PHYSICS RESEARCH AND TECHNOLOGY

PHYSICS AND MECHANICS OF NEW MATERIALS AND THEIR APPLICATIONS

PHYSICS RESEARCH AND TECHNOLOGY

MATERIALS SCIENCE AND TECHNOLOGIES

Additional books in this series can be found on Nova's website
under the Series tab.

Additional e-books in this series can be found on Nova's website
under the e-books tab.

PHYSICS RESEARCH AND TECHNOLOGY

PHYSICS AND MECHANICS OF NEW MATERIALS AND THEIR APPLICATIONS

IVAN A. PARINOV
AND
SHUN-HSYUNG CHANG
EDITORS

nova publishers
New York

For permission to use material from this book please contact us:
Telephone 631-231-7269; Fax 631-231-8175
Web Site: http://www.novapublishers.com

Library of Congress Cataloging-in-Publication Data

ISBN 978-1-62618-535-7

Published by Nova Science Publishers, Inc. † New York

CONTENTS

PREFACE

The advanced materials and composites based on nanotechnology approaches, modern piezoelectric techniques, and also using last achievements of Material Science, Condensed Matter Physics and Mechanics of Deformable Solids have found broad applications in the modern science, techniques and technologies. Tremendous interest is connected with fast development of theoretical, experimental and numerical methods which ensure to obtain new knowledge and are capable to control and give forecast on development of critical phenomena and improvement of very fine processes, in particular nano- and microstructure transformations taking place during processing, loading and working modern materials and composites, which demonstrate a wide spectrum of properties in scales from nanometers up to macroscopic range. Numerous devices with their very high accuracy, longevity and extended possibilities to work in wide temperature and pressure ranges, aggressive media, etc. demonstrate characteristics directly defined by used materials and composites. It is obvious a tremendous role of processing techniques rapidly developed into framework of modern science.

This collection of 30 papers presents selected reports of the Russian-Taiwanese Symposium "Physics and Mechanics of New Materials and Their Applications" (PMNM-2012), which has been taken place in Rostov-on-Don, Russia, 4-6 June, 2012 (http://pmnm.math.rsu.ru/) and sponsored by Russian Foundation for Basic Research, National Science Council of Taiwan, South Scientific Center of Russia Academy of Sciences, New Century Education Foundation (Taiwan), Unity Opto Technology Co., Formosa Epitaxy Inc., ChenFull International Co., Genius Net Co., AREA Energy Group Inc., EPOCH Energy Technology Corp., Fair Well Fishery Corp., South Russian Regional Centre for Preparation and Implementation of International Projects, Ltd. This edited book continues the author's series of edited books, devoted to piezoelectrics, related materials and their applications, which have been published together with Nova Science Publishers in the last years.

The presented papers are divided into four scientific directions: (i) processing techniques of new materials, (ii) physics of new materials, (iii) mechanics of new materials, and (iv) applications of new materials.

Into framework of first theme are considered, in particular the nanotechnology approaches and novel methods in processing of wide class of piezocomposites and multiferroics. Moreover, there are also presented investigations of Cu_2ZnSnS_4 solar cell material manufactured by using multistage techniques, based on the nanotechnology achievements, and also β-Ga_2O_3 epilayers grown by MOCVD.

Second direction is opened by theoretical studies of Hall effect and inhomogeneous magnetic fields in ferromagnetic samples, investigations of electromechanical properties of relaxor ferroelectrics, dielectric retardation and relaxation in multiferroic ceramics. There are also presented studies of physical properties of the quantum resonant materials and structures, structure of dilute nitride GaAsSbN. Finally, there are discussed the methods of electrochemical prevention of biofouling and also novel optical means and methods for measurement of displacements in technical diagnostics of various materials and goods.

From viewpoint of mechanics are studied dislocations and disclinations defining nonlinear deformation of thin-film materials, fracture and damage processes in advanced materials on the base of acoustic emission diagnostics, numerical approaches to define different properties of piezocomposites with regular structure and porous piezoceramics, and also contact problems for coatings of complicated structure and functionally-graded materials.

Applications are presented by results of modeling and optimal design of power high stroke piezoelectric actuators for rotorcraft applications; R&D of new piezo- and pyroelectric materials and piezo-generators for energy harvesting devices, studied as by finite element methods as using natural experiments; development of superconducting magnetic energy storage for railway traction applications; hydrophone based on epitaxial PZT film for acoustic devices application; devices of underwater acoustic multimedia communication based on the piezoelectric technologies; and finally investigations of light-emitting diodes. We also accounted to be very useful to finish this collection by the paper which generalizes the results of Taiwanese experience aimed to bring Science, Technology and Society, inter-disciplinary studies to engineering education.

The book is addressed to students, post-graduate students, scientists and engineers taking part in R&D of nano-materials, ferro-piezoelectrics and related materials, and also different devices based on them having broad applications in different areas of modern science and technique.

Ivan A. Parinov (Ed.)

Shun-Hsyung Chang (Ed.)

July, 2012

I. Processing Techniques of New Materials

In: Physics and Mechanics of New Materials ...
Editors: Ivan A. Parinov and Shun-Hsyung Chang
ISBN: 978-1-62618-535-7

Chapter 1

NANOPARTICLES TRANSPORT IN CERAMIC MATRIXES: A NOVEL APPROACH FOR CERAMIC MATRIX COMPOSITES FABRICATION

***Andrey N. Rybyanets*[1,*] *and Anastasia A. Naumenko*[2,†]**
[1]Institute of Physics, Southern Federal University, Rostov-on-Don, Russia
[2]Physical Faculty, Southern Federal University, Rostov-on-Don, Russia

ABSTRACT

A novel approach for fabrication of nano- and microporous piezoceramics as well as ceramic matrix piezocomposites was proposed. The technique is based on nanoparticles transport in ceramic matrixes using a polymer nanogranules coated or filled with a various chemicals, with successive porous ceramics fabrication processes. Resulting ceramic matrix piezocomposites are composed by superlattices of closed pores filled or coated by nanoparticles of metals, oxides, ferromagnetic, etc. embedded in piezoceramic matrix. New family of nano- and microporous piezoceramics and ceramic matrix piezocomposites are characterized by a unique spectrum of the electrophysical properties unachievable for standard PZT ceramic compositions and fabrication methods, including giant dielectric relaxation, giant piezoelectric and electrocaloric effects as well as the possibility of controllable changes of the main properties within a wide range.

1. INTRODUCTION

In recent years, multiphase ceramic composites (nanocomposites, porous ceramics, composites of ceramics/ceramics, and ceramics/crystals types) are widely used for industrial and ultrasonic transducer applications [1-4]. Numerous technologies based on incorporating of nanoquantities of functional ceramics into structural ones and vice-versa have been

* E-mail: arybyants@gmail.com.
† E-mail: charliekorablik@gmail.com.

developed and the novel design idea has been applied in the field of functional ferroelectric ceramics [2, 5].

Nanoparticles are perfect building blocks offering a wide variety of compositions, structures and properties, ideally suited to designing functional nanomaterials and nanodevices. Nanoparticles can be embedded in various matrixes. A major technological problem is spontaneous aggregation of nanoparticles which, as a result, lose their unique properties. One possible solution to this problem is to use supports interacting with nanoparticles. Many organic polymers have been used in various nanoparticles surface engineering approaches [6].

Recently polymer nano- and microgranules filled with nanoparticles (magnetic, metal, oxides, etc.) started widely to use as a delivery means of nanosubstances in medical, biotechnology and chemical applications [7]. Very recently, there was a tendency to fix (immobilize) 2-10 nm particles on the surface of spherical polymeric microgranules (typically 0.1 - 20 μm in size) [8, 9]. Such composite micro-nano systems offer a number of significant advantages. When fixed to a surface or embedded in microgranules, nanoparticles lose their tendency to readily aggregate but retain their reactivity and, for the most part, their physical properties. Besides, microgranules are easier to manipulate then nanoparticles. Micro-granules, coated or filled with nanoparticles can be used to produce "homogeneous" disperse systems, such as sols and aerosols, and fabricate films, coatings or bulk materials. The preparation and properties of microgranules for different applications as well as the properties of particular types of nanoparticles were described elsewhere [6-9].

Intensive research and technological works as well as improvements of fabrication methods have allowed large-scale manufacture of porous piezoelectric ceramics with reproducible and controllable porosity and properties [5, 10, 11]. Closed cell porous ceramics are usually formed by adding porosifiers to ceramics powders (burnable plastic spheres) [5]. A universal manufacturing method suitable for mass production of a wide class of nano- and microporous piezoceramics and ceramic matrix piezocomposites (CMPC) were described in [5, 12, 13].

In this paper we propose a novel approach for fabrication of nano- and microporous piezoceramics as well as CMPC. The technique is based on nanoparticles transport in ceramic matrixes using polymer nano- and microgranules coated or filled by nanoparticles of various chemicals, with successive porous ceramics fabrication processes.

2. Nanoparticles Transport Concept

The nanoparticles transport concept (NPT) proposed in this paper is a nontrivial combination of manufacture technologies of porous ceramics and ceramic matrix composites [12, 13] and microcapsules transport techniques used in biomedical applications for targeting nanoparticles drugs delivery [6].

Over the past years, considerable advances were made to improve the physical, electrical and functional properties of piezoelectric ceramics using composite approaches [5]. Numerous composite technologies were developed and a novel design idea was applied to develop functional piezoelectric ceramics [2, 12, 13].

The original microstructural design concept (MSD) for ceramic matrix composites was proposed in [2]. The MSD concept offers a brand-new range of polymer-free piezocomposite materials with parameters that are adjustable over a wide range. The MSD concept is based on controllable substitution during composite formation of separate crystallites or crystallite groups making a polycrystal by pores, crystallites with other composition and/or structure, or amorphous substances, all with preliminary choice of chemically, thermally and technologically compatible components and with FEM modeling of polycrystalline composite properties. The distinctive peculiarities of the MSD concept is the transition to a micro- or nanoscale level of separate crystallites and the use of the full set of technological approaches at particular stages of synthesis, preparation of initial materials (powders, solutions, suspensions, gels), granulation and compacting of a green bodies, sintering and the-post sintering treatment of a composites.

The MSD concept offers an innovative range of polymer-free piezocomposite materials (micro-, mesa- and macroporous piezoelectric ceramics, piezocomposites of ceramics /ceramics and ceramic/crystal types, and CMPC) with properties combining better parameters for PZT, PMN, PN and PT type ceramics and 1-3 composites [4, 10, 13].

Porous piezoceramic is composite material composed of two phases: a piezoelectrically active (piezoceramic) phase and a passive (porous) phase. The piezoelectrically active phase forms a three-dimensional matrix, in which the passive phase exists in the form of isolated and/or interconnected inclusions. The case of isolated inclusions corresponds to 3-0 connectivity while the interconnected inclusions case, to 3-3 connectivity type (closed and open porosity, respectively) [12]. The microstructure of porous ceramics is unequivocally defined by the fabrication method and by the initial chemical composition of the piezoceramic phase. The structure and properties of porous ceramics are controlled by their processing.

Figure 1 shows micrographs illustrating the marginal cases of a micro-structural variety of porous ceramic materials. The resulting elastic, dielectric, and piezoelectric properties of porous piezoceramics are defined by the properties of its piezoceramic matrix: that is, the porosity, type of connectivity, shape, and size of pores (and, in the case of irregular shaped pores, by the degree of their primary orientation and arising anisotropy) [5, 10, 12].

All existing methods of porous ceramics fabrication can be divided into two basic approaches: subtractive processes and processing limitations. In subtractive processes, certain elements of an original structure are selectively removed to create pores. In processing limitations, technological processing regimes are modified to form porous structures [12]. Porous ceramics with an open cell structure are usually formed by replicating a polymer foam/sponge. Closed cell porous ceramics are usually formed by adding porosifiers to ceramics powders.

A simpler technique for production of highly porous compacts was proposed in [11] with sintered mixtures of burnable plastic spheres (PMMA) and PZT powders. A series of further process for the preparation of porous ceramics and composites with 3-3 connectivity were subsequently introduced [2, 10, 12].

The universal manufacturing method suitable for mass production of a wide range of porous piezoceramics and piezocomposite materials based on burning out of burnable particles (polymer, organic substances or salts) proposed in [5, 12] was used in the NPT concept.

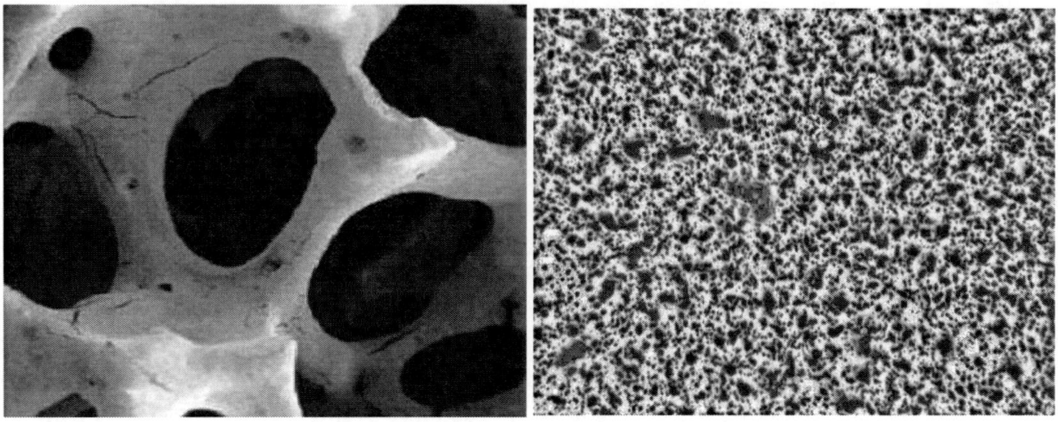

a b

Figure 1. SEM micrographs of different porous ceramics structures: (a) open cell PZT ceramics, fabricated by polymer foam method; (b) closed cell PZT ceramics, obtained by burning out polymer granules.

Nanoparticles transport techniques using microgranules or microcapsules as a delivery means are intensively developed and widely used now for targeting drugs delivery and other biomedical and chemical applications [6, 7]. A fabrication techniques of polymer microgranules coated or filled (microcapsules) by nanoparticles of a various chemicals are well developed and different types of microgranules are commercially available [6]. The preparation and properties of microgranules for different applications as well as the properties of particular types of nanoparticles were described elsewhere [6-9].

In the NPT method, proposed in this paper, polymer nano- and microgranules coated or filled with nanoparticles of various chemicals are used as a porosifiers agents for modified porous ceramics fabrication process. Thereby, polymer nano- and microgranules act as a delivery means for targeting transport of nanosubstances into ceramic matrixes with closed nanopores.

The NPT concept comprises:

- nanoparticles, nano- and microgranules preparation;
- filing or coating of nano- and microgranules with nanoparticles;
- characterization of the nanoparticles and microgranules;
- porous ceramics and CMPC fabrication process;
- resulting porous ceramics and CMPC characterization.

The full scale use of the NPT concept will results in a new generation of nano- and microporous ceramics and composite materials derived from technologically simple processing, whilst offering a unique combination of piezoelectric, magnetic and electrets properties along with a possibility of controllable changes of the main properties.

3. Experimental Scenario

3.1. Polymer Microgranules and Nanoparticles Preparation

Four types of polymer microgranules were used for "pilot" porous ceramics and CMPC samples fabrication:

- pure polystyrene microgranules (PS);
- polystyrene microgranules (PS) filled with magnetite ($FeO{\bullet}Fe_2O_3$) and NiO nanoparticles;
- pure polytetrafluoroethylene nanogranules (PTFENG); and
- polytetrafluoroethylene nanogranules (PTFENG) coated with metal-containing (Fe, Co, Ni, Cu, Pt and Pd) nanoparticles (MCNP).

Magnetite ($FeO{\bullet}Fe_2O_3$) nanoparticles of 10-40 nm in diameter were prepared by chemical deposition of solid phase from Fe-salts solutions. NiO nanoparticles of 10-30 nm in diameter were prepared by thermal decomposition of and Ni-salts ($NiCO_3$). PS microgranules of 80-100 nm in diameter were prepared from polystyrene/magnetite or polystyrene/NiO colloidal fluids with successive polymerization [7].

PTFENG microgranules of 150–500 nm in diameter were fabricated by the thermal gas-dynamic method [8]. MCNP nanoparticles of 3.5 – 6.5 nm in diameter have been synthesized and immobilized on the surface of polytetrafluoroethylene nanogranules via thermal decomposition of metal-containing precursors on the surface of PTFENG in mineral oil, as described in [9].

To produce MCNP from MR_n precursors (M = Co, Fe, Cu, Ni, Pd; R = CO, HCOO, CH_3-COO), well-developed thermal decomposition processes have been used [9]. At PTFENG sizes from 100 to 500 nm, a fluidized bed of microgranules was formed over the surface of heated oil. This behavior of microgranules was used to immobilize MCNP. In the course of deposition, the nanogranules became progressively heavier, and some of them left the fluidized bed and settled in the oil. As a result, the MCNP on their surfaces stopped growing.

3.2. Characterization of the Nanoparticles and Microgranules

Polystyrene microgranules filled with magnetite or Ni nanoparticles were examined using transmission and scanning electron microscopy (TEM and SEM).

Figure 2 shows a typical TEM image of magnetite nanoparticles and their size distribution. Nanoparticles have average diameter 10 nm at size variation 9.6%. Exemplary micrographs of polystyrene microgranules containing magnetite nanoparticles are shown on Figure 3.

The MCNP nanoparticles were examined by X-ray diffraction (XRD), transmission and scanning electron microscopy (TEM and SEM), EXAFS spectroscopy, X-ray emission and Mössbauer spectroscopy [8, 9]. The results demonstrate that the nanoparticles with average size of 3.5–6.5 nm have a complex structure, are isolated from one another and strongly bonded to the surface of the nanogranules.

The metal content of the samples was determined by elemental analysis. TEM images were taken on a JEOL JEM-100B at an accelerating voltage of 100 kV.

Figure 4 shows a representative TEM image of as-prepared PTF-ENG/MCNP [9].

As seen in Figure 4, MCNP are distributed rather evenly throughout the PTFENG surface. It easily to see that all the nanogranules are covered with nearly monodisperse MCNP. From the TEM image in Figure 4, the average size of the MCNP [c-Fe_2O_3 prepared from $Fe(CH_3\text{-}COO)_3$] was determined to be ~ 5 nm, with a narrow size distribution. The average sizes of Ni- and Pd-containing nanoparticles were 3.5 and 7.5 nm, respectively. Although the number of MCNP surrounding each PTFENG varied from sample to sample, we were able to control the total amount of the deposited metal per nanogranules by varying the precursor concentration in the solution.

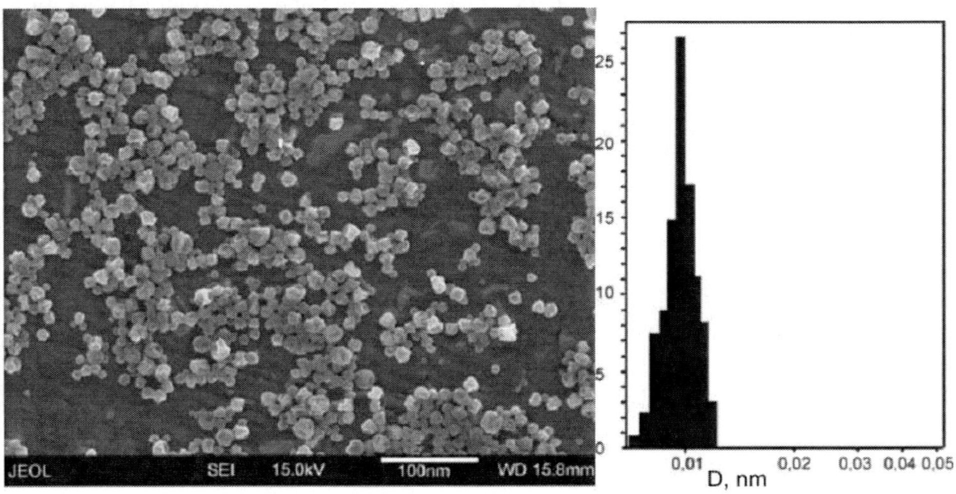

a b

Figure 2. SEM micrographs of magnetite nanoparticles (a) and their size distribution (b).

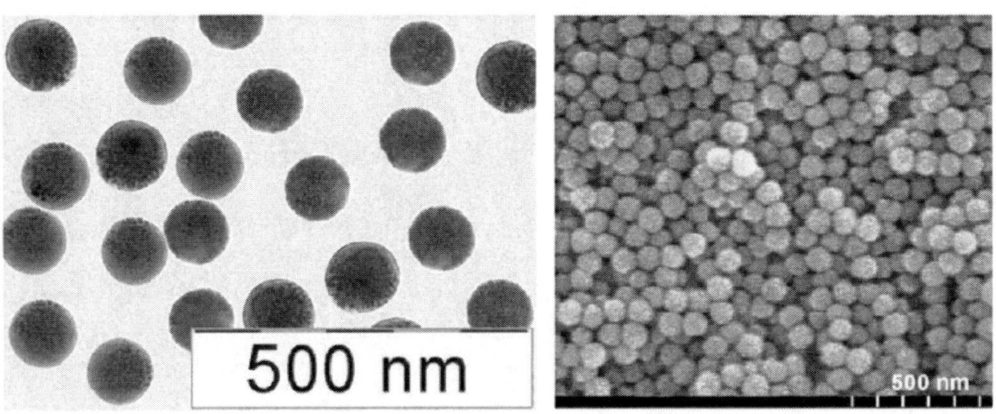

a b

Figure 3. Micrographs of polystyrene microgranules containing magnetite nanoparticles: (a) TEM, (b) SEM. Magnetite mass content ~ 40 %, average diameter of microgranules ~ 100 nm.

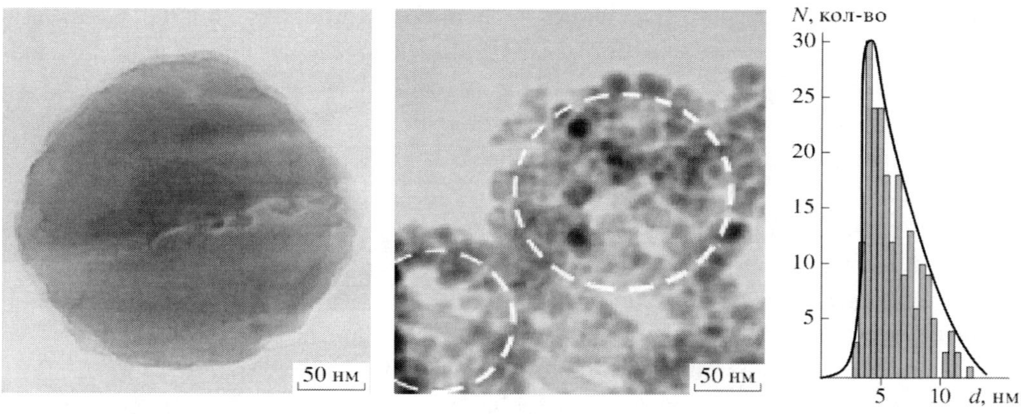

a b c

Figure 4. Exemplary TEM image of initial PTFENG (a), as-prepared PTFENG/MCNP (b) and their size distribution (c).

3.3. Porous Ceramics and CMPC Fabrication Processes

Schematic illustration of the technological processes used for porous ceramics and CMPC fabrication are shown In Figure 5.

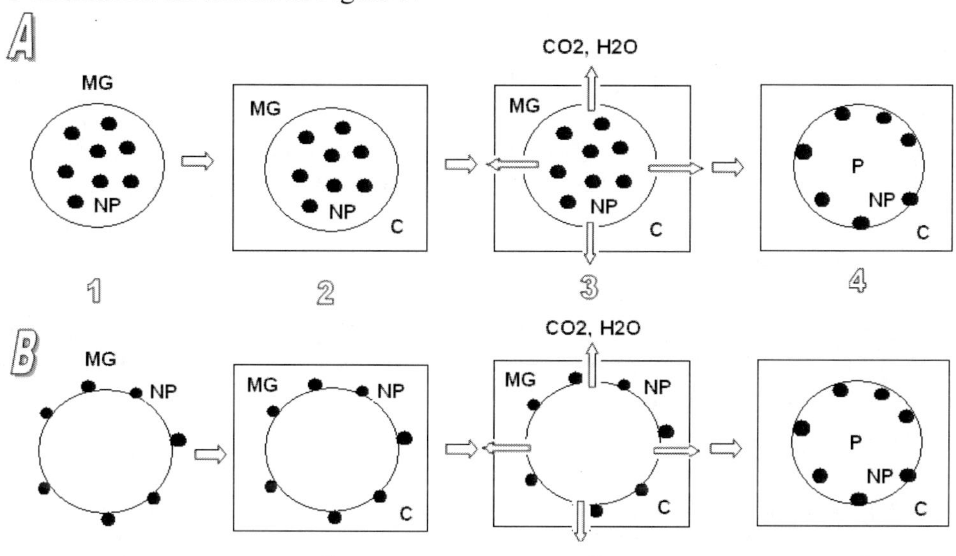

Figure 5. Schematic technological process for nano- and microporous piezoceramics and CMPC fabrication using polymer microgranules filled (A) and coated (B) with nanoparticles:
1 - preparation of microgranules (MG) filled (A) and coated (B) with nanoparticles (NP);
2 - embedding of microgranules (MG) in ceramic matrix (C) by means of mixing and compacting;
3 - burning out (thermal decomposition) of polymer microgranules (MG) and deposition of nanoparticles (NP) on a pore surface;
4 - sintering of porous ceramic matrix (P) and covering internal pores surface with nanolayers of metals or oxides, local doping or modification of stoichiometric composition of surface layer of piezoceramics (C).

Standard sequence of porous ceramics technological processes (mixing, compacting, burning out and sintering) were used for CMPC fabrication. Polymer nanogranules, coated or filled with a various chemicals, were used as porosifiers agents. Resulting CMPC were composed by super lattices of closed pores filled or coated by nanoparticles of metals, oxides, ferromagnetic, etc. embedded in piezoceramic matrix.

Special regimes of wet and dry mixing were used for homogeneous distribution of polymer nanogarnules in piezoceramic raw powders. Burning out was fulfilled at 700 °C to provide full decomposition of polymer components. Sintering was carried out at optimal temperature for each piezoceramic composition.

Special anti-oxidation nanoparticles additives (SiC and C, as an examples) were embedded in polymer nanogranules simultaneously with metal or oxide nanoparticles in proper mass proportion, to prevent undesirable oxidation of metal nanoparticles or to deoxidize metal oxides during ceramics sintering process.

4. Experimental Results

4.1. Method of Measurements

The basic techniques for finding material constants of piezoelectric materials are outlined in the IEEE Standard on Piezoelectricity (1987) [14]. These methods work for many of the most widely used commercial piezoceramics based on lead titanate-zirconate (PZT) compositions that are high-QM and high-coupling coefficient piezoelectric materials. However, there is a general agreement that their use in many new piezoelectric materials such as porous ceramics, piezoelectric polymers or piezoelectric composites may lead to significant errors. Furthermore, the current IEEE Standard does not comprehensively account for the complex nature of material coefficients, as it uses only the dielectric loss factor (tan δ) and the mechanical quality factor (Q_M) to account for loss.

Numerous techniques using complex material constants have been proposed to take into account losses in low-Q_M materials and to overcome limitations in the IEEE Standard [15-18]. Iterative methods [19-21] provide a means to accurately determine the complex coefficients in the linear range of poled piezoceramics from complex impedance resonance measurements.

The Piezoelectric Resonance Analysis Program (PRAP) automatic iterative method [21] was proposed for complete complex characterization of a wide range of materials with very high and moderate loss factors. The PRAP software analyses impedance spectra to determine complex material properties. This software uses a generalized form of Smits's method to determine material properties for any common resonance mode, and a generalized ratio method for the radial mode valid for all material Q's. By analyzing on each harmonic, PRAP can determine complex material properties as a function of frequency. The software always generates an impedance spectrum from the determined properties to indicate validity of the results.

New nanoporous piezoceramics and CMPC are lossy and direct use of IEEE Standards for material constant evaluation can leads to significant errors, therefore the PRAP automatic iterative measurement method [21-23] was applied to the full set of standard geometries and resonance modes needed to complete complex characterization of the pilot samples.

Measurements of electric parameters were made using the Solartron Impedance/Gain-Phase Analyzer SL 1260 and Agilent 4294A Impedance Analyser. Pulse-echo and through-transmit measurements of ultrasonic transducers were made using LeCroy digital oscilloscope and Olympus pulser/receiver. Piezoelectric modulus was measured using APC d_{33} tester. The microstructure of polished, chemically etched, and chipped surfaces of composite samples was observed with optical (NeoPphot-21) and scanning electron microscopes (SEM, Karl Zeiss).

4.2. Microstructure of Nanoporous Ceramics and CMPC

Pilot samples of nano and microporous ceramics and CMPC were fabricated using different piezoceramics compositions (PZT, lead-potassium niobate and lead titanate) as a ceramic matrix and PS and PTFENG microgranules as porosifiers agents.

Figure 6 shows SEM micrograph of grinded surface of PZT nanoporous piezoceramics prepared as a result of pure PS microgranules burning out and sintering. The resulting porous ceramics demonstrate mixed inter- and intracrystalline closed porosity. Notable variation of intra- and intercrystalline pore sizes is caused by partial agglomeration of PS microgranules during mixing and compacting processes.

Figure 7 shows SEM micrograph of chipped surface of PZT nanoporous piezoceramics prepared as a result of pure PTFENG microgranules burning out and sintering. It is obvious that PTFENG microgranules are prevent compacting of PZT powder and lead to intercrestalline nanopores formation.

Figure 6. SEM micrograph of grinded surface of PZT piezoceramics with mixed inter- and intracrystalline nanoporosity.

Figure 7. SEM micrograph of chipped surface of PZT nanoporous piezoceramics with closed intercrystalline nanoporosity.

Figure 8 shows optical micrograph of lead-potassium niobate (PKN) ceramics microstructure with closed intra-crystallite nanopores filled by magnetite nanoparticles formed as a result of PS-magnetite microgranules burning out and sintering. Resulting piezomagnetic CMPC demonstrate notable magnetic properties at retained piezoelectric characteristics of initial PKN ceramics.

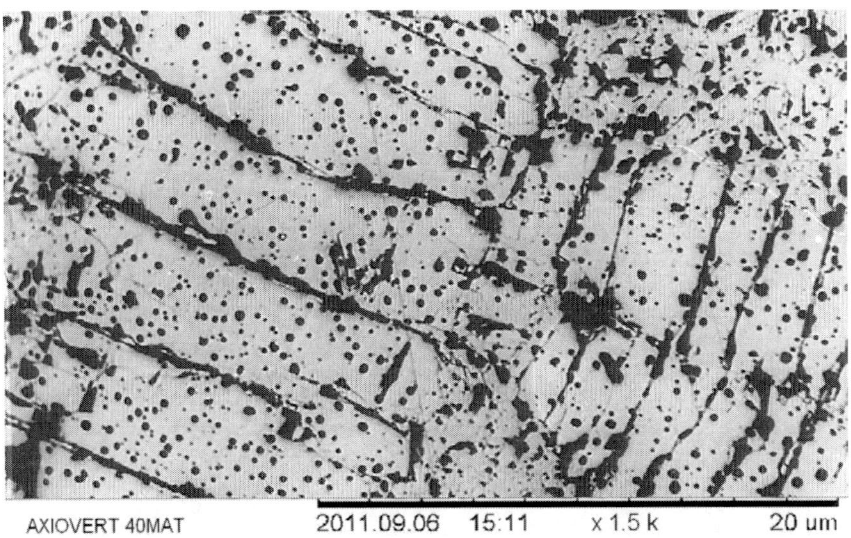

Figure 8. Optical micrograph of PKN piezoceramics with closed intracrystalline nanoporosity filled by magnetite nanoparticles.

Figure 9. SEM micrograph of internal surface of closed pore in PZT piezoceramics covered by Pd nanoparticles.

Figure 9 shows SEM micrograph of internal surface of closed pore in PZT piezoceramics covered by Pd nanoparticles formed as a result of PTFENG-Pd microgranules burning out and sintering. Figure 9 clearly demonstrates a possibility of pin-point delivery of definite number of nanoparticles into target ceramic matrix.

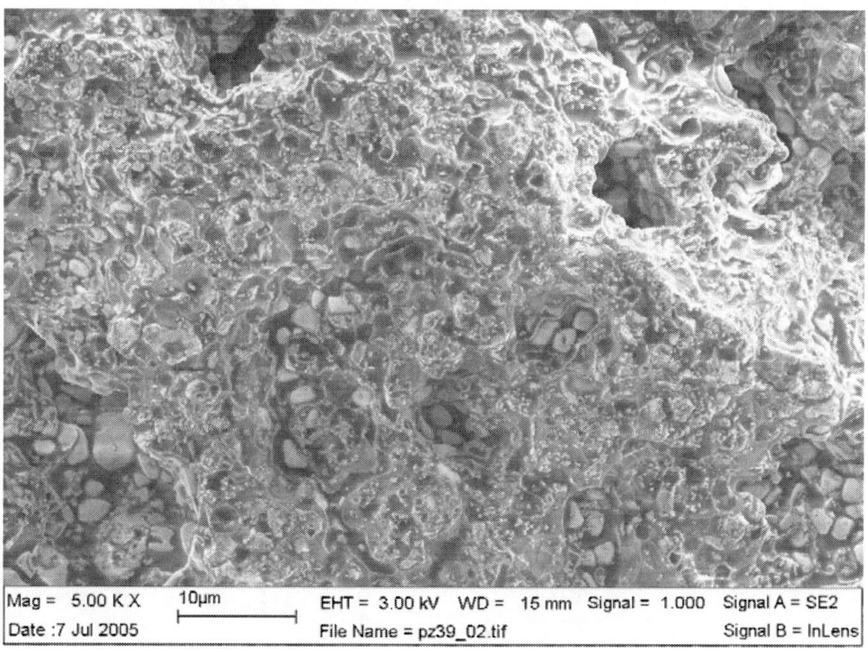

Figure 10. SEM micrograph of internal surface of closed pore in lead titanate piezoceramics covered by Ni nanoparticles layer.

Figure 10 shows SEM micrograph of internal surface of closed micropores in PZT piezoceramics covered by Ni nanoparticles layer (covering rate 80%) after PS-NiO+SiC microgranules burning out and sintering.

4.2. Dielectric and Piezoelectric Properties

The distinctive peculiarity of a new family of nano- and microporous piezoceramics and CMPC are the possibility of controllable changes of the main properties within a wide range along with a possible combination of piezoelectric, magnetic and electrets properties:

(i) acoustic impedance $0.1Z_{AC} \leq Z_A \leq 2Z_{AK}$, where Z_{AC} is the impedance for initial ceramics;
(ii) dielectric constant $0.1\varepsilon_{33}{}^{T}{}_{C} \leq \varepsilon_{33}{}^{T} \leq 5\varepsilon_{33}{}^{T}{}_{C}$, where $\varepsilon_{33}{}^{T}{}_{C}$ is the dielectric constant for initial ceramics;
(iii) electromechanical coupling factors $0.1k_{pC} \leq k_p \leq k_{pC}$, $k_{tC} \leq k_t \leq 1.5\ k_{tC}$, where k_{pC} and k_{tC} are the radial and thickness for initial ceramics;
(iv) piezoelectric modulus $d_{33C} \leq d_{33} \leq 2d_{33C}$, $0.1d_{31C} \leq d_{31} \leq d_{31C}$, where d_{33C} and d_{31C} are the longitudinal and thickness for initial ceramics.

Figure 11 shows impedance spectra (impedance modulus $|Z|$ and phase angle φ as function of frequency f) for nanoporous PZT piezoceramics at different relative porosity P. It can be seen from Figure 11 that the electromechanical activity of porous ceramics (difference between resonant f_r and antiresonant frequency f_a) grows with the relative porosity P increase.

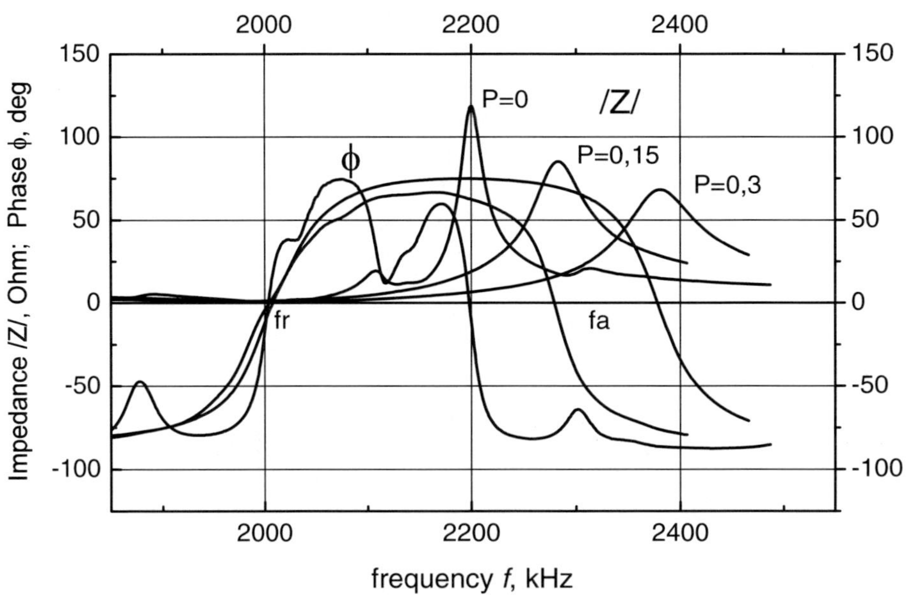

Figure 11. Impedance spectra for nanoporous PZT piezoceramics at different relative porosity P.

Table 1 shows dielectric and piezoelectric properties of lead metaniobate $PbNb_2O_6$ and lead potassium niobate (Pb, K)Nb_2O_6 nanoporous piezoceramics. It is obvious that the nanoporous ceramics technology results in increasing of piezoelectric anisotropy, removal of internal mechanical stress, increasing of mechanical durability, preventing of cracking and finally, allows production of stable in time elements of such “technologically difficult” ceramics as lead metaniobate with excellent and reproducible properties.

Table 1. Dielectric and piezoelectric properties of $PbNb_2O_6$ and (Pb,K)Nb_2O_6 nanoporous piezoceramics

Parameter/Material	$PbNb_2O_6$ P=15%	(Pb,K)Nb_2O_6 P=15%
Electromechanical coupling factor for planar mode, k_p	0	0
Electromechanical coupling factor for thickness mode, k_t	0.42	0.45
Piezoelectric modulus, d_{33} (10^{-12}), C/N	75	90
Dielectric constant, $\varepsilon_{33}{}^{T}/\varepsilon_0$	160	160
Dielectric loss tan, tan δ, %	0.6	0.8
Mechanical quality factor for thickness mode, $Q_M{}^{t}$	< 10	< 10
Density, ρ (10^3), g/cm^3	5.7	5.6
Curie temperature, T_C, °C	560	550
Frequency constant, N_t, kHz·mm	1300	1380

Dielectric and piezoelectric parameters of piezoceramic-metal nanoporous CMPC (lead titanate piezoceramics with closed nanoporous covered by Ni nanoparticles layers) are shown in Table 2.

Table 2. Piezoelectric properties of piezoceramic-metal nanoporous CMPC

Parameter/Material	$PbTiO_3$/Ni P=20%	$PbTiO_3$/Ni P=40%
Electromechanical coupling factor for planar mode, k_p	0	0
Electromechanical coupling factor for thickness mode, k_t	0.55	0.6
Piezoelectric modulus, d_{33} (10^{-12}), C/N	150	275
Dielectric constant, $\varepsilon_{33}{}^{T}/\varepsilon_0$	250	480
Dielectric loss tan, tan δ, %	2	4
Mechanical quality factor for thickness mode, $Q_M{}^{t}$	< 12	< 10
Density, ρ (10^3), g/cm^3	6.4	4.8
Curie temperature, T_C, °C	490	480
Frequency constant, N_t, kHz·mm	1200	800

We can see from Table 1 that electromechanical and piezoelectric activity as well as dielectric constant of nanoporous CMPC “piezoceramic-metal” increase with relative porosity, and consequently with metal content grows. This behavior is in good conformity with the theoretical and FEM modeling results [12, 24, 25] predicting grows of dielectric constant and piezoelectric modulus in the vicinity of dielectric-metal percolation threshold.

The electromechanical behavior of CMPC with hollow metal spheres (pore with metalized internal surface) is drastically differs from the solid metal sphere case, because of practically infinite elastic compliance of pores.

Preliminary experiments on nanoporous CMPC "piezoceramic-metal" in low frequency range $10^{-6} - 10^{4}$ Hz confirmed also the theoretical predictions of giant dielectric relaxation and giant piezoelectric effect resulting from Maxwell-Wagner relaxation (Figure 12), as was foreshown in [24, 25].

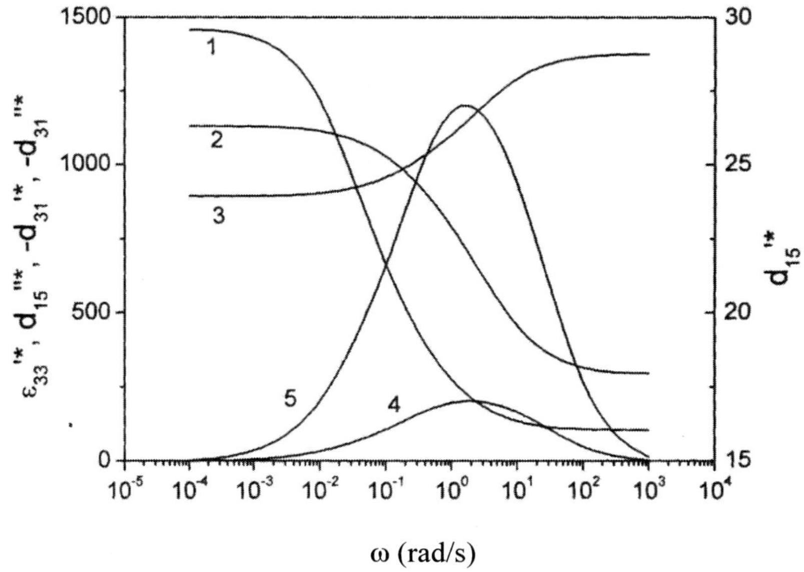

Figure 12. Frequency dependencies of real parts of dielectric constants ε_{33}'* (1), piezoelectric moduli $-d_{31}$'*, 10^{-14} C/N (2), d_{15}'*, 10^{-12} C/N (3), as well as imaginary parts of $-d_{31}$''*, 10^{-14} C/N (4), d_{15}''*, 10^{-15} C/N (5) for CMPC.

References

[1] *Piezoelectric Materials in Devices*. N. Setter (Ed.). Swiss Institute of Technology, Switzerland, 2002; pp 1 – 518.

[2] Rybjanets, A. N.; Nasedkin, A. V.; Turik A. V. *Integrated Ferroelectrics*. 2004, *vol. 63*, 179 – 182.

[3] Xiang, P.-H.; Dong, X.-L; Chen, H.; Zhang Z.; Guo, J.-K. *Ceramics International*. 2003, *vol. 29*, 499 – 503.

[4] Rybianets, A .N. *Ferroelectrics*. 2007, *vol. 360*, 84 – 89.

[5] Rybyanets, A. N. In *Piezoceramic Materials and Devices*, Ivan A. Parinov (Ed.). Nova Science Publishers Inc. NY: 2010, Chapter 3, pp 113 – 174.

[6] Donbrow, Max. *Microcapsules and Nanoparticles in Medicine and Pharmacy;* CRC Press, London, 1992, pp 1 – 368.

[7] *Magnetic Nanoparticles*, Sergey P. Gubin (Ed.), Manheim, Willey-VCH, 2009. pp 1 – 283.

[8] Gubin, S. P.; Yurkov, G. Yu; Korobov, M. S.; et. al. *Acta Materialia*, 2005, *vol.53*, 1407 – 1413.

[9] Gubin, S. P.; Yurkov, G. Yu; Kataeva, N. A. *Inorganic Materials*, 2005, *vol. 41*, 1017 – 1032.
[10] Rybjanets, A. N.; Razumovskaja, O. N.; Reznitchenko, L. A.; Komarov, V. D.; Turik, A. V. *Integrated Ferroelectrics.* 2004. *vol. 63*, 197 – 200.
[11] Shrout, T. R.; Shulze, W. A.; Biggers J. V. *Mat. Res. Bull.* 1979, *vol. 14,* 1553 – 1559.
[12] Rybyanets, A. N. *IEEE Trans.UFFC.* 2011, *vol. 58*, 1492 – 1507.
[13] Rybyanets, A. N.; Rybyanets, A. A. *IEEE Trans. UFFC.* 2011, *vol. 58*, 1757–1774.
[14] *IEEE Standard on Piezoelectricity*. ANSI/IEEE Std. 1987, 176.
[15] Alguero, M.; Alemany, C.; Pardo, L. Gonzalez, A. M. *J. Am. Ceram. Soc.* 2004, *vol. 87*, 209 – 212.
[16] Holland, R. *IEEE Trans.Sonics Ultrason.* 1967, *vol. SU-14*, 18 – 24.
[17] Sherrit, S.; Wiederick H. D., Mukherjee, B.K. *Ferroelectrics.* 1992., *vol. 134*, 111 – 115.
[18] Sherrit, S.; Wiederick H. D., Mukherjee, B.K. *Ultrasonic Transducer Engineerinq. SPIE Proceedinqs.* 1997, *vol. 3037*, 158 – 163.
[19] Sherrit, S.; Wiederick H. D., Mukherjee, B. K. *Smart Materials. Structures, and Integrated Svstems, SPIE Proceedings.* 1997, *vol. 3241*, 327 – 333.
[20] Smits, J.G. *IEEE Trans. Sonics Ultrason.* 1976, *vol. SU-23*, 393 – 402.
[21] PRAP (Piezoelectric Resonance Analysis Program). TASI Technical Software Inc. www.tasitechnical.com.
[22] Rybianets, A. N.; Nasedkin A. V. *Ferroelectrics.* 2007, *vol. 360*, 57 – 62.
[23] Rybianets, A. N.; Tasker R. *Ferroelectrics.* 2007, *vol. 360*, 90 – 95.
[24] Turik, S. A.; Reznitchenko, L. A.; Rybjanets A. N.; Dudkina, S. I.; Turik A. V.; Yesis A. A. *J. Appl. Phys.* 2005, *vol. 97*, 064102.
[25] Turik, A. V.; Chernobabov, A. I.; Radchenko, G. S.; Turik. S. A. *Sov. J. Solid State Physics.* 2004, *vol. 46*, 2139 – 2142.

In: Physics and Mechanics of New Materials ...
Editors: Ivan A. Parinov and Shun-Hsyung Chang
ISBN: 978-1-62618-535-7

Chapter 2

Cu_2ZnSnS_4 Solar Cell Material Prepared by Spin-Coating Metal Chloride Precursors and Treated with Post-Synthesizing and Annealing

Min Yen Yeh*[a], Yu-Fong Huang[a], Cheng-Liang Huang[a] and Dong-Sing Wuu[b]

[a]Department of Microelectronics Engineering,
National Kaohsiung Marine University, Kaohsiung, Taiwan
[b]Department of Materials Science and Engineering,
National Chung Hsing University, Taichung, Taiwan

Abstract

Cu_2ZnSnS_4 (CZTS) thin films applied for solar cells were prepared by sol-gel spin-coated method. The as-coated films were treated with post-synthesizing and annealing processes. CZTS precursors were prepared by using solutions of copper (II) chloride, zinc (II) chloride, tin (IV) chloride, and thiourea. The post-annealing process was found to have great impacton the crystalline properties of CZTS films. It was found that in addition to CZTS phase SnS and SnS_2 were also observed in as-annealed films. The precursors were prepared in the ratio of $CuCl_2 : ZnCl_2 : SnCl_4 : CH_4N_2S = 4 : 1 : 1 : 8$, where $SnCl_4$ is of 1 mol/l. The SEM images show uniform surface morphology for as deposited CZTS films. An average absorption coefficient of 3×10^4 cm^{-1} and an optical energy gap of ~1.65 eV were obtained resulting from a cupper-poor composition CZTS.

* E-mail: minyen@webmail.nkmu.edu.tw, Department of Microelectronics Engineering National Kaohsiung Marine University, 142, Haijhuan Rd., Nanzih District, Kaohsiung 81157, Taiwan

1. INTRODUCTION

Kesterite structure Cu_2ZnSnS_4 with a direct optical band gap energy of ~1.5 eV is considered as potential material for photo-voltaic devices. It has good properties of low cost, non-toxicity, abundance on Earth. With a stannite structure similar to $CuIn_xGa_{1-x}Se_2$ (CIGS) and large absorption coefficients of over 10^4 cm^{-1} in the visible band [1], it is a suitable absorption layer for applications in thin-film solar cells. With these properties, CZTS is thought to be an attractive replacement of CIGS ($Cu(In, Ga)S_2$) based solar cells, though CIGS has achieved 20.3 % efficiency [2]. Up to date, the best convert efficiency for CZTS is 6.8 % [3] and 9.6 % for $Cu_2ZnSn(S, Se)_4$ [4]. CZTS can be prepared by various methods, such as spray-pyrolysis deposition [5], electron-beam evaporation [6], radio-frequency magnetron sputtering deposition [7], electroplated deposition [8], pulsed laser deposition [9], and sol-gel deposition [10]. Most of the CZTS films have sulfur-poor problems, and they were furher treated through post-sulfurization using hydrogen sulfide (H_2S) [6-7, 11-12]. In this work, CZTS thin films were prepared by sol-gel spin-coating deposition [13], and then treated with post annealing processes.

2. EXPERIMENTAL METHODS

In this work, soda lime glass (SLG) with 2 × 2 cm^2 dimension was used as substrates. The precursors prepared by using copper (II) chloride, zinc (II) chloride, tin (IV) chloride, and thiourea in concentrations of 2 *M*, 1 *M*, 1 *M*, and 8 *M*, respectively, were added to a 50% ethanol-water mixture, stirred at 40 °C for 30 min and then spinning coated on SLG substrates. SLG substrates were ultrasonically cleaned in acetone and D.I. water, respectively for a couple times. Two-step spin-coating was performed at 200 rpm for 10 sec and then 1000 rpm for 60 sec. The as spin-coated films were baked on a hot plate at 100 °C for 10 min to remove residual solvent, and then synthesized at the temperature of 280 °C. The route of spin-coating and synthesizing processes were repeated several times to obtain a favorable thickness of CZTS. The as-synthesized CZTS films were post-annealed in nitrogen ambiance for 60 min. Surface morphology of the CZTS was examined by scanning electron microscopy (SEM). Crystallinity of the CZTS films was characterized by an X-ray diffractometer (XRD). Analyzer 1280 from the n&k Technology was used to measure UV-visible spectra from 190 to 1000 nm. Compositions of the CZTS thin films were evaluated by energy dispersive spectrometry (EDS). Electrical characteristics of the CZTS were examined by the four-point probe method. Raman scattering measurement was conducted under exciting from a 532 nm LASER line.

3. RESULTS AND DISCUSSION

The surface morphology and the cleavage of the CZTS were examined by scanning electron microscope (SEM). The cleavage images of as obtained CZTS layers are solid, continuous and uniform. Figure 1 shows the surface images of CZTS thin films (~ 2.9 μm) prepared without annealing and with annealing processes at 300, 350, and 400 °C, respectively. The CZTS films reveal voids over surface texture structures. The number of voids seems to shrink with increasing annealing temperature. A bit of cracks on surface resulting from subsequent annealing processes are found in Figure 1(b)-(d).

The crystallinity of the CZTS was characterized by an X-ray diffract meter (XRD). The synthesized process at different temperatures was then examined. It was observed that degree of crystallinity of the CZTS increases with synthesizing temperatures. At temperature below 280 °C, the CZTS may not have a complete synthesization, and phases other than CZTS appeared. The CZTS reflects the kesterite structure with the major XRD diffraction from (112), (200), (220) and (312) planes as the synthesizing temperature is higher than 240 °C. In this work, the as spin-coated CZTS was synthesized at 280 °C. In order to further investigate the influence of annealing process at higher temperatures on the properties of as-synthesized CZTS, we carried out the processes in a furnace filled with nitrogen ambiance.

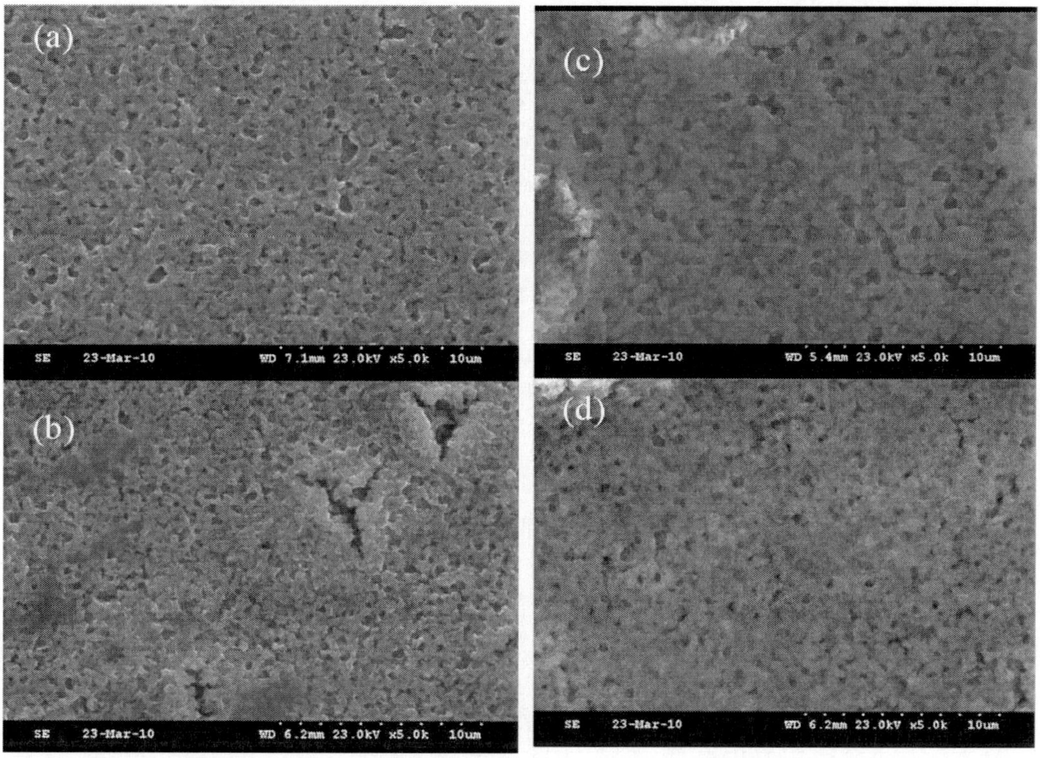

Figure 1. SEM photographs of CZTS films (a) as synthesized at 280 °C, and then treated with post-annealing at (b) 300 °C, (c) 350 °C, (d) 400 °C.

From Figure 2, we can see that degree of crystallinity of the CZTS decreases with annealing temperatures. The degrading crystallinity of CZTS resulting from the annealing process at a temperature 400 °C was observed. The phenomena imply that CZTS can't withstand temperature treatment higher than ~ 400 °C without sulfur supply. The degrading CZTS crystallinity at ~ 400 °C may be attributed to changing chemical composition of sulfur content due to outgassing.

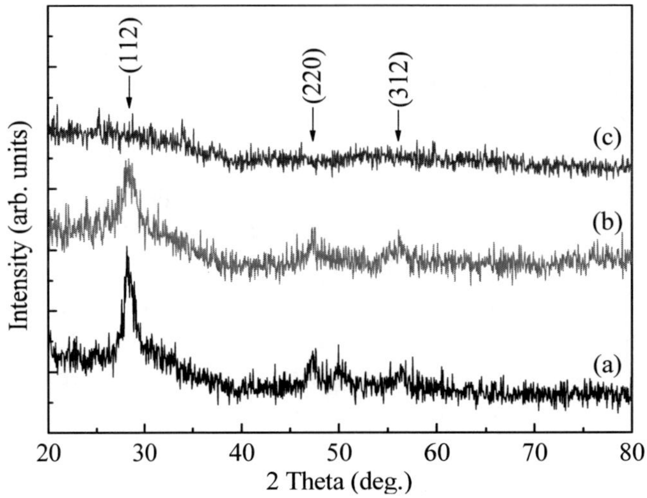

Figure 2. XRD patterns of CZTS films treated with post-annealing at (a) 300 °C, (b) 350 °C, (c) 400 °C.

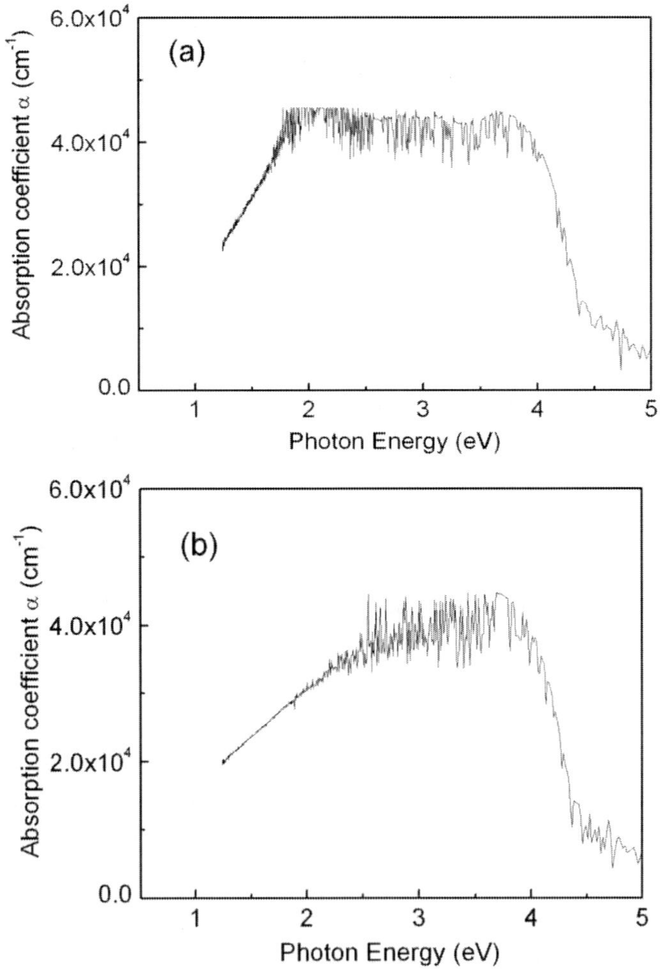

Figure 3. (continued)

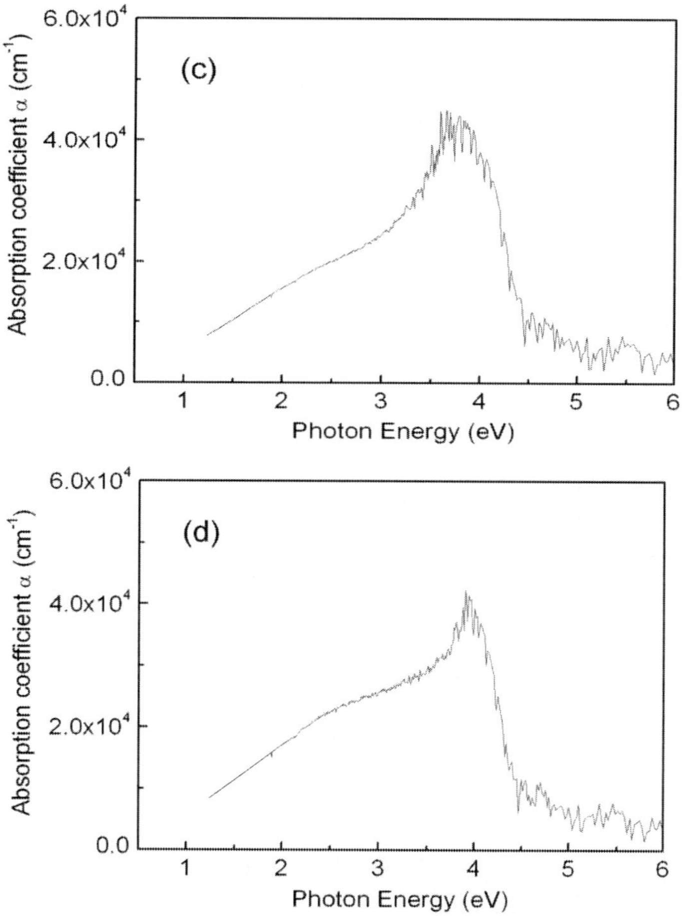

Figure 3. Absorption coefficient spectrum of CZTS films, (a) as synthesized at 280 °C, and then treated with post-annealing at (b) 300, (c) 350, (d) 400 °C.

The transmittance and reflectance of the CZTS films were examined using UV-visible spectra from 190 to 1000 nm. Optical absorption coefficients were then calculated [14] as

$$\alpha = \frac{1}{t}[\ln(1 - R_\lambda) - \ln T_\lambda], \tag{1}$$

where t is the film thickness, R_λ and T_λ are transmittance and reflectance coefficients, respectively.

Figure 3 shows absorption coefficient spectrum of as-synthesized CZTS at 280 °C and further post-annealed CZTS at 300, 350 and 400 °C, respectively. It is obvious that all of absorption coefficients of CZTS samples are higher than 1×10^4 cm^{-1} over the solar radiation band. A high absorption property over the solar radiation band is thought to be an attractive candidate for solar cells applications. Figure 3(a) reveals that before annealing treatment, the as-synthesized CZTS has a high uniform absorption spectrum over a larger photon energy region. As annealing temperature increases, its absorption properties deteriorate apparently for low photon energy. The phenomenon of degrading absorption properties of CZTS in lower photon energy region may relate to the deviation of chemical composition through annealing processes.

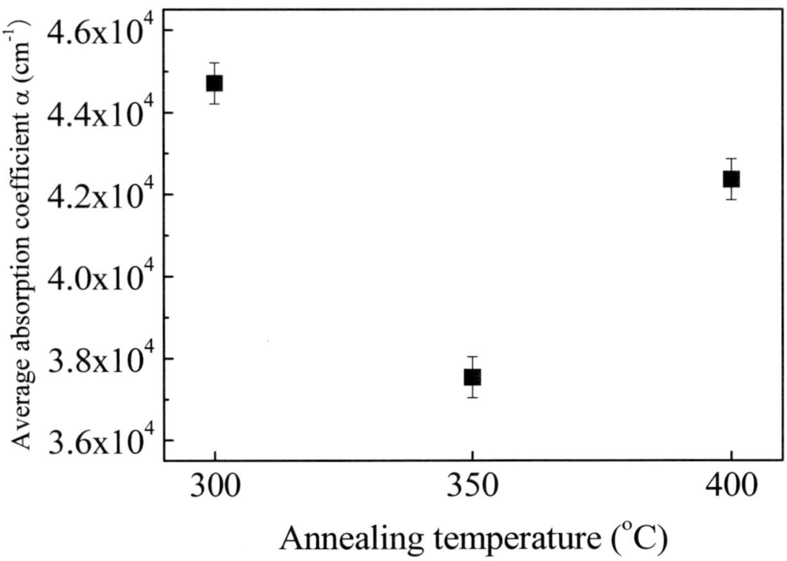

Figure 4. Average absorption coefficients of CZTS films as a function of annealing temperatures.

As we know, the major power of solar radiation is in the range of 400 ~ 700 nm (1.8 ~ 3.1 eV). High average absorptions over the range of 1.8 to 3.1 eV are essential requirement for using in solar cell absorption layer. Figure 4 shows average optical absorption coefficients of CZTS resulting from different annealing temperatures. The average absorption coefficients of all the CZTS samples are higher than 3×10^4 cm^{-1} in the major range of solar radiation spectrum. In this work, despite the CZTS samples were annealed at ~ 400 °C, they still have high optical-absorption properties suitable for solar cell application.

From [14], we know that the absorption coefficient related to direct transitions has following spectral dependence:

$$\alpha = A(h\nu - E_g)^{1/2}, \tag{2}$$

where A is a constant, $h\nu$ is the photon energy and E_g is the energy band gap. A direct transition can then be found by the linear dependence of $(\alpha h\nu)^2$ vs. $h\nu$. Thus optical-energy gap can be determined by extrapolating the linear part of the spectrum of the square absorption coefficients of the CZTS to zero absorption coefficient as a function of photon energy. As shown in Figure 5, the CZTS optical energy gap is ~ 1.65, ~ 1.85, and ~ 1.97 eV for annealing at 300, 350 and 400 °C, respectively. The as-obtained CZTS optical energy gap reveals different values through annealing at different temperatures. It may result from different compositions in the as-annealed CZTS.

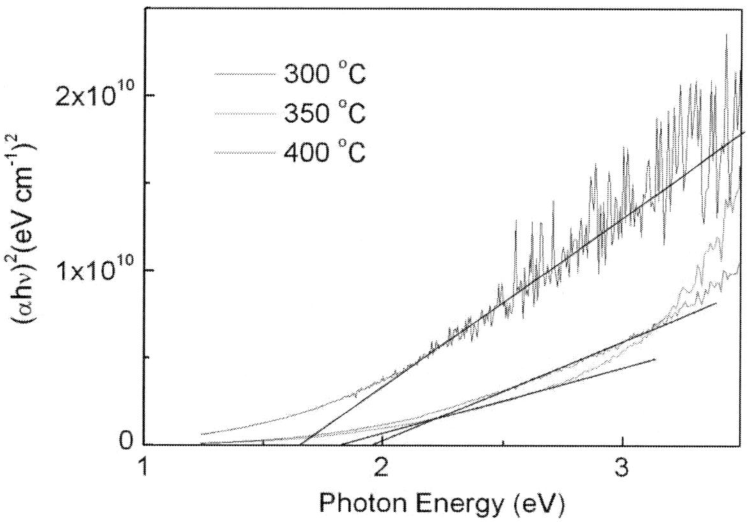

Figure 5. $(\alpha h\nu)^2$ as a function of photon energy for as-annealed CZTS at 300, 350 and 400 °C.

Table 1. CZTS composition ratio with respect to annealing temperature.

	Composition Ratio			
Annealing Temperature	Cu/(Zn+Sn)	Zn/Sn	S/metal*	O/total**
300°C	0.5	0.31	1.49	0.42
350°C	0.77	0.53	0.58	0.6
400°C	0.18	0.49	0.68	0.75

*metal: Cu + Zn + Sn, **total: Cu + Zn + Sn + S + O

The influence of annealing processes on CZTS properties was also evaluated from chemical composition ratios. In this work, the annealing process was conducted in an open outlet furnace made of tube quartz, in which nitrogen gas was introduced from inlet of furnace. Evaluation of CZTS treated with annealing processes at high temperature may give valuable information for CZTS solar cells when situated at a high thermal environment. Composition ratios in the CZTS thin films with respect to annealing temperature were examined by EDS tabulated in Table 1. Figure 6 shows chemical compositions of constituents in the CZTS as a function of annealing temperature. The composition of Cu:Zn:Sn:S = 14:7:22:57 for annealing at 300 °C reveals a cupper-poor CZTS. The chemical composition of sulfur in the CZTS decreases with annealing temperature. This may result from outgassing of sulfur at high temperature. Also we can see that oxygen was incorporated in a great quantity into the CZTS through the annealing processes using the semi-open quartz furnace. The composition ratios of oxygen obviously increase with annealing temperatures. The phenomenon of oxygen incorporation in CZTS samples resulting from annealing processes is worth further investigating.

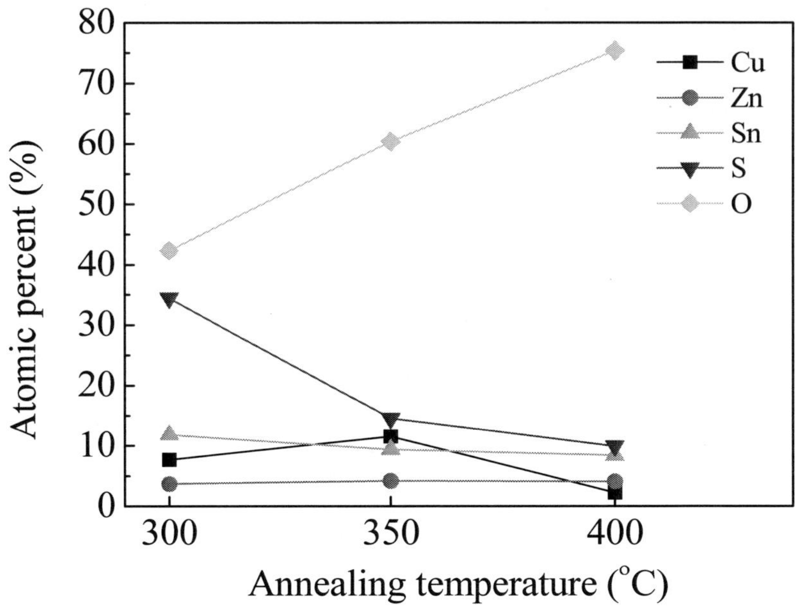

Figure 6. Composition ratio of CZTS films as a function of annealing temperatures.

4. CONCLUSIONS

Kesterite structure Cu_2ZnSnS_4 with direct optical band gap energy of ~ 1.65 eV was obtained by synthesizing and annealing the sol-gel spin-coated precursors. In this work, solutions of copper (II) chloride, zinc (II) chloride, tin (IV) chloride, and thiourea were used as the precursors for preparing CZTS thin films. The CZTS thin films were successfully prepared on SLG substrates by so-gel spin-coated deposition. The surface morphology of the CZTS reveals texture structures. The CZTS films synthesized at 280 °C and then treated with annealing reveal kesterite structures as examined from XRD. The annealing treatment of CZTS reveals useful information on CZTS composition ratio. The optical-energy gap of the CZTS varies with annealing temperature, and 1.65 eV is obtained through 300 °C annealing. The average optical-absorption coefficient of the CZTS is over ~ 3×10^4 cm^{-1}, and the absorption band covers the major solar irradiation spectrum. In this work, we have successfully prepared crystalline CZTS with high optical-absorption coefficients.

ACKNOWLEDGMENTS

The authors would like to express much appreciation to the National Science Council of the Republic of China under the contract, No. NSC 97-2221-E-022-002 and No. NSC 99-2221-E-022-005, and National Kaohsiung Marine University under the project, No. RND-98-012 for supporting this work.

REFERENCES

[1] Benedetto, F. D.; Bernardini, G. P.; Borrini, D.; Lottermoser, W.; Tippelt G.; Amthauer, G. *Phys. Chem. Minerals 2005, vol. 31*, 683-690

[2] Jackson, P.; Hariskos, D.; Lotter, E.; Paetel, S.; Wuerz, R.; Menner, R.; Wischmann, W.; Powalla, *M. Prog. Photovoltaics Res. Appl. 2011, vol. 19*, 894–897

[3] Katagiri, H.; Jimbo, K.; Yamada, S.; Kamimura, T.; Maw, W. S.; Fukano, T.; Ito, T.; Motohiro, *T. Appl. Phys., Express 2008, vol. 1*, 041201-041202

[4] Todorov, T. K.; Reuter, K. B.; Mitzi, D. B. *Adv. Mater. 2010, vol. 22*, E156-E159

[5] Kamoun, N.; Bouzouita, H.; Rezig, B. *Thin Solid Films. 2007, vol. 515*, 5949-5952

[6] Katagiri, H. *Thin Solid Films. 2005, vol. 480*, 426-432

[7] Jimbo, K.; Kimura, R.; Kamimura, T.; Yamada, S.; Maw, W.S.; Araki, H.; Oishi, K.; Katagiri, H. *Thin Solid Films. 2007, vol. 515*, 5997-5999

[8] Ennaoui,A.; Lux-Steiner, M.; Weber, A.; Abou-Ras, D.; Kotschau, I.; Schock, H.W.; Schurr, R.; Holzing, A.; Jost, S.; Hock, R.; Vo, T.; Schulze, J.; Kirbs, A. *Thin Solid Films. 2009, vol. 517*, 2511-2514

[9] Pawar,S.M.; Moholkar, A.V.; Kim, I.K.; Shin, S.W.; Moon, J.H.; Rhee, J.I.; Kim, J.H. *Curr. Appl. Phys. 2010, vol. 10*, 565-569

[10] Tanaka, K.; Moritake, N.; Uchiki, H. *Sol. Energy Mater. Sol. Cells. 2007, vol.91*, 1199-1201

[11] Katagiri, H.; Ishigaki, N.; Ishida, T.; Saito, K. *Jpn. J. Appl. Phys. 2001, vol.40*, 500-504

[12] Kobayashi, T.; Jimbo, K.; Tsuchida, K.; Shinoda, S.; Oyanagi, T.; Katagiri, H. *Jpn. J. Appl. Phys. 2005, vol. 44*, 783-787

[13] Yeh, M.Y.; Lee, C.C.; Wuu, D.S. *J. Sol-Gel Sci. Technol. 2009, vol. 52*, 65-68

[14] Pankove, J. I. *Optical Processes in Semiconductors*; Dover Publication: New York, 1975; pp 34-95

In: Physics and Mechanics of New Materials … ISBN: 978-1-62618-535-7
Editors: Ivan A. Parinov and Shun-Hsyung Chang

Chapter 3

EFFECTS OF ANNEALING ON STRUCTURAL AND OPTICAL PROPERTIES OF SINGLE CRYSTALLINE GA_2O_3 EPILAYERS

Dong-Sing Wuu*[*a,b], Ray-Hua Horng[c,d], Parvaneh Ravadgar[d] and Sin-Liang Ou[a]

[a]Department of Materials Science and Engineering,
National Chung Hsing University, Taichung, Taiwan
[b]Department of Materials Science and Engineering,
Da-Yeh University, Changhua, Taiwan
[c]Graduate Institute of Precision Engineering,
National Chung Hsing University, Taichung, Taiwan
[d]Department of Photonics & Advanced Optoelectronic Technology Center,
National Cheng Kung University, Tainan, Taiwan

ABSTRACT

The (-201) oriented single crystalline β-Ga_2O_3 epilayers were deposited on (0001) sapphire at 450 °C by metalorganic chemical vapor deposition, and then annealed at 800°C in atmosphere. From the analyses by transmission electron microscopy and X-ray photoelectron spectroscopy, it was evident that the oxygen vacancies existed in the as-deposited epilayer and reduced after annealing. Due to the decrease in oxygen vacancy, the crystal quality of Ga_2O_3 epilayer was improved. Photoluminescence results indicated that the defect emissions in the as-deposited epilayer could result from surface states and point defects. The defect emissions decreased apparently after annealing, causing by the recombination of oxygen atoms with gallium atoms inside the Ga_2O_3.

Moreover, the metal-semiconductor-metal photodetectors were fabricated with the as-deposited and annealed Ga_2O_3 epilayers, respectively. Current-voltage measurements revealed that the photodetector with annealed Ga_2O_3 epilayer had a lower dark current with 2-3 orders of magnitude than that fabricated with as-deposited Ga_2O_3 epilayer at the

* dsw@dragon.nchu.edu.tw, [a]Department of Materials Science and Engineering, National Chung Hsing University, No.250, Kuo Kuang Road, Taichung 40227, Taiwan, [b]Department of Materials Science and Engineering, Da-Yeh University, No.168, University Road, Changhua 51591, Taiwan

same bias voltage. The improved performance is attributed to the decrease in free electrons resulted from oxygen vacancies, leading to a lower current leakage.

1. INTRODUCTION

Recently, monoclinic structure gallium oxide (β-Ga_2O_3) has attracted great interest because of its wide bandgap (~ 4.8-4.9 eV) [1]. Moreover, Ga_2O_3 is a thermally and chemically stable transparent conductive oxide over a wide range including the visible and ultraviolet wavelength [2]. Such properties indicate that β-Ga_2O_3 has high potential used in plenty of optoelectronic devices consisting of gas sensors, flat-panel displays, phosphors and transparent electronic devices [1, 2]. Lately, there have been a number of investigations about bulk- and thin-film-typed β-Ga_2O_3 deep ultraviolet (DUV) photodetectors [3, 4]. Solar-blind photodetectors exhibit the high sensitivity to DUV light with wavelength shorter than 280 nm, i.e., they have the ability to confine the photo-response to DUV region even under sun or room illuminations. Therefore, DUV photodetectors could be applied for various devices, such as flame detector, missile plume sensor, and ozone holes monitor [5, 6].

As we known, dark current and photoresponsivity are the main characteristics of β-Ga_2O_3 photodetector performance, which are influenced by surface states and point defects in β-Ga_2O_3. Surface states could be resulted from the adsorption of oxygen ions at the surface, leading to the attraction of free electrons to the oxygen molecules and forming disordered surface. On the other hand, point defects are usually induced by the formations of oxygen vacancies, oxygen interstitials, gallium vacancies and gallium interstitials [7, 8]. Therefore, the present challenge in designing β-Ga_2O_3 films is the cure for surface states and point defects, and the performance of their optoelectronic devices will be improved.

Various methods have been proposed for the growth of β-Ga_2O_3 thin films, such as sputtering, molecular beam epitaxy, solution-phase process, pulsed laser deposition, and metalorganic chemical vapor deposition (MOCVD) [9-13]. Up to now, there is little research on the pure single oriented β-Ga_2O_3. Ga_2O_3 with high-quality single crystal is important for fabrication of high performance solar-blind photodetector. After Ga_2O_3 growth, the epilayers were annealed to cure surface states and point defects. Furthermore, the as-deposited and annealed β-Ga_2O_3 epilayers were used to fabricate photodetectors.

2. EXPERIMENTAL

In this chapter, Ga_2O_3 epilayers were grown on (0001) sapphire substrates by MOCVD system (Emcore, D-180). In order to minimize the gas phase reaction which can induce the formation of particles and thus degrade the Ga_2O_3 quality, organometallic sources and oxygen were introduced separately into the reactor and then mixed immediately before entering into the growth chamber. Trimethylgallium (TMGa) and pure oxygen (99.999%) were applied for Ga and O sources, respectively. Pure Ar (99.999%) was used as the carrier gas, passed through the TMGa bubbler to deliver TMGa vapor into the reactor. Substrate temperature and chamber pressure for the Ga_2O_3 growth were kept at 450 °C and 15 torr, respectively. The flows of TMGa and oxygen were maintained at 200 and 50 sccm, respectively. The thickness of Ga_2O_3 epilayer was fixed at 500 nm. After Ga_2O_3 growth, the epilayers were annealed at 800 °C for 30 min in air ambient.

The crystalline quality and orientation of Ga_2O_3 epilayers were analyzed by X-ray diffraction (XRD) (PANalytical, X'Pert Pro MRD). The Cu K_α line (λ = 1.5418 Å) was employed as the source and Ge (220) was used as the monochromator for XRD. The microstructure of Ga_2O_3 epilayer was taken with transmission electron microscopy (TEM) (JEOL JEM-2100F) in which the electron diffraction pattern of the epilayer was also analyzed. Chemical bonding states in the films were characterized by X-ray photoelectron spectroscopy (XPS, ULVAC-PHIPHI 5000). Al K_α (1486.7 eV) radiation was used as the excitation source. All obtained spectra were calibrated to a C 1*s* electron peak at 284.6 eV. Photoluminescence (PL) measurement with a 325 nm He–Cd laser was performed to investigate the optical properties of epilayers. For metal-semiconductor-metal (MSM) photodetector device process, Ni/Au (50/100 nm) film deposited onto the sample by thermal evaporation was used as the Schottky contact metal. The interdigitated contact electrodes were prepared by standard photolithography and wet etching. The size of interdigitated pattern on the shadow mask was 2 mm wide and 2.2 mm long. The finger width and interspacing were the same as 100 μm. Figure 1 shows a schematic diagram of the fabricated Ga_2O_3 photodetector. Current–voltage (I-V) characteristics of the MSM photodetector were investigated by a HP 4156 semiconductor analyzer at room temperature in dark environment.

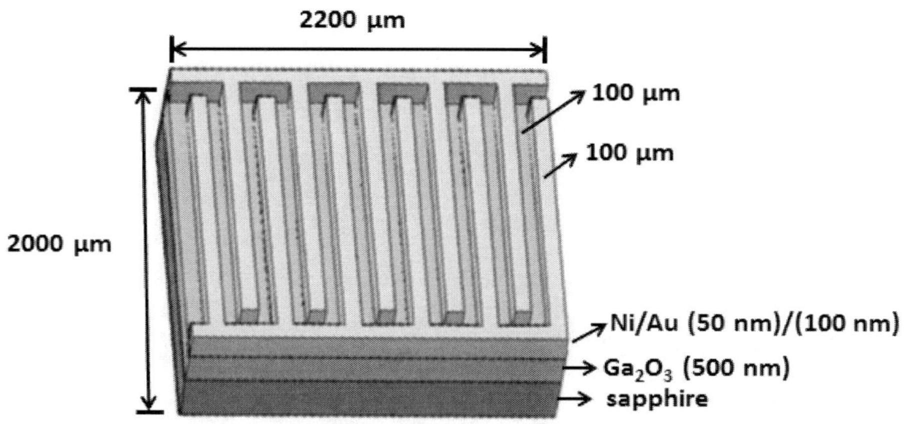

Figure 1. Schematic diagram of the fabricated Ga_2O_3 photodetector.

3. Analysis of Experimental Data

Figure 2(a) exhibits the XRD patterns of the Ga_2O_3 epilayers grown at 450, 500 and 600°C, respectively. We found that all the films show the diffraction peaks that belong to the (-201) plane family of β-Ga_2O_3, i.e. (-201), (-402) and (-603) planes of single crystalline β-Ga_2O_3 [14]. It is worthy to mention that the single crystalline β-Ga_2O_3 films have been reported seldom. Figure 2(b) shows the XRD rocking curve of (-402) reflection for β-Ga_2O_3 epilayers grown at 450-600 °C. It can be seen that the full width at half maximum (FWHM) of rocking curve at (-402) plane of these films deposited at 450, 500 and 600 °C were 366, 363 and 368 arcsec, respectively. Obviously, the FWHM results indicate that the Ga_2O_3

epilayers prepared at 450-600 °C all have high crystal quality. Furthermore, we chose the Ga_2O_3 epilayer grown at 450 °C to perform the annealing process at 800 °C and compare the characteristics with as-deposited epilayer. Figure 2(c) shows the XRD rocking curve of (-402) reflection for as-deposited and 800 °C-annealed β-Ga_2O_3 epilayers grown at 450 °C, the FWHM values are determined to be 366 and 350 arcsec, respectively. This reveals that the crystal quality of β-Ga_2O_3 epilayer is improved after annealing at 800 °C.

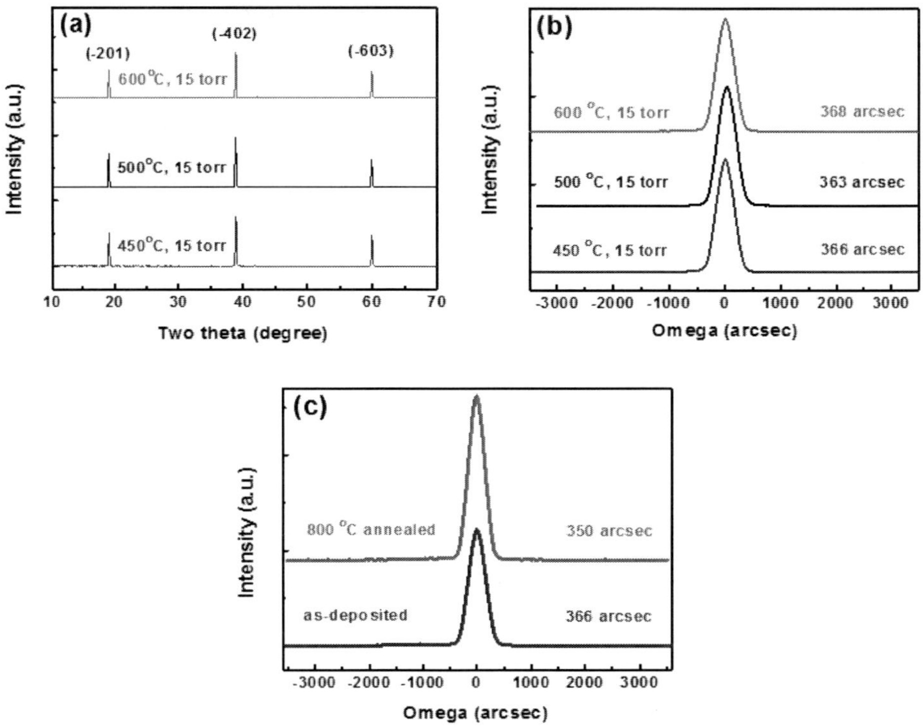

Figure 2. (a) XRD patterns of Ga_2O_3 films grown at 450-600 °C, (b) XRD rocking curves for (-402) diffraction peak of Ga_2O_3 grown at 450-600 °C and (c) XRD rocking curves for (-402) diffraction peak of as-deposited and 800 °C-annealed Ga_2O_3 grown at 450 °C.

The high-resolution cross-sectional TEM images of Ga_2O_3/sapphire structures and the corresponding electron diffraction patterns of Ga_2O_3 epilayers in as-deposited and annealed states are shown in Figures 3(a) and 3(b), respectively. The *d* spacings of Ga_2O_3 films in as-deposited and annealed states are 4.82 and 4.7 Å, respectively, while the standard *d* spacing of β-Ga_2O_3 (-201) is 4.67 Å according to the JCPDS card. Based on the XRD results shown in Figure 2(a), the *d* spacing can be indexed to (-201) plane. In the past research, the β-Ga_2O_3 thin film which consisted of (-201) oriented planes could be resulted from the similar oxygen atoms arrangements at the interface between a *c*-plane sapphire substrate and a β-Ga_2O_3 (-201) plane [15]. From our calculation, the corresponding diffraction patterns taken at the Ga_2O_3 epilayers have the same zone axis of [102]. It proves that the epilayers are both β-Ga_2O_3 phase again. However, except for the β-Ga_2O_3 planes (marked with green circles) in the Figure 3(a), the other diffraction dots marked with yellow circles are also observed. It indicates that the diffraction dots marked with yellow circles are probably resulted from the formation of oxygen vacancies in Ga_2O_3 epilayer, leading to the distortion and enlargement in *d* spacing. After annealing at 800 °C, the diffraction pattern of epilayer shows a clear β-Ga_2O_3 phase, as shown in Figure 3(b). Obviously, the annealing process can provide the oxygen atoms to Ga_2O_3 epilayer and eliminate oxygen vacancies, which cause a close *d* spacing (4.7 Å) to the standard *d* spacing (4.67 Å).

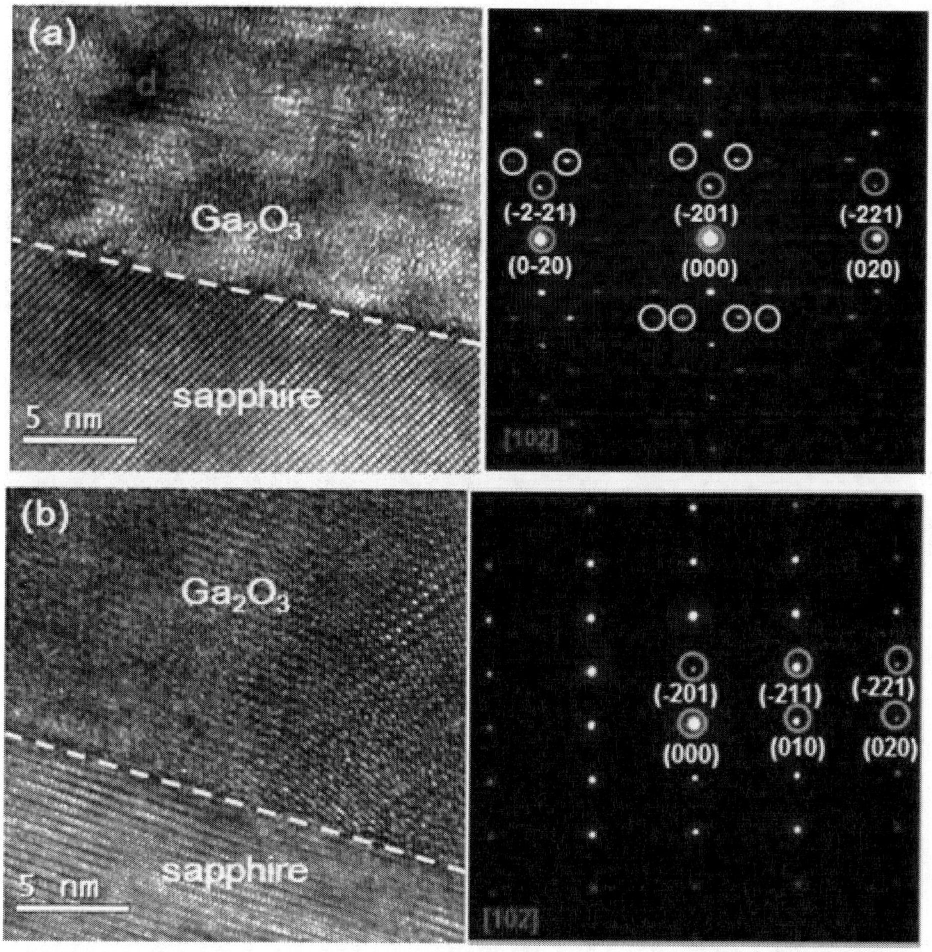

Figure 3. High-resolution cross-sectional TEM photograph of Ga_2O_3/sapphire structure and the corresponding electron diffraction pattern of Ga_2O_3 epilayer in (a) as-deposited and (b) annealed states.

Figure 4 displays a schematic diagram of the epitaxial relationship between β-Ga_2O_3 (-201) plane (as-deposited and annealed states) and Al_2O_3 (0006) plane. According to our calculations for the electron diffraction patterns shown in Figures 3(a) and 3(b), the lattice distances of [102] and [010] directions in Ga_2O_3 epilayer were decreased to the standard values after annealing at 800 °C. As we discussed, the decrease in lattice distance is attributed to the elimination of oxygen vacancies during annealing. Moreover, the lattice mismatch between Ga_2O_3 epilayer and Al_2O_3 substrate can be determined by further calculations. The lattice mismatch between as-deposited Ga_2O_3 epilayer and Al_2O_3 substrate was calculated to be about 10.1%. After 800 °C annealing, the lattice mismatch between these two materials can be reduced to 7.8%. It is well known that a lower lattice mismatch would improve the structural property of deposited film. The results are in good agreement with XRD rocking curves for (-402) diffraction peak of as-deposited and annealed Ga_2O_3 films, as shown in Figure 2(c).

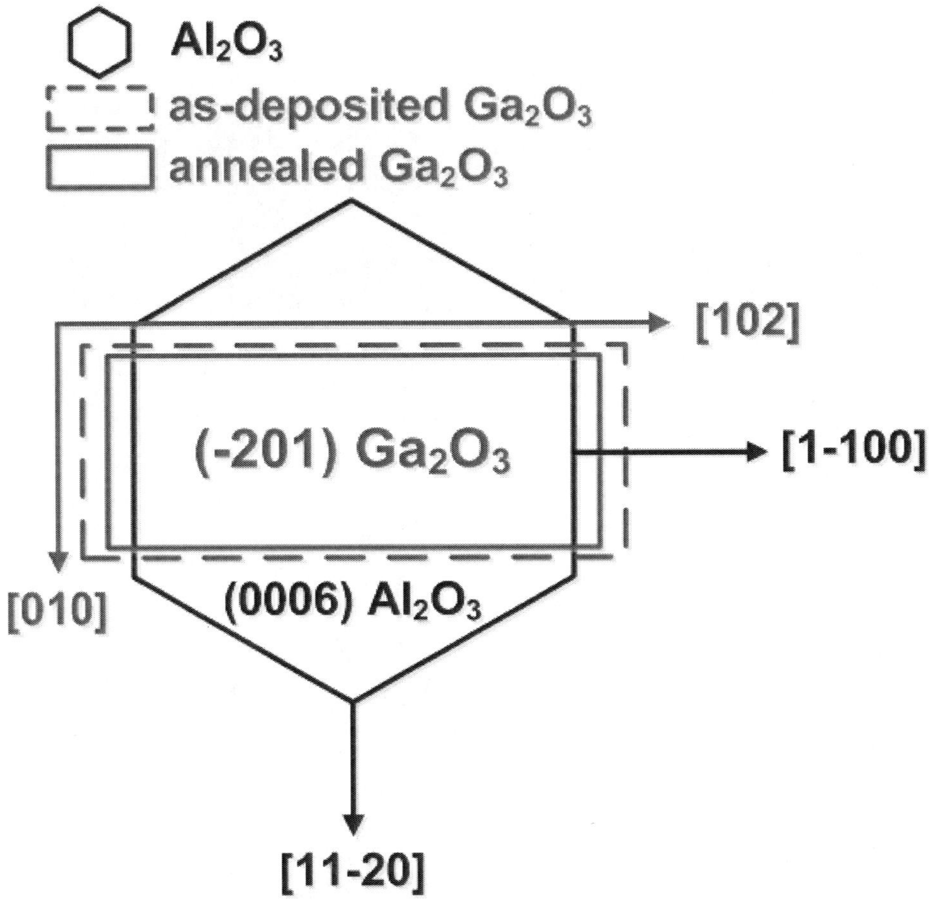

Figure 4. Schematic diagram of the epitaxial relationship between the β-Ga_2O_3 (-201) plane (as-deposited and annealed states) and the Al_2O_3 (0006) plane.

Figures 5(a) and 5(b) exhibit the O1s and $Ga2p^3$ core level XPS spectra of as-deposited and annealed Ga_2O_3 epilayers, respectively. All XPS spectra of the samples were fitted well with Gaussian functions. We observed that O1s peaks of as-deposited and annealed Ga_2O_3 epilayers centered at 531 and 531.5 eV, respectively. In general, the O1s peak at around 531 eV can be ascribed to the O^{2-} ions in the oxygen-deficient regions [16], resulting in the non-stoichiometric oxides. After annealing at 800 °C, a shift of O1s binding energy in the direction toward higher energy side was found, indicating the number of oxygen atom in the film was increased. It is suggested that the increment in the number of oxygen atom would decrease the oxygen vacancies in Ga_2O_3 epilayer, improving the stoichiometric composition [17]. In Figure 5(b), the peak positions of $Ga2p^3$ are both higher than 1117.5 eV. As compared with the peak for elemental gallium at the binding energy of 1116.6 eV, the core level of gallium has a positive shift from elemental gallium [18]. It reveals that the gallium atoms in Ga_2O_3 epilayer are indeed in compound state.

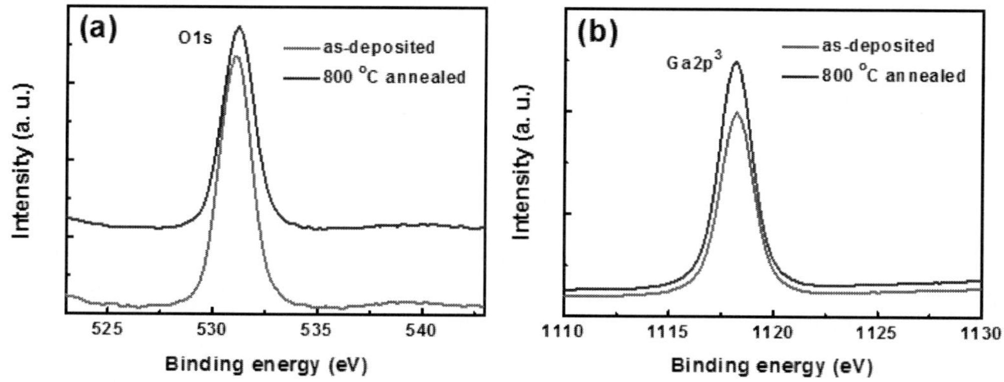

Figure 5. XPS spectra of (a) O1s core level and (b) Ga2p[3] core level of as-deposited and annealed Ga_2O_3 epilayers.

Figures 6(a) and 6(b) exhibit the PL spectra of as-deposited and 800 °C annealed Ga_2O_3 epilayers, respectively. The measured wavelength is increased from 350 to 600 nm. In Figure 6(a), it was found that two emissions centered at 408 and 526 nm, which resulted from the defects. We believed that the PL defect emissions of Ga_2O_3 epilayer are induced by the formations of adsorbed oxygen at the film surface, oxygen vacancies, gallium vacancies and gallium–oxygen vacancy pairs inside Ga_2O_3. The defect emissions probably generated from surface states and point defects in the form of donor levels to acceptor levels (electron-hole recombination). After annealed at 800 °C, the PL defect emissions were cured apparently, which may be resulted from the recombination of oxygen atoms in atmosphere with gallium atoms inside the Ga_2O_3. Oxygen recombination in Ga_2O_3 epilayer would decrease the oxygen vacancies and lower the influence of band bending caused by surface defects and point defects.

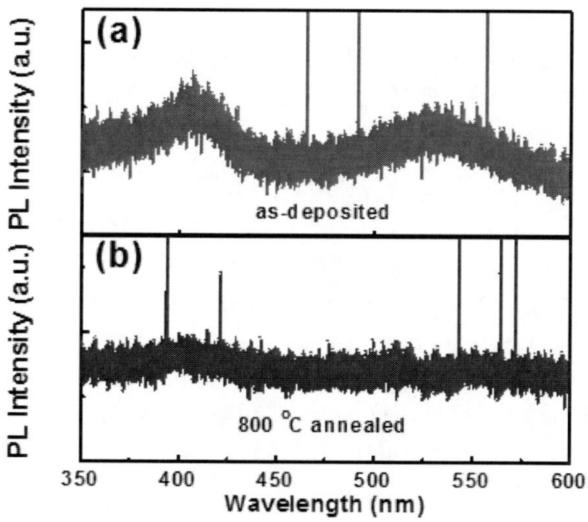

Figure 6. PL spectra of (a) as-deposited and (b) 800 °C annealed Ga_2O_3 epilayers.

To investigate the effect of annealing on device performance, the as-deposited and 800°C annealed Ga_2O_3 epilayers were fabricated the photodetectors and analyzed their dark current characteristics. The room temperature I–V curves measured in dark of Ga_2O_3 photodetectors are exhibited in Figure 7. The inset in Figure 7 is the cross sectional schematic diagram of Ga_2O_3 photodetector. We found that the dark current of photodetector fabricated with as-deposited Ga_2O_3 epilayer was obviously higher than that with 800 °C annealed Ga_2O_3 epilayer. Under the same bias voltage, there exists a large difference in dark current about 2-3 orders of magnitude between these two photodetectors. The result suggested that the high leakage current in as-deposited Ga_2O_3 epilayer was resulted from surface states and oxygen vacancies. The oxygen vacancies in Ga_2O_3 epilayer would lead to a lot of free electrons from gallium atoms. However, the free electrons can be decreased by the oxygen recombination with Ga_2O_3 after annealing, which induced a significant improvement in current leakage.

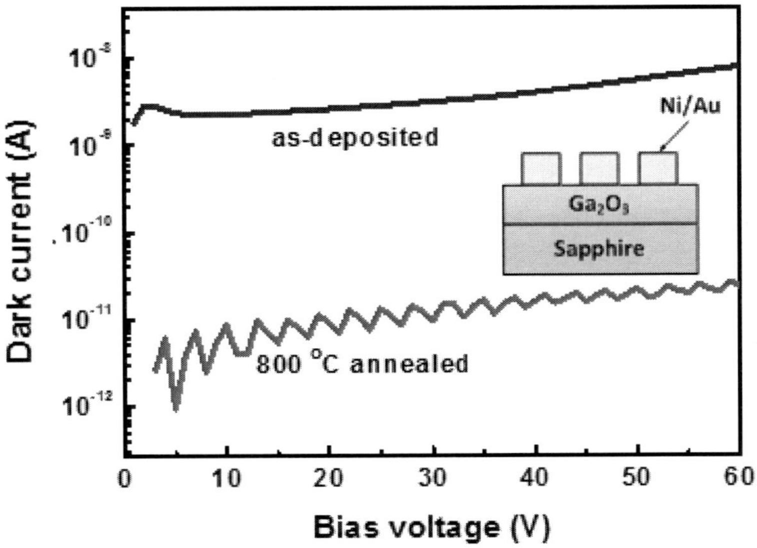

Figure 7. Current-voltage measurements of the β-Ga_2O_3 photodetectors. Inset is the cross sectional schematic diagram of β-Ga_2O_3 photodetector.

4. Conclusion

In summary, the (-201) oriented single crystal β-Ga_2O_3 epilayers have been grown on (0001) sapphire at 450 °C by MOCVD system. The crystal quality of Ga_2O_3 epilayer can be improved after annealing at 800 °C due to a decrease in oxygen vacancy formed in the as-deposited epilayer. The defect emissions in PL spectrum of as-deposited Ga_2O_3 were arisen from the formations of surface states and point defects, and the emissions were cured after annealing. The photodetectors were fabricated with the as-deposited and annealed Ga_2O_3 epilayers. From current-voltage measurements, the photodetector with annealed Ga_2O_3 epilayer exhibited an obviously lower dark current than that with as-deposited Ga_2O_3 epilayer. It was attributed to the decrease in free electrons resulted from oxygen vacancies, which leaded to a lower current leakage.

REFERENCES

[1] Varley, J. B.; Weber, J. R.; Janotti, A.; Van de Walle, C. G. *Appl. Phys. Lett.* 2010, *vol. 97*, 142106-1-142106-3.

[2] King, P. D. C.; McKenzie, I.; Veal, T. D. *Appl. Phys. Lett.* 2010, *vol. 96*, 062110-1-062110-3.

[3] Suzuki, R.; Nakagomi, S.; Kokubun, Y.; Arai, N.; Ohira, S. *Appl. Phys. Lett.* 2009, *vol. 94*, 222102-1-222102-3.

[4] Oshima, T.; Okuno, T.; Arai, N.; Suzuki, N.; Ohira, S.; Fujita, S. *Appl. Phys. Exp.* 2008, *vol. 1*, 011202-1-011202-3.

[5] Razeghi, M.; Rogalski, A.; *J. Appl. Phys.* 1996, *vol. 79*, 7433–7473.

[6] Hirano, A.; Pernot, C.; Iwaya, M.; Detchprohm, T.; Amano, H.; Akasaki, I. *Phys. Stat. Sol. (a)* 2001, *vol. 188*, 293-296.

[7] Vanithakumari, S. C.; Nanda, K. K. *Adv. Mater.* 2009, *vol. 21*, 3581-3584.

[8] Wang, T.; Radovanovic, P. V. *Chem. Commun.* 2011, *vol. 47*, 7161-7163.

[9] Baban, C. I.; Toyoda, Y.; Ogita, M. *Jpn. J. Appl. Phys.* 2004, *vol. 43*, 7213-7216.

[10] Tsai, M. Y.; Bierwagen, O.; White, M. E.; Speck, J. S. *J. Vac. Sci. Technol. A* 2010, *vol. 28*, 354-359.

[11] Kokubun, Y.; Miura, K.; Endo, F.; Nakagomi, S. *Appl. Phys. Lett.* 2007, *vol. 90*, 031912-1-031912-3.

[12] Matsuzaki, K.; Yanagi, H.; Kamiya, T.; Hiramatsu, H.; Nomura, K.; Hirano, M.; Hosono, H. *Appl. Phys. Lett.* 2006, *vol. 88*, 092106-1-092106-3.

[13] Kim, H. W.; Kim, N. H. *Appl. Surf. Sci.* 2004, *vol. 230*, 301-306.

[14] Gottschalch, V.; Mergenthaler, K.; Wagner, G.; Bauer, J.; Paetzelt, H.; Sturm, C.; Teschner, U. *Phys. Stat. Sol. (a)* 2009, *vol. 206*, 243-249.

[15] Oshima, T.; Okuno, T.; Fujita, S. *Jpn. J. Appl. Phys.* 2007, *vol. 46*, 7217-7220.

[16] Xiao, Z.; Liu, Y.; Zhang, J.; Zhao, D.; Lu, Y.; Shen, D.; Fan, X. *Semicond. Sci. Technol.* 2005, *vol. 20*, 796-800.

[17] Kang, S. J.; Joung, Y. H.; Shin, H. H.; Yoon, Y. S. *J. Mater. Sci: Mater. Electron.* 2008, *vol. 19*, 1073-1078.

[18] Briggs, D.; Seah, M. P. *Practical Surface Analysis*, Wiley, New York, 1979.

In: Physics and Mechanics of New Materials ...
Editors: Ivan A. Parinov and Shun-Hsyung Chang

ISBN: 978-1-62618-535-7

Chapter 4

EFFECT OF MODIFYING LITHIUM CARBONATE AND MANGANESE OXIDE ON THE STRUCTURE, MICROSTRUCTURE, FERROELECTRIC AND PIEZOELECTRIC PROPERTIES OF THE $PbFe_{1/2}Nb_{1/2}O_3$

A. V. Pavlenko**, *I. A. Verbenko, S. I. Shevtsova, L. A. Shilkina and L. A. Reznichenko
Research Institute of Physics, Southern Federal University, Rostov-on-Don, Russia

ABSTRACT

Uncontaminated ceramic samples Lead ferroniobate, $PbFe_{1/2}Nb_{1/2}O_3$ (PFN), pure (PFN) and modified as 1 wt. % Lithium carbonate (I) (PFNL) oxide and manganese (IV) (PFNM) were obtained. The introduction of modifiers has led to changes in the kinetics of sintering of ceramics, in their crystal structure, and as a consequence, the dielectric and piezoelectric properties. The observed effects are due to the formation of liquid phases and disorder in the cation sublattice. These also lead to a decrease in the Curie temperature, smearing of the ferroelectric paraelectric transition and increased electrical resistance. The modified materials have good piezoelectric properties and high dielectric constant. It makes it possible to use the modified ceramics in the low-frequency devices.

1. INTRODUCTION

Multiferroics are currently one of the most promising materials for microelectronics, spintronics, sensor and microwave technologies [1]. One of these materials, lead ferroniobate $PbFe_{1/2}Nb_{1/2}O_3$ (PFN), belongs to the set of ferroelectromagnets with perovskite structure, the magnetic properties of which are related to the presence of Fe^{3+} ions in octahedral positions [2]. The transitions to the antiferromagnetic and polar phases in this compound occur,

* E-mail: tolik_260686@mail.ru, Research Institute of Physics, Southern Federal University, 194, pr. Stachki, 344090 Rostov-on-Don, Russia

respectively, at the Neél temperature ($T_N \sim 140 - 170$ K) and the Curie temperature ($T_C \sim 368 - 385$ K), which depend, as well as ferroelectric and piezoelectric characteristics of the thermodynamic history of the samples [3, 4]. The stabilization of PFN properties is often achieved by the introduction of modifiers [5-8]. Among the latter, and at the same time, the best ones, as shown in [6-8], include lithium carbonate (I) and manganese oxide (IV). In this sense, it seems urgent to study the influence of Lee, and LT on the regularities of the structural, ferroelectric and piezoelectric properties of PFN ceramics, together with a detailed analysis of their microstructure, which was the subject of this paper.

2. Sample Preparation and Experimental Technique

The samples were synthesized by solid-phase reactions from high-purity oxides PbO, Fe_2O_3, Nb_2O_5, MnO_2 and Li_2CO_3 carbonate (of pure and analytical grade) by two stage firing and intermediate grinding at temperatures $T_1 = T_2 = 1123$ K and exposure times $\tau_1 = \tau_2 = 4$ h. MnO_2 and Li_2CO_3 was added in excess of stoichiometry in the stage of their synthesis in an amount of 1 wt.%. The ceramic billets were sintered at the temperature $T_{sint} = (1333 - 1373)$ K, depending on the composition, for 2.5 h.

Phase composition and synthesis completeness were checked by X-ray diffraction analysis. A precise study was performed on fine-grained ceramic cakes by powder X-ray diffraction on DRON-3 diffractometer using the filtered $Co_{K\alpha}$ radiation (Bragg–Brentano focusing). The linear (a), angular (α), and volume (V, perovskite cell volume) parameters were calculated by the standard technique. The experimental (ρ_{exp}) sample density was found by hydrostatic weighing in octane; the X-ray density ($\rho_{X\text{-}ray}$) was calculated from the formula $\rho_{X\text{-}ray} = 1.66M/V$, where M is the weight unit in grams and V is the perovskite cell volume in E_3; and the relative density ρ_{rel} was calculated from the formula $\rho_{rel} = (\rho_{exp}/\rho_{X\text{-}ray})\times 100\%$.

The complex dielectric permeability ($\varepsilon^* = \varepsilon' + i\varepsilon''$, where ε' and ε'' are the real and imaginary parts, respectively) was measured in the temperature range of 300 – 700 K at frequencies from 10 Hz up to 10^5 Hz using HIOKI 3522-50 LCR meter

Hysteresis loops in the temperature range of (300 – 450) K at the frequency of 50 Hz were obtained using an oscilloscope setup Sawyer - Tauera. We evaluated the residual polarization, P_R, and the coercive field, the E_C samples.

The microstructure of the sintered ceramic samples was studied by means of scanning electron microscope with the help of a CAMEBAX-MICRO microanalyzer.

Dielectric, piezoelectric and elastic parameters of ceramics were measured using the resonance–antiresonance method. The following parameters were determined: the relative dielectric permittivity of poled ($\varepsilon'^T_{33}/\varepsilon_0$) and unpoled ($\varepsilon'/\varepsilon_0$) samples, the piezoelectric modulus (d_{ij}), the electromechanical coupling, factor of planar vibration mode (K_p), the mechanical quality factor (Q_m).

3. Experimental Results and Discussion

The introduction of Li_2CO_3 and MnO_2 reduced the optimum sintering temperature, T_{sin}, ceramics from 1373 K (PFN) down to 1353 K (PFNL) and 1333 K (PFNM). The experimental density, $\rho_{exp,}$ also decreased from 8.07 g/cm^3 down to 7.76 g/cm^3 for the PFNL and increased from 8.07 g/cm^3 up to 8.08 g/cm^3, for the PFNM. However, the relative density remained above 90% (Table. 1). The results obtained are acceptable for conventional ceramic technology. X-ray studies show that the objects possess rhombohedral distorted perovskite structure and contain no extraneous phases. The modification results to a slight increase in the structural parameters: a linear cell parameter (a), unit cell volume (V) and the rhombohedral angle cells (α_{Rh}) (Table 1).

Table 1. X-ray data for PFN, PFNL and PFNM samples at room temperature.

Compounds	Symmetry	α, Å	α, angle degrees	V, Å^3	ρ_{exp} g/cm^3	ρ_{rel}, %
PFN	Rh	4.0141	89.92	64.68	8.07	95
PFNL	Rh	4.0149	89.93	64.72	7.76	91
PFNM	Rh	4.0153	89.93	64.74	8.08	96

Figure 1 shows images of the fragments of the cleaved ceramics studied.

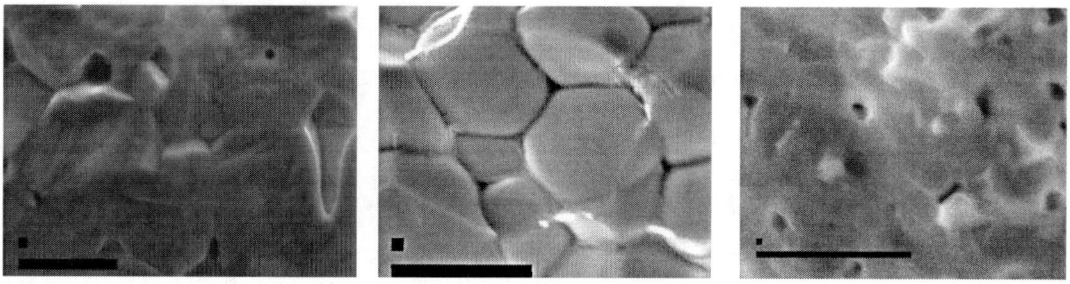

a b c

Figure 1. Pictures of areas of the cleaved ceramics PFN (a), PFNL (b) and PFNM (c) (Markers in fragments correspond to 10 μm).

In all cases, the dense structures of the crystallites in the shape of irregular polyhedral crystals are formed. The sizes of the grains in all objects are varied greatly, and the average grain size was in PFN (4 – 5) μm, in PFNL (10 – 16) μm, in PFNM (4 – 8) μm. Inhomogeneity of intragranular structure of ceramic raw material is pronounced. It manifests itself in the presence of dark and light areas in the image. This may indicate both the complex morphology of the surface, formed by cleavage, and the fluctuations of the volume of grains. In the study were found PFNL intergranular layers that were missing from in the images of the PFN and the PFNM. Figure 1 shows that the modifiers have different effects on the kinetics of sintering the ceramics. Thus, during sintering, PFN (Figure 1, b) is cleaved both through grains, and on their boundaries. This means that the strength of the grains and their boundaries are approximately the same. In PFNL cleavage takes place only at the grain boundaries. This means a higher strength of the grains. In PFNM the cleavage takes place only through grains that indicates growth strength at the grain boundaries.

Figure 2 shows the hysteresis loops of the objects at room temperature. In all cases, at the saturation are formed loops, which to be typical for classical ferroelectrics.

The P_R and E_C of PFN loops were ~ 0.2 C/m^2 and ~ 1.6×10^5 V/m, respectively. When modifying Li_2CO_3, they increased up to values ~ 0.22 C/m^2 and ~ 1.8×10^5 V/m, and when modifying MnO_2, they slightly decrease down to values ~ 0.175 C/m^2 and ~ 1.55×10^5 V/m, respectively. It was also found that the modified ceramics possess a greater electrical resistance than the unmodified. In PFNL it is increased from 2.9×10^7 Ω·m up to 2.6×10^{10} Ω·m and in PFNM it is increased from 2.9×10^7 Ω·m up to 7.2×10^9 Ω·m.

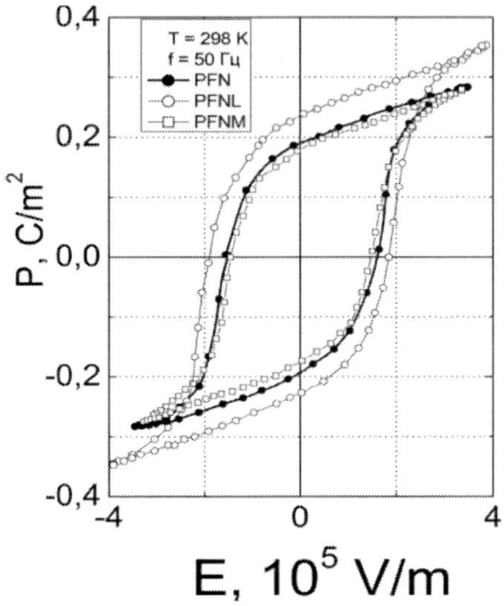

Figure 2. Hysteresis loops of ceramics PFN, PFNL and PFNM.

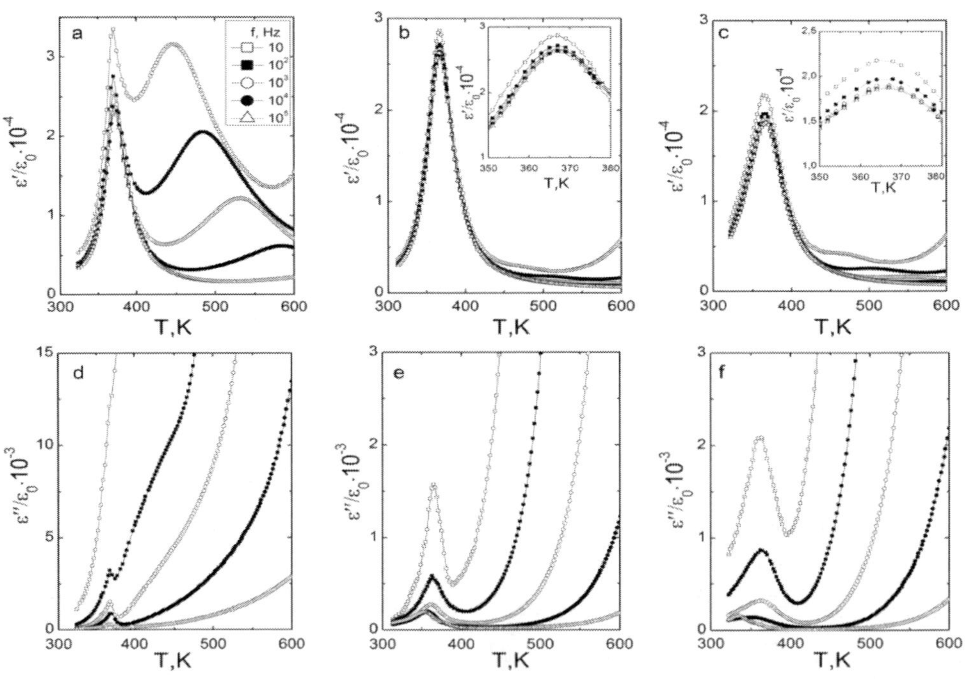

Figure 3. Temperature dependences of ε'/ε$_0$ and ε''/ε$_0$ PFN (a, d), PFNL (b, f) and PFNM (c, g) in the temperature range (300 – 600) K at frequencies f = 10^2, 10^3, 10^4 and 10^5 Hz.

Figure 3 shows the temperature dependences of $\varepsilon'/\varepsilon_0$ and $\varepsilon''/\varepsilon_0$ for the ceramics RFN (a, d), and RFNL (b, f), RFNM (c, g) in the temperature range (300 – 700) K at frequencies of 10, 10^2, 10^3, 10^4, 10^5 Hz, obtained in the cooling mode. Dispersions of $\varepsilon'/\varepsilon_0$ in the PFNL and PFNM are less than in PFN both at room temperature and at a temperature of the ferroelectric transition. It was also found that the PFNL and PFNM have lesser T_C (369 K and 370 K, respectively) than the T_C of PFN and T_C of these compounds is independent of frequency in the investigated frequency range. Moreover, the extremes on the curves obtained for the PFN are lesser visible than those on the curves obtained for PFNL and PFNM. These extremes, as in the case of PFN [9], are the result of peculiarities of the real (defect) structure of the ceramics, which leads to the appearance of Maxwell-Wagner polarization and relaxation. Similar patterns are observed in the dependences $\varepsilon''/\varepsilon_0$ (T). But the temperature of maximum $\varepsilon''/\varepsilon_0(T)$ depends of frequency.

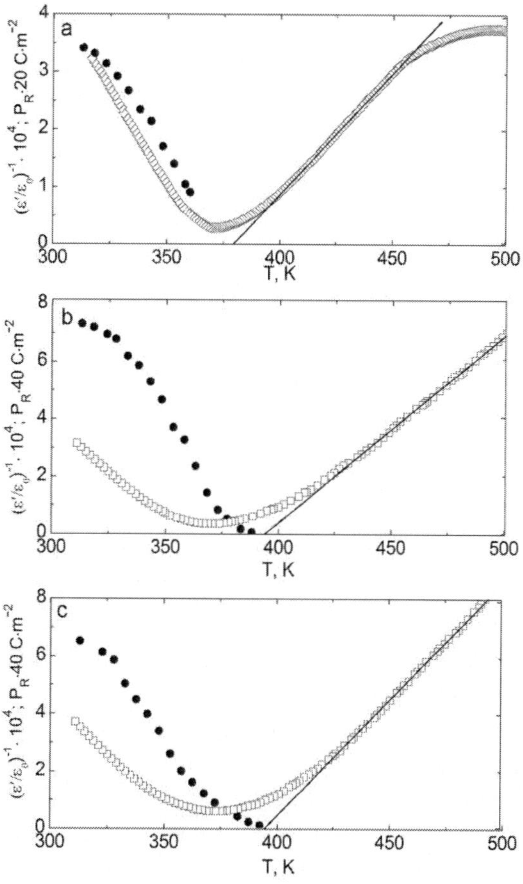

Figure 4. Dependencies $(\varepsilon'/\varepsilon_0)^{-1}$ ($T, f = 10^5$ Hz) (open symbols) и $P_R(T)$ (dark symbols) PFN (a), PFNL (b) и PFNM (c) in the temperature range (20 – 650) K. The solid lines present illustration of the implementation of the Curie - Weiss law.

The values of the widths of the peaks in the area of T_C have been calculated on half of maximum $\varepsilon'/\varepsilon_0$ at the frequency 10^5 Hz. In comparing values of the widths of the peaks of $\varepsilon'/\varepsilon_0$ in the T_C area, we obtained an increasing set: PFN (ΔT = 29 K) → PFNL (ΔT = 41 K) →

PFNM ($\Delta T = 54$ K). This indicates the increasing blurring of the phase transition of the FE → PE with the introduction of modifiers. This conclusion is confirmed by the analysis of dependencies $(\varepsilon'/\varepsilon_0)^{-1}(T)$ at $f = 10^5$ Hz, and $P_R(T)$ of the objects that are presented in Figure 4.

In PFN value P_R have been calculated only up to $T = 350$ K, which is associated with an increase in electrical conductivity of ceramics at these temperatures and, therefore, it is impossible to obtain the dielectric hysteresis loops. The P_R of PFNL and PFNM is monotonically decreases with increasing temperature and approaches zero at $T = 388$ K and ~ 393 K, respectively. These temperatures are higher than T_c of these compounds (see Figure 3.4). All objects for $T > T_C$ showed performance of the Curie - Weiss law (see Figure 4, solid lines):

$$\varepsilon' / \varepsilon_0 \sim \frac{C}{T - Q}, \qquad (2)$$

where C is Curie constant, Q is temperature in Curie - Weiss law.

The calculated values of Q (see Table 2) were greater than T_C. At the same time in increasing order of magnitude $Q - T_C$ compounds can be arranged in the following sequence: PFN → PFNL → PFNM. In our opinion the above arguments mean that the objects under study are ferroelectrics with diffuse phase transition.

Table. 2 shows the basic ferroelectric and piezoelectric characteristics of the studied ceramics.

Table. 2. Ferroelectrics and piezoelectric characteristics of ceramics PFN, PFNL and PFNM.

	$\varepsilon'_{33}{}^T/\varepsilon_0$	$\varepsilon''_{33}{}^T/\varepsilon_0$	K_p	Q_M	$\lvert d_{31}\rvert$, pC·N^{-1}	d_{33}, pC·N^{-1}	Q, K	T_C, K
PFN	2740	130	0.13	240	34	101	378	371
PFNL	2235	36	0.30	262	82	230	394	369
PFNM	2712	60	0.20	223	59	150	397	370

The effects caused by the introduction of modifiers can be explained as follows. Reducing the sintering temperature (particularly strong in the case of modification of MnO_2) is a consequence of the changing nature of sintering: from solid-phase sintering to one in the presence of liquid phase [10]. This is confirmed by the presence of "double" the boundaries and intergranular layers in PFNL, as well as strengthening the attachment between grains in PFNM. During sintering PFNL and PFNM the sources of liquid phases are Li_2CO_3, which has a low melting point [11], and MnO_2 [12], that were involved in the formation of the final product.

Based on the fact that grain growth in PFNL increases sharply, one can assume that the Li-containing liquid phase primarily serve as a transport medium. This favors the mass transfer and diffusion in the sintering process. The increase in the average grain size and decrease their total surface leads to a weakening of grain boundaries. This is reflected in the character of the cleavage and the deconsolidation of ceramics, as mentioned above. In the case of PFNM, we clearly not find the presence of such layers. This may be due to manifestations of "cementing" the role of Mn-containing liquid phases. They are tightly associated with each other grains and an excess squeezed out of the structure in the form of a small amount of impurities amorphized.

Described above, the unusual behavior of the structural parameters of PFNL and PFNM may be due to incomplete embedding (or not embed) modifiers in the crystal structure of the lead compound. Following [13], Li and Mn would have to take the *B*-position in the structure, that is to replace Fe or Nb (Although the Li is placed in *A*-positions of perovskite compounds, some of it can be embedded in a *B*-positions or interstitial positions) [14, 15]. The volume of the unit cell would have either an increase (in the case of Li), or a decrease (in the case of Mn), following the logic of change in ionic radii of the elements of a "host" and modifiers: Pb^{2+} (1.26) → Fe^{3+} (0.67) → Nb^{5+} (0.66) → Li^{1+} (0.68) → Mn^{4+} (0.52) (in parentheses are the values of ionic radii measured in Å [16] for coordination number 6).

Conclusions

These results should be used when creating multifunctional materials based on ferroniobate lead with a high dielectric constant for use in low-frequency devices.

Acknowledgements

This work was carried out with partial financial support from the Ministry of Education and Science of the Russian Federation (State contract № 16.513.11.3032).

References

[1] Zvezdin, A. K.; Loginov, A. S.; Meshkov, G. A.; Pyatakov, A. P.. *Izv. Ross. Akad. Nauk, Ser. Fiz.*, 2007, *vol. 71*, 1604-1605.
[2] Venevtsev, Yu. N.; Gagulin, V. V.; Lyubimov, V. N. *Segnetomagnetiki (Ferroelectromagnets),* Nauka, Moscow, 1982; pp. 1 – 254.
[3] Malyshkina, O. V.; Baranova, E. V.; Gavrilova, N. D.; Lotonov, A. M.. *Pis'ma Zh. Tekh. Fiz.*, 2007, *vol. 33(*18*)*, 70-75.
[4] Yang, Y.; Lui, J. M; et al. *Phys. Rev. B.* 2004, *vol. 70*, 132101-1-132101-5
[5] Bochenek, D.; Guzdek, P.. *J. Magnetism and Magnetic Materials.* 2011, *vol. 323*, 369-374.
[6] Bochenek, D.; Surowiak, *Z. Phys. Status Solidi A*, 2009, *vol. 206(*12*)*, 2857-2865.
[7] Raevski, I. P.,;Kubrin, S. P.; et al. *Ferroelectrics.* 2010. *vol. 397*. 96-101.
[8] Raevski, I. P.; Kirilov, S. T; Malittskaya, M. A.; et al. *Inorganic Materials*, 1988, *vol. 24*(2), 286-291.
[9] Turik, A. V.; Pavlenko, A. V.; et al. *Phys. Solid State*, 2012, *vol. 54*, 947–950.
[10] Geguzin, Y. E. *Physics of Sintering.* Nauka, Moscow, 1984; pp. 1 – 312.
[11] Reznitchenko, L. A.; Razumovskaya, O. N.; Shilkina, L. A.; et al. *Proceedings of the 7th International Workshop on the Physics of Ferroelectric Semiconductors. Rostov-on-Don.* 1996. pp. 149-150.

[12] Reznitchenko, L. A.; Razumovskaya, O. N.; Klevtsov, A. N.; et al. *Proceedings of International Scientific-Practical Conference "PEZOTEHNIKA-99".* Rostov-on-Don, 1999, pp. 268-275.
[13] Fesenko, E. G. *Perovskite Family and Ferroelectricity*, Atomizdat, Moscow, 1972; pp. 1 – 248.
[14] Kuzminov, Y. S. *Electro-Optical and Nonlinear Optical Crystal of Lithium Niobate.* Nauka, Moscow, 1987; pp. 1 – 262.
[15] Sidorov, N. V.; Wolf, T. R.; Mavrin, B. N.; Kalinnikov, V. T. *Lithium Niobate. Defects. Photo-refraction. The Vibration Spectrum. Polaritons.* Nauka, Moscow, 2003; pp. 1 – 255.
[16] Poray-Koshits, M. A. *Practical Course of X-ray Analysis.* Vol. 2. Boky G. B. (Ed.). Publishing House of Moscow University, Moscow, 1960; pp. 1 – 632.

II. Physics of New Materials

In: Physics and Mechanics of New Materials ...
Editors: Ivan A. Parinov and Shun-Hsyung Chang

ISBN: 978-1-62618-535-7

Chapter 5

Hall Effect in Inhomogeneous Magnetic Field

***V. K. Ignatjev*[*] *and A. A. Orlov*[†]**
Volgograd State University, SEC "Radiophysics", Volgograd, Russia

Abstract

The development of modern electronics is moving towards the introduction of new physical principles in the design of circuits. One of the most promising for the application of physical phenomena are the spin and magnetostatic waves. The need for miniaturization of devices, magnetic electronics leads the developer to analyze the behavior of a complex problem in an inhomogeneous magnetoactive media field. This is especially important in the calculation of spin-wave electronics devices.

An example of such a calculation is made for the heterostructure AsIn sensor of magnetic field into a voltage. To characterize the heterogeneity within the measurement method is used the system (instrument) functions that characterize the action of the field on a small portion of the converter. An approximate method of numerical calculation of system functions is to be through the Green's function. A relationship is obtained for the system function galvanomagnetic transducer in a linear approximation:

$$u_H(\mathbf{r}) = \frac{i}{en}\int_D \mathbf{B}(\mathbf{r}')\mathbf{W}(\mathbf{r}-\mathbf{r}')d^3r', \text{ where } \mathbf{W}(\mathbf{r}) = \lim_{i\to 0}\frac{[\mathbf{j}_1(\mathbf{r})\times\mathbf{j}_2(\mathbf{r})]}{i^2}.$$

Here the measured voltage u_H ($\mathbf{r}$) is a function of magnetic field $\mathbf{B}$, with a minimized system function $\mathbf{W}(\mathbf{r})$. Here $\mathbf{j}_1$ is the current density by passing a current through contacts 1 and 3, and measuring the voltage on pins 2 and 4, and the current density $\mathbf{j}_2$ by passing current through the contacts 2 and 4 and measuring the voltage on pins 1 and 3, i is the current transmitted through contacts, e is the electron charge, n is the concentration of carriers.

It is carried out the experimental verification of reciprocity galvanomagnetic transducer in an inhomogeneous field and measuring the distortion of magnetic field

* E-mail: ignatjev@vlpost.ru.
† E-mail: orlwork@inbox.ru.

distribution caused by the reduction of the instrumental function of the converter. The experimental setup is a group of parallel thin (90 mm) wires and switching current system, controlled by PC. Measurement of the voltage on the converter is carried out using the analog-digital unit of the Hall magnetometer, developed in the SEC, "Radiophysics."

The reduced theoretical material makes the proof of reciprocity transducer in a nonuniform field (divergence of the data was the noise level meter of the magnetic field, the standard deviation of which was 5 mkT), as well as the consistency of the experimental and theoretical data for measurements of the inhomogeneous distribution of the magnetic field.

1. Introduction

The development of modern electronics is moving towards the introduction of new physical principles in the design of circuits. One of the most promising for the application of physical phenomena are the spin and magnetostatic waves. Studies are under way to replace energy-consuming way to pass the signal through an electrical current to the spin currents, which require little or no energy cost. At the present time we can talk about the appearance of the whole area of functional electronics - Magnetic microelectronics, the problem of creation of electronic components and integrated circuits for the spin currents and other magnetic effects [1].

The transformation of the magnetic field in the measured voltage can be carried out in various ways. The most promising in this case is the Hall effect. In this semiconductor device can be viewed as a system for which the magnetic field and the field current is the input parameters and the voltage across the potential contacts of the resulting value.

One of the fundamental physical principles is the Onsager reciprocity relation. It argues that the kinetic coefficients of a thermodynamic system in equilibrium and the magnetic field **H** is symmetric under inversion of the magnetic field. However, the proof of the Onsager relations assumed homogeneity of the field **H**, it does not guarantee reciprocity in an inhomogeneous magnetic field. To better understand the processes occurring under the influence of a strongly inhomogeneous magnetic field, it is necessary to show the conditions under which the reciprocity of a physical system is preserved, and if violated, then what is the order of the inhomogeneity, which leads to a significant violation of reciprocity.

To characterize the heterogeneity of the measured value within the measurement method is used the system (instrument) functions that characterize the action of the field on a small section of the converter [2].

2. Direct Magnetostatic Problem

Consider the generalized structure of magnetoactive device (Figure 1) as the volume of the four contact element D with two current (S_1 and S_3) and two potential (S_2 and S_4) contacts. Ohm's law for an isotropic conductor in a magnetic field is

$$\mathbf{j} = \sigma \mathbf{E} + \frac{\sigma}{en}[\mathbf{j} \times \mathbf{B}''],$$

where σ is the conductivity, n is the concentration of charge carriers, **B**'' is the full magnetic field in a conductor can be written in matrix form as

$$A \cdot \mathbf{j} = \sigma \cdot \mathbf{E},$$

where

$$\beta = \frac{\sigma}{en},\ A = \begin{pmatrix} 1 & -\beta \cdot B''_z & \beta \cdot B''_y \\ \beta \cdot B''_z & 1 & -\beta \cdot B''_x \\ -\beta \cdot B''_y & \beta \cdot B''_x & 1 \end{pmatrix}.$$

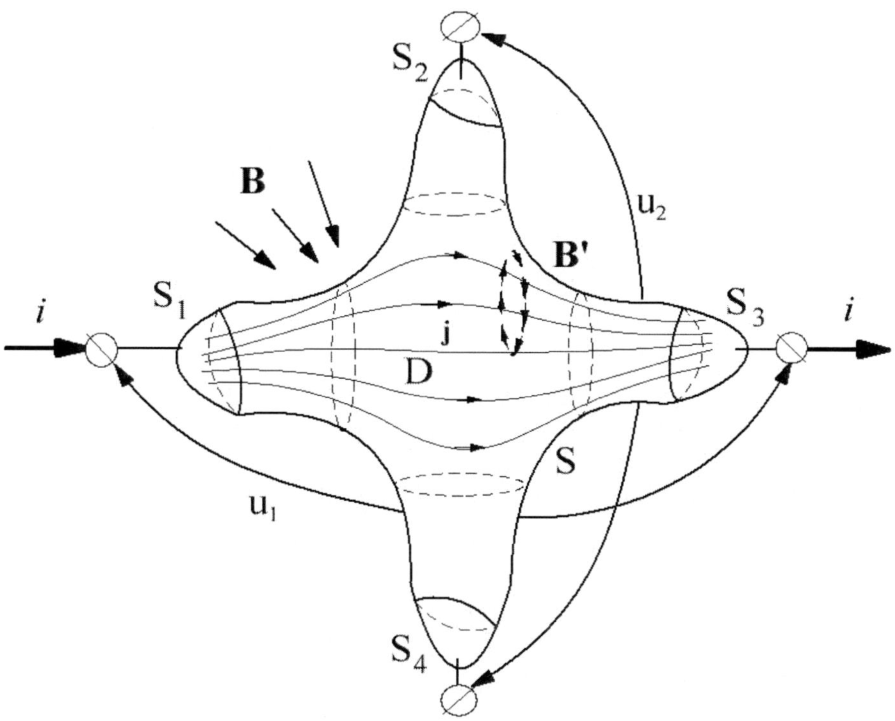

Figure. 1. The structure of the magnetoelectronic device.

We can obtain relative current density in the form

$$\mathbf{j} = \sigma\ A^{-1}\ \mathbf{E},$$

where

$$A^{-1}=\frac{1}{1+(\beta\mathbf{B}'')^2}\begin{pmatrix}1+\beta^2B_x''^2 & -\beta B_z''-\beta^2B_x''B_y'' & \beta B_y''+\beta^2B_z''B_x''\\ \beta B_z''+\beta^2B_x''B_y'' & 1+\beta^2B_y''^2 & -\beta B_x''-\beta^2B_z''B_y''\\ -\beta B_y''-\beta^2B_z''B_x'' & \beta B_x''+\beta^2B_z''B_y'' & 1+\beta^2B_z''^2\end{pmatrix}=$$

$$=\frac{A_0^{-1}+\beta A_1^{-1}+\beta^2A_2^{-1}}{1+(\beta\mathbf{B}'')^2},$$

$$A_0^{-1}=\begin{pmatrix}1&0&0\\0&1&0\\0&0&1\end{pmatrix};\ A_1^{-1}=\begin{pmatrix}0&-B_z''&B_y''\\B_z''&0&-B_x''\\-B_y''&B_x''&0\end{pmatrix};$$

$$A_2^{-1}=\begin{pmatrix}B_x''^2&-B_x''B_y''&B_z''B_x''\\B_x''B_y''&B_y''^2&-B_z''B_y''\\-B_z''B_x''&B_z''B_y''&B_z''^2\end{pmatrix}.$$

In weak fields, when $|\mathbf{B}''|\beta \ll 1$ in the linear magnetic field approximation, we obtain

$$\mathbf{j}=\sigma\left(A_0^{-1}+\beta A_1^{-1}\right)=\sigma\mathbf{E}+\frac{\sigma^2}{en}\left[\mathbf{E}\times\mathbf{B}''\right].\tag{1}$$

Divide the total magnetic field **B**" into the field **B** generated by external sources and the field **B**' of the currents flowing through the sensor: **B**" = **B** + **B**'. For a homogeneous and non-magnetic material from Equation (1) and the equations of magnetostatics

$\operatorname{div}\mathbf{j}=0$, $\operatorname{rot}\mathbf{B}=0$, $\operatorname{rot}\mathbf{B}'=\mu_0\mathbf{j}$, $\operatorname{rot}\mathbf{E}=0$, $\mathbf{E}=-\operatorname{grad}\varphi$

we obtain

$$\Delta\varphi=-\frac{\sigma^2\mu_0}{en}(\nabla\varphi)^2,\tag{2}$$

$$\operatorname{rot}\mathbf{B}'=\mu_0\sigma\left\{\mathbf{E}+\frac{\sigma}{en}\left[\mathbf{E}\times(\mathbf{B}+\mathbf{B}')\right]\right\}.\tag{3}$$

The right-hand side of Equation (2) describes the self-action of the current flowing through the sensor, i.e. the pinch effect.

Let the domain D, bounded by surface S, the current i flowing through the current pin 1 (the surface S_1), and follows through the current pin 3 (surface S_3), through the rest of the

surface S is not current (condition of unloaded contacts). The condition for the normal component of current density on the boundary of S has the form:

$$j_n\big|_S = i\left[f_3(\mathbf{r}) - f_1(\mathbf{r})\right],$$

where $f_m(\mathbf{r})$ is the function – carrier of surface S_m:

$$\begin{cases} \oint\limits_{S_m} f_m(\mathbf{r})ds = 1; \\ f_m(\mathbf{r}) = 0, \mathbf{r} \notin S_m. \end{cases} \tag{4}$$

From the constitutive Equation (1) in this case, the potential distribution is the solution of the mixed problem when the boundary is given a linear combination of normal and tangential derivatives of the potential

$$\left.\frac{\partial\varphi}{\partial n}\right|_S + \frac{\sigma}{en}\left([\nabla\varphi \times \mathbf{B}]\mathbf{n}\right)\Big|_S + \frac{\sigma}{en}\left([\nabla\varphi \times \mathbf{B}']\mathbf{n}\right)\Big|_S = i\frac{f_3(\mathbf{r}) - f_1(\mathbf{r})}{\sigma}, \tag{5}$$

where $\mathbf{n}$ is the external normal to surface S.

At the same time, the measured voltage is the potential difference between contacts 2 and 4:

$$u_2 = \varphi(\mathbf{r}_4) - \varphi(\mathbf{r}_2).$$

The points $\mathbf{r}_4$ and $\mathbf{r}_2$ are located on the contact surfaces of S_4 and S_2, respectively. In real magnetoeletronic device the contacts always have a nonzero size. Taking into account Formula (4), we obtain

$$u_2 = \varphi(\mathbf{r}_4)\oint\limits_{S_4} f_4(\mathbf{r}')ds' - \varphi(\mathbf{r}_2)\oint\limits_{S_2} f_2(\mathbf{r}')ds'.$$

After applying the mean-value theorem, assuming equipotential contacts, we can introduce the potential under the integral sign:

$$u_2 = \oint\limits_{S_4} \varphi(\mathbf{r}) f_4(\mathbf{r}')ds' - \oint\limits_{S_2} \varphi(\mathbf{r}) f_2(\mathbf{r}')ds'.$$

Then by using the property of the function of the carrier (4), we finally obtain an expression for the measured voltage:

$$u_2 = \oint_S \varphi(\mathbf{r})\left(f_4(\mathbf{r}') - f_2(\mathbf{r}')\right)ds'. \tag{6}$$

We take a zero mean value of the potential in the region D, and seek a joint solution of Equations (2) and (3) with boundary conditions (5) by using the method of successive approximations, setting

$$\varphi(\mathbf{r}) = u_1\psi(\mathbf{r}),\ \mathbf{B}(\mathbf{r}) = \mu_0\sigma u_1\mathbf{b}(\mathbf{r}),\ \mathbf{B}'(\mathbf{r}) = \mu_0\sigma u_1\mathbf{b}'(\mathbf{r}),$$

$$\psi(\mathbf{r}) = \psi_0(\mathbf{r}) + \alpha\psi_1(\mathbf{r}) + \ldots,\ \mathbf{b}'(\mathbf{r}) = \mathbf{b}_0(\mathbf{r}) + \alpha\mathbf{b}_1(\mathbf{r}),$$

where $u_1 = \varphi(\mathbf{r}_1) - \varphi(\mathbf{r}_3)$ is the voltage drop across the magnetoactive device, $\alpha = \dfrac{\sigma^2\mu_0}{en}u_1$, $|\alpha| << 1$. Then Equations (2), (3), (5) are equivalent to a chain of boundary value problems for the Neumann, Laplace and Poisson equations

$$\begin{aligned}
&\Delta\psi_0 = 0,\\
&\left.\frac{\partial\psi_0}{\partial n}\right|_S = \frac{i}{\sigma u_1}\left[f_3(\mathbf{r}) - f_1(\mathbf{r})\right],\\
&\operatorname{rot}\mathbf{b}_0 = -\nabla\psi_0,\quad \operatorname{rot}\mathbf{b} = 0,
\end{aligned} \tag{7}$$

$$\begin{aligned}
&\Delta\psi_1 = -(\nabla\psi_0)^2,\\
&\left.\frac{\partial\psi_1}{\partial n}\right|_S = [\mathbf{b}\times\nabla\psi_0]\mathbf{n} + [\mathbf{b}_0\times\nabla\psi_0]\mathbf{n}.
\end{aligned} \tag{8}$$

Let $G(\mathbf{r}, \mathbf{r}')$, where $\mathbf{r} \in D$, $\mathbf{r}' \in S$ is the second Green's function of the Neumann problem for the domain D, bounded by surface S. Then the solution of problem (7) has the form

$$\psi_0(\mathbf{r}) = \frac{i}{\sigma u_1}\oint_S G(\mathbf{r},\mathbf{r}')\left[f_3(\mathbf{r}') - f_1(\mathbf{r}')\right]ds', \tag{9}$$

$$\mathbf{b}_0(\mathbf{r}) = -\frac{1}{4\pi}\int_D \frac{\partial\psi_0(\mathbf{r}')/\partial\mathbf{r}'\times(\mathbf{r}-\mathbf{r}')}{|\mathbf{r}-\mathbf{r}'|^3}d^3r'. \tag{10}$$

and the solution of (8) has the form

$$\psi_1(\mathbf{r}) = \int_D G(\mathbf{r},\mathbf{r}')\left(\frac{\partial\psi_0(\mathbf{r}')}{\partial\mathbf{r}'}\right)^2 d^3r' +$$

$$+\oint_S G(\mathbf{r},\mathbf{r}')\left[\mathbf{b}_0(\mathbf{r}')\times\frac{\partial\psi_0(\mathbf{r}')}{\partial\mathbf{r}'}\right]d\mathbf{s}' + \oint_S G(\mathbf{r},\mathbf{r}')\left[\mathbf{b}(\mathbf{r}')\times\frac{\partial\psi_0(\mathbf{r}')}{\partial\mathbf{r}'}\right]d\mathbf{s}'. \tag{11}$$

We transform the second term in equation (9), taking into account the third equation of system (7)

$$\oint_S G(\mathbf{r},\mathbf{r}')\left[\mathbf{b}_0(\mathbf{r}')\times\frac{\partial\psi_0(\mathbf{r}')}{\partial\mathbf{r}'}\right]d\mathbf{s}' = \int_D \frac{\partial}{\partial\mathbf{r}'}\left\{G(\mathbf{r},\mathbf{r}')\left[\mathbf{b}_0(\mathbf{r}')\times\frac{\partial\psi_0(\mathbf{r}')}{\partial\mathbf{r}'}\right]\right\}d^3r' =$$

$$= \int_D \left[\mathbf{b}_0(\mathbf{r}')\times\frac{\partial\psi_0(\mathbf{r}')}{\partial\mathbf{r}'}\right]\frac{\partial G(\mathbf{r},\mathbf{r}')}{\partial\mathbf{r}'}d^3r' + \int_D G(\mathbf{r},\mathbf{r}')\frac{\partial}{\partial\mathbf{r}'}\left[\mathbf{b}_0(\mathbf{r}')\times\frac{\partial\psi_0(\mathbf{r}')}{\partial\mathbf{r}'}\right]d^3r' =$$

$$= \int_D \mathbf{b}_0(\mathbf{r}')\left[\frac{\partial\psi_0(\mathbf{r}')}{\partial\mathbf{r}'}\times\frac{\partial G(\mathbf{r},\mathbf{r}')}{\partial\mathbf{r}'}\right]d^3r' - \int_D G(\mathbf{r},\mathbf{r}')\left[\frac{\partial\psi_0(\mathbf{r}')}{\partial\mathbf{r}'}\right]^2 d^3r'.$$

For the third term, respectively, we obtain

$$\oint_S G(\mathbf{r},\mathbf{r}')\left[\mathbf{b}(\mathbf{r}')\times\frac{\partial\psi_0(\mathbf{r}')}{\partial\mathbf{r}'}\right] = \int_D \mathbf{b}(\mathbf{r}')\left[\frac{\partial\psi_0(\mathbf{r}')}{\partial\mathbf{r}'}\times\frac{\partial G(\mathbf{r},\mathbf{r}')}{\partial\mathbf{r}'}\right]d^3r'.$$

Thus, in view of (9) and (10), we have

$$\varphi(\mathbf{r}) = \varphi_0(\mathbf{r}) + \frac{\sigma}{en}\int_D \mathbf{B}(\mathbf{r}')\left[\frac{\partial\varphi_0(\mathbf{r}')}{\partial\mathbf{r}'}\times\frac{\partial G(\mathbf{r},\mathbf{r}')}{\partial\mathbf{r}'}\right]d^3r' +$$

$$+\frac{\sigma^2\mu_0}{en}\int_D\int_D\left[\frac{\partial\varphi_0(\mathbf{r}')}{\partial\mathbf{r}'}\times\frac{\mathbf{r}'-\mathbf{r}''}{|\mathbf{r}'-\mathbf{r}''|^3}\right]\left[\frac{\partial\varphi_0(\mathbf{r}')}{\partial\mathbf{r}'}\times\frac{\partial G(\mathbf{r},\mathbf{r}')}{\partial\mathbf{r}'}\right]d^3r'd^3r''.$$

The output voltage u_2 can be expressed as a value proportional to the external magnetic field **B** and the Hall voltage u_H does not depend on the external magnetic field **B** voltage unbalance (asymmetry) u_A, which can be divided into linear and nonlinear part of the current i. Then we have

$$u_{A1} = \frac{i}{\sigma}\oint_S\oint_S G(\mathbf{r},\mathbf{r}')(f_4(\mathbf{r}) - f_2(\mathbf{r}))\cdot(f_3(\mathbf{r}') - f_1(\mathbf{r}'))ds\,ds', \tag{12}$$

$$u_{A2} = \frac{\mu_0 i^2}{4\pi en} \int\limits_D \int\limits_D \left[\left\{ \oint\limits_S \frac{\partial G(\mathbf{r}'', \mathbf{r})}{\partial \mathbf{r}''} \cdot (f_3(\mathbf{r}) - f_1(\mathbf{r}))ds \right\} \times \frac{(\mathbf{r}' - \mathbf{r}'')}{|\mathbf{r}' - \mathbf{r}''|^3} \right] \times$$

$$\times \left[\left\{ \oint\limits_S \frac{\partial G(\mathbf{r}'', \mathbf{r}''')}{\partial \mathbf{r}'''} [f_3(\mathbf{r}''') - f_1(\mathbf{r}''')] ds''' \right\} \times \left\{ \oint\limits_S \frac{\partial G(\mathbf{r}, \mathbf{r}')}{\partial \mathbf{r}'} [f_4(\mathbf{r}) - f_2(\mathbf{r})] ds \right\} \right] d^3 r' d^3 r'',$$

(13)

$$u_H = \frac{i}{en} \oint\limits_S (f_4(\mathbf{r}) - f_2(\mathbf{r})) \times$$

$$\times \int\limits_D \mathbf{B}(\mathbf{r}') \left[\left\{ \oint\limits_S \frac{\partial G(\mathbf{r}', \mathbf{r}'')}{\partial \mathbf{r}'} [f_3(\mathbf{r}'') - f_1(\mathbf{r}'')] ds'' \right\} \times \frac{\partial G(\mathbf{r}, \mathbf{r}')}{\partial \mathbf{r}'} \right] d^3 r' ds,$$

By changing the order of integration, we obtain:

$$u_H = \frac{i}{en} \int\limits_D \mathbf{B}(\mathbf{r}') \left[\left\{ \oint\limits_S \frac{\partial G(\mathbf{r}', \mathbf{r}'')}{\partial \mathbf{r}'} [f_3(\mathbf{r}'') - f_1(\mathbf{r}'')] ds'' \right\} \times \left\{ \oint\limits_S \frac{\partial G(\mathbf{r}, \mathbf{r}')}{\partial \mathbf{r}'} [f_4(\mathbf{r}) - f_2(\mathbf{r})] ds \right\} \right] d^3 r'.$$

(14)

Under the condition of uniqueness of the Green's function, from Neumann equation

$$\oint\limits_S G(\mathbf{r}, \mathbf{r}') dS_{\mathbf{r}} = 0.$$

one is symmetric [3]:

$$G(\mathbf{r}'', \mathbf{r}') = G(\mathbf{r}', \mathbf{r}''). \quad (15)$$

Note that the expression (9) represents the potential distribution in the flow of current, without the influence of the magnetic field. Then, using (1) by putting $\mathbf{B}'' = 0$, as well as the notation $\mathbf{E} = -\operatorname{grad} \varphi$ and (9), we obtain an expression for the distribution of current density:

$$\mathbf{j}_1(\mathbf{r}) = i \oint\limits_S \frac{\partial G(\mathbf{r}, \mathbf{r}')}{\partial \mathbf{r}} [f_3(\mathbf{r}') - f_1(\mathbf{r}')] ds',$$

$$\mathbf{j}_2(\mathbf{r}) = i \oint\limits_S \frac{\partial G(\mathbf{r}, \mathbf{r}')}{\partial \mathbf{r}} [f_4(\mathbf{r}') - f_2(\mathbf{r}')] ds'.$$

Here $\mathbf{j}_1(\mathbf{r})$ is the current density in the absence of an external magnetic field by passing a current i through terminals 1 and 3, $\mathbf{j}_2(\mathbf{r})$ is the current density in the absence of an external magnetic field by passing a current i through the terminals 2 and 4. Then Equation (14) can be written as follows:

$$u_H = \frac{1}{eni} \int_D \mathbf{B}(\mathbf{r}')[\mathbf{j}_1(\mathbf{r}') \times \mathbf{j}_2(\mathbf{r}')] d^3 r', \tag{16}$$

Equation (16) shows that the Onsager reciprocity relations are valid in the nonuniform field. The basic approach here is the equipotentiality of contacts and Equation (6) for the voltage across the potential contacts of the sensor.

In Equation (16), the coordinates are measured from the center of device. If we consider the output voltage of magnetoelectronic device as a function of the coordinates of its center, Equation (16) can be transformed into a convolution:

$$u_H(\mathbf{r}) = i \cdot R_H \int_D \mathbf{B}(\mathbf{r}') \mathbf{W}(\mathbf{r} - \mathbf{r}') d^3 r', \tag{17}$$

where $\mathbf{r}$ are the coordinates of the center of the sensor, $R_H = 1/en$ is Hall coefficient, $\mathbf{W}(\mathbf{r})$ is the vector volume system function of magnetoelectronic device (SFMD):

$$\mathbf{W}(\mathbf{r}) = \lim_{i \to 0} \frac{[\mathbf{j}_1(\mathbf{r}) \times \mathbf{j}_2(\mathbf{r})]}{i^2}. \tag{18}$$

The approximation of small current in (18) is required for strict compliance with the conditions of the first approximation of the problem (8) into the right side which includes the value of the potential at zero magnetic field.

In most cases, magnetoeletronic device is a thin film of semiconductor material, so it makes sense to reduce the problem to a two-dimensional one. Suppose that D is a cylinder of height d with base C, lying in the plane xy, then the film is thin enough that the distribution of current and magnetic field along the z axis be considered uniform. Then the vectors $\mathbf{j}_1(\mathbf{r})$ and $\mathbf{j}_2(\mathbf{r})$ can be considered as lying in the plane xy, and their vector product is directed along the axis z. The triple integral in (16) can be taken first on the axis z, which is equivalent to multiplication by d double integral over the base S.

We introduce the relative linear current density of $\boldsymbol{\lambda}$ in the absence of a magnetic field

$$\boldsymbol{\lambda}(x, y) = \lim_{i \to 0} \left[\frac{1}{i} \int_0^d \mathbf{j}(x, y, z) dz \right] = \lim_{i \to 0} \left[\frac{d}{i} \mathbf{j}(x, y, 0) \right]$$

and rewrite (16) as

$$u_H = \frac{iR_H}{d} \int_C B_z(x', y') [\lambda_{1x}(x', y') \lambda_{2y}(x', y') - \lambda_{2x}(x', y') \lambda_{1y}(x', y')] dx' dy'. \tag{19}$$

Then Equations (17) and (18) can be written as follows:

$$u_H(x,y)=i\cdot R_H\int_C B_n(x',y')w(x-x',y-y')\cdot dx'\cdot dy', \tag{20}$$

$$w(x,y)=\frac{\lambda_{1x}(x,y)\cdot\lambda_{2y}(x,y)-\lambda_{2x}(x,y)\cdot\lambda_{1y}(x,y)}{d}. \tag{21}$$

3. Finite-Element Modeling of Magnetoelectronic Devices on the Example of the Hall Transducer

Obtainment of SFMD by using formula (21) through the distribution of currents in the conductor film is analytically possible only for simple geometries of the sensor. For real magnetoactive device system function, it is advisable to find the numerical methods, such as physical modeling environment COMSOL.

To verify these relations construct a system function such as the Hall transducer (HT) PHE602817A [4], whose geometry is shown in Figure 2. In the transition to two-dimensional problem it was assumed that the density of flow into the region of current converter is uniform along the border contacts. The film thickness sensor is $d = 10$ mm.

Numerical simulation was performed by using the program of physical modeling environment COMSOL [5]. It allowed one to use the grid method by using an adaptive grid, with a maximum size of an element of 9 μm. The total number of elements was 158394. The calculation was performed on the PC with a Pentium (R) Dual-Core T4300 and 3GB of RAM. Two-dimensional SFMD is shown in Figure. 3.

To measure SFMD depending on two coordinates, it is necessary to create the magnetic field distribution, at both the local coordinates. This is possible in principle (for example, the small size of the solenoid), but practically difficult to be done. For experimental verification are more comfortable one-dimensional system functions. They are obtained if in equation (20) put $B_n(x', y') = B_n(x')$:

$$u_H(x)=i\cdot R_H\int_{x-x_{\min}}^{x-x_{\max}} B_n(x')\int_{y-y_{\min}}^{y-y_{\max}} w(x-x',y-y')\cdot dy'\cdot dx'.$$

Here $x_{\max}$, $x_{\min}$; $y_{\max}$, $y_{\min}$ are the coordinates that define the dimensions of the sensor for the respective axes. Then, by introducing one-dimensional SFMD for x and y:

$$w_x(x)=\int_{y_{\min}}^{y_{\max}} w(x,y')\cdot dy', \quad w_y(y)=\int_{x_{\min}}^{x_{\max}} w(x',y)\cdot dx', \tag{22}$$

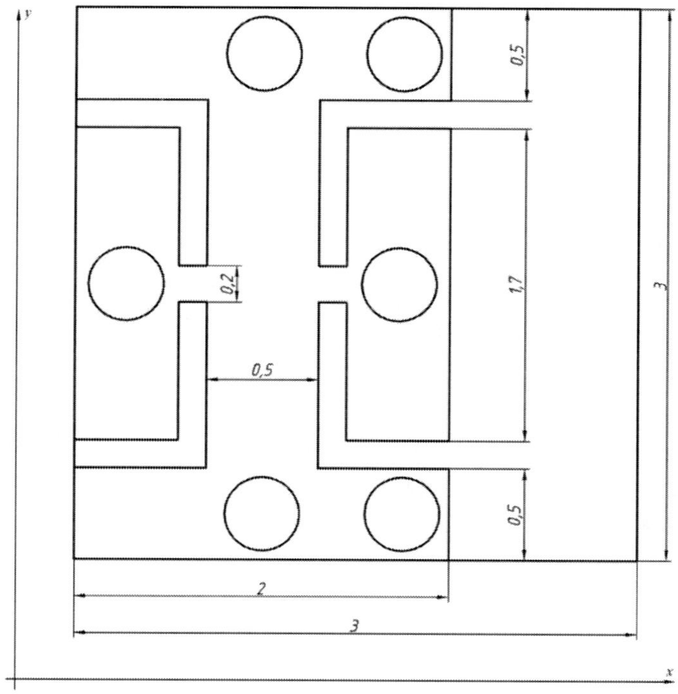

Figure 2. The converter design PHE602817A.

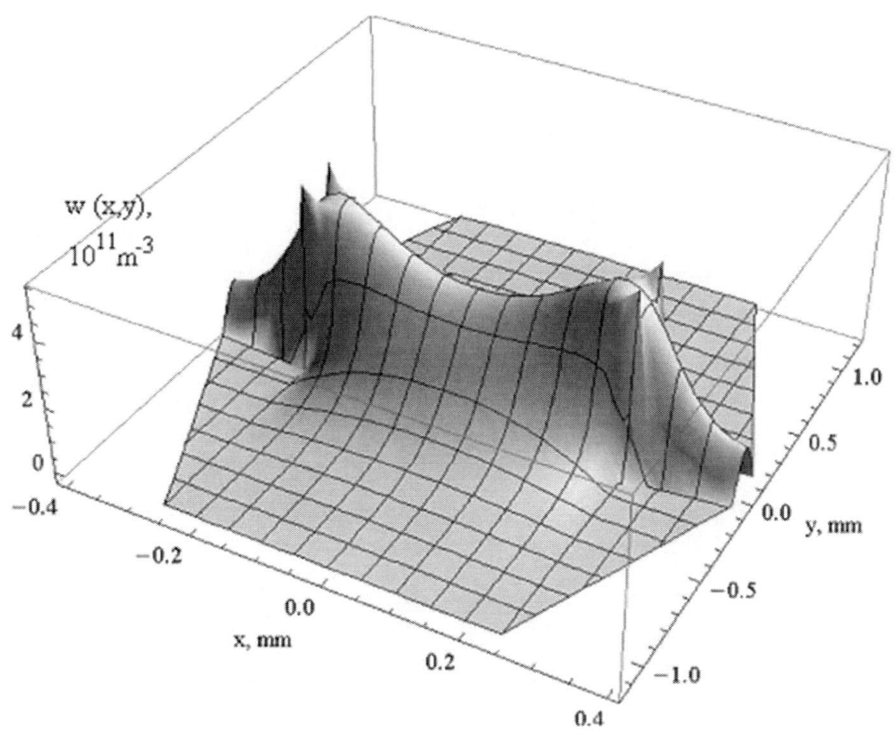

Figure 3. Graphic two-dimensional SFMD PHE602817A.

We obtain expressions for the Hall voltage:

$$u_H(x) = i \cdot R_H \int_{x-x_{\min}}^{x-x_{\max}} B_n(x') w_x(x-x') \cdot dx';$$
$$u_H(y) = i \cdot R_H \int_{y-y_{\min}}^{y-y_{\max}} B_n(y') w_y(y-y') \cdot dy'. \quad (23)$$

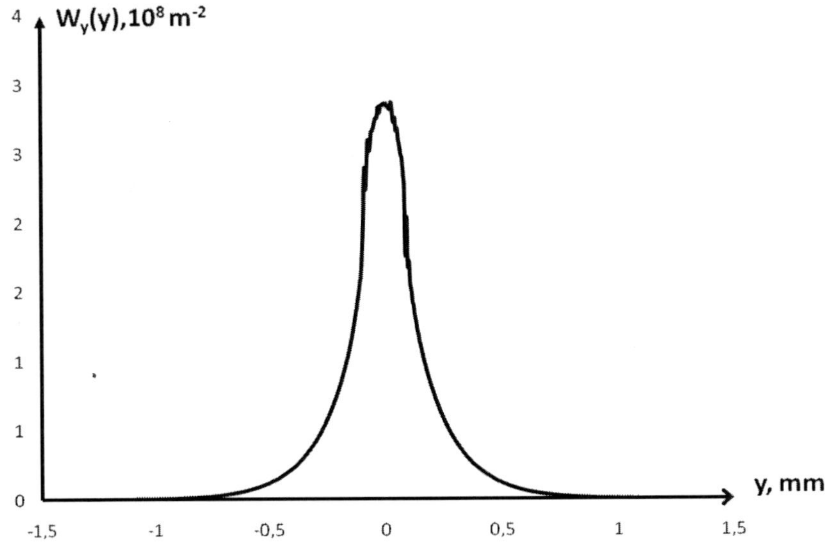

Figure 4. One-dimensional SFMD PHE602817A, with a uniform field along the x axis.

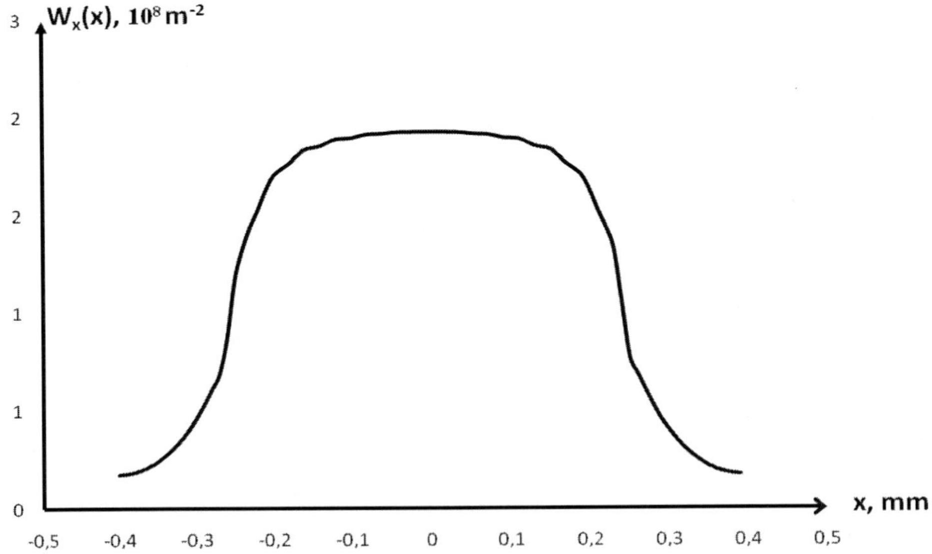

Figure 5. One-dimensional SFMD PHE602817A, with a uniform field along the y axis.

For one-dimensional SFMD, two-dimensional instrument function was integrated according to (22). Figures 4 and 5 show the one-dimensional SFPH for homogeneity in different coordinates.

4. EXPERIMENTAL VERIFICATION OF RECIPROCITY IN NON-HOMOGENIOS FIELD

To check relation (16), it was developed a special setup whose block-scheme shown in Figure 6.

PC interface software provides control and timing measurements, and the installation of the current distribution. With the converter interface (IC) code sequence is transmitted to the microcontroller (MC), which manages the team of the PC unit current switching (SUC). Source current (CS) is a support for the SUC, switching current carrying conductors in the system (SC). On the SC fixed HT study, the Hall is connected to a digital magnetometer [6] (DHM) - it measures the magnetic field and sends the data to a PC interface program which writes to a file for further processing. DHM allows measurement of the magnetic in different modes: the current through the sensor can be passed through the current as well as through the potential contacts. In this case the Hall voltage measurements carried out on the remaining two free contacts.

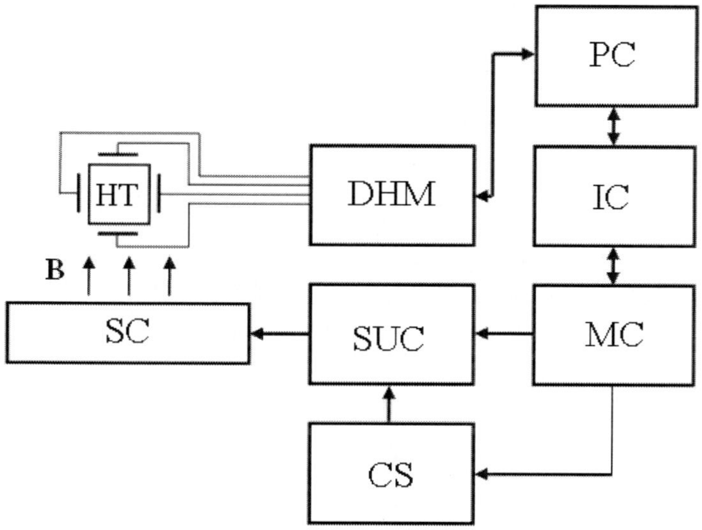

Figure 6. Block-scheme of experimental test setup for the calculation of SFMD.

The system of conductors presents itself 64 parallel wires with a diameter of 90 mm, located in one plane (Figure 7). With its production on the plastic cylinder of 20 mm in diameter through a pad of paper condenser coil to the coil was wound with wire sizing with BF-6, diluted with alcohol in a ratio of 1:6. After this, the wire was cut perpendicular to the coils, thus obtained a system of a large number of parallel conductors. This system was placed on an aluminum plate to provide the required level of heat dissipation. Wires soldered to the two contact cards, which were glued to the aluminum plate with glue "Monolith".

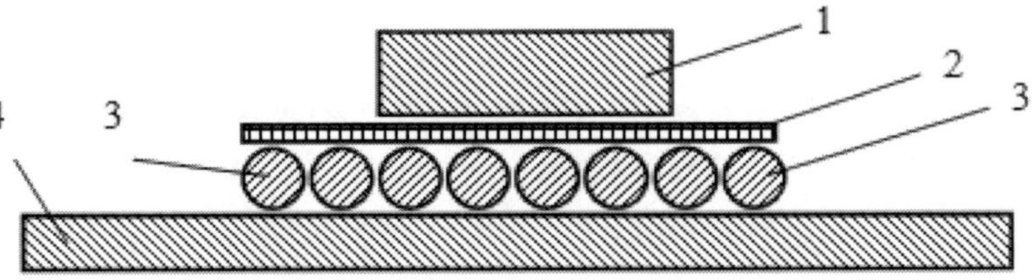

Figure 7. System of conductors: 1 – HT, 2 - insulator, 3 - copper wire, 4 - aluminum base.

With the help of obtained from (23) one-dimensional SFMD, the reaction of HT was restored to the currents in the conductors. The reconstructed and measured sensor response to an external inhomogeneous field is shown in Figure 8.

Experimental results are superimposed on the curve obtained by taking into account SFPH, indicating the reliability of the method of calculating the measurement error. The points corresponding to different modes of measurement fields are not shown because they lie within the measurement error, which is characterized by the standard deviation of ≈ 4 μT,

The apparatus function is an important parameter of any measurement system. The introduction of such a function in the state of the art design allows the device by choosing the parameters of converters more consciously and specifically for a particular task with minimal material costs. The study and, especially, the obtained formula (16) for SFMD allow its easy evaluation and analysis of physical properties.

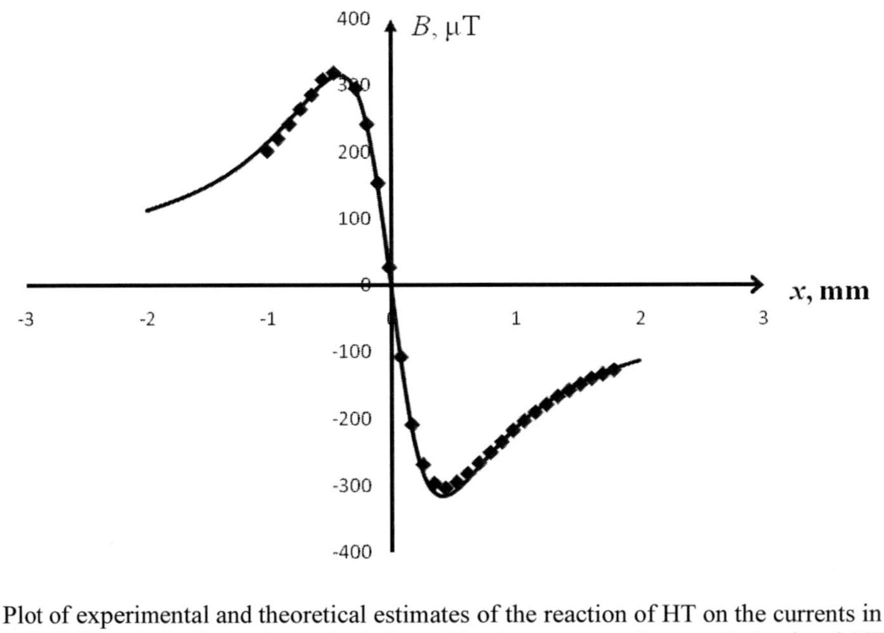

Figure 8. Plot of experimental and theoretical estimates of the reaction of HT on the currents in the conductors. Solid line is the theoretical calculation taking into account the one-dimensional SFMD; the points show the corresponding experimental values.

The reciprocity of converters in an inhomogeneous field indicates that the Onsager relations [7] can be extended to some systems with a high magnetic field inhomogeneity. In addition, these expressions for the stress components are independent of the field, can take into account the parasitic parameters of the converter at the design stage, which allow increasing the output of HT to a new level.

Application of this method of calculation and accounting SFMD can be recommended for enterprises that produce magnetoactive devices in order to obtain products having the best technical and economic characteristics.

ACKNOWLEDGEMENTS

This work was supported by the Federal Target Program “Scientific and Pedagogical Manpower” for 2009-2013 (Project № 14.740.11.0830)

REFERENCES

[1] Ignatiev, V. K.; Liashenko, A. A. *Geteromagnitnaya Microelectronics. Microactive Type*; Science: Moscow, 2006; pp 1 – 877 (in Russian).

[2] Ignatjev, V.K.; Protopopov, A. G. *Proceedings of the Universities. Instrument*. 2003. *vol. 46(3)*. 38 – 44 (in Russian).

[3] Sveshnikov, A. G.; Bogolyubov, A. N.; Kravtsov V. V. *Lectures on Mathematical Physics: A Handbook. Allowance*; Moscow State University Press: Moscow, 1993; pp 1 – 352 (in Russian).

[4] *Specifications of Converters Hall Series PCE*. Electronic Resource: http://www. nvlab.spb.ru/Hall% 20sensors.pdf.

[5] *Official Website of the Software COMSOL Multiphysic*. Electronic resource: http://www.comsol.com/

[6] Golubev, A. A.; Ignatjev, V. K.; Nikitin, A. V. A *High-Precision Magnetometer Instruments and Experimental Techniques*. 2008. *vol.* 51(5). 753-758 *(in Russian).*

[7] *Lars Onsager. Physycal Review. 1931. vol. 37 (February 15).*

In: Physics and Mechanics of New Materials ... ISBN: 978-1-62618-535-7
Editors: Ivan A. Parinov and Shun-Hsyung Chang

Chapter 6

NONLINEAR RECIPROCAL RELATIONS OF PLASMA-LIKE MEDIA

V. K. Ignatjev* and S. V. Perchenko†
Volgograd State University, Institute of Physics and Technology, Volgograd, Russia

ABSTRACT

Onsager's reciprocal relations are obtained in the linear approximation, based on the invariance of the macroscopic motion under time reversal and the assumption that the average relaxation of spontaneous fluctuations in the system is in accordance with the macroscopic laws. This paper shows that in the particular case of plasma-like media, to which the staff of the charge carriers in a semiconductor is referred, it is possible to obtain reciprocal relations for the nonlinear case.

OBTAINMENT OF NONLINEAR RECIPROCAL RELATIONS OF PLASMA-LIKE MEDIA

Phenomena, in which there is interaction between the cross transport processes, are studied for a long time. In 1854, Kelvin studied the thermoelectric effect, which arises while the flow of electric current and heat, and were deduced first reciprocal relations based on thermodynamic arguments. Such relations are deduced by Helmholtz in 1876 for transport processes in electrolytes, and later by Eastman (1926) for diffusion and heat fluxes [1]. The general theory was formed by Lars Onsager in 1931, for which he was awarded the Nobel Prize in 1968.

Onsager reciprocity relations are fundamental and valid not only in thermodynamics. For example, in mechanics, they are expressed in the Maxwell's law of reciprocal deflections, according to which for a linearly deformable body moving σ_{ik} in the point of application first

* ignatjev@vlpost.ru.
† sergrf1104@mail.ru.

state single force P_k in the direction of its action, caused by any other second single force P_i, is equal to moving σ_{ki} the point of force application P_i in the direction of its action on the unit force P_k [2]. The principle of reciprocity is widely used in acoustics [3], and also it is used for calculation and measurement of the characteristics of receiving antennas [4].

Onsager relations of symmetry of kinetic coefficients [5] are obtained in the linear approximation, based on the invariance of the macroscopic motion under time reversal and the assumption that the average relaxation of spontaneous fluctuations in the system is in accordance with the macroscopic laws. In the thermodynamics of nonequilibrium processes, the reciprocal relations for linear systems are postulated [6] and sometimes regarded as the fourth law of thermodynamics. Common statistical techniques of reciprocal relations argumentation are generally inapplicable when there are nonlinear relationships between forces and flows generated by these forces [7]. In a number of papers [8], the higher-order reciprocal relations for special cases are given without derivations, but they have not been confirmed in some approximation by a precise and general method (or experimental), and the theory, which is built on it, is empty formalism [7]. Influence of magnetic field on the higher-order reciprocal relations, unlike to the linear theory of Onsager, generally has not been investigated.

In the particular case of plasma-like media, to which relate the collective of the charge carriers in semiconductors, the construction of a theory of nonlinear permeability based on the solution of kinetic equations by expansion in powers of the electromagnetic field is developed enough in a contemporary nonlinear electrodynamics [9], but the reciprocal relations for the obtained susceptibilities are not considered.

Consider the physical causes of nonlinearity in the plasma-like media. Transport processes in the relaxation approximation are described by the kinetic Boltzmann equation, which for the medium in the electric and magnetic fields, takes the form Vlasov equation [9]:

$$\frac{\partial f}{\partial t}+\mathbf{v}\frac{\partial f}{\partial \mathbf{r}}+q\left(\mathbf{E}+\frac{1}{c}[\mathbf{v}\times\mathbf{B}]\right)\frac{\partial f}{\partial \mathbf{p}}=-\frac{f-f_0}{\tau}. \tag{1}$$

There $f(t, \mathbf{r}, \mathbf{p})$ is the distribution function of the charge carriers (electrons or holes) over coordinates and impulses, $f_0(W)$ is the equilibrium distribution function, $W(\mathbf{p})$ is the carrier energy, q is the electrical charge (of electron or hole), τ is the ensemble average of relaxation time.

The equilibrium distribution function does not contribute to the current density:

$$\int v f_0(t,r,p)d^3p=0;\quad j(t,r)=q\int v f_1(t,r,p)d^3p=0. \tag{2}$$

Denote $f_1(t, \mathbf{r}, \mathbf{p}) = f(t, \mathbf{r}, \mathbf{p}) - f_0(W)$. Then (1) takes the form:

$$\frac{\partial f_0}{\partial t}+\frac{\partial f_1}{\partial t}+\mathbf{v}\frac{\partial f_0}{\partial \mathbf{r}}+\mathbf{v}\frac{\partial f_1}{\partial \mathbf{r}}+q\left(\mathbf{E}+\frac{1}{c}[\mathbf{v}\times\mathbf{B}]\right)\left(\frac{\partial f_0}{\partial \mathbf{p}}+\frac{\partial f_1}{\partial \mathbf{p}}\right)=-\frac{f_1}{\tau}. \tag{3}$$

In the steady state, the distribution function does not depend on time:

$$\frac{\partial f_0}{\partial t}=0;\frac{\partial f_1}{\partial t}=0.$$

Consider the case that the semiconductor itself and the temperature distribution in it are homogeneous. Then the spatial dependence of the distribution function is determined only by the inhomogeneity of the external fields. If the electric and magnetic fields vary insignificantly on the free path length, then in equation (3) we can ignore the derivative of the distribution function with respect to coordinate:

$$\frac{\partial f_0}{\partial \mathbf{r}}\approx 0;\frac{\partial f_1}{\partial \mathbf{r}}\approx 0.$$

Reconstruct a derivative of the equilibrium distribution function with respect to impulse:

$$\frac{\partial f_0}{\partial \mathbf{p}}=\frac{df_0}{dW}\frac{\partial W}{\partial \mathbf{p}}=\mathbf{v}\frac{df_0}{dW}.$$

With use of the imposed assumptions, Equation (3) takes the following form:

$$e(\mathbf{Ev})\frac{df_0}{dW}+e\mathbf{E}\frac{\partial f_1}{\partial \mathbf{p}}+\frac{q}{c}[\mathbf{v}\times\mathbf{B}]\frac{\partial f_1}{\partial \mathbf{p}}=-\frac{f_1}{\tau}. \tag{4}$$

We seek the solution of Equation (4) in the form of a series of functions as

$$f_1=f^{(1)}+f^{(2)}+f^{(3)}+\ldots, \tag{5}$$

where $f^{(1)}$ is the the function, linear by electrical field E, the function $f^{(2)}$ proportional E^2, the function $f^{(3)}$ proportional E^3 [5]. Substituting (5) in equation (4) and collecting terms of the same order of E, we obtain

$$\begin{aligned}
&q(\mathbf{Ev})\frac{df_0}{dW}+\frac{q}{c}[\mathbf{v}\times\mathbf{B}]\frac{\partial f^{(1)}}{\partial \mathbf{p}}=-\frac{f^{(1)}}{\tau},\\
&q\mathbf{E}\frac{\partial f^{(1)}}{\partial \mathbf{p}}+\frac{q}{c}[\mathbf{v}\times\mathbf{B}]\frac{\partial f^{(2)}}{\partial \mathbf{p}}=-\frac{f^{(2)}}{\tau},\\
&q\mathbf{E}\frac{\partial f^{(2)}}{\partial \mathbf{p}}+\frac{q}{c}[\mathbf{v}\times\mathbf{B}]\frac{\partial f^{(3)}}{\partial \mathbf{p}}=-\frac{f^{(3)}}{\tau}.
\end{aligned} \tag{6}$$

For isotropic semiconductor $\mathbf{v}=\mathbf{p}/m^*$, where m^* is the effective mass of charge carriers, we can introduce averaged drift mobility of charge carrier:

$$\mu=\frac{q\tau}{m^*}.$$

Taking it into account the first equation of system (6) can be rewritten as:

$$\mu(\mathbf{Ep})\frac{df_0}{dW}+\frac{\mu}{c}[\mathbf{p}\times\mathbf{B}]\frac{\partial f^{(1)}}{\partial\mathbf{p}}=-f^{(1)}. \qquad (7)$$

In order for the distribution function $f^{(1)}$ depends on the magnetic field B, it is necessary that $\frac{\partial f^{(1)}}{\partial\mathbf{p}}\neq 0$. This condition is satisfied by the solution $f^{(1)}=\mathbf{Ap}$. As a basis for the vector **A** we choose the vectors **E**, **B** и [**B**×**E**], then the solution will be in the form:

$$f^{(1)}=\mathrm{I}(\mathbf{Ep})+\mathrm{J}(\mathbf{Bp})+\mathrm{K}([\mathbf{B}\times\mathbf{E}]\mathbf{p}). \qquad (8)$$

Substituting (8) into (7) and collecting terms with identical vectors, we obtain expressions for the coefficients I, J and K:

$$\begin{cases} I=\frac{\mu}{c}KB^2-\mu\frac{df_0}{dW}; \\ K=-\frac{\mu}{c}I; \\ J=-\frac{\mu}{c}K(\mathbf{BE}). \end{cases} \qquad (9)$$

Solving system of equations (9) we find the coefficients:

$$\begin{cases} I=-\mu(B^2)\frac{df_0}{dW}; \\ K=\mu(B^2)\frac{\mu}{c}\frac{df_0}{dW}; \\ J=-\mu(B^2)\left(\frac{\mu}{c}\right)^2(\mathbf{BE})\frac{df_0}{dW}, \end{cases}$$

where $\mu(B^2)=\frac{\mu}{1+(\mu B/c)^2}$.

Substituting these coefficients into equation (8) we obtain solution of equation (7):

$$f^{(1)}=-\mu(B^2)\left\{\mathbf{E}+\frac{\mu}{c}[\mathbf{E}\times\mathbf{B}]+\left(\frac{\mu}{c}\right)^2\mathbf{B}(\mathbf{BE})\right\}\mathbf{p}\frac{df_0}{dW}. \qquad (10)$$

In the crossed fields, the third term in the curly brackets in (10) is equal to zero. Then the second equation of system (6) takes the form:

$$-\mu(B^2)\left(\frac{df_0}{dW}+\frac{p^2}{m^*}\frac{d^2 f_0}{dW^2}\right)qE^2+\frac{q}{c}[\mathbf{v}\times\mathbf{B}]\frac{\partial f^{(2)}}{\partial\mathbf{p}}=-\frac{f^{(2)}}{\tau}. \tag{11}$$

Direct substitution is easy way to see that the equation (11) has following solution:

$$f^{(2)}=\mu(B^2)\left(\frac{df_0}{dW}+\frac{p^2}{m^*}\frac{d^2 f_0}{dW^2}\right)q\tau E^2. \tag{12}$$

Substituting (12) into the third equation of system (6), we obtain for the function $f^{(3)}$ an equation similar to the first equation of system (6) for $f^{(1)}$. Accordingly, the solution of this equation is similar to the expression (10). For case of the crossed fields, we obtain:

$$f^{(3)}=-q\tau\mu^2(B^2)E^2\left(3\frac{d^2 f_0}{dW^2}+\frac{p^2}{m^*}\frac{d^3 f_0}{dW^3}\right)\left\{\mathbf{E}+\frac{\mu}{c}[\mathbf{B}\times\mathbf{E}]\right\}\mathbf{p}. \tag{13}$$

As it follows from the second equation (2), even the impulse function $f^{(2)}$ of the form (12) does not contribute to the current density. Therefore, in the second formula (2) we substitute the sum of the distribution function (10) and (13):

$$f_1=-\mu(B^2)\left\{\frac{df_0}{dW}+q\tau\mu(B^2)E^2\left(3\frac{d^2 f_0}{dW^2}+\frac{p^2}{m^*}\frac{d^3 f_0}{dW^3}\right)\right\}\left\{\mathbf{E}+\frac{\mu}{c}[\mathbf{B}\times\mathbf{E}]\right\}\mathbf{p}. \tag{14}$$

Integration of (2) with the distribution functions (10) and (13) holds for a spherical energy surface $W = \mathbf{p}^2/(2m^*)$ and the Boltzmann equilibrium distribution $f_0 = \exp(-W/\theta)$, where $\theta = k_B T$ is the temperature in energy units. For the distribution function $f^{(1)}$, the current density is given by:

$$\mathbf{j}^{(1)}(t,\mathbf{r})=\mu(B^2)\left\{\mathbf{E}+\frac{\mu}{c}[\mathbf{E}\times\mathbf{B}]\right\}\frac{q}{m^*\theta}\int_{-\infty}^{\infty}\int_{-\infty}^{\infty}\int_{-\infty}^{\infty}p^2\exp\left(-\frac{p^2}{2m^*\theta}\right)dp_x dp_y dp_z\ .$$

We pass from the integration in Cartesian coordinates to spherical:

$$I = \int_{-\infty}^{\infty}\int_{-\infty}^{\infty}\int_{-\infty}^{\infty} p^2 \exp\left(-\frac{p^2}{2m^*\theta}\right) dp_x dp_y dp_z = \int_0^{2\pi} d\varphi \int_0^{\pi} \sin\theta d\theta \int_0^{\infty} \rho^4 \exp\left(-\frac{\rho^2}{2m^*\theta}\right) d\rho =$$

$$= 4\pi \int_0^{\infty} \rho^4 \exp\left(-\frac{\rho^2}{2m^*\theta}\right) d\rho,$$

where $\rho = \sqrt{p_x^2 + p_y^2 + p_z^2}$.

The obtained expression is convenient to integrate after differentiation of the integrand by the independent parameter $\alpha = \dfrac{1}{2m^*\theta}$:

$$\frac{\partial^2}{\partial\alpha^2} \exp\left(-\alpha\rho^2\right) = \rho^4 \exp\left(-\alpha\rho^2\right) \Rightarrow$$

$$I = 4\pi \frac{\partial^2}{\partial\alpha^2} \int_0^{\infty} \exp\left(-\alpha\rho^2\right) d\rho = 2\pi\sqrt{\pi} \frac{\partial^2}{\partial\alpha^2} \alpha^{-\frac{1}{2}} =$$

$$= \frac{3}{2}\pi^{\frac{3}{2}}\alpha^{-\frac{5}{2}} = \frac{3}{2}\pi^{\frac{3}{2}}\left(2m^*\theta\right)^{\frac{5}{2}} = 3m^*\theta\left(2\pi m^*\theta\right)^{\frac{3}{2}}.$$

In summary

$$\mathbf{j}^{(1)} = 3q\mu(B^2)\left\{\mathbf{E} + \frac{\mu}{c}[\mathbf{E}\times\mathbf{B}]\right\}\left(2\pi m^*\theta\right)^{\frac{3}{2}}.$$

In the same manner for $\mathbf{j}^{(3)}$ we have

$$\mathbf{j}^{(3)} = -3\frac{q^2\tau\mu^2(B^2)E^2}{m^*\theta^2}\left\{\mathbf{E} + \frac{\mu}{c}[\mathbf{B}\times\mathbf{E}]\right\} \int_{-\infty}^{\infty} p^2 \exp\left(-\frac{p^2}{2m^*\theta}\right) d^3 p =$$

$$-9\frac{q^2\tau\mu^2(B^2)E^2}{\theta}\left\{\mathbf{E} + \frac{\mu}{c}[\mathbf{B}\times\mathbf{E}]\right\}\left(2\pi m^*\theta\right)^{\frac{3}{2}}.$$

The full current density:

$$\mathbf{j} = \mathbf{j}^{(1)} + \mathbf{j}^{(3)} = 3q\mu(B^2)\left(1 - 3\frac{q\tau\mu(B^2)E^2}{\theta}\right)\left\{\mathbf{E} + \frac{\mu}{c}[\mathbf{E}\times\mathbf{B}]\right\}\left(2\pi m^*\theta\right)^{\frac{3}{2}} =$$

$$= \sigma\left(E^2, B^2\right)\left\{\mathbf{E} + \frac{\mu}{c}[\mathbf{E}\times\mathbf{B}]\right\},$$

where quasi-linear conductivity $\sigma\left(E^2, B^2\right)$ is given as

$$\sigma\left(E^2, B^2\right) = 3q\mu(B^2)\left(1 - 3\frac{q\tau\mu(B^2)E^2}{\theta}\right)\left(2\pi m^*\theta\right)^{\frac{3}{2}}.$$

It is obvious that the expression for a quasi-linear conductivity is symmetric about magnetic induction. Assuming that the magnetic field is directed along the axis z, and the electric field and current density lie in the plane *xy*, we obtain expressions for the currents j_x and j_y:

$$j_x = \sigma\left(E^2, B^2\right)\left\{E_x + \frac{\mu}{c}E_y B_z\right\}; j_y = \sigma\left(E^2, B^2\right)\left\{E_y + \frac{\mu}{c}E_x B_z\right\} \quad (15)$$

Equations (15) allows us to record steady-state constitutive relations of plasma-like medium in the form of quasi-linear Ohm's law:

$$j_i = \Sigma_{ij}(E^2, B^2)E_j. \quad (16)$$

Here, repeated indices imply summation. The matrix of the quasi-linear conductivity Σ_{ij} has the form:

$$\Sigma_{ij}\left(E^2, B^2\right) = \sigma\left(E^2, B^2\right)\begin{bmatrix} 1 & \frac{\mu}{c}B_z \\ -\frac{\mu}{c}B_z & 1 \end{bmatrix}. \quad (17)$$

It is easy to see that the matrix elements (17) satisfy the classical symmetry relations:

$$\Sigma_{ij}(B) = \Sigma_{ji}(-B).$$

ACKNOWLEDGEMENTS

This work was supported by the Federal Target Program "Scientific and Pedagogical Manpower" for 2009-2013 (project № 14.740.11.0830).

REFERENCES

[1] Petrov, N.; Brankov, J. *Modern Problems of Thermodynamics*. Mir, Moscow, 1986, pp 1 - 288.

[2] Gantmacher, F. R.. *Lectures in Analytical Mechanics*. Nauka, Moscow, 1966, pp 1 – 300.

[3] Lyamshev, L.M. *Reports of the USSR Academy of Sciences*. 1959, *vol. 125(6)*.

[4] Makarov, G. T. *Antennas*. Energia, Moskow, 1975, pp 1 – 528.

[5] Onsager, L. *Physical Review*, 1931, *vol. 37*, 405 – 426.

[6] DeGroot, S. R. *The Thermodynamics of Irreversible Processes.* GITTL, Moskow, 1956, pp 1 - 281.

[7] Djarmati, I. *Non-equilibrium Thermodynamics. Field Theory and Variational Principles.* Mir, Moskow, 1974, pp 1 – 304.

[8] Fain, V. M., Hanin J. I. *Quantum Radiophysics.* Sovetskoe Radio, Moskow, 1965, pp 1 – 606.

[9] Aleksandrov, A. F.; Bogdankevich, L. S.; Ruhadze, A. A. *Fundamentals of Plasma Electrodynamics*. Vishaya Shkola, Moskow, 1978, pp 1 – 407.

In: Physics and Mechanics of New Materials …
Editors: Ivan A. Parinov and Shun-Hsyung Chang
ISBN: 978-1-62618-535-7

Chapter 7

MAGNETOSTATICS OF CYLINDRICAL FERROMAGNETIC SAMPLES

V. K. Ignatjev*, A. A. Orlov and D. A. Stankevich
Volgograd State University, Institute of Physics and Technology, Volgograd, Russia

ABSTRACT

The paper describes a method of recovering magnetization of a ferromagnetic cylinder caused by magnetic field, measured at least at three points around the sample. The method of recovering parameters of the magnetic dipoles, as well as a way of checking gauges of the magnetic moment are presented.

1. INTRODUCTION

A variety of properties of the different type steels and goods can be described by their different microstructures. The study of the microstructure allows one to state the stage of early failure of the material, to prevent the failure propagation and its catastrophic consequences. The analysis allows one to study the metal microstructure during fatigue processes or the instability of various defects. The applied methods of microstructure analysis are based on correlations between physicochemical, mechanical and structure properties of the material and its magnetic characteristics.

The actual problem of magnetic methods is to detect violations of the material microstructure, such as defects of continuity, the assessment of location and geometric parameters that are used to diagnose the technical state of the product. Elucidation of the influence of microstructure and external electromagnetic fields on the magnetic properties of solids is important for modern functional electronics, based on the propagation of spin waves in magnetically ordered heterostructures, as well as for analysis of nanomaterial structure.

---

* e-mail: ignatjev@vlpost.ru

The priority in the field of magnetic methods for microstructure analysis is the creation of computer diagnostics.

The development of computer technology makes this problem to be solvable in principle for the corresponding increase in information content of measurements due to a substantial increase in the number of measurement points and a spatial region in which these measurements are made with the required accuracy and spatial resolution.

The mathematical estimation of the microstructure parameters on the measured magnetic field belongs to a class of ill-posed inverse problems, the use of simplified models does not lead to satisfactory estimates of the parameters of actual microstructures, even if we use powerful mathematical methods of solution [1]. It should be noted that the inverse problem for any solving algorithm is incorrect, and a small error in the input data can cause large errors in the control output data. Analysis of gradient methods for solving the inverse problem of magnetometry showed that the main source of the error is the numerical calculation of derivatives of the magnetic induction measured by its values at closely disposed points. The instability of the algorithm of numerical differentiation is much higher than the measurement error of the field.

Therefore, the method, used in geophysics for calculation of the magnetic dipole tensor spatial derivatives of the magnetic field components [2], has little application in the magnetic microstructure analysis, which should be carried out by using microtopologic integral, but no differential magnitudes of the magnetic field.

The use of modern methods of structural and parametric identification of dynamic objects and robust statistics allows one to obtain mathematical models of the required accuracy. The presence of sufficiently accurate models and constant monitoring of changes in the parameters of the object provide diagnostics and prognosis of its behavior with a high degree of reliability. The main problem in achieving the necessary metrological characteristics of magnetometric methods of the microstructure analysis is the requirement to simultaneously measure with high accuracy the magnetic fields in several closely disposed sites and to process in real time a large volume of data at scanning of the specimen surface, followed by solving the inverse problem.

Difficulties in interpreting the results of magnetic structurescopy arise from the fact that the dependence of the mechanical and magnetic properties on solid microstructure are now theoretically explained in the case of weak magnetic fields, whose energy can be considered as a perturbation of the model Hamiltonian in magnetically ordered materials [3]. The magnetic structural analysis is performed in a state of technical saturation, where the magnetostriction creates elastic stresses, which significantly distort the current microstructure of the material.

Waiver of technical saturation complicates the identification of a defect microstructure, depending on the orientation of the object relatively to the control vector of the geomagnetic field. Its magnetization and stray field can vary considerably, because the weak magnetization hysteresis depends strongly on the magnetic history.

On the other hand, in weak fields, comparable to Earth's field, the nonlinearity of the magnetization curve leads to bulk magnetic charges and additional fields. As a result, for a defect related to the microstructure, a magnetic dipole is drawn out, sometimes hundreds of times along the direction of magnetization and near the boundary of the specimen [4], that improves the resolution of internal defects with micron sizes at the necessary sensitivity of the magnetic measurements.

2. Method of Recovering Magnetization of Ferromagnetic Cylinder

We consider the cylindrical geometry of the magnetostatic problem taking place in magnetostructure analysis of long objects (wires, cables, pipes, etc.) with a weak magnetization, for example, in the geomagnetic field (Figure 1).

Let us suppose that in a uniform cylindrical rod the magnetization $M_0(z')$ is constant over the cross-section and varies only along the longitudinal coordinate z, there is a defect of microstructure at r_d that distorts the magnetization in a small region of v, bounded by the surface s. We denote the field, produced by the defects, as $M_d(r')$. In addition, we take into account the random component of the magnetization $M_n(r')$, created by the magnetic moments localized near the grain boundaries, i.e., we present

$$M(r') = M_0(z') + M_d(r') + M_n(r') \quad (1)$$

both on the surface s, and everywhere $M_d(r') \equiv 0$ for $r' \notin v$. Then the magnetic potential φ can be represented as the sum of the potentials φ_0, φ_d and φ_n, created by magnetizations M_0, M_d and M_n, respectively.

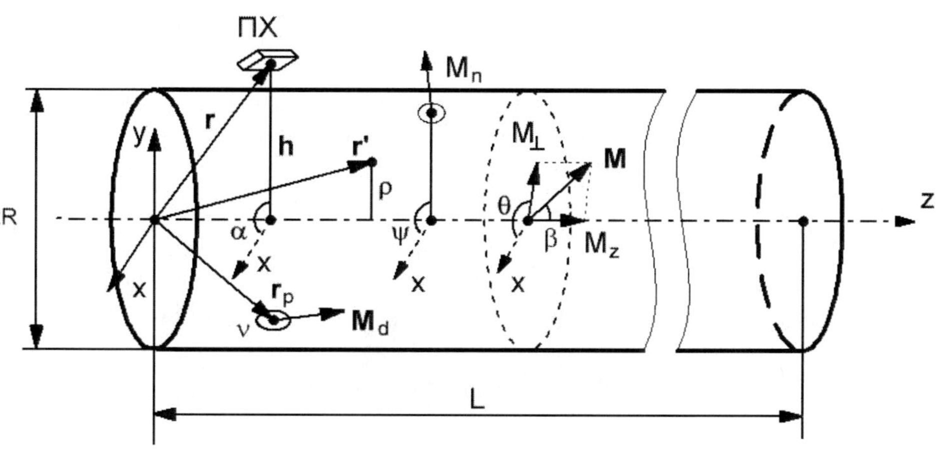

Figure 1. The geometry of the magnetostatic problem.

We denote $M_z = M_0\cos(\beta)$, $M_\perp = M_0\sin(\beta)$, $M_x = M_\perp\cos(\theta)$, $M_y = M_\perp\sin(\theta)$ (see Figure 1). The point r, in which the field is the measured, we describe by the cylindrical coordinates (z, h, α). Since the components of the magnetization in the formula (1) provide an independent contribution to the measured field, then we solve the direct problem of finding a field for each separate component of the magnetization.

We use the representation of the magnetic field through the potential φ_0 to find the magnetic field, produced by magnetization M_0 as

$$H_0(r) = -\nabla\varphi_0(r).$$

The solution is a formula expressing the potential of the magnetic field through the divergence of the magnetization inside the body and the normal to the boundary of the magnetization on the surface:

$$\varphi_0(\mathbf{r}) = \frac{1}{4\pi}\int_{V-S} \frac{-\operatorname{div}_{\mathbf{r}'}\mathbf{M}(\mathbf{r}')}{|\mathbf{r}-\mathbf{r}'|} d^3r' + \frac{1}{4\pi}\oint_S \frac{M_n(\mathbf{r}')}{|\mathbf{r}-\mathbf{r}'|} ds'. \quad (2)$$

This expression arises in the limit near the boundary of the body where there is a jump in the magnetization, and the operation of divergence is not defined. Using the formula (2), taking into account the geometry of the problem and the notation adopted, we can write the expression for the potential created by the magnetization M_0 as

$$\varphi_0(\mathbf{r}) = \frac{1}{4\pi}\int_0^L M_\perp(z')\int_{-\pi}^{\pi} \frac{R\cos[\theta(z')-\psi]d\psi}{\sqrt{(z-z')^2 + (h\cos(\alpha) - R\cos(\psi))^2 + (h\sin(\alpha) - R\sin(\psi))^2}} dz' -$$

$$-\frac{M_z(0)}{4\pi}\int_0^R\int_{-\pi}^{\pi} \frac{\rho d\rho d\psi}{\sqrt{(z)^2 + (h\cos(\alpha) - \rho\cos(\psi))^2 + (h\sin(\alpha) - \rho\sin(\psi))^2}} +$$

$$+\frac{M_z(L)}{4\pi}\int_0^R\int_{-\pi}^{\pi} \frac{\rho d\rho d\psi}{\sqrt{(z-L)^2 + (h\cos(\alpha) - \rho\cos(\psi))^2 + (h\sin(\alpha) - \rho\sin(\psi))^2}} -$$

$$-\frac{1}{4\pi}\int_0^L \frac{M_z(z')}{dz'}\int_0^R\int_{-\pi}^{\pi} \frac{\rho d\rho d\psi}{\sqrt{(z-z')^2 + (h\cos(\alpha) - \rho\cos(\psi))^2 + (h\sin(\alpha) - \rho\sin(\psi))^2}} dz'.$$

The first term, here describes the contribution of the normal component of the magnetization on the lateral surface of the cylinder, while the second and third ones are responsible for the contributions of the side surfaces of the cylinder. The fourth term describes the magnetic field produced by the divergence of the magnetization in the thickness of the cylinder. We transform this expression by integrating the fourth integral part on the longitudinal coordinate z, at given the second and third terms:

$$\varphi_0(\mathbf{r}) = \frac{1}{4\pi}\int_0^L R\left[M_x(z')\cos(\alpha) + M_y(z')\sin(\alpha)\right] \times$$

$$\times \int_{-\pi}^{\pi} \frac{\cos(\psi)d\psi}{\sqrt{(z-z')^2 + h^2 + R^2 - 2Rh\cos(\psi)}} dz' + \quad (3)$$

$$+\frac{1}{4\pi}\int_0^L M_z(z')\int_0^R\int_{-\pi}^{\pi} \frac{\rho(z-z')d\rho d\psi}{\left[(z-z')^2 + h^2 + \rho^2 - 2\rho h\cos(\psi)\right]^{3/2}} dz'.$$

We calculate the normal component of the field at the point of measurement. In order to make it we differentiate the potential φ_0 and obtain

$$H(\alpha,z)=\int_0^L M_z(z')K(z-z')dz'+\int_0^L [M_x(z')\cos(\alpha)+M_y(z')\sin(\alpha)]P(z-z')dz', \tag{4}$$

where the functions $K(z)$ and $P(z)$ are the system functions of the magnetization:

$$K(z)=\frac{3}{4\pi}\int_0^R\int_{-\pi}^{\pi}\frac{\rho[h-\rho\cos(\psi)]z}{\left[z^2+h^2-2\rho h\cos(\psi)+\rho^2\right]^{5/2}}d\rho d\psi\ , \tag{5}$$

$$P(z)=\frac{1}{4\pi}\int_{-\pi}^{\pi}\frac{R[h-R\cos(\psi)]\cos(\psi)}{\left[z^2+h^2-2Rh\cos(\psi)+R^2\right]^{3/2}}d\psi\ . \tag{6}$$

The system uses magnetic microstructure analysis of the magnetic field with help of measuring devices, each of which is located at a fixed angle α (see Figure 1). We can use a combination of several meters for information on the magnetization by using the formula (4). Schemes of three- and four-sensor systems with devices measuring spatial magnetic field are shown in Figure 2.

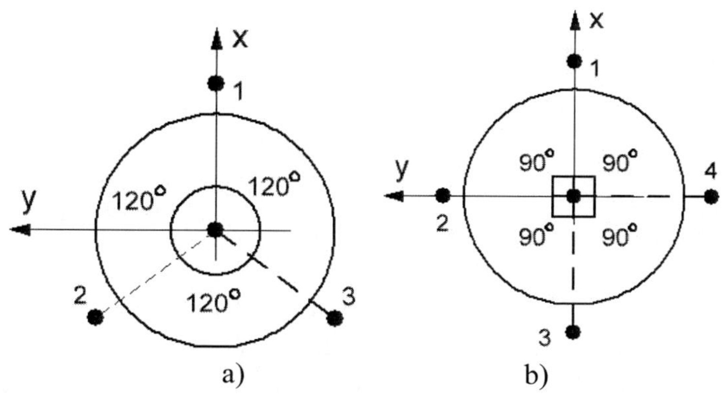

Figure 2. The location of the magnetic field measuring devices: (a) three-sensor scheme, (b) four-sensor scheme.

For the normal component of the magnetic field at points 1, 2 and 3 (H_1, H_2 and H_3) with $\alpha = 0$, $\alpha = 2\pi/3$ and $\alpha =-2\pi/3$ we obtain the convolution equations:

$$H_1(z)=\int_0^L \{M_z(z')K(z-z')+M_x(z')P(z-z')\}dz'\ ;$$

$$H_2(z)=\int_0^L \left\{M_z(z')K(z-z')+\left[-\frac{1}{2}M_x(z')+\frac{\sqrt{3}}{2}M_y(z')\right]P(z-z')\right\}dz'\ ;$$

$$H_3(z)=\int_0^L \left\{M_z(z')K(z-z')+\left[-\frac{1}{2}M_x(z')-\frac{\sqrt{3}}{2}M_y(z')\right]P(z-z')\right\}dz'\ .$$

The inverse problem of finding the magnetization of a defect-free rod measured on the surface by using normal values of the magnetic field, provided that all $H_i(-\infty) = 0$, has the form:

$$H_1(z)+H_2(z)+H_3(z)=3\int_0^L M_z(z')K(z-z')dz';$$

$$2H_1(z)-H_2(z)-H_3(z)=3\int_0^L M_x(z')P(z-z')dz'; \tag{7}$$

$$H_2(z)-H_3(z)=\sqrt{3}\int_0^L M_y(z')P(z-z')dz'.$$

For the four-sensor scheme (Figure 2,b) with points 1 – 4 at $\alpha = 0$, $\alpha = \pi/2$, $\alpha = \pi$ and $\alpha = 3\pi/2$, we obtain

$$H_1(z)=\int_0^L \{M_z(z')K(z-z')+M_x(z')P(z-z')\}dz';$$

$$H_2(z)=\int_0^L \{M_z(z')K(z-z')+M_y(z')P(z-z')\}dz';$$

$$H_3(z)=\int_0^L \{M_z(z')K(z-z')-M_x(z')P(z-z')\}dz';$$

$$H_4(z)=\int_0^L \{M_z(z')K(z-z')-M_y(z')P(z-z')\}dz'.$$

The inverse problem for the four-sensor system has the form:

$$H_1(z)+H_2(z)+H_3(z)+H_4(z)=4\int_0^L M_z(z')K(z-z')dz';$$

$$H_1(z)-H_3(z)=2\int_0^L M_x(z')P(z-z')dz'; \tag{8}$$

$$H_2(z)-H_4(z)=2\int_0^L M_y(z')P(z-z')dz'.$$

By comparing the relationships for the inverse problems (7) and (8), it is easy to see that fourth-sensor data are redundant in order to find the magnetization vector of the rod. In turn, two magnetic field sensors are not allowed one to share the magnetization and to obtain the inverse problems like (7) and (8). Thus, the three-sensor scheme is optimal measuring one to recover slowly varying magnetization.

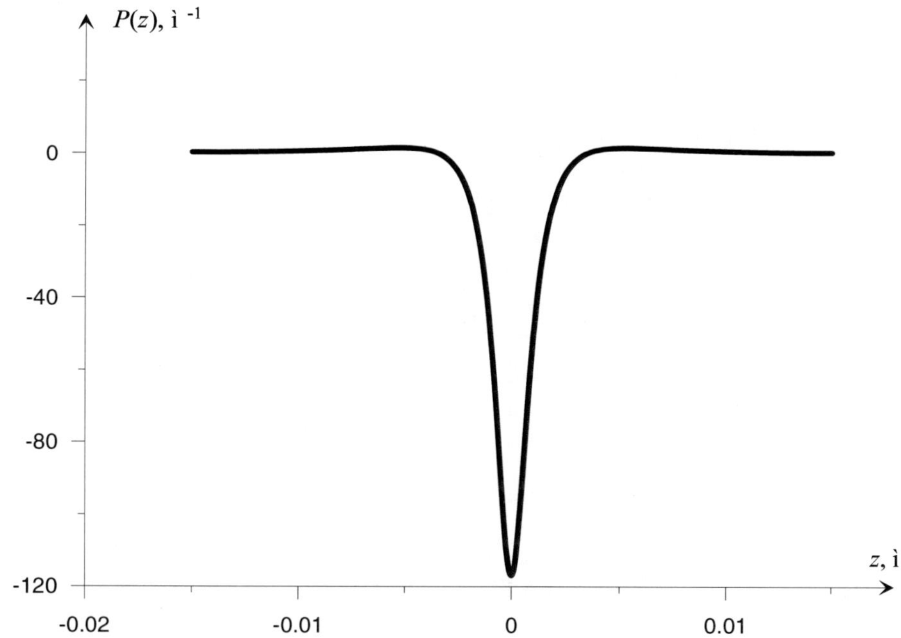

Figure 3. System function $P(z)$ ($R = 2$ mm, $h = 3$ mm).

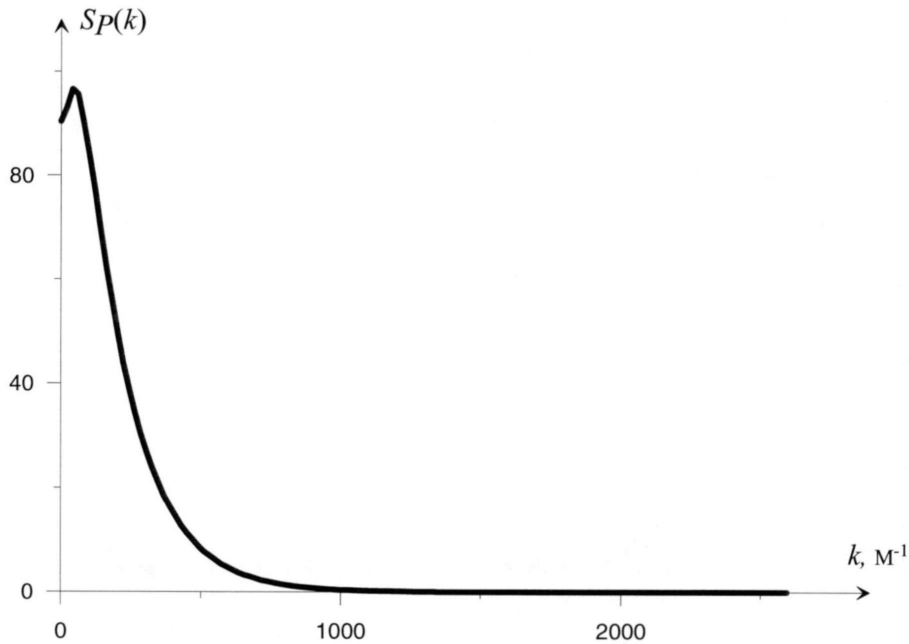

Figure 4. Spectrum of the system function $S_P(k)$ ($R = 2$ mm, $h = 3$ mm).

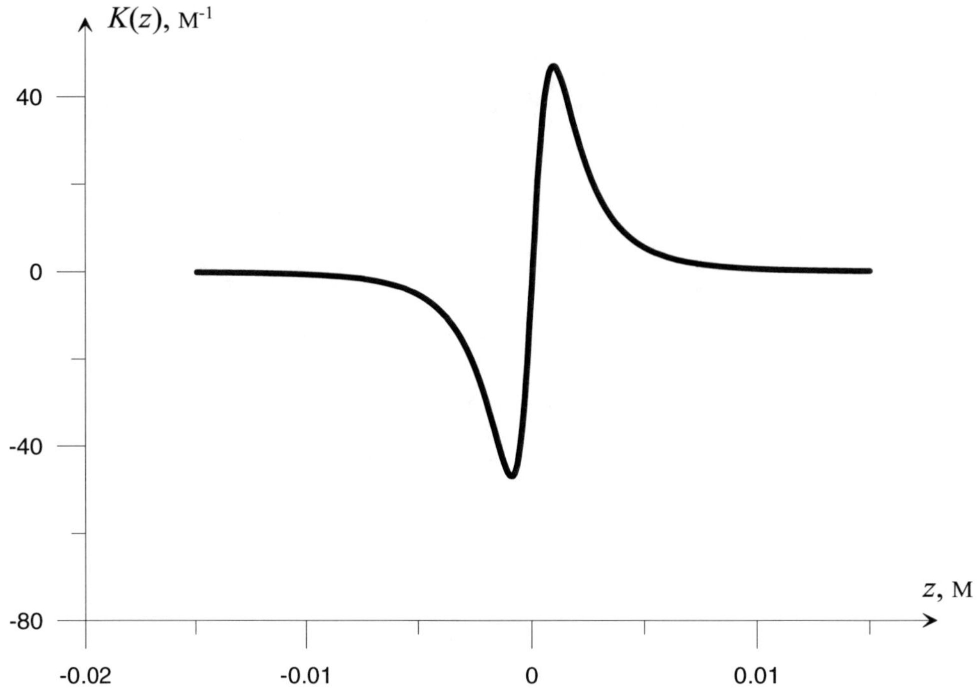

Figure 5. System function $K(z)$ ($R = 2$ mm, $h = 3$ mm).

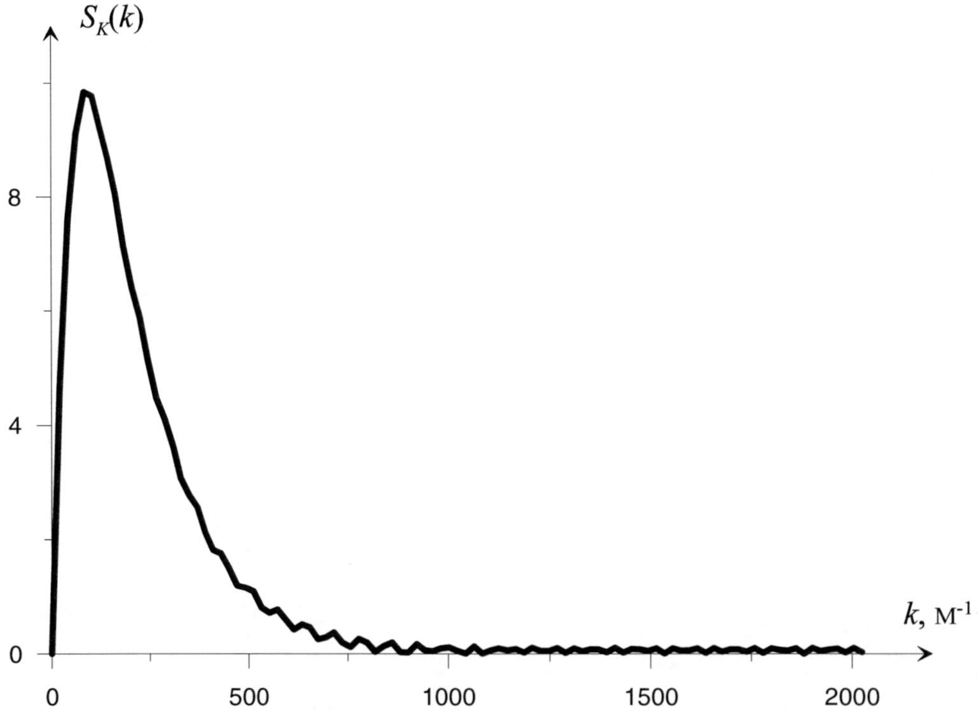

Figure 6. Spectrum of the system function $S_K(k)$ ($R = 2$ mm, $h = 3$ mm).

Let us analyze expressions (7) on components. They are the equations of convolution type and can be solved using the Fourier transform and the transition from the convolution of the product of their spectra. In this case, an absence of zeros in the spectra of the system functions $K(z)$ and $P(z)$ is necessary.

The analysis was based on calculation of the functions $K(z)$ and $P(z)$ by using Equations (5) and (6), respectively, and their spectra (see Figures 3-6), for almost reasonable geometry parameters.

Figure 6 shows that spectrum of the function $S_P(k)$ is zero at zero frequency, indicating inability to resolve the inverse problem for the first equation of systems (7) by the spectral method. On the other hand, the solution of the equations for M_x and M_y can be built as a spectral one, and any other way. This is due to the fact that the system function $P(z)$ is the even one (see Figure 3) and has no zeros in its spectrum (Figure 4).

For the potential created by the defect microstructure φ_d, using the small size of the region v compared with the distance r, we expand the denominator of the integrand in a Taylor series:

$$\varphi_d(\mathbf{r}) = \frac{1}{4\pi}\int_v \frac{-\operatorname{div}_{\mathbf{r}'}\mathbf{M}_d(\mathbf{r}')}{|\mathbf{r}-\mathbf{r}'|}d^3r' = -\frac{1}{4\pi r}\int_v \operatorname{div}_{\mathbf{r}'}\mathbf{M}_d(\mathbf{r}')d^3r' +$$
$$+\frac{1}{4\pi}\int_v \operatorname{div}_{\mathbf{r}'}\mathbf{M}_d(\mathbf{r}')\left(\mathbf{r}'\frac{\partial}{\partial\mathbf{r}}\right)\frac{1}{r}d^3r' + \cdots \quad .$$

The first term is zero due to $M_d(r'\in s) \equiv 0$. The second term is a dipole, and the next ones present the multipole potentials. By using the formula for the potential of the magnetic moment of expectation m_d:

$$\varphi_d(r) = (m_d\nabla)(1/r),$$

we obtain

$$\mathbf{m}_d = \frac{1}{4\pi}\int_v \mathbf{r}'\operatorname{div}_{\mathbf{r}'}\mathbf{M}_d(\mathbf{r}')d^3r'.$$

For the cylindrical geometry, the dipole potential has the form:

$$\varphi_d(\mathbf{r}) = \left(\mathbf{m}_d\frac{\partial}{\partial\mathbf{r}}\right)\frac{1}{|\mathbf{r}-\mathbf{r}_d|} = -\frac{(x-x_d)m_x + (y-y_d)m_y + (z-z_d)m_d}{\left[(x-x')^2+(y-y')^2+(z-z')^2\right]^{3/2}} =$$
$$= -\frac{(h\cos(\alpha)-\rho\cos(\psi))m_x + (h\sin(\alpha)-\rho\sin(\psi))m_y + (z-z_d)m_z}{\left[(z-z_d)^2 + (h\cos(\alpha)-\rho\cos(\psi))^2 + (h\sin(\alpha)-\rho\sin(\psi))^2\right]^{3/2}}.$$

For the normal component of the magnetic field, $H_d = -\partial\varphi_d/\partial h$, we obtain

$$H_d(z)=\frac{m_d\sin(\beta)[\cos(\alpha)\cos(\gamma)+\sin(\alpha)\sin(\gamma)]}{\left[(z-z_d)^2+(h\cos(\alpha)-\rho\cos(\psi))^2+(h\sin(\alpha)-\rho\sin(\psi))^2\right]^{3/2}}+$$
$$+\frac{3m_d\{[h\cos(\alpha)-\rho\cos(\psi)]\cos(\alpha)+[h\sin(\alpha)-\rho\sin(\psi)]\sin(\alpha)\}}{\left[(z-z_d)^2+(h\cos(\alpha)-\rho\cos(\psi))^2+(h\sin(\alpha)-\rho\sin(\psi))^2\right]^{5/2}}\times \quad (9)$$
$$\times\{[[h\cos(\alpha)-\rho\cos(\psi)]\cos(\gamma)+[h\sin(\alpha)-\rho\sin(\psi)]\sin(\gamma)]\sin(\beta)+(z-z_d)\cos(\beta)\}.$$

Here (z_d, ρ, ψ) are the cylindrical coordinates of the defect, β and γ are the polar and azimuthal angles of the magnetic dipole moment.

We assume that the random component of the magnetic potential φ_n, created by uniformly distributed over the volume of the rod magnetic dipoles with random magnetic moments m_i, is defined as

$$\varphi_n(\mathbf{r})=\sum_{\mathbf{r}_i\in V}\left(\mathbf{m}_i\frac{\partial}{\partial\mathbf{r}}\right)\frac{1}{|\mathbf{r}-\mathbf{r}_i|}=w\iiint_V\left(\mathbf{m}(\mathbf{r}')\frac{\partial}{\partial\mathbf{r}}\right)\frac{d^3r'}{|\mathbf{r}-\mathbf{r}'|}=$$
$$=-w\iiint_V\frac{(x-x')m_x(x',y',z')+(y-y')m_y(x',y',z')+(z-z')m_z(x',y',z')}{\left[(x-x')^2+(y-y')^2+(z-z')^2\right]^{3/2}}dx'dy'dz'.$$

Here w is the density of random dipoles, m_x, m_y, and m_z are the random functions with zero mean value and normal distribution, which can be regarded as white noise.

For the cylindrical geometry of the problem (see Figure 1), we obtain

$$\varphi_n(z)=-w\int_0^L\int_0^R\int_{-\pi}^{\pi}\frac{(h\cos(\alpha)-\rho\cos(\psi))m_x+(h\sin(\alpha)-\rho\sin(\psi))m_y+(z-z')m_z}{\left[(z-z')^2+(h\cos(\alpha)-\rho\cos(\psi))^2+(h\sin(\alpha)-\rho\sin(\psi))^2\right]^{3/2}}\rho d\rho d\psi dz'$$

and for the normal component of the random magnetic field, $H_n = -\partial\varphi_n/\partial h$, we have

$$H_n(z)=w\int_0^L\int_0^R\int_{-\pi}^{\pi}\frac{(\cos(\alpha)m_x+\sin(\alpha)m_y)\rho d\rho d\psi dz'}{\left[(z-z')^2+(h\cos(\alpha)-\rho\cos(\psi))^2+(h\sin(\alpha)-\rho\sin(\psi))^2\right]^{3/2}}-$$
$$-3w\int_0^L\int_0^R\int_{-\pi}^{\pi}\frac{(h\cos(\alpha)-\rho\cos(\psi))m_x+(h\sin(\alpha)-\rho\sin(\psi))m_y+(z-z')m_z}{\left[(z-z')^2+(h\cos(\alpha)-\rho\cos(\psi))^2+(h\sin(\alpha)-\rho\sin(\psi))^2\right]^{5/2}}\times \quad (10)$$
$$\times\{[h\cos(\alpha)-\rho\cos(\psi)]\cos(\alpha)+[h\sin(\alpha)-\rho\sin(\psi)]\sin(\alpha)\}\rho d\rho d\psi dz'.$$

We find the autocorrelation function of fluctuations of the magnetic field, assuming a delta-correlated random magnetization as

$$\langle m_x(z,\rho,\psi)m_x(z_1,\rho_1,\psi_1)\rangle=\langle m_y(z,\rho,\psi)m_y(z_1,\rho_1,\psi_1)\rangle=\langle m_z(z,\rho,\psi)m_z(z_1,\rho_1,\psi_1)\rangle=$$
$$=C\delta(z-z_1)\delta(\rho-\rho_1)\delta(\psi-\psi_1),$$
$$\langle m_x(z,\rho,\psi)m_y(z_1,\rho_1,\psi_1)\rangle=\langle m_y(z,\rho,\psi)m_z(z_1,\rho_1,\psi_1)\rangle=\langle m_x(z,\rho,\psi)m_z(z_1,\rho_1,\psi_1)\rangle=0.$$

We assume that the observation point z disposes far enough from boundaries of the rod, so that the outer limits of the integral (10) can be replaced by $\pm\infty$.

Then, denoting $G_0 = Cw^2$, we obtain

$$W(p) = \langle H_n(z)H_n(z-p)\rangle =$$

$$= G_0 \int_{-\infty}^{\infty}\int_0^R\int_{-\pi}^{\pi} \frac{\rho^2 d\rho d\psi dz}{\left[z^2+h^2+\rho^2-2\rho h\cos(\psi)\right]^{3/2}\left[(z+p)^2+h^2+\rho^2-2\rho h\cos(\psi)\right]^{3/2}} +$$

$$+9G_0 \int_{-\infty}^{\infty}\int_0^R\int_{-\pi}^{\pi} \frac{\left[z(z+p)+h^2+\rho^2-2\rho h\cos(\psi)\right]\left[h+\rho\cos(\psi)\right]^2 \rho^2 d\rho d\psi dz}{\left[z^2+h^2+\rho^2-2\rho h\cos(\psi)\right]^{5/2}\left[(z+p)^2+h^2+\rho^2-2\rho h\cos(\psi)\right]^{5/2}}.$$

The expression for the spectral intensity of fluctuations of the magnetic field at the same time takes the following form:

$$G(k) = F[W(p)] = \int_{-\infty}^{\infty} W(p)e^{-ikp}dp = G_0\int_0^R\int_{-\pi}^{\pi}\left|F\left[\frac{\rho}{\left(p^2+h^2+\rho^2-2h\rho\cos(\psi)\right)^{3/2}}\right]\right|^2 d\rho d\psi +$$

$$+9G_0\int_0^R\int_{-\pi}^{\pi}\left|F\left[\frac{\rho\left[h+\rho\cos(\psi)\right]}{\left(p^2+h^2+\rho^2-2h\rho\cos(\psi)\right)^{5/2}}\right]\right|^2 \left[h^2+\rho^2-2\rho h\cos(\psi)\right]d\rho d\psi +$$

$$+9G_0\int_0^R\int_{-\pi}^{\pi}\left|F\left[\frac{p\rho\left[h+\rho\cos(\psi)\right]}{\left(p^2+h^2+\rho^2-2h\rho\cos(\psi)\right)^{5/2}}\right]\right|^2 d\rho d\psi.$$

By denoting the modified Bessel function as $K_\nu(x)$, we rewrite this expression as

$$G(k) = \frac{\pi G_0 k^2}{2\Gamma^2(3/2)}\int_0^R\int_{-\pi}^{\pi}\frac{\rho^2}{h^2+\rho^2-2h\rho\cos(\psi)}K_1^2\left[k\sqrt{h^2+\rho^2-2h\rho\cos(\psi)}\right]d\rho d\psi +$$

$$+\frac{9\pi G_0 k^4}{8\Gamma^2(5/2)}\int_0^R\int_{-\pi}^{\pi}\frac{\rho^2\left[h+\rho\cos(\psi)\right]^2}{h^2+\rho^2-2h\rho\cos(\psi)}K_2^2\left[k\sqrt{h^2+\rho^2-2h\rho\cos(\psi)}\right]d\rho d\psi +$$

$$+\frac{9\pi G_0 k^4}{2\Gamma^2(5/2)}\int_0^R\int_{-\pi}^{\pi}\frac{\rho^2\left[h+\rho\cos(\psi)\right]^2}{h^2+\rho^2-2h\rho\cos(\psi)}K_{-1}^2\left[k\sqrt{h^2+\rho^2-2h\rho\cos(\psi)}\right]d\rho d\psi.$$

Analysis of the correlation shows that each sensor measures the field, which is a superposition of the four fields of quasi-stationary and one non-stationary sources. The stationary sources are:

1) the geomagnetic field slowly varying in the space and the field produced by surrounding ferromagnetic objects of the control object,
2) slowly varying in space the field generated by permanent magnetization of the object under test,

3) the rapidly changing field in space caused by the microstructure of the test object;
4) the field of defect, which in first approximation can be regarded as a dipole field.

As a non-stationary sources, the various types of magnetic interference are treated, which should be avoided producing a series of experiments. Since the magnetic fields, produced by the stationary sources, depend on the permanent magnetization of the control object, the problem of the magnetic microstructure analysis should not be regarded as a measurement of intensity or induction fields, but a measurement of the topology of constant magnetization.

The primary data for the magnetic structurescopy will be the values of the magnetic field at equidistant points:

$$H_i[k] = H_i(z = k\Delta z),$$

where $i = 1, 2, 3$ is the number of sensors, Δz is the step of the spatial discretization. Then the inverse problem (7) can be formulated through discrete convolution equations as

$$H_1[k] + H_2[k] + H_3[k] = 3\Delta z \sum_{n=0}^{N-1} M_z[n] K'[k-n];$$

$$2H_1[k] - H_2[k] - H_3[k] = 3\Delta z \sum_{n=0}^{N-1} M_x[n] P'[k-n];$$

$$H_2[k] - H_3[k] = \sqrt{3}\Delta z \sum_{n=0}^{N-1} M_y[n] P'[k-n].$$

Here, N is the number of counts along the length of the sample ($L = N\Delta z$), $M_x[n]$, $M_y[n]$, $M_z[n]$ are the components of the magnetization of samples, $K'[k]$ and $P'[k]$ are the system functions (5) and (6), respectively, taken at points of $k\Delta z$. The sampling counts of slowly varying magnetization, in general, may differ from the spatial discretization of the original data, but without loss of generality we can put them the same. The next stage of micromagnetic multiparametric microstructure analysis is the determination of the dipoles on the distribution of the magnetic field.

The physically noticeable object, associated with the defect microstructure, is a magnetic dipole. The magnetic microstructure analysis of weakly magnetized materials is required to determine the presence of magnetic dipoles with the values of the magnetic moment of the order of 10^{-6} A·m^2. Currently, the verification scheme of measurement allows one to reproduce the magnetic moment with value in the range from 10^{-4} up to 10 A·m^2 [5]. Therefore, to calibrate the magnetic structurescopy as a physical model, a point magnetic dipole ring is used with the current supplied to it by using a twisted pair of wires, for which the magnetic fields of currents cancel each other.

Structurally, the model is a plastic cylinder, inside of which the multiple twisted pairs of wires PELSHO with diameter of 0.17 mm are stretched. At one end of the cylinder, the twisted pairs form a circular loop wrapped around glass cylinder with diameter of 1.5 mm. The loops and wires are attached to both ends of a plastic cylinder by using glue. We produced two different models of the dipole, namely: z- and xy-dipole.

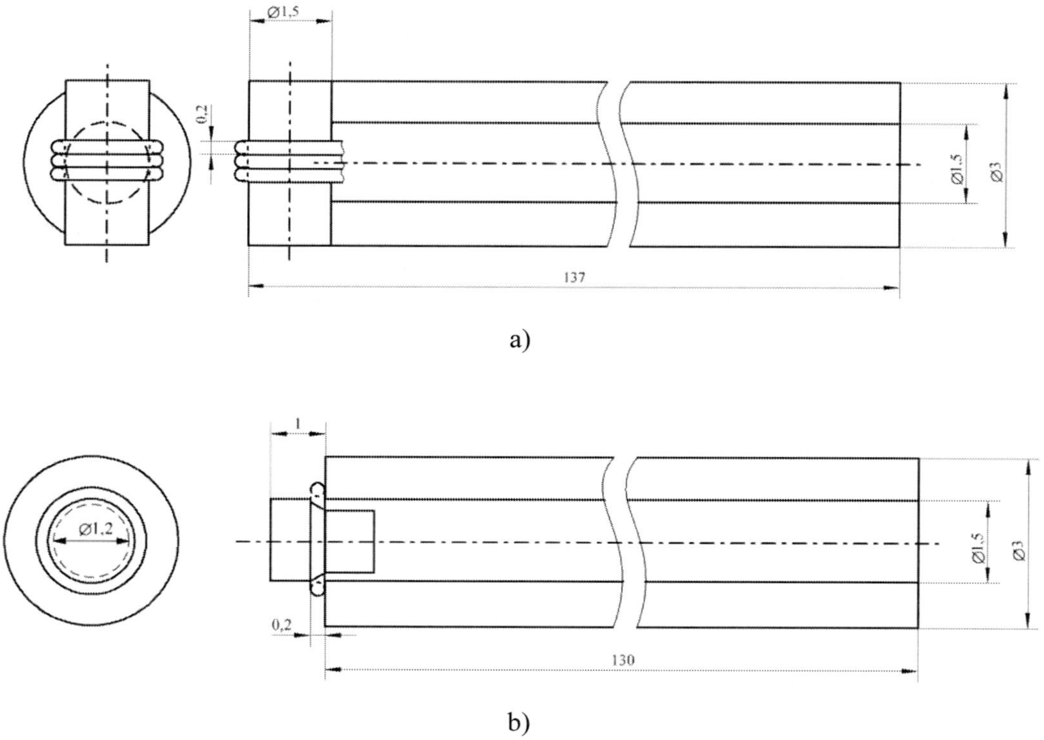

Figure 7. Sketches of the physical models of a dipole: (a) *xy*-dipole, and (b) *z*-dipole. Lead in wires in the form of twisted pairs, as well as the glued joints are not shown.

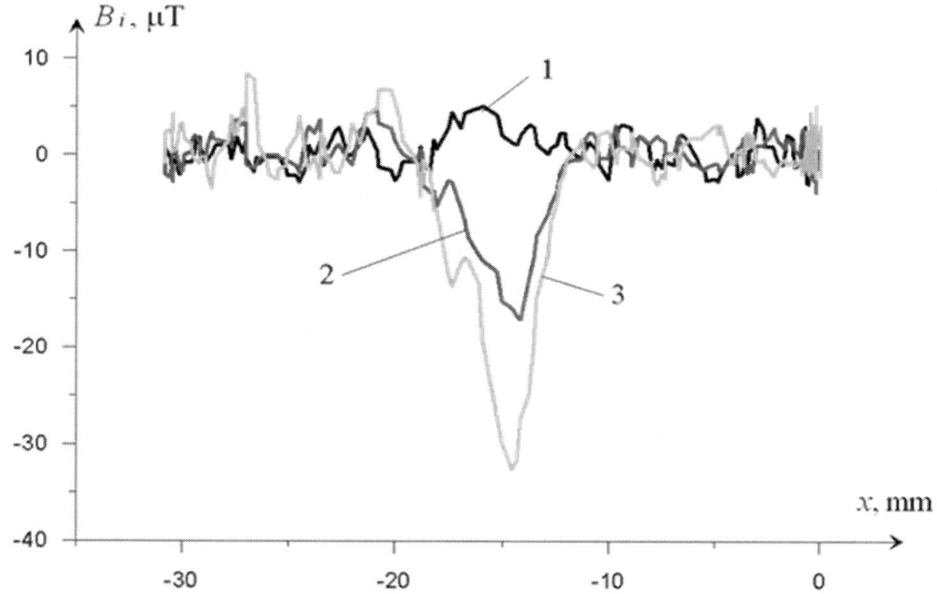

Figure 8. The distribution of the magnetic field of the point *xy*-dipole with current $i = 1$ A and excluded external geomagnetic field. Digitals correspond to the numbers of transducer graphs.

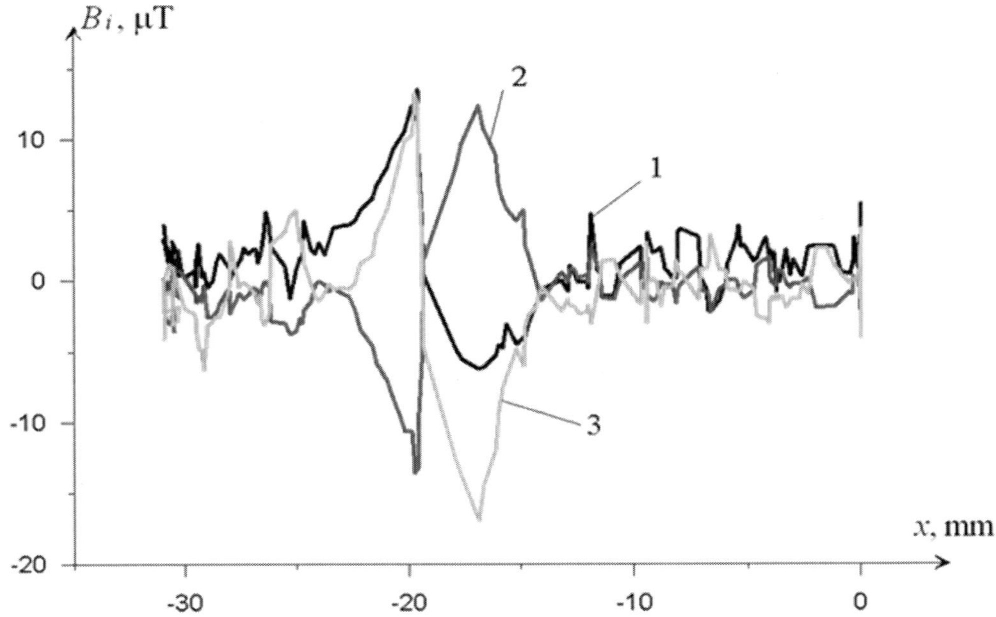

Figure 9. The distribution of the magnetic field of the point *z*-dipole with current $i = 1$ A and excluded external geomagnetic field. Digitals correspond to the numbers of transducer graphs.

In the case of the *z*-dipole, as magnetic dipole we define a vector which is directed along the axis of the sensor. At the same time, the magnetic *xy*-dipole vector is perpendicular to the sensor. The sketches of these dipoles are shown in Figure 7.

The distribution of the magnetic field, produced by the loop with a current, is the same as the dipole field having dipole moment $m = iS$, where i is the current through the loop, S is the area of the loop.

The area of the coil defined by the dipole was 2.79 mm^2. When a current passes 1 A, the dipole moment is 2.79×10^{-6} A·m^2.

After exclusion of the external field and the calibration sensor, the field distribution is obtained near the magnetic dipoles. The Figures 8 and 9 suggest the presence of reliable fixation of the dipoles with the magnetic moment of order of 3×10^{-6} A·m^2. The use of mathematical algorithms allows one to obtain the parameters of the dipole according to the distributions of the field. Thus, the sensitivity of the three-sensor scheme is of the order of 10^{-6} A·m^2. The experiment using Hall transducers and measurement algorithm have been described in [6 – 8].

The parameters of the defect are determined by fitting the parameters of the theoretical model that adequately describes the actual field defect, or by solving a semi-empirical equations for the parameters of the defect by using numerical minimization algorithms [9]. Let us suppose that in a uniform cylindrical rod, the magnetization is constant over the cross section and varies only along the longitudinal coordinate z, and r_d states a defect site that distorts the magnetization in a small area. In this case, the normal component of the magnetic field has the form (9). We rewrite this expression as

$$H_d(z) = \frac{3m_d\left[h - \rho\cos(\alpha - \psi)\right]\left\{\left[h\cos(\alpha - \gamma) - \rho\cos(\gamma - \psi)\right]\sin(\beta) + (z - z_d)\cos(\beta)\right\}}{\left[(z - z_d)^2 + h^2 + \rho^2 - 2h\rho\cos(\alpha - \psi)\right]^{\frac{5}{2}}} - \frac{m_d \sin(\beta)\cos(\alpha - \gamma)}{\left[(z - z_d)^2 + h^2 + \rho^2 - 2h\rho\cos(\alpha - \psi)\right]^{\frac{3}{2}}}.$$

Here, as before, (z_d, ρ, ψ) are the cylindrical coordinates of the defect, β and γ are the polar and azimuthal angles of the magnetic dipole moment m_d (see Figure 1). Measurement of the normal component of the magnetic field is produced by three sensors located on the cylinder surface at the angle of 120° (see Figure 2,a).

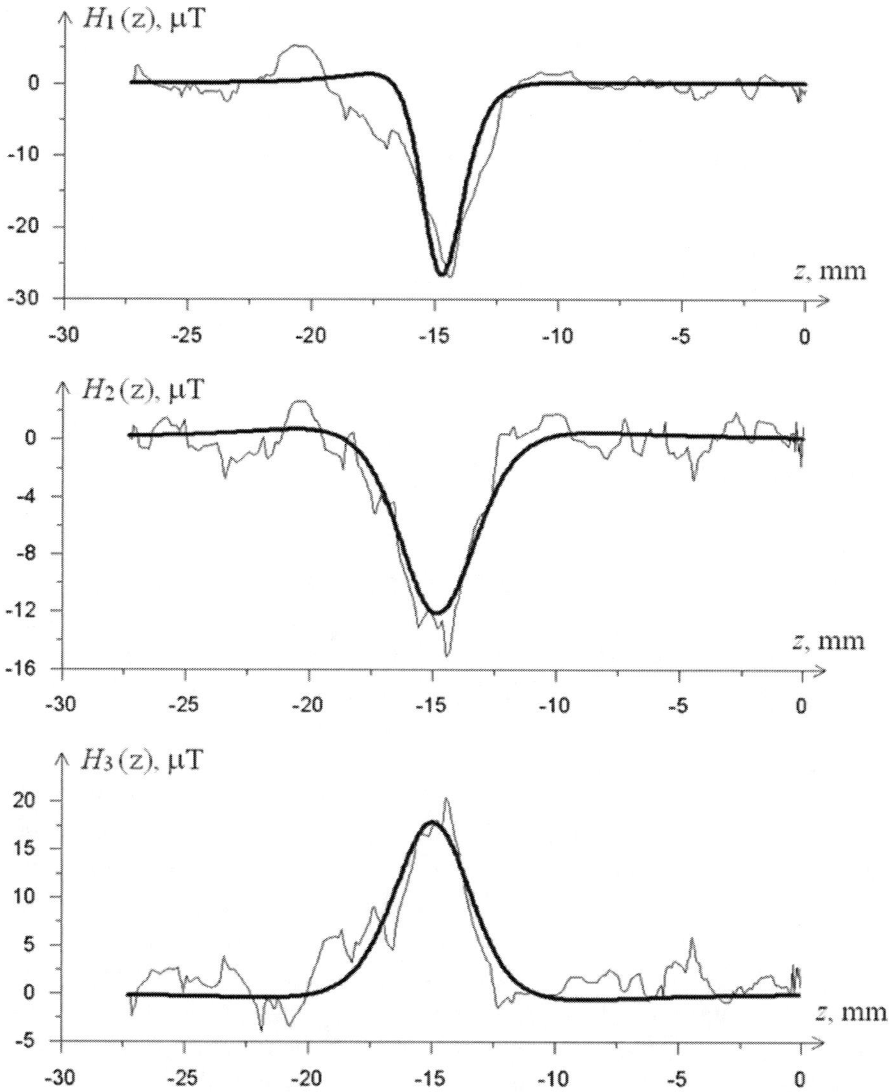

Figure 10. Experimental (thin) and calculated by the gradient method (bold) curves defining the dependences of $H_k(z)$ for point dipole ($i = 1$ A, $\beta = 90°$).

Thus, the defect model is described by six variables whose values can be obtained by using algorithms to minimize the mean square error functional:

$$E(z_d, \rho, \psi, \beta, \gamma, m_d) = \frac{1}{I}\sum_{i=0}^{I}\left(F[i] - H(i\Delta z)\right)^2,$$

where Δz is the step in the z-axis during the measurements, $F[i]$ is the result of the i-th measurement, corresponding to coordinate $i\Delta z$, $H(i\Delta z)$ is the value of the model function along the corresponding coordinate, I is the number of measurements. By using the gradient method [9] at each step, it is necessary to calculate the derivatives of the error as

$$\frac{\partial E}{\partial \varepsilon_m} = \frac{\partial E}{\partial H}\frac{\partial H}{\partial \varepsilon_m} = \frac{2}{I}\sum_{i=0}^{I}\left(H(i\Delta z) - F[i]\right)\frac{\partial H(i\Delta z)}{\partial \varepsilon_m},$$

where ε_m is any of the six variables of the function error ($m = 1, ..., 6$).

In accordance with parameters adjusted by the gradient method we obtain [10]

$$\varepsilon_m(k+1) = \varepsilon_m(k) - \frac{\partial E}{\partial \varepsilon_m(k)}\frac{\Delta l}{\Theta}; \quad \Theta = \frac{1}{6}\sqrt{\sum_{m=1}^{6}\varepsilon_m^2(k)},$$

where Δl is the step approach, Θ is the vector normalization approach.

Figure 10 shows the dependence of $H_k(z)$ for point dipole, and the error of recovering parameters of the dipole does not exceed of 17%.

CONCLUSION

The developed optimization algorithm allowed one to gradually solve the problem of previously considered intractable localization of magnetic dipoles in a material with unknown magnetic characteristics. The use of integral equations of convolution type with the calculated instrumental function of the magnetometric sensors and the developed sustainable regularization algorithm for solving the inverse problem allowed us to eliminate the influence of magnetic history on the results obtained. The microstructure analysis was conducted for weak magnetic fields, when the magnetic energy of the residual magnetization is much smaller than the structure energy and can be considered as a weak perturbation. It allowed one to provide the necessary precision, sensitivity and resolution of magnetic measurements to eliminate the theoretical and experimental problems associated with the technical saturation and to obtain objective information on the local and distributed defect microstructure.

The conducted field experiments confirmed the high metrological characteristics of the algorithm. Thus, the position of the magnetic dipole was determined with an accuracy of 0.6 mm, which is sufficient for the tasks of micromagnetic multiparametric microstructure analysis.

ACKNOWLEDGMENTS

This work was supported by the Federal Target Program "Scientific and Pedagogical Manpower" for 2009-2013 (Project № 14.740.11.0830).

REFERENCES

[1] Pechenkov, A. N. *Russian Journal of Nondestructive Testing.* 2005, No 11, 20 - 24.
[2] Semenov, V. G., Dzaria G. M. *Metrology*. 1977, No 12, 29 - 34.
[3] Izyumov, Yu. A, Katsnelson, M. I.; Skryabin, Yu. N. The Magnetism of Itinerant Electrons. *Fizmatlit,* Moscow, 1994, pp 1 – 368.
[4] Zolotovitskii, A. B.; Shur, M. L.; Zagidulin, R. V. *Russian Journal of Nondestructive Testing*. 1989, No 5, 42 - 50.
[5] Methods of Calibration Standard and Working Means of Measurements of the Magnetic Moment of the MI 191-79. *Publishing House of Standards*, Moscow, 1980, pp 1 – 19.
[6] Golubev, A. A.; Ignatjev, V. K.; Nikitin, A. V. *Instruments and Experimental Techniques*. 2008, No 5, 123 - 128.
[7] Perchenko, S. V.; Stankevich, D. A. *Bulletin of the VolSU.* 2010, No. 13, 118.
[8] Ignatjev, V. K.; Nikitin, A. V.; Perchenko, S. V.; Stankevich, D. A. *Technosphere Technology Security*, 2011, No. 6 (40).
[9] Ignatjev, V. K,; Stankevich, D. A. *Journal of Engineering Don*, 2012. No 2.
[10] Il'in, V.A.; Poznyak, E. G. Principles of Mathematical Analysis: Textbook. for Universities. *Part I. Fizmatlit*, Moscow, 2002, pp 1 – 648.

In: Physics and Mechanics of New Materials …
Editors: Ivan A. Parinov and Shun-Hsyung Chang
ISBN: 978-1-62618-535-7

Chapter 8

FEATURES OF ELECTROMECHANICAL PROPERTIES OF RELAXOR FERROELECTRICS BASED CERAMICS FOR USE IN MEMS

***M. V. Talanov*[1] *and L. A. Reznichenko*[2]**
[1]Research Institute of Physics, Russia
[2]Southern Federal University, Rostov-on-Don, Russia

ABSTRACT

Electric-field-induced strain of piezoceramic materials based on a multicomponent system mPbMg$_{1/3}$Nb$_{2/3}$O$_3$ (PMN) - nPbNi$_{1/3}$Nb$_{2/3}$O$_3$ (PNN) - yPbZn$_{1/3}$Nb$_{2/3}$O$_3$ (PZN)-xPbTiO$_3$ (PT), modified by barium was investigated in the range of electric fields 0 – 12 kV/cm. It is established that with increasing content of lead titanate the behavior of electric-field-induced strain varies from typical for relaxor ferroelectrics to classical ferroelectrics. In addition, the ceramics containing more than $x = 0.3$ of lead titanate are characterized by significant influence of the electric field applying mode on the maximum deformation of the specimen. The observed features are associated with the formation of domain structure with increasing content of lead titanate, and, as a consequence, with increasing the contribution of piezoelectric components in the induced strain.

1. INTRODUCTION

Relaxor ferroelectrics (RF) are the basis for both existing and prospective high perfomance piezoelectric and electrostrictive materials. They are characterized by ultra-high values of piezoelectric coefficients [1], electromechanical coupling factors [2] and electrostriction [3], which makes it possible to use them in different kinds of actuators, and piezomotors [4-6], as well as advanced MEMS-devices [7]. The basis of the materials for these applications are well-known RF: PbMg$_{1/3}$Nb$_{2/3}$O$_3$ (PMN), PbZn$_{1/3}$Nb$_{2/3}$O$_3$ (PZN), PbNi$_{1/3}$Nb$_{2/3}$O$_3$ (PNN), and their solid solutions (SS) with the classical ferroelectric PbTiO$_3$

(PT), which is devoted to the study of rather extensive bibliography (see, for example [8] and references therein).

It is believed, that the electric field E applied to RF can transform the latter to "macro-domain" ferroelectric (FE) state [9] which can also appear as system disorder decreases due to the substitution of the $B^{2+}{}_{1/3}Nb^{5+}{}_{2/3}$ complex with Ti^{4+} ions [10]. In this case, the formation of the "normal" domain in RFs should be reflected in their physical properties, in particular, in the dependences of ξ_3 on E.

It is well known fact [11], that these high perfomance materials undergo an irreversible electric-field-induced phase transition, which affects the character of the manifestations of electromechanical properties. Thus, in some studies [12, 13] was reported about the significant differences between the behavior of ξ_3 in electric fields applied in various modes (monopolar and bipolar) to the crystals PMN-30PT, PZN-8PT . The purpose of this work was to study the behavior of electric-field-induced strain measured in different modes of ceramics based on PMN, PZN, PNN and PT.

2. EXPERIMENTAL

The objects of this study were multicomponent system ceramics $m\text{PbMg}_{1/3}\text{Nb}_{2/3}\text{O}_3$ (PMN) – $n\text{PbNi}_{1/3}\text{Nb}_{2/3}\text{O}_3$ (PNN) – $y\text{PbZn}_{1/3}\text{Nb}_{2/3}\text{O}_3$ (PZN) –$x\text{PbTiO}_3$ (PT) ($x = 0.25 - 0.4$, $y = 0.1130 - 0.0842$, $m = 0.4844 - 0.1298$, $n = 0.1266 - 0.4726$), modified by barium. Samples for this investigation were prepared by solid-state reactions with elements of a columbite route. The limiting solid solutions were prepared in one step by reacting PbO, TiO_2, $BaCO_3$ and niobates ($ZnNb_2O_6$, $MgNb_2O_6$, and $NiNb_2O_6$) at 950°C for 4 h. Zinc, magnesium, and nickel niobates were synthesized in two steps: t_1 = 1000°C (6 h), t_2 = 1100°C (4 h) for $ZnNb_2O_6$ and $MgNb_2O_6$; t_1 = 1000°C (6 h), t_2 = 1240°C (2 h) for $NiNb_2O_6$.

Ceramics were prepared in the form of disks of 1 mm thick and 10 mm in diameter coated with silver-electrode end faces.

A specially designed stand was used to measure the longitudinal strain ξ_3 induced by the discretely increasing or discretely decreasing E applied to the samples [3]. The effective piezoelectric modulus was determined by the equation: $d_{33} = \xi_3/E$.

3. RESULTS AND DISCUSSION

Figure 1 (a) shows behavior of ξ_3 (E) for the studied ceramics with different contents of the PT (x) obtained under monopolar mode. All of the presented half-cycles of electromechanical hysteresis (ξ_3 (E)-curve) go through a similar sequence of processes: with increasing E initially observed relatively weak, almost linear strain, then pass through the area of the sharp growth, and, finally, the growth rate of ξ_3 with increasing E decreases again. By reducing E the sequence of stages is repeated in reverse order, but the ξ_3 (E)-curve is higher. It is evident that the increase in the content x leads to a lowering of the maximum values of strain, since the concentration $x = 0.275$. The evolution of the shape of the curves is similar to previously found in the study of PMN-PT [14]. The ξ_3 (E)-curves of ceramics with a low content of PT (< 0.325) are characterized by the presence of kinks typical for SS of RF and

PT. Analogous kinks on ξ_3 (E)-curves of crystals PMN-PT and PZN-PT, are usually associated with electric-field-induced phase transitions sequence [14].

Figure 1 (b) shows behavior of $d_{33}(E)$ for the studied ceramics with different contents of the PT (x) obtained from ξ_3 (E)-curves. The maximum values of d_{33}, as well as ξ_3, were obtained when the content of $x = 0.275$. The increase of x leads to a gradual smoothing of the d_{33} maximum.

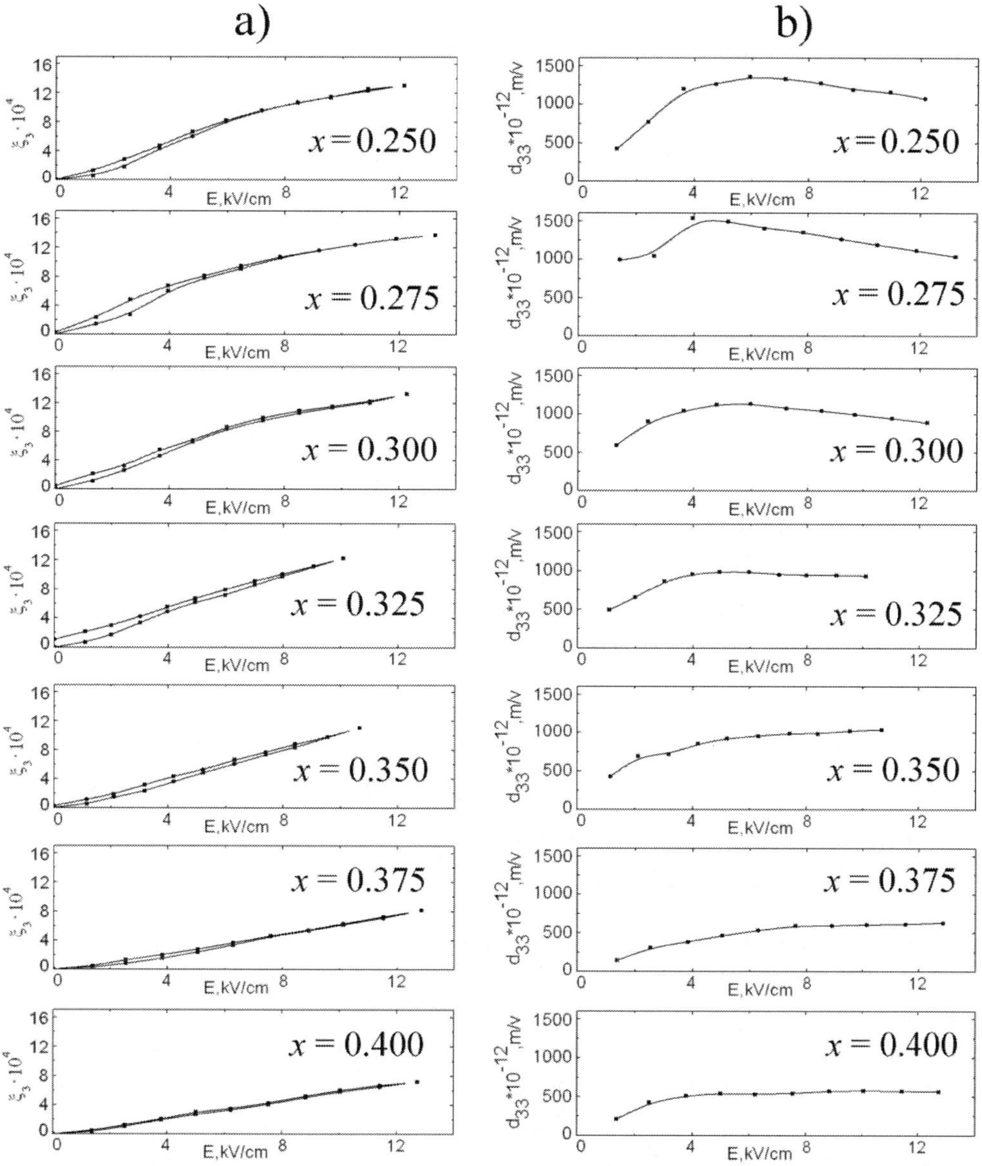

Figure 1. Dependences of ξ3(E) (a) and d33 (E) (b) for the studied ceramics varying with different contents of x, measured under monopolar mode.

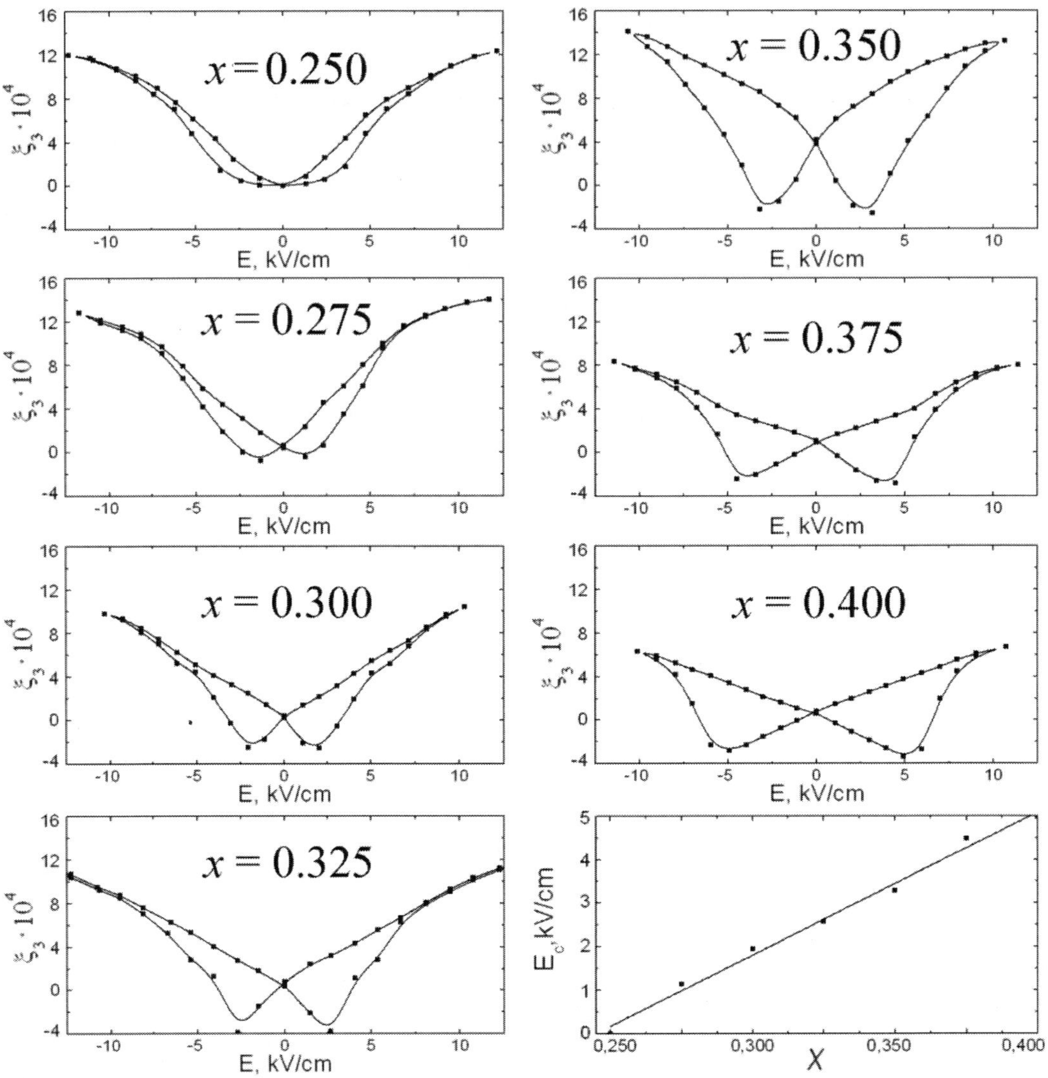

Figure 2. Dependences of $\xi_3(E)$ for the studied ceramics with different contents of x measured under bipolar mode. The inset shows the dependence of the coercive field (E_c) on the content of x.

Figure 2 shows behavior of ξ_3 (E) for the studied ceramics with different contents of the PT (x) obtained under bipolar mode. The ξ_3 (E)-curves of the ceramics with low content of PT are characterized by almost no hysteresis behavior, which is typical for RF. However, with increasing of PT, the ξ_3 (E)-curves changes to the "butterfly"-like behaviors, which are typical for classical ferroelectrics. Furthermore, the increase in x (PT) leads to an almost linear increase in coercive field (inset in Figure 2) defined by the minima of ξ_3.

Figure 3 shows the maximum values of ξ_3 and d_{33}, and the remanent strain ξ_{rem}, obtained for the ceramics with different content of x (PT) under various modes. Lines in the figure indicate the boundaries of stability regions of the cubic, pseudocubic and tetragonal phases, as well as the region of phase coexistence [15]. It was established that SS with $x \leq 0.3$ are characterized by close values of maximum ξ_3 and d_{33}, obtained both under bipolar and bipolar mode. In ceramics with $x > 0.35$, the hysteresis increases and maximum values of ξ_3 are observed under bipolar mode. We assume that these features are associated with the evolution of the dominant contributions to ξ_3 (E): from electrostrictive to piezoelectric one with increasing content of the PT. We suggest that with increasing of x the domain structure is formed, as a consequence, of increasing the piezoelectric contribution to the induced deformation.

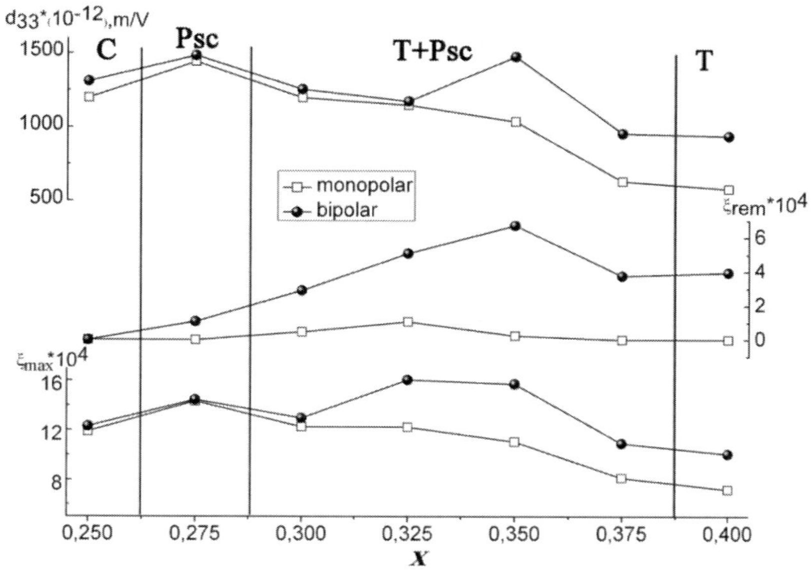

Figure 3. d_{33}, ξ_{rem} and ξ_{max} obtained under different conditions as a function of PT (x) concentration.

This feature can be used to improve the efficiency of piezoelectric actuators, such as used in MEMS devices. Thus, the maximum value of d_{33} heterophase samples can be increased by (15-20)% under bipolar mode (compared with monopolar mode).

ACKNOWLEDGEMENTS

This work was supported by the Russian State Contract № 16.513.11.3032.

REFERENCES

[1] Park, S. - E.; Shrout, T. R. *J. Appl. Phys.* 1997, *vol. 82*, 1804-1811.
[2] Paik, D. - S.; Park, S. - E.; Wada, S.; Liu, S. - F.; Shrout, T. R. *J. Appl. Phys.* 1999, *vol. 85*, 1080-1083.
[3] Turik, A. V.;Yesis, A. A.; Reznitchenko, L. A. *J. Phys.: Cond. Matt.* 2006, *vol. 18*, 4839-4843.
[4] Blackwood, G. H.; Yaley, M. A. *Smart Mater. Struct.* 1993, *vol. 2*, 124-133.
[5] Morita, T. *Sensors and Actuators A.* 2003, *vol. 103*, 291-300.
[6] Uchino, K. *J. Electroceram.* 2008, *vol. 20*, 301-311.
[7] Spearing, S. M. *Acta Mater.* 2000, *vol. 49*, 179-196.

[8] Bokov, A. A.; Ye, Z.-G. *Journal of Materials Science.* 2006, *vol. 41*, 31-.52
[9] Ye, Z.-G.. *Ferroelectrics.* 1993, *vol. 140*, 319-326.
[10] Colla, E. V.; Yushin, N. K.; Vieland, D. *J. Appl. Phys.* 1998, *vol. 83*, 3298-3304.
[11] Bai, F.; Wang, N.; Li, J.; et al. *J. Appl. Phys.* 2004, *vol. 96*, 1620 - 1627.
[12] Viehland, D.; Li, J. *J. Appl. Phys.* 2002, *vol. 92*, 7690 - 7692.
[13] Viehland, D. *J. Appl. Phys.* 2000, *vol. 88*, 4794 - 4806.
[14] Davis, M.; Damjanovic, D.; Setter, N. *Phys. Rev. B.* 2006, *vol. 73*, 014115(1 – 16).
[15] Talanov, M. V.; Verbenko, I. A.; Shilkina, L. A.; Reznichenko, L. A. *Inorg. Mat.* 2012, *vol. 48*, 455 – 459.

In: Physics and Mechanics of New Materials ...
Editors: Ivan A. Parinov and Shun-Hsyung Chang

ISBN: 978-1-62618-535-7

Chapter 9

DIELECTRIC RETARDATION AND RELAXATION IN $PbFe_{1/2}Nb_{1/2}O_3$ CERAMICS

A.V. Turik**, *A.V. Pavlenko and L. A. Reznitchenko
Research Institute of Physics, Southern Federal University
Rostov-on-Don, Russia

ABSTRACT

Temperature and frequency dependences of the real and imaginary parts of the complex dielectric permittivity ε^* of the $PbFe_{1/2}Nb_{1/2}O_3$ ceramics have been studied. It has been shown that, in the paraelectric phase, the relaxation extrema are formed in the dependences. A deviation of the ratio of the average retardation and relaxation time constants from the Debye value $\tau_\varepsilon / \tau_M = \varepsilon_s / \varepsilon_\infty$ has been examined. This relation is strictly carried out only for Debye dielectrics, semiconductors and ferrites. However for non-Debye materials the relation can appear incorrect and is to be checked. An adequate description of the experimental spectra of ε^* of the $PbFe_{1/2}Nb_{1/2}O_3$ ceramics is possible using the model of an insulator with the distribution function of the relaxation times in the shape of a rectangle.

1. INTRODUCTION

Lead ferroniobate $PbFe_{1/2}Nb_{1/2}O_3$ (PFN) is a multiferroic in which the electric and magnetic orders coexist [1, 2]. Earlier, when studying the PFN ceramics, we detect anomalous behavior of the relative dielectric permittivity and the dielectric loss in the vicinity of the Neél ($T_N = -(100 - 130)$ °C) and Curie ($T_C \sim 98$ °C) temperatures [3]. The aim of this work is to state the regularities of forming the dielectric properties of the PFN ceramics in the temperature range 20 – 400 °C based on detailed studying the dielectric spectra in the frequency range of $1 - 2 \cdot 10^6$ Hz.

* Correspondence: 194, Stachki Street, 344090 Rostov-on-Don, Russia; 194, Stachki Street, 344090 Rostov-on-Don, Russia

For a linear medium, it is accepted to distinguish between dielectric retardation and dielectric relaxation. This difference is due to the specific features of the behavior of the complex dielectric permittivity $\varepsilon = \varepsilon' - i\varepsilon''$ under applied step voltage (retardation) and complex electric modulus $M = 1/\varepsilon$ under applied step charge (relaxation) [4, 5]. The relation between the average retardation and relaxation time constants $\tau_\varepsilon / \tau_M = \varepsilon_s/\varepsilon_\infty$, where ε_s and ε_∞ are the static and high-frequency dielectric permittivity, respectively. The relation is precisely given only for Debye dielectrics, semiconductors and ferrites. However for non-Debye materials the relation can appear incorrect and is to be checked.

2. THEORY

At the assumption that the electric field E existed infinitely long up until the considered instant t, it is possible to obtain the electric induction D by means of the superposition principle [4] as

$$D(t) = \varepsilon_\infty E(t) + \int_{-\infty}^{t} E(u)\alpha(t-u)du\,. \tag{1}$$

We will consider that the function $\alpha(t)$, describing decay of the electric induction D or polarization P with time, has the Debye type (an exponent with the characteristic time τ which does not depend on time, but can depend on the temperature):

$$\alpha(t) \propto e^{-t/\tau}. \tag{2}$$

Differentiation of D with respect to time t allows transforming the integral equation (1) to the differential one:

$$\tau \frac{d}{dt}(D - \varepsilon_\infty E) + (D - \varepsilon_\infty E) = \tau\alpha(0)E \tag{3}$$

To define $\alpha(0)$, i.e. $\alpha(t)$ at initial time, it is possible using a special case of balance in constant electric field. Thus

$$\frac{d}{dt}(D - \varepsilon_\infty E) = 0;\quad D = \varepsilon_s E,\ \alpha(0) = \frac{\varepsilon_s - \varepsilon_\infty}{\tau}. \tag{4}$$

It allows receiving the equation:

$$\tau \frac{d}{dt}(D - \varepsilon_\infty E) + (D - \varepsilon_\infty E) = (\varepsilon_s - \varepsilon_\infty)E \tag{5}$$

at the assumption that the function $\alpha(t)$ looks like

$$\alpha(t) = \frac{\varepsilon_s - \varepsilon_\infty}{\tau} e^{-t/\tau}. \tag{6}$$

Let us consider now solutions of Equation (5) for the following two cases.

1) A Constant Pressure on Plates of a Condenser Filled with a Dielectric (Retardation)

Thus $dE/dt = 0$, $E = E_0$ and according to (5) we obtain

$$\tau_\varepsilon \frac{dD}{dt} + D = \varepsilon_s E_0, \ (\mathrm{D} - \varepsilon_s E_0) \propto e^{-t/\tau_\varepsilon}; \ \ \tau_\varepsilon = \tau, \tag{7}$$

i.e. the charge on the condenser plates exponentially rises with time.

2) A Constant Charge on Plates of the Condenser (Relaxation)

Thus $dD/dt = 0$, $D = D_0$ and according to (5) we obtain

$$\tau_M \frac{dE}{dt} + E = D_0 / \varepsilon_s; \ \ -(D_0 / \varepsilon_s - E) \propto e^{-t/\tau_M}; \ \ \tau_M = \frac{\varepsilon_\infty}{\varepsilon_s} \tau, \tag{8}$$

($M = 1/\varepsilon$), i.e. the voltage on the condenser plates exponentially decreases with time.

Thus, the relation $\tau_\varepsilon / \tau_M = \varepsilon_s/\varepsilon_\infty$, characterizing delaying the retardation processes in comparison with the relaxation ones, can be strictly proved only for Debye dielectrics with one retardation (relaxation) time and exponential function of decaying the electric induction (or the electric field) with time. For non-Debye dielectrics transition from the integrated equation (1) to the differential equation (3) and obtaining the relation $\tau_\varepsilon / \tau_M = \varepsilon_s/\varepsilon_\infty$ are impossible, and for each kind of distribution function of relaxation times a special consideration is required.

3. Sample Preparation and Experimental Technique

The PFN ceramic samples were synthesized by the method of solid-phase reactions using high purity oxides PbO, Fe_2O_3, and Nb_2O_5, and by annealing for two stages with intermediate grinding at the temperatures $T_1 = T_2 = 850$ °C and holding times $t_1 = t_2 = 4$ h. The ceramic tablets were sintered at the temperature $T_{\sin} = 1100$ °C for 2 h.

The X-ray diffraction studies were performed on ground ceramic sinters by the method of X-ray powder diffraction on a DRON-3 diffractometer using the filtered CoK_α radiation (Bragg–Brentano focusing). According to the X-ray phase and microstructure analyses, the impurity-free ceramic PFN samples prepared were single-phase (with the rhombohedral symmetry at room temperature) with practically a non-porous fine-grain structure. The grain structure of the samples was examined by using Hitachi TM-1000 scanning electron microscope.

The complex dielectric permittivity $\varepsilon^* = \varepsilon' - i\varepsilon''$ (ε' and ε'' are the real and imaginary parts of ε^*, respectively) was measured at frequencies $f = 1 - 2{\cdot}10^6$ Hz using a Novo-control ALPHA High-Resolution Dielectric Analyzer equipped by a system for high-temperature measurements and LCR-meter HIOKI 3522-50 and Agilent 4285A. The specific conduction was calculated by the formula $\gamma^* = i\omega\varepsilon^* = \gamma' + i\gamma''$, where a circular frequency $\omega = 2\pi f$. The temperature dependences of $\varepsilon'/\varepsilon_0$ and $\varepsilon''/\varepsilon_0$ (ε_0 is the dielectric constant) are depicted in Figure 1. The maxima of $\varepsilon'/\varepsilon_0$ and $\varepsilon''/\varepsilon_0$ at the temperature $T = T_C$ are related to a transition from the ferroelectric to paraelectric phase [3] (in the PFN ceramics studied in [6, 7], $T_C \approx 110$ °C). At $T > T_C$, in the paraelectric phase, additional extreme (maxima and minima) form in the temperature dependences of $\varepsilon'/\varepsilon_0$ and $\varepsilon''/\varepsilon_0$; the extreme are shifted to higher temperatures and decrease with increasing f in the temperature range of 130 – 400 °C. A characteristic feature of the $\varepsilon''(T)/\varepsilon_0$ dependence at $T > T_C$ is a fast increase in $\varepsilon''/\varepsilon_0$ at temperatures with increasing f. This fact is caused by an increase in the dc-conductivity that makes a significant contribution to $\varepsilon''/\varepsilon_0$. As a result of the increase in the conductivity, there are no maxima of $\varepsilon''(T)/\varepsilon_0$ at $T = T_C$ at low frequencies.

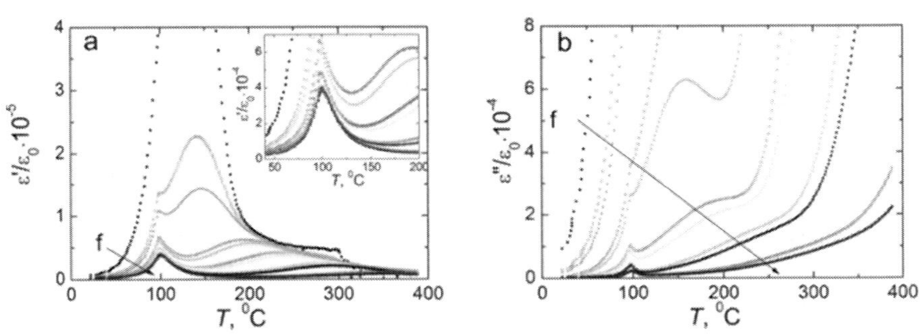

Figure 1. Dependences (a) $\varepsilon'(T)/\varepsilon_0$ and (b) $\varepsilon''(T)/\varepsilon_0$ of the PFN ceramics measured at $T = 20$ –400 °C and $f = 1, 6, 10, 60, 10^2, 6{\times}10^2, 10^3, 6{\times}{\cdot}10^3, 10^4, 6{\times}{\cdot}10^4$, and 10^5 Hz. The arrows indicate the direction of increasing f. The inset shows the region near $T_C = 98$ °C in more detail.

4. Retardation

The $\varepsilon'(\omega)/\varepsilon_0$, $\varepsilon''(\omega)/\varepsilon_0$ and $\varepsilon''/\varepsilon_0(\varepsilon'/\varepsilon_0)$ (Cole–Cole diagrams) measured at temperature of 250 °C and frequencies of $(5 - 1.2{\times}10^6)$ Hz are shown in Figure 2. The $\varepsilon'(\omega)/\varepsilon_0$ spectra have a plateau at low and high frequencies, whereas the $\varepsilon''(\omega)/\varepsilon_0$ spectra demonstrate a monotonic increase as ω decreases (bright symbols); this increase is due to very high dc-conductivity of the PFN ceramics at these temperatures (inset in Figure 2a). To clearly select the relaxation maxima of $\varepsilon''(\omega)/\varepsilon_0$, we exclude, from the imaginary part of the permittivity $\varepsilon''/\varepsilon_0$, the singular term $\gamma/(\varepsilon_0\omega)$ determined by the contribution of the dc-conductivity. As a result, the maxima appear in the $\varepsilon''(\omega)/\varepsilon_0$ spectra; the maxima are shifted to higher frequencies and decrease as temperature increases. In addition the arcs, which were not observed initially, appear in the Cole–Cole diagrams at T = 250 °C (Figure 2c, dark symbols). This fact testifies that at the temperature, the relaxation process occurs in the PFN ceramics. The shape of the $\varepsilon''/\varepsilon_0(\varepsilon'/\varepsilon_0)$ dependences indicates that the non-Debye relaxation is observed in the PFN ceramics.

We approximated the experimental relaxation spectra $\varepsilon'(\omega)$ and $\varepsilon''(\omega)$ by formulae [4]

$$\varepsilon' = \varepsilon_\infty + (\varepsilon_s - \varepsilon_\infty)\int_0^\infty \frac{f(\tau)d\tau}{1+(\omega\tau)^2};\ \varepsilon'' = (\varepsilon_s - \varepsilon_\infty)\int_0^\infty \frac{\omega\tau\ f(\tau)d\tau}{1+(\omega\tau)^2};\ \int_0^\infty f(\tau)d\tau = 1 \tag{9}$$

using the simplest model for an insulator with the distribution function of relaxation times $f(\tau)$ in the shape of a rectangle: $f(\tau)$ = const in the range $\tau_1 < \tau < \tau_2$, while $f(\tau) = 0$ at $\tau < \tau_1$ and $\tau > \tau_2$ [8]. In such insulator, the relaxation oscillator on microscopic scale cannot be described by the model of deep potential well with two equilibrium positions (the Fröhlich relaxation oscillator [4] with the distribution function of relaxation times $f(\tau)$ in the form of a hyperbole). Figure 3 shows the results of fitting the $\varepsilon'(\omega)/\varepsilon_0$, $\varepsilon''(\omega)/\varepsilon_0$ and $\varepsilon''/\varepsilon_0(\varepsilon'/\varepsilon_0)$ performed by relationships (9) for T = 250 °C; $f(\tau)$ = const was calculated by the latter formula (9).

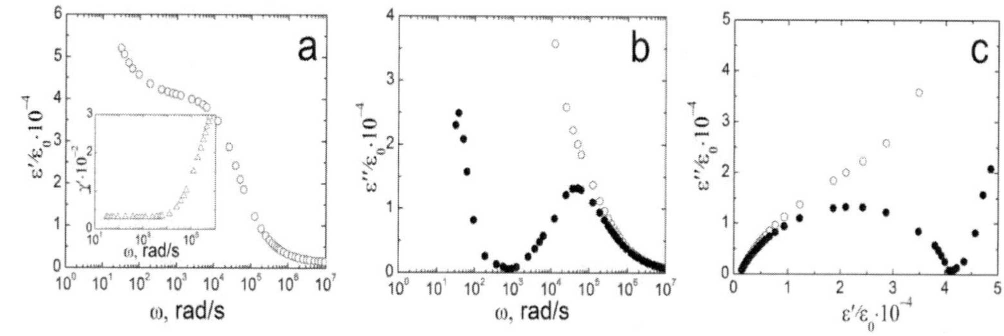

Figure 2. Dependences (*a*) $\varepsilon'(\omega)/\varepsilon_0$, (*b*) $\varepsilon''(\omega)/\varepsilon_0$, $\gamma'(\omega)$ (the inset in figure (*a*)), and (*c*) $\varepsilon''/\varepsilon_0(\varepsilon'/\varepsilon_0)$ of the PFN ceramics measured at temperature of 250 °C. The dark symbols indicate the values of $\varepsilon''/\varepsilon_0$ after excluding the singular (divergent at $\omega \to 0$) terms.

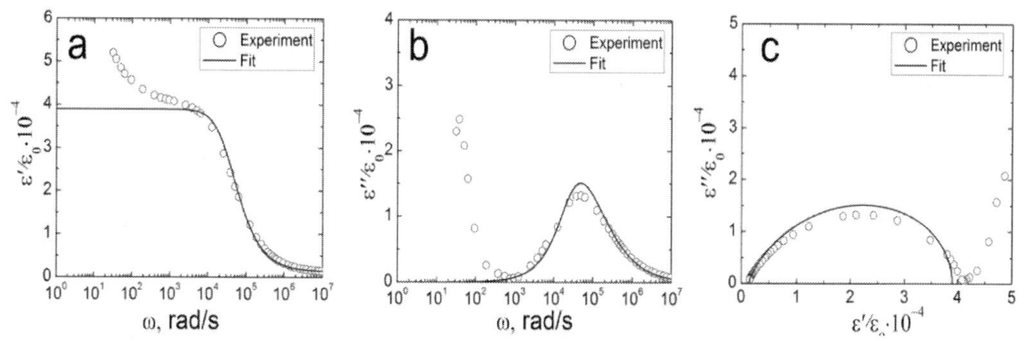

Figure 3. Fitting (*a*) $\varepsilon'(\omega)/\varepsilon_0$, (*b*) $\varepsilon''(\omega)/\varepsilon_0$ and (*c*) $\varepsilon''/\varepsilon_0(\varepsilon'/\varepsilon_0)$ experimental dependencies of the PFN ceramics measured at 250 °C. Solid lines are the results of calculations for τ_1 = 0 s, $\tau_2 = 7.5\times10^{-5}$ s, ε_s = 42000, and ε_∞ = 3000.

It is seen that the chosen model of the equally probable distribution of the relaxation times allows adequate approximation of the experimental data. According to [9], the physical basis of the model can be the Maxwell–Wagner polarization [10] and relaxation in an electrically inhomogeneous matrix system with cells consisting of ceramic grains with approximately the same diameters surrounded with thin layers (shells) having a small [9] or high [6, 7] conductivity and the dielectric permittivity different than that of grains. The variation of the permittivities, conductivities, and thicknesses of grains and shells leads to a wide distribution of the relaxation times and provides high values of the permittivity and conductivity and the non-Debye relaxation in the PFN ceramics. It is obvious that, along with the equally probable distribution of the relaxation times in the case of the Maxwell–Wagner relaxation, the relaxation time distribution is possible around the most probable and the least probable values. The simplest variant of the latter distribution, that is a parabola with a minimum at $\tau_0 = (\tau_1 + \tau_2)/2$ in the middle of the interval of the relaxation time distribution, provide the best approximation of the experimental data for the PFN ceramics.

It is necessary to notice that, along with investigated by us high-frequency retardation, one more retardation takes place in PFN ceramics at the temperature $T = 250$ °C: it is a low-frequency (at $\omega < 10^2$ rad/s) retardation, clearly observed in dependence $\varepsilon''(\omega)$. In this work we do not study the low-frequency retardation which should be a subject of special research. Of even greater Maxwell–Wagner relaxation is observed in PFN ceramics at lower temperatures (Figure 4 at the temperature of 175 °C). The wide and low ε'' maximum testifies the non-Debye character of the dielectric spectrum.

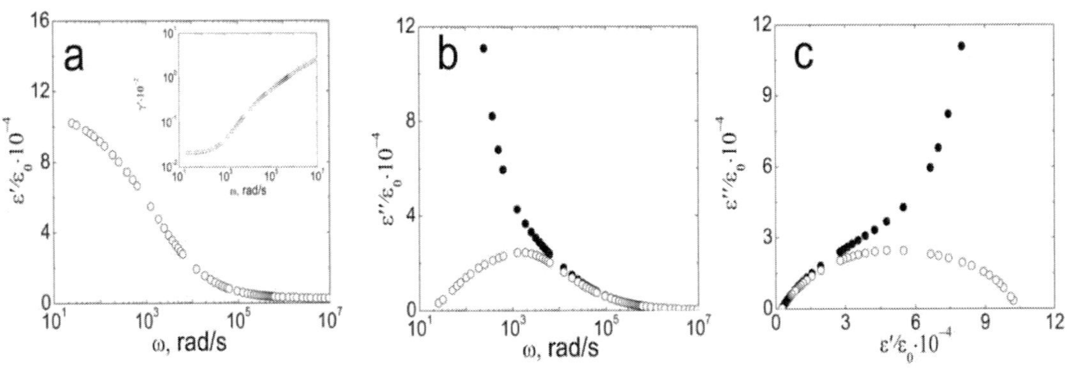

Figure 4. Dependences (*a*) $\varepsilon'/\varepsilon_0(\omega)$, (*b*) $\varepsilon''/\varepsilon_0(\omega)$, $\gamma'(\omega)$ (the inset in figure (*a*)), and (*c*) $\varepsilon''/\varepsilon_0(\varepsilon'/\varepsilon_0)$ of the PFN ceramics measured at temperature of 175 °C. The dark symbols indicate the values of $\varepsilon''/\varepsilon_0$ after excluding the singular (divergent at $\omega \to 0$) terms.

5. RELAXATION

Experimental dependences of $M'(\omega)$ and $M''(\omega)$ on the circular frequency ω are depicted in Figure 5. Comparison of Figures 2 and 5 shows, that the main difference of the relaxation from the retardation is very big distinction of their characteristic frequencies: $\omega_r^M \approx 7\times10^6$ rad/s >> $\omega_r^\varepsilon \approx 5\times10^4$ rad/s. Unlike $\varepsilon''(\omega)$, experimental dependence of $M''(\omega)$ has well-defined relaxation maximum. Great values of relaxation frequencies provide weak influence singular members on dielectric spectra.

And the Cole-Cole diagram for $M''(M')$ dependence looks more convincingly, than in $\varepsilon''(\varepsilon')$ case. From Figure 5, it is visible, that approximation of experimental data using the simplest model for an insulator with the distribution function of relaxation times $f(\tau)$ in the shape of the rectangle allows one to describe well frequency dependences of M', and M''.

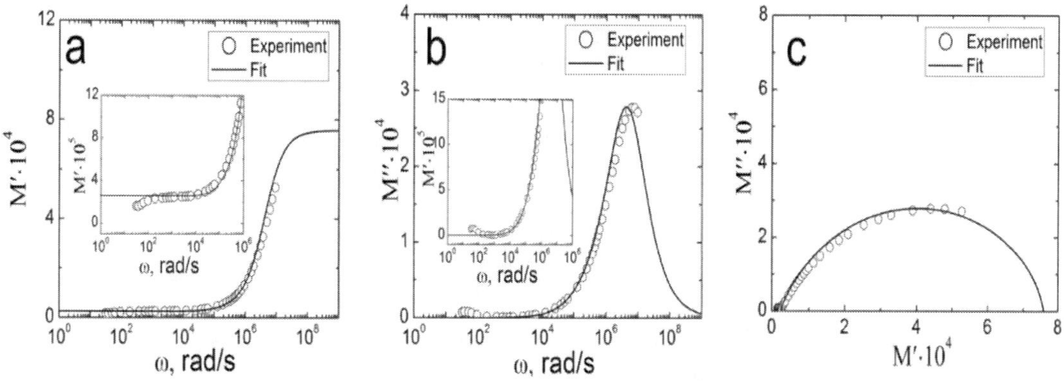

Figure 5. Dependences (*a*) $M'(\omega)$, (*b*) $M''(\omega)$ and (c) $M''(M')$ of the PFN ceramics measured at temperature of 250 °C. The solid lines are the results of calculations for $\tau_1 = 0$ s, $\tau_2 = 7.5\times10^{-5}$ s, $\varepsilon_s = 42000$, and $\varepsilon_\infty = 3000$.

More well-defined experimental $M''(\omega)$ maximum and more expressed the non-Debye spectrum were observed at temperature of 175 °C (Figure 6). The Cole-Cole diagram also well show that we deal with a non-Debye dielectric. In this case the very big distinction between characteristic frequencies ω_r^M and ω_r^ε has been observed, too.

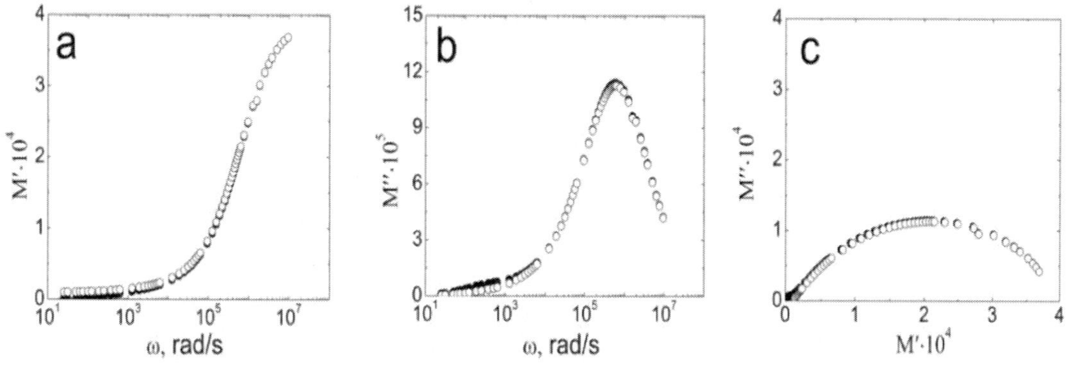

Figure 6. Experimental dependences (*a*) $M'(\omega)$, (*b*) $M''(\omega)$, and (*c*) $M''(M')$ of the PFN ceramics measured at temperature of 175 °C. The dark symbols indicate the values of *M* after excluding the singular (divergent at $\omega \to 0$) terms.

Properties of PFN ceramics in strong electric fields are shown in Figure 7. Differential dielectric permittivity $\varepsilon = \partial D/\partial E$ and susceptibility $\chi = \partial P/\partial E$ (*P* is the polarization) close $E = 0$ decrease in process of increase in amplitude of variable electric field. Dielectric permittivity of PFN ceramics depends on a magnetic field, too (the magnetodielectric effect [11], Figure

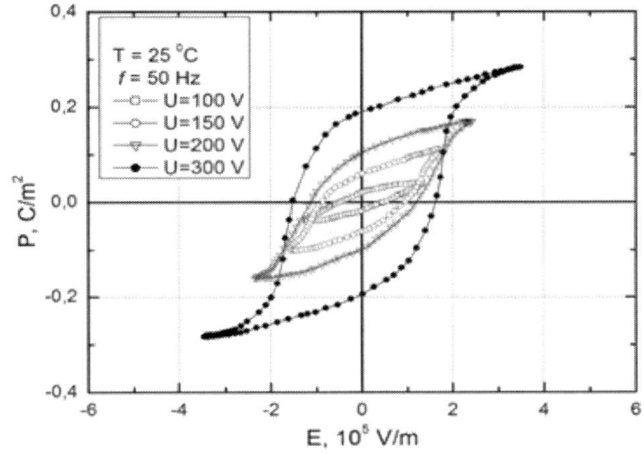

Figure 7. Family of hysteresis loops for the PFN ceramics at the frequency of 50 Hz and at room temperature for different voltage amplitudes $U = (1)$ 100, (2) 150, (3) 200, and (4) 300 V. The sample size is ∅11.7 × 1.2 mm.

8). The most probable model for the magnetodielectric effect in the PFN ceramics is associated with the shift ΔT_C of the ferroelectric–paraelectric phase transition temperature under an applied magnetic field. The observed enhancement of the magnetodielectric effect in the PFN ceramics is associated with the spontaneous magnetization M_s revealed at a temperature $T > 300$ K, which is explained by the manifestation of a weak ferromagnetism due to the presence of ferromagnetic inclusions. In the ceramic samples studied in our work, the electron-probe microanalysis revealed the presence of ferromagnetic inclusions of the Fe_2O_3 magnetite, which had sizes of several micrometers and were located between grains of the parent phase. The volume fraction of these inclusions, which was determined from the cleavage and metallographic section of the sample, did not exceed 0.2%.

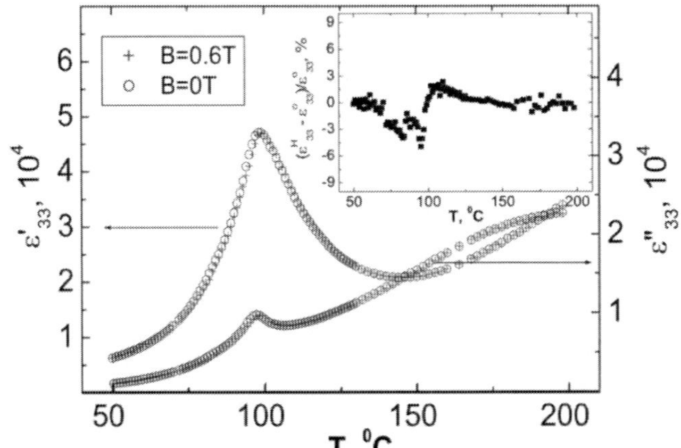

Figure 8. Temperature dependences of the real and imaginary parts of the dielectric constants ε_{33}^{0} in absence and ε_{33}^{H} in presence of a magnetic field H of the PFN ceramics at the frequency of 1 kHz for mutually perpendicular orientation of the measuring electric and constant magnetic fields. The inset shows the temperature variations in the magnetodielectric effect for the real parts of the dielectric constantε_{33}.

CONCLUSION

It has been established that, in the paraelectric phase of the PFN ceramics, the $\varepsilon'/\varepsilon_0(T)$, $\varepsilon''/\varepsilon_0(T)$ dependences have the extrema that are a result of the non-Debye relaxation and that are shifted to higher temperatures as the frequency f of the measuring electric field increases. The retardation and relaxation are explained using the model of the Maxwell–Wagner polarization and relaxation in an electrically inhomogeneous matrix system with cells consisting of ceramic grains with approximately the same diameters surrounded with thin layers (shells) with a small [9] or high [6, 7] conductivity and the permittivity different than that of grains.

ACKNOWLEDGEMENTS

This work was carried out with partial financial support from the Ministry of Education and Science of the Russian Federation (State contract № 16.513.11.3032).

REFERENCES

[1] Smolenskii, G. A; Chupis, I. E. *Sov. Phys. - Usp.* 1982, *vol. 25*, 475-493.
[2] Schmid, H. *J. Phys.: Condens. Matter.* 2008, *vol. 20*, 434201-1 - 434201-24.
[3] Pavlenko, A. V.; Turik, A. V.; Reznitchenko, L. A.; et al. *Phys. Solid State,* 2011, *vol. 53,* 1872–1875.
[4] Fröhlich, H. *Theory of Dielectrics: Dielectric Constant and Dielectric Loss*; Clarendon: Oxford, 1958; pp 1-252.
[5] Jäckle, J.; Richert, R. *Phys. Rev. E.* 2008, *vol. 77,* 031201-1 - 031201-6.
[6] Raymond, O.; Font R.; Suárez-Almodovar, N.; Portelles, J.; Siqueiros, J.M. *J. Appl. Phys.* 2005, *vol. 97,* 084107-1-084107-8.
[7] Raymond, O.; Font R.; Suárez-Almodovar, N.; Portelles, J.; Siqueiros, J.M. *J. Appl. Phys.* 2005, *vol. 97,* 084108-1-084108-8.
[8] Turik, A. V.; Rodinin, M. Yu. *Tech. Phys. Lett.* 2010, *vol. 36(1)*, 16-18.
[9] Raevski, I. P.; Prosandeev, S. A.; Bogatin, A. S.; et al. *J. Appl. Phys.* 2003, *vol. 93,* 4130-4136.
[10] Catalan, G. *Appl. Phys. Lett.* 2006, *vol. 98,* 102902-1-102902-3
[11] Turik, A. V.; Pavlenko, A. V.; Andryushin, K. P.; et al. *Phys. Solid State,* 2012, *vol. 54,* 947–950.

In: Physics and Mechanics of New Materials ...
Editors: Ivan A. Parinov and Shun-Hsyung Chang

ISBN: 978-1-62618-535-7

Chapter 10

SHORT RANGE STRUCTURE OF DILUTE NITRIDE GAASSBN

Hao-Hsiung Lin[1,2,3,*], Chian-Lin Chiou[3], Yan-Ting Lin[2], Ta-Chun Ma[2], Jian-Sheng Wu[3] and Zhe-Chuang Feng[1,3]

[1]Department of Electrical Engineering, National Taiwan University, Taipei, Taiwan
[2]Graduate Institute of Electronics Engineering, National Taiwan University, Taiepi, Tawan
[3]Graduate Institute of Photonics and Optoelectronics, National Taiwan University, Taipei, Taiwan

ABSTRACT

We employ N K-edge X-ray absorption near edge structure (XANES) and Sb K-edge extended X-ray absorption find structure (EXAFS) to investigate the short range chemical structure of dilute nitride $GaAs_{0.91}Sb_{0.07}N_{0.02}$. The GaAsSbN sample was prepared by RF plasma-assisted gas-source molecular-beam epitaxy. The experimental XANES result is compared with simulations on a series of N-centered atomic clusters with different short range configurations. To construct the clusters, we used valence force field (VFF) model to calculate the lattice structure of 256-atom zincblende GaAsSbN supercells with different N and Sb configurations. The supercells were then stacked periodically to form 381-atom N-centered spherical clusters. Then we used FEFF9 code to calculate their XANES spectra. After comparing the experimental and simulation XANES results, we found that the best cluster is with a central NN1 nitrogen pair surrounding by aggregate Sb atoms. The two N atoms of the pair are located at the center and in the second shell of the cluster, respectively. The Sb atoms aggregate in the second shell of the cluster. The GaSb bond length, determined from EXAFS, of the GaAsSbN sample is 2.582 Å. After inspecting the average GaSb bond lengths of the supercells with different short range configurations, we found that the bond length of the cells with Sb aggregation is longer than those of the cells with randomly distributed Sb atoms and is

* E-mail: hhlin@ntu.edu.tw.

close to the experimental length. These findings suggest the existence of NN1 pair and Sb aggregation in $GaAs_{0.91}Sb_{0.07}N_{0.02}$.

1. INTRODUCTION

GaAsSbN can be lattice-matched to GaAs and with a tunable energy gap ranging from 0.8 to 1.42 eV [1], which makes it a promising material for multi-junction solar cells [2-3] and optoelectronic devices [4]. Its narrow energy gap results from the tiny amounts of N and Sb incorporated in the host GaAs. The electronegativity values of these two atoms are very different from that of As atom, leading to localized energy states embedded in the GaAs energy bands. The anti-crossing between the localized states and the bands results in the narrow energy gap [5-8]. A band anti-crossing model, considering the repulsive effects has well explained the energy gap reduction in GaAsSbN [8]. In this manner, the local environments of N and Sb, resulting from specific growth conditions or thermal treatment conditions, are very crucial to the energy gap of the dilute nitride. Dilute nitrides are grown at extremely low temperature to enhance the incorporation of nitrogen. However, the growth condition usually leads to non-radiative defects. Thermal annealing is commonly used to restore the optical properties of the materials [9]. Short range atomic exchange accompanies the annealing has been reported for InGaAsN [10-12]. The exchange leads to a blue shift in optical absorption edge and photoluminescence, which is an undesirable behavior for long wavelength applications. Raman and IR absorption measurement were used to show that the change of group-3 atoms neighboring N atoms is responsible for the unwanted blue shift [10-12].

Because dilute nitrides GaAsN and GaAsSbN consist of only one group-3 element – Ga, atomic change in the first neighboring atoms of N becomes impossible. However, the change in the second neighboring atoms is also able to affect the electronic structure. Kent has calculated the electronic structures of a variety of short range configurations, including Ga-centered tetrahedrons with one to four N in host GaP and in host GaAs and N chains with 3 to 6 N extended along [110] direction in host GaP, using empirical pseudopotential method [13]. For the case in host GaAs, the energy level of tetrahedron with only one N, i. e., isolated N, is above the conduction band edge. While for the cases of tetrahedrons with two N (pair), three N (triplet), and four N (quadruplet), the levels are below the conduction band edge. When the N mole fraction exceeds a critical value of 0.6%, the edge of the perturbed host state sweeps and passes the localized states. Since the localized level of N pair is lower than that of isolated N, the perturbed host state of GaAsN containing N pairs is lower as compared with that of GaAsN containing only isolated N.

In literature, strong blue shift in absorption edge has been observed in GaAsN and GaAsSbN, when the annealing temperature is higher than 750°C [8, 14-16]. Several reports attributed this behavior to the dissociation of N pairs in thermal annealing [8, 16-17]. The theory assumes the existence of N pairs in as-grown GaAsN and GaAsSbN, leading to a low energy gap. Thermal annealing dissociates the pairs into isolated N, resulting in the blue shift in absorption edge. The dissociation, in fact, is just an atomic change of the second nearest neighboring atom.

X-ray absorption near edge structure (XANES) and extended X-ray absorption fine structure (EXAFS) are powerful tools capable of probing the short range local environment around the absorbing atoms [18]. In the measurements of these two structures, specific atom absorbs X-ray and releases core electrons to high-lying continuum states where they are backscattered by neighboring atoms, leading to resonance in absorption spectrum. Basically, f the excited electrons in the region of XANES has small kinetic energy and can be backscattered by atoms within a long range. In contrast, the excited electrons in the EXAFS region are mainly backscattered by the nearest neighboring atoms.

Gupta et al. reported their observation of N pairs in GaAsN [19]. They used density function theory to generate supercells and then used FEFF8 code to simulate XANES spectra, which were compared with experimental results. Ciatto et al. used XANES and EXAFS to study the short range structures of GaAsSbN and GaAsN and suggested that the breaking of the N pairs during the thermal annealing is responsible for the unusual blue shift observed in these alloys [16]. The mechanism has been further confirmed by Lin et al. using photo-reflectance spectroscopy [20].

In this paper, we report on the short range configuration of N atoms in GaAsSbN. The alloys were grown on GaAs using gas-source molecular-beam epitaxy (GSMBE). N source was from a RF plasma cell. N K-edge XANES and Sb K-edge EXAFS were measured on the samples in National Synchrotron Radiation Research Center (NSRRC), Taiwan. The results were then compared with simulation curves for GaAsSbN supercells with different short range arrangements. The lattice structures of the supercells were calculated using valence force field (VFF) model [21-22]. FEFF9 code was used to simulate the XANES curves [23]. After comparing the experimental and simulation XANES curves, we found that GaAsSbN grown with atomic-N rich source contains N pairs surrounded by Sb atoms. The GaSb bond lengths obtained from EXAFS are also close those determined from EXAFS. Details of the measurement of EXAFS and XANES as well as the structure of the supercells are presented.

2. Experimental Procedures

One-μm-thick $GaAs_{0.91}Sb_{0.07}N_{0.02}$ sample was deposited on semi-insulating (100) GaAs wafers at 490°C by a VG V80H GSMBE system. Ga beam was supplied by an EPI SUMO cell. Its flux was calibrated using an ion gauge to keep the growth rate at 1 μm/hr. As_2 molecular was cracked from AsH_3 precursor at 1000°C. Sb source was supplied by an EPI cracking cell. Sb_4 flux from the reservoir zone was decomposed at the cracking zone at 1050°C into a mixed beam of Sb_2 dimer and Sb monomer. N species was generated by injecting N_2 gas to an EPI uni-bulb RF plasma K-cell operating at a radio frequency of 13.56 MHz. The aperture plate of the plasma cavity contains 253 holes, each with a diameter of 0.2 mm. A PBN shutter was placed in front of the K-cell to reduce the N flux and ionized species. In this manner, the N-radicals reached the samples must undergo multiple reflections, leading to the relaxation of excited radicals. Since relaxed nitrogen molecule is inactive in the growth, the source became atomic N rich [24].

The compositions of the GaAsSbN were determined by a JEOL JAX-8200 electron probe microanalysis (EPMA) system. XANES measurements were performed on the branch line (20A) of the NSRRC in Hsinchu, Taiwan. The spot size of X-ray is 1.5×1 mm^2, and the incident beam is perpendicular to the sample surface. Signals were from fluorescence yield of the sample and collected by a micro-channel plate detector. Nitrogen K-edge of the 2 μm thickness GaAsSbN samples was acquired from 395 eV to 430 eV. Step size of every point was 0.25 eV and the measurement time was 50 s for each point. Before the measurement, the N K-edge of a standard BN samples was used to calibrate the monochromator.

Sb K-edge EXAFS measurement was performed with fluorescence yield mode at beam line 01C of the NSRRC in Hsinchu, Taiwan. In the measurement, the incident angle was 45°. The flux and spot size were 3×10^{10} photons/sec and 0.9×0.2 mm, respectively. The double crystal monochromator at beam-line 01C is equipped with Pt-coated mirrors. The fluorescence was detected by a Lytle detector.

3. SIMULATIONS

The simulation has two steps. In the first step, VFF model is used to calculate the coordinates of the atoms in periodic supercells with different short range arrangements. The supercells for VFF calculation consist of $3 \times 3 \times 3$ zincblende unit cells. In the second step, for each short range arrangement, we stack the supercell periodically and select atoms to form an N-centered spherical cluster for the XANES simulation using FEFF9 code [23]. The spherical cluster used for FEEF9 calculation consists of 381 atoms. The size of the cluster and the number of atoms was determined from test results which will be described later.

The supercells mainly consist of Ga and As atoms. The numbers of N atoms and Sb atoms are chosen in accordance with their corresponding compositions determined from EPMA. The central N atom is placed at (0, 0, 0). For the case of nitrogen pair, hereafter referred to as NN1, the second N atom is placed at (0.5*a*, 0.5*a*, 0), where *a* is the lattice constant of the zincblende supercells. Since the N composition determined from EPMA is 2%, a third N atom is placed at (*a*, 1.5*a*, 1.5*a*), roughly on the midway between the centered N and its counterpart in the diagonally adjacent supercell. For the case of isolated nitrogen, besides the central N atom, the second N atom is placed at (*a*, 1.5*a*, 1.5*a*).

In order to understand the aggregation effect of Sb atoms, there are three arrangements for the positions of Sb atoms. In the first one, all the Sb atoms are randomly distributed over the anion sublattice of the supercell with the number coinciding with the Sb composition determined from EPMA measurement. We use a random number to select the position for each atom. In the second one, a Sb cluster structure is established. We place all Sb atoms in the second shell to the central N atom. In the third one, partial Sb atoms are placed in the second shell to the central N atom, and the rest atoms are randomly distributed in the supercell.

We use VFF model to calculate the distortion energy of the aforementioned supercells following Keating and Martin's formula for zincblende crystal [21-22]. Figure 1 shows the schematic diagram of a bond in a zincblende lattice, each bond. The bond connects a group-3 atom and a group-5 atom, which are also the central atoms of the two tetrahedral unit cells shown in the figure. The formula for distortion energy U is

$$U = \sum_{l} \sum_{m=1}^{4} \frac{3}{8d_m^{\,2}} \alpha_m \left[\vec{r}_m(l,s=1) \cdot \vec{r}_m(l,s=1) - d_m^{\,2} \right]^2$$
$$+ \sum_{s=1}^{2} \sum_{l} \sum_{m=1}^{3} \sum_{n=m+1}^{4} \frac{3}{8d_m d_n} \beta_{mn} \left[\vec{r}_m(l,s) \cdot \vec{r}_n(l,s) + \frac{d_m d_n}{3} \right]^2, \quad (1)$$

where d_m is the equilibrium interatomic distance, $s = 1, 2$ denotes group 3 and group 5 atoms, $\vec{r}_m(l,s)$ is the vector connecting atom s in tetrahedral unit cell l to its m-th nearest neighbor, and α_m and β_{mn} are the bond stretching and bond-bending force constants, respectively. The first term represents the stretching energy of the bonds connecting to the first nearest neighboring atoms and sum over only group-3-atom ($s = 1$) centered unit cells. Note that in each tetrahedral unit cell, there are 4 bonds. Since each bond contains a group 3 atom, to sum over the group-3-atom centered unit cells has counted the stretching energy of all bonds in the supercell. The second item corresponds to the bond bending energy which counts the scalar product of two vectors connecting atom s and its mth and n^{th} nearest neighboring atoms. Figure 1 shows a bond, the hatched one, and its six adjacent bonds. Three adjacent bonds are in the group-3-atom centered unit cell and share the central group 3 atom with the hatched bond. The other three adjacent bonds are in the group-5 atom centered unit cell and share the central group 5 atom with the hatched bond. Therefore, we have to sum over both group-5-atom centered unit cells and group-3-atom centered unit cells to count the bending energy of all bonds. The force constants, α and β, the equilibrium bond length of three bonds are listed in Table 1.

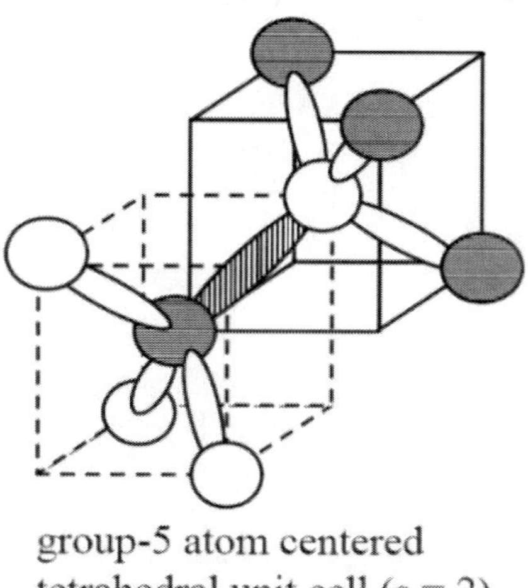

Figure 1. A bond (the hatched one) and its six adjacent bonds in a zincblende lattice. Three adjacent bonds share the group-3 atom (white circle) of the bond and, together with the bond, comprise the group-3 atom centered tetrahedral unit cell. The other three adjacent bonds share the group-5 atom (hatched circle) of the bond and, together with the bond, comprise the group-5 atom centered tetrahedral unit cell.

Table 1. Force constants and bond length used in VFF calculation

Bond	α	β	d	Reference
GaN	95760 dyn/cm	17150 dyn/cm	1.9500 Å	[25]
GaAs	41190 dyn/cm	8950 dyn/cm	2.4478 Å	[22]
GaSb	33160 dyn/cm	7220 dyn/cm	2.6396 Å	[22]

The force constants of GaN are taken from Ref. [25], and those of GaAs and GaSb are taken from Ref. [22]. β_{mn} is taken to be the geometric mean of the bond connecting atom s and its m-th neighbor and the bond connection atoms and its n-th neighbor.

Once the supercell has been constructed, we sequentially adjust the position of each atom to minimize the distortion energy of the whole supercell. The convergent limit for each atomic site is 10^{-8} of the lattice constant. After 40 cycles, fairly good convergence can be obtained. We then stack the supercells periodically to select the spherical cluster for FEFF9 calculation. In the simulation, the size of the spherical cluster is 381 atoms.

In order to correctly and accurately compare the experimental result and the simulation curves, we used a high quality (0001) GaN, grown on sapphire by metal-organic chemical vapor deposition, as a reference. The lattice constant a and c of the reference sample were determined from high resolution X-ray diffraction. A wurtzite GaN supercell was then constructed according to the obtained lattice constants for the FEFF9 simulation. Finally, the major peak energy of the FEFF9 simulation curve was compared with the experimental XANES result of the reference sample. The difference in energy obtained from the comparison was used to correct the energy of the XANES curves of other GaAsSbN samples measured in the same batch.

4. Results and Discussion

4.1. XANES Simulation Using FEFF9 Code

Before starting the simulation for dilute nitride, we performed tests to understand the effect of the cluster size on the XANES spectrum. Figure 2 shows the simulation curves for N-centered spherical clusters with different sizes. The largest one consists of 418 atoms including a central N atom and 15 randomly distributed Sb atoms. The smaller clusters, in fact, are taken from the 418-atom cluster and have the same configuration with the 418-atom cluster in the common shells. The number indicated for each curve in the figure represents the atom number of the cluster, i. e., the cluster size in number of atoms. As can be seen, when the number increases, in the region slight above the absorption edge, the line-shape dramatically changes from a broad multi-peaks band to a single main peak. In this region, the electron, excited from the core level of the central N atom, is with low kinetic energy. Because the wavelength of the excited electron is long, atoms in long distance are still able to affect its resonance. As a result, the line-shape in this region changes with the increasing cluster size. The broad band of the curve with a cluster size of 5 atoms probably resembles the multiple peaks line-shape observed for GaN. The line-shape then gradually evolves to a major peak, which was named single white line in literature [18], and becomes steady when the number of atoms is larger than 329 (18 shells). Finally, we chose 381 for the later FEFF9 simulation.

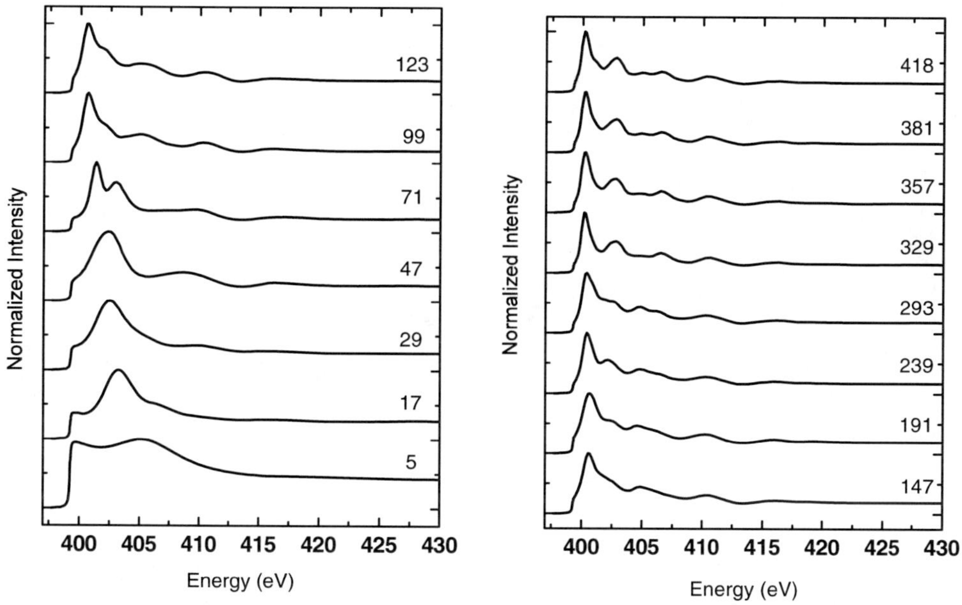

Figure 2. XANES simulation curves of a spherical GaAsSbN cluster. The center of the cluster is an isolated N atom, while the Sb atoms randomly distribute in the cluster. The number indicated for each curve is the number of atom included in the FEFF9 simulation.

Effect of the distances between the central N atom and the surrounding atoms on the white line position and intensity was also investigated. The 381-atom spherical cluster consisting of central isolated N atom and 15 randomly distributed Sb atoms was used for simulation. We intentionally shrank or expanded distance between the 4 first neighboring Ga atoms and the central N atoms. The rest of the cluster was kept unchanged. The results are shown in Figure 3. As can be seen, the white line position shifts to lower energy with the increasing average Ga-N bond length. In addition, the lower intensity peaks at 402 eV, 406.5 eV and 410 eV are also shifted to higher energy with the increasing bond length. Another noticeable point is that the intensity of the peak at 402 eV virtually unchanged. We also show the effects of the 12 As atoms, the second neighbors surrounding the central N atom and the Ga atoms in the first shell, in Figure 4. Opposite to that of the first neighboring Ga atoms, change the distance between the atoms in the second shell and the central N atom does not shift the peaks of the XANES spectra. However, shrinks the distance does enhance the intensities of the peaks at 402 eV, 406.5 eV and 410 eV. Besides, we can see an obvious increase of the peak signal at 402 eV when the As-N distance shrinks. Having understood how the atoms in the first and second shells affect the XANES line-shape, we are able to configure possible N short range structures for the fitting of the N K-edge XANES.

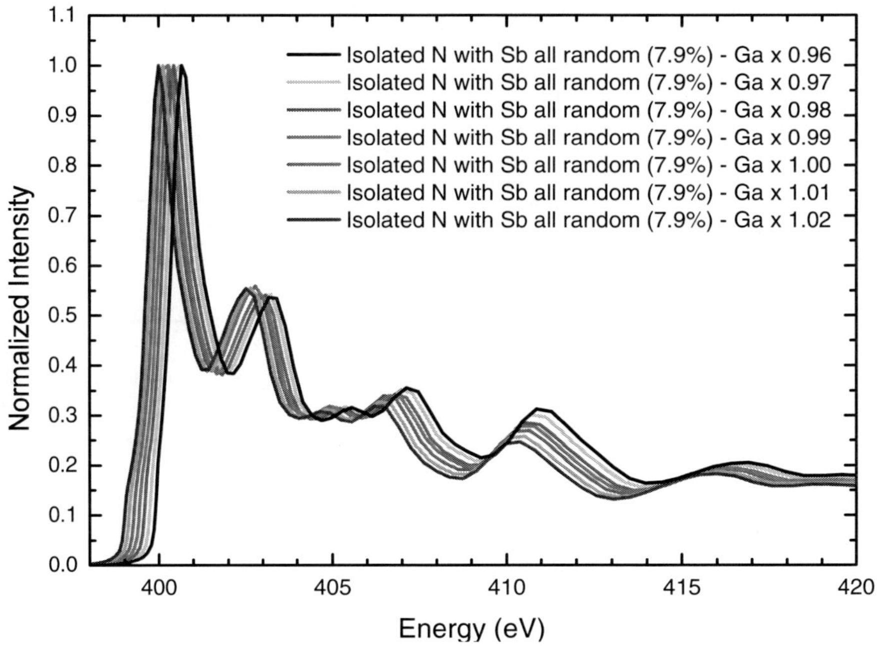

Figure 3. XANES simulation curves of spherical GaAsSbN clusters with different distance between central atom and its first nearest neighboring atoms. The cluster consists of a central isolated N atom and 15 randomly distributed Sb atoms. The number indicated for each curve is the ratio of the distance to the distance obtained from VFF calculation. Only the positions of the 4 first nearest neighbors are changed. The rest of the cluster is kept unchanged.

4.2. Short Range Configuration of GaAsSbN

In this section, we study the short range configuration of a $GaAs_{0.91}Sb_{0.07}N_{0.02}$ sample. The electronic characteristics of the sample have been investigated using photo-reflectance (PR) spectroscopy [20]. Through the Kramers-Kronig modulus of the PR signal, multi peaks relevant to transition from different N pairing and isolated N were resolved. In here, we performed N K-edge XANES to investigate the short range configuration of this sample. Basically, EXAFS is used to acquire the bond lengths. Unfortunately, the K-edge absorption edge of the next atom, oxygen, is 530 eV, very close to that of nitrogen. The energy difference is too small to provide enough room for a good N K-edge EXAFS. In addition, dilute nitride needs longer integral time to obtain reliable data due to the low N composition. XANES is used in this work because its spectral range is shorter than EXAFS. The shorter spectral range not only fits the spacing in energy but also allows longer integral time. In addition, XANES results from the back scattering from atoms in a long range and contains more information than EXAFS. However, its spectrum is quite complicated, requiring simulation to elucidate the physical meaning. In the following, we compare the experimental XANES result with the simulation results for a series of N-centered spherical clusters with different short range configurations to determine the most likely structure in the GaAsSbN sample.

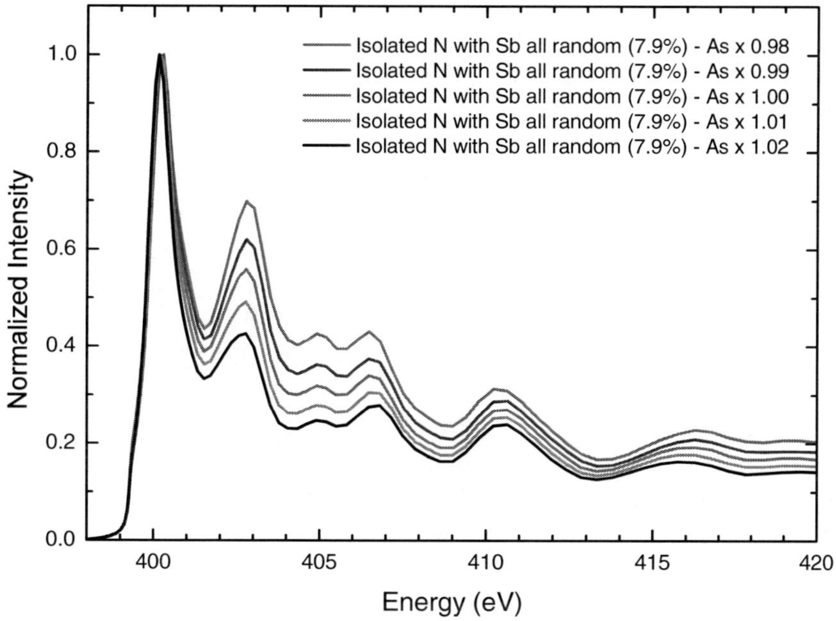

Figure 4. XANES simulation curves of spherical GaAsSbN clusters with different distance between central atom and its second nearest neighboring atoms. The cluster consists of a central isolated N atom and 15 randomly distributed Sb atoms. The number indicated for each curve is the ratio of the distance to the distance obtained from VFF calculation. We changed the position of the 12 second nearest neighbors only. The rest of the cluster was kept unchanged.

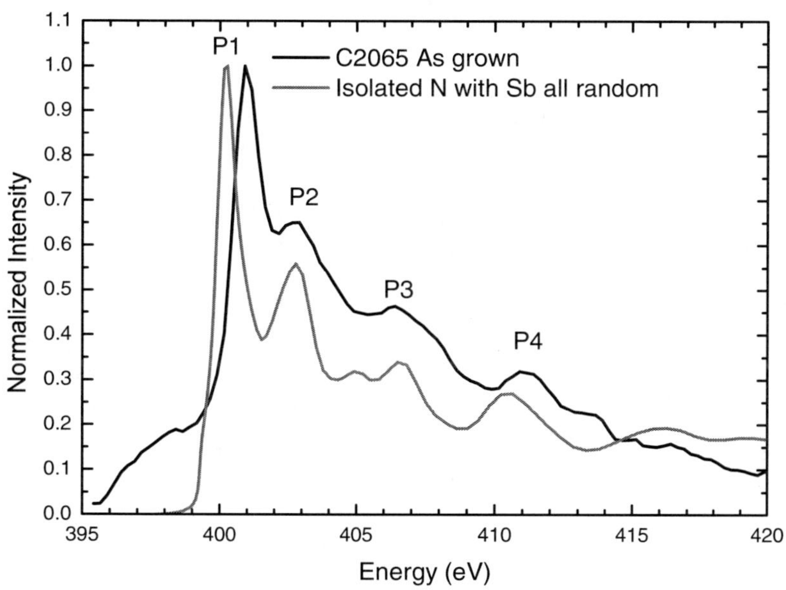

Figure 5. Black and red curves represent the experimental XANES spectrum of $GaAs_{0.91}Sb_{0.07}N_{0.02}$ and the simulation curve of a spherical GaAsSbN cluster, respectively. The cluster consists of a central isolated N atom and 15 randomly distributed Sb atoms.

Figure 5 shows the measured XANES curve of $GaAs_{0.91}Sb_{0.07}N_{0.02}$ and the simulation curve for the N-centered cluster studied in Section 4.1. The cluster contains a central N atom and 15 randomly distributed Sb atoms. Obviously, there are discrepancies in the energy of major white lines (denoted as P1 in the figure) and in the intensities of the second and third peaks (denoted as P2 and P3 in the figure). A comparison between the experimental XANES and the simulation curve for a cluster containing central NN1 pair and 15 randomly distributed Sb atoms is shown in Figure 6. As can be seen, the discrepancies are reduced, which can be attributed to the NN1 pair. For comparison, Table 2 summarizes the structure, the distances of the first shell and the second shell, the white line (P1) energy and the P2 intensity of the clusters studied in this section. The white line (P1) energy and P2 intensity of the experimental result are also listed in the table. Note that the NN1-pair-centered cluster is identical to the isolated-N-centered cluster except the second N atom in the second shell. Because the bond length in an alloy resembles its original length in binary and GaN bond is the shortest among the three bonds in GaAsSbN, the incorporation of N in the second shell reduces the average distance of the second shell, as evidenced in Table 2. According to our finding in section 4.1, the reduction in the distance of second shell will enhance the intensity of P2. As we can see in Figure 6, the P2 intensity of the simulation curve indeed increases and becomes slightly higher than that of the experimental one. The increase of P2 intensity resulting from NN1 pair was also reported for GaAsN in Ref. [19].

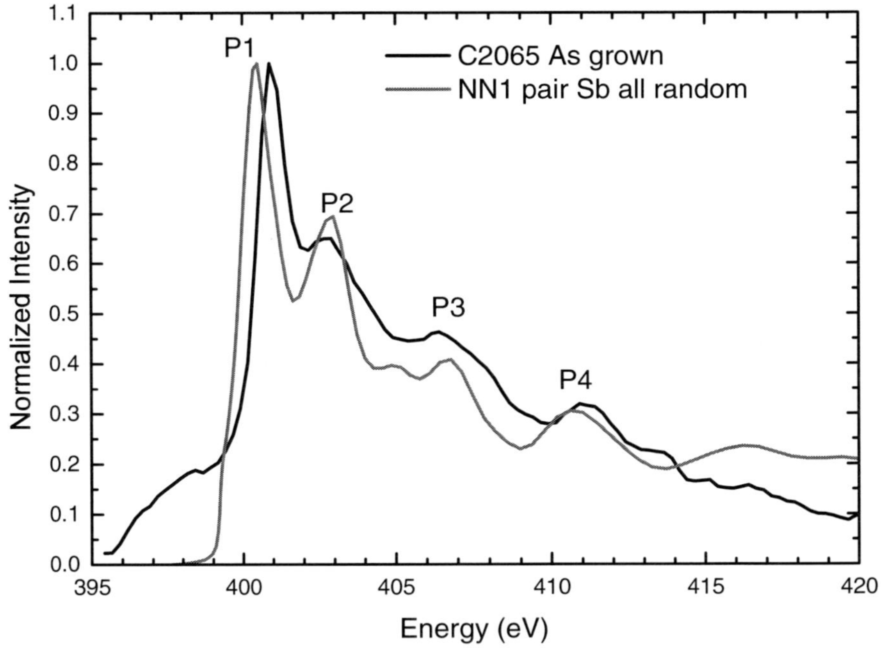

Figure 6. Black curve is the experimental XANES spectrum of $GaAs_{0.91}Sb_{0.07}N_{0.02}$ and the red curve is the simulation curve of a spherical GaAsSbN cluster. The center of the cluster is a NN1 pair. There are 15 Sb atoms randomly distributed in the cluster. Note that the positions of the 15 Sb atoms are identical to those in the isolated-N-centered cluster whose simulation curve shown in Figure 5.

As can be seen from Table 2, the white line (P1) energy of the NN1-centered cluster is higher than that of isolated-N-centered cluster by 0.2 eV. The first shell distance of the NN1-centered cluster, however, is longer than that of the isolated N-centered cluster. The longer distance is ascribed to the N atom in the second shell. This N atom connects a Ga atom in the first shell and pulls the Ga atom out of the center, leading to the increase of the average distance between the first shell and the central atom. The blue shift of the white line appears to be in conflict with the finding in Figure 3. However, in the case of Figure 3, the change in the structure is simply in the distance of first shell and the effect of atom is not considered. Blue shift in white line energy is due to the second N atom in the second shell. The blue shift of NN1-centered cluster was also observed in the simulation for GaAsN in Ref. [19].

Table 2. Summary of cluster structure, the distances of the first shell and the second shell to the cluster center, the P1 energy and the P2 intensity determined from the simulation

Cluster	N	Sb in 2nd shell	1st shell distance	2nd shell distance	P1 energy (eV)	P2 intensity
1	isolated	0 (random)	2.026 Å	3.849 Å	400.27	0.559
2	NN1	0 (random)	2.036 Å	3.810 Å	400.47	0.694
3	NN1	11	2.006 Å	3.918 Å	400.70	0.596
4	NN1	8	2.010 Å	3.886 Å	400.81	0.645
5	isolated	12	1.995 Å	3.961 Å	400.51	0.476
$GaAs_{0.91}Sb_{0.07}N_{0.02}$ XANES			--	--	400.91	0.651

The P1 energy and P2 intensity determined from experimental result are also listed.

Although NN1 pair structure enhances P2 intensity, the white line position cannot fit the experimental result. In here, we intentionally concentrate the Sb atoms in the second shell of a new spherical cluster. In this NN1-pair-centered cluster, all the atoms in the second shell except the N atom are replaced with Sb atoms. The simulation results as well as the experimental spectrum are shown in Figure 7. The white line energy and P2 intensity of the simulation curve are close to those of experimental curve. Inspect Table 2 and we can find that the Sb aggregation greatly reduces the distance of the first shell and increases the distance of the second shell. Because GaSb bond is the longest among the bonds in GaAsSbN, the aggregation of Sb atoms in the second shell will elongate the distance of the second shell. At the same time, the Ga atoms in the first shell will be compressed to the cluster center to accommodate the long GaSb bonds. According to our findings in Section 4.1, the shrinkage of the first shell will result in a blue shift in white line and the elongation of the second shell will weaken the intensity of P2, which explains the observation in Figure 7.

To fine tune the simulation curve, we reduce the number of Sb atoms in the second shell. As can be seen in Figure 7, the P2 intensity of the simulation curve is slightly lower than that of the experimental result, indicating that the second shell distance is too long. Reduce the number of Sb atoms is able to shorten the distance and enhance the P2 intensity. After testing, we keep 8 Sb atoms in the second shell and select new position for the rest 3 Sb atoms by random numbers. The simulation result of the new Sb-aggregate cluster is shown in Figure 8. For comparison, the experimental result is also depicted in the figure. As can be seen, both white line (P1) energy and P2 intensity are in fairly good agreement. However, there are discrepancies in the minor peaks. The P3 of the experimental curve is slightly weaker the simulation one, which could be attributed to errors in measurement and data processing. The P4 of the experimental curve at 411 eV is slightly higher than the P4 of simulation curve at 410 eV. The energy difference could be due to errors in Hedin-Lundqvist self-energy [19, 26].

This self-energy correction is more important at high wave-vectors and gives only a smaller effect on the white line. In addition, the simulation curve shows a broad band at 417 eV, which is not observed in the experimental curve. This broad band is indistinct in the simulation of clusters with randomly distributed Sb atoms and was not observed in previous report for GaAsN [19]. The band is clear due to the aggregation of Sb atoms. Its absence in experimental curve is not clear yet and needs more studies. Nonetheless, the success in the fitting of major peaks suggests not only the existence of NN1 pair in the dilute nitride and but also the Sb surrounding of the pair.

In order to confirm the existence of NN1 pair, we generate an isolated-N-centered spherical cluster with its second shell filled with 12 Sb atoms. The simulation results and the experimental XANES spectrum are shown in Figure 9. Although the cluster has shortest first shell distance, as evidenced in Table 2, its white line shifts to 400.51 eV, deviating the experimental white line. This finding suggests again that the white line position is not solely determined by the distance of the first shell. The atoms in the second shell certainly also play important roles. Also shown in Table 2, the distance of the second shell is the longest one. This is not surprising because the second shell is filled with Sb atoms. The long second shell distance retards the intensity of P2. The deviation in P1 position and P2 intensity suggests that the isolated-N-centered cluster is not able to well fit the experimental result.

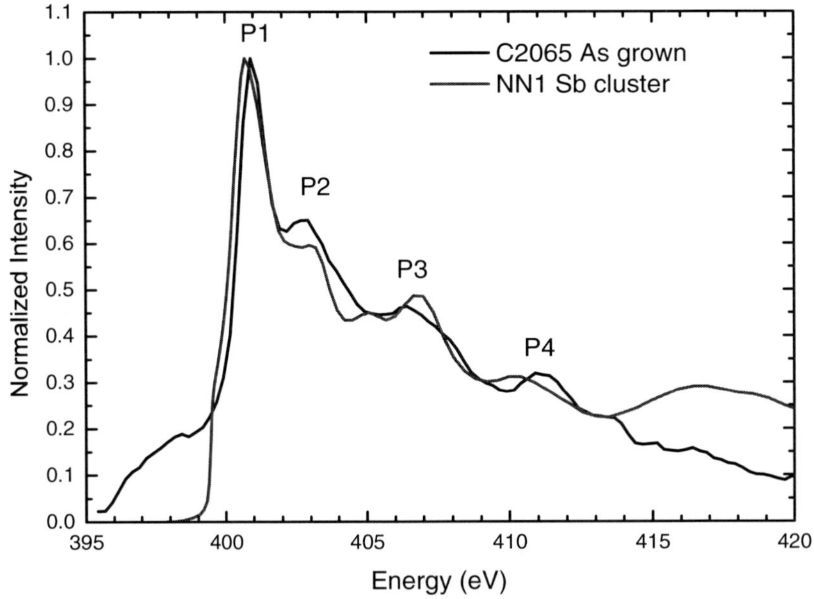

Figure 7. Black curve is the experimental XANES spectrum of $GaAs_{0.91}Sb_{0.07}N_{0.02}$ and the blue curve is the simulation curve of a NN1-centered spherical GaAsSbN cluster. Eleven Sb atoms aggregate in the second shell of the cluster and leave a position to the second N atom of the NN1 pair.

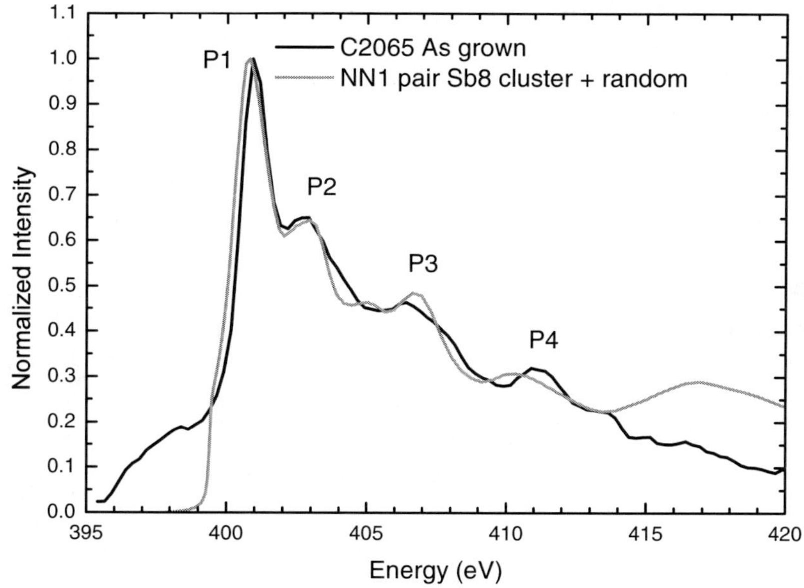

Figure 8. Black curve is the experimental XANES spectrum of $GaAs_{0.91}Sb_{0.07}N_{0.02}$ and the green curve is the simulation curve of a NN1-centered spherical GaAsSbN cluster. The number of Sb atoms aggregate in the second shell is reduced to eight. Three Sb atoms randomly distribute in other shells.

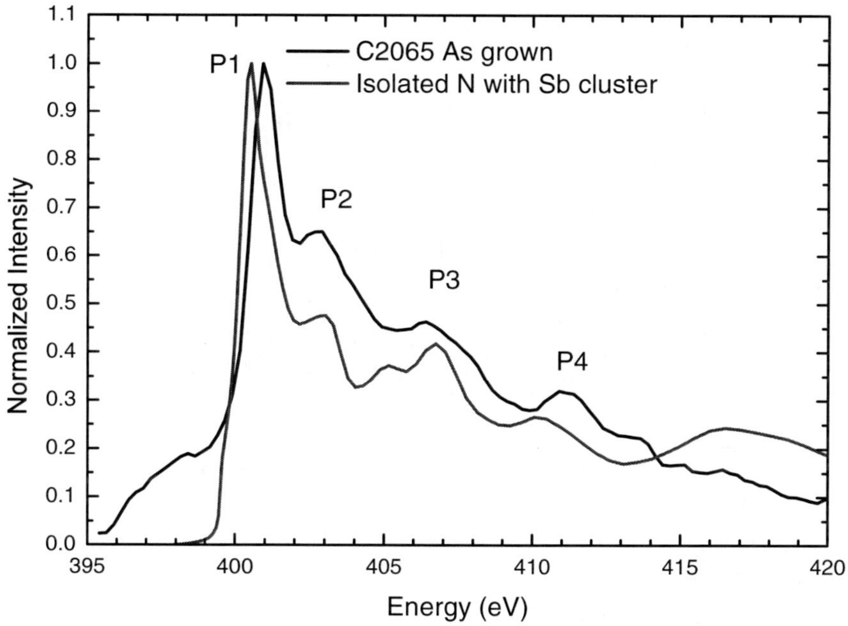

Figure 9. Black curve is the experimental XANES spectrum of $GaAs_{0.91}Sb_{0.07}N_{0.02}$ and the blue curve is the simulation curve of an isolated-N-centered spherical GaAsSbN cluster. The second shell of the cluster is filled with Sb atoms.

On the ground of above findings, we conclude the existence of NN1 pairs in the GaAsSbN sample, which coincides with our previous report based on the observation from PR [20]. A more striking finding is the aggregate Sb atoms that surround the NN1 pair. This structure was not observed previously and not considered in the theoretical studies on the electronic structure of GaAsSbN yet. The Sb aggregation could complicate the electronic structure of GaAsSbN. However, its understanding could elucidate the peculiar annealing effect on the optical and electronic properties of GaAsSbN [20].

4.3. Bond Length of GaSb in GaAsSbN

The Sb K-edge extended X-ray absorption fine structure (EXAFS) signal of the as-grown $GaAs_{0.97}Sb_{0.07}N_{0.02}$ is depicted in Figure 10. The signal is weak and noisy, which is due to the low photon flux of the beam line and the low Sb composition in the $GaAs_{0.91}Sb_{0.07}N_{0.02}$ sample. We used ATHENA code to remove the background and extract the EXAFS oscillations from the k-space signal [27]. The R-space signal is the Fourier transform of the k-space signal and is shown in Figure 11. Only one distinct peak is observed. Obviously, this peak represents the distance between the Sb atom and its first nearest neighboring atoms. Then we fit the EXAFS data with ARTEMIS program set [27]. The best fitting result, represented by the white circles in Figure 11, is in agreement with the peak of the experimental signal. In the fitting of the Sb K-edge EXAFS data, only one path was considered because the four nearest neighbors of the Sb atom are all Ga atoms. The interatomic distance and the relevant parameters of the best fitting are summarized in Table 3. From the table, the GaSb bond length is 2.582 Å.

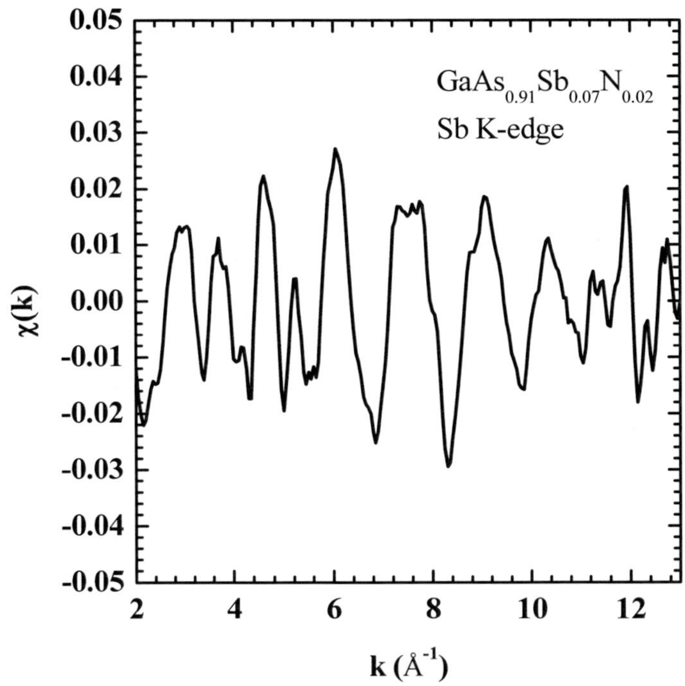

Figure 10. Sb K-edge extended X-ray absorption fine structure oscillation of $GaAs_{0.91}Sb_{0.07}N_{0.02}$.

Table 4 lists the average GaSb bond length in supercells with different short range configuration. The bond length determined from EXAFS measurement for the $GaAs_{0.91}Sb_{0.07}N_{0.02}$ sample and the bond length reported for GaSb binary [22] are also listed in the table. As can be seen, the bond lengths of the three clusters with aggregate Sb atoms are close to the experimental one. The bond lengths of the other two clusters with randomly distributed Sb atoms are clearly shorter than the experimental result. The comparison obviously suggests the existence of Sb aggregation in the $GaAs_{0.91}Sb_{0.07}N_{0.02}$ sample and is in consistent with the finding we obtained in Section 4.2.

In order to further elucidate the effect of Sb aggregation, the average bond lengths of Sb atoms in the second shell and in other shells are listed in the table. As can be seen, the average bond length relevant to Sb atoms in the second shell is obviously longer than that relevant to Sb atoms in other shells. Because the short GaN bonds in the cluster center leave space to the GaSb bonds relevant to the Sb atoms in the second shell, the stretching of the GaSb bonds is mitigated, resulting in longer bond length as compared with the GaSb bonds relevant to Sb atoms in other shells.

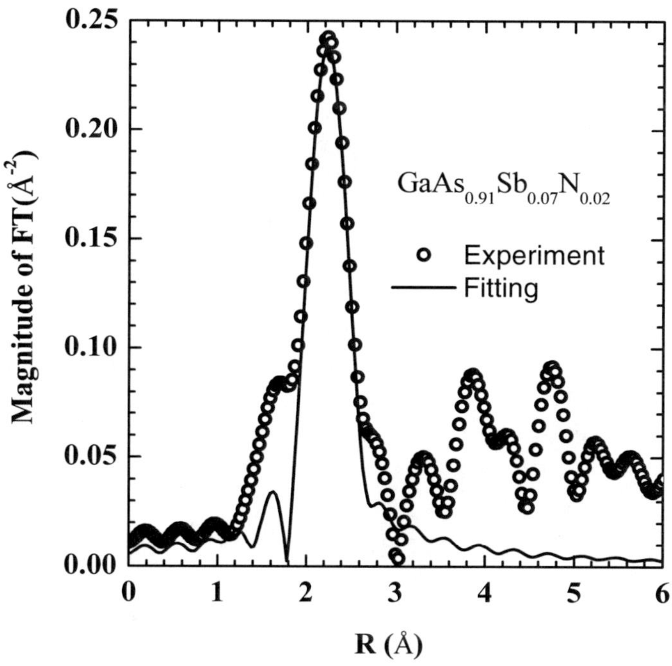

Figure 11. Fourier transformed extended X-ray absorption fine structure signals for $GaAs_{0.91}Sb_{0.07}N_{0.02}$.

Table 3. Fitting parameters for ARTEMIS code

Item	Value
Amplitude	0.81
Debye-Waller's factor, σ^2	0.001 Å
Range, *R*	2.582 Å
R-factor	0.002
χ^2	196

Table 4. Summary of cluster structure and average GaSb bond length calculated using valence force field model

Cluster	N	Sb in 2nd shell	Average bond length for all Sb	Average bond length for Sb in 2nd shell	Average bond length for Sb in other shells
1	isolated	0 (random)	2.5713 Å	--	2.5713 Å
2	NN1	0 (random)	2.5723 Å	--	2.5723 Å
3	NN1	11	2.5806 Å	2.5823Å	2.5611 Å
4	NN1	8	2.5800 Å	2.5809Å	2.5765 Å
5	isolated	11	2.5788 Å	2.5814Å	2.5643 Å
GaSb binary			2.6396 Å	--	--
$GaAs_{0.91}Sb_{0.07}N_{0.02}$ EXAFS			2.582 Å	--	--

The experimental result was determined from Sb K-edge extended X-ray absorption fine structure.

CONCLUSION

N K-edge XANES and Sb K-edge EXAFS have been used to study the short range configuration in $GaAs_{0.91}Sb_{0.07}N_{0.02}$ grown by GSMBE. The experimental XANES result is compared with simulations on the ground of VFF model and FEFF9 code. A series of 381-atom N-centered spherical clusters with different short range configurations are simulated. After comparing the white line energy and the intensity of peak P2, the best cluster is with a central NN1 nitrogen pair surrounding by aggregate Sb atoms. The GaSb bond length determined from EXAFS is 2.582 Å. After inspecting the average GaSb bond lengths of the supercells, solved by VFF model, we found that the bond length of the cells with Sb aggregation is longer than those of the cells with randomly distributed Sb atoms and is close to the experimental result. These findings suggest the existence of NN1 nitrogen pair and Sb aggregation in the as-grown $GaAs_{0.91}Sb_{0.07}N_{0.02}$ sample.

ACKNOWLEDGEMENTS

This work was supported by the National Science Council, R. O. C., under contract number NSC 99-2221-E-002-199- MY3. The authors would like to thank Dr. Jin-Ming Chen, spokesperson of beam line 20A, NSRRC, and Dr. Ting-Shan Chan, spokesperson of beam line 01C, NSRRC, for their valuable discussion in XANES and EXAFS measurements.

REFERENCES

[1] Ma, T. C.; Lin, Y. T.; Lin, H. H. *J. Cryst. Growth.* 2008, *vol. 310*, 2854 -2858.

[2] Yuen, H. International solar energy technology conference, ISETC 2011, Santa Clara, USA.

[3] Jackrel, D. B.; Bank, S. R.; Yuen, H. B.; Wistey, M. A.; Harris, J. S.; Ptak, A. J.; Johnston, S. W.; Friedman, D. J.; Kurtz, S. R. *J. Appl. Phys*. 2007, *vol. 101*, 114916.
[4] Wicaksono, S.; Yoon, S. F.; Loke, W. K.; Tan, K. H.; Ng, B. K. *J. Appl. Phys*. 2006, *vol. 99*, 104502.
[5] Shan, W.; Walukiewicz, W.; Ager, J. W., III; Haller, E. E.; Geisz, J. F.; Friedman, D. J.; Olson, J. M.; Kurtz, S. R. *Phys. Rev. Lett*. 1999, *vol. 82*, 1221-1224.
[6] Wu, J.; Shan, W.; Walukiewicz, W. *Semicond. Sci. Technol*. 200, *vol. 17*, 860-869.
[7] Alberi, K.; Wu, J.; Walukiewicz, W.; Yu, K. M.; Dubon, O. D.; Watkins, S. P.; Wang, C. X.; Liu, X.; Cho, Y. J.; Furdyna, J. *Phys. Rev.* B 2007, *vol. 75*, 045203.
[8] Lin, Y. T.; Ma, T. C.; Chen, T. Y.; Lin, H. H. *Appl. Phys. Lett*. 2008, *vol. 93*, 171914.
[9] Bian, L. F.; Jiang, D. S.; Tan, P. H.; Lu, S. L.; Sun, B. Q.; Li, L. H.; Harmand, J. C. *Solid State Communications*. 2004, *vol. 132*, 707-711.
[10] Kondow, M.; Kitatani, T.; Shirakata, S. *Jpn. J. Appl. Phys*. 2003, *vol. 42*, 4286-4287.
[11] Kitatani, T.; Kondow, M.; Kudo, M. *Jpn. J. Appl. Phys*. 2001, *vol. 40*, L750-L752.
[12] Kurtz, S.; Webb, J.; Gedvilas, L.; Friedman, D.; Geisz, J.; Olson, J.; King, R.; Joslin, D.; Karam, N. *Appl. Phys. Lett*. 2001, *vol. 78*, 748-750.
[13] Kent, P. R. C.; Zunger, A. *Phys. Rev. B* 2001, *vol. 64*, 115208.
[14] Grenouillet, L.; Bru-Chevallier, C.; Guillot, G.; Gilet, P.; Ballet, P.; Duvaut, P.; Rolland, G.; Million, A. *J. Appl. Phys*. 2002, *vol. 91*, 5902-5904.
[15] Wicaksono, S.; Yoon, S. F.; Loke, W. K. Proceeding of 2005 International Conference on Indium Phosphide and Related Materials, pp. 421-423.
[16] Ciatto, G.; Harmand, J. C.; Glas, F.; Largeau, L.; Le Du, M.; Boscherini, F.; Malvestuto, M.; Floreano, L.; Glatzel, P.; Alonso Mori, R. *Phys. Rev. B* 2007, *vol. 75*, 245212.
[17] Jenichen, A.; Engler, C. *Phys. Stat. Sol. (b)* 2004, *vol. 241*, 1883-1895.
[18] Koningsberger, D. C.; Prins, R. *X-Ray Absorption: Principles, Applications, Techniques of EXAFS, SEXAFS and XANES*; John Wiley and Sons, New York, 1988.
[19] Gupta, J. A.; Dharma-wardana, M. W. C.; Jürgensen, A.; Crozier, E. D.; Rehr, J. J.; Prange, M. *Solid State Comm*. 2005, *vol. 136*, 351-355.
[20] Lin, Y. T.; Ma, T. C.; Lin, H. H.; Wu, J. D.; Huang, Y. S. *Appl. Phys. Lett*. 2010, *vol. 96*, 011903.
[21] Keating, P. N. *Phys. Rev.* 1966, *vol. 145*, 637-645.
[22] Martin, R. M. *Phys. Rev. B* 1970, *vol. 1*, 4005-4011.
[23] Rehr, J. J.; Albers, R. C. *Rev. Mod. Phys*. 2000, *vol. 72*, 621-654.
[24] Ma, T. C.; Lin, Y. T.; Lin, H. H. *J. Cryst. Growth*. 2011, *vol. 318*, 363-366.
[25] Verma, A. S. *Phys. Lett. A* 2008, *vol. 372*, 7196-7198.
[26] Ankudinov, A. L.; Ravel, B.; Rehr, J. J.; Conradson, S. D. *Phys. Rev. B* 1998, *vol. 58*, 7565-7576.
[27] Ravel, B.; Newville, M. *J. Synchro. Rad.* 2005, *vol. 12*, 537-541.

In: Physics and Mechanics of New Materials …
Editors: Ivan A. Parinov and Shun-Hsyung Chang

ISBN: 978-1-62618-535-7

Chapter 11

Influence of Parasitic Effect between Frequency and Quantum Impedance for Quantum Resonant Materials and Structures

Chih Chin Yang[1,*], Chong Yan Chen[1], Jenny Chih Yu Lee[1], Ming Lung Hsieh[1], Shau Jie Shia[1], Shun Hsyung Chang[1], Ian Yi-Yu Bu[1] and Ivan A. Parinov[2]

[1]Department of Microelectronics Engineering, National Kaohsiung Marine University, Kaohsiung City, Taiwan, Republic of China

[2]Vorovich Mechanics and Applied Mathematics Research Institute, Southern Federal University, Rostov-on-Don, Russia

Abstract

The variety of impedance of resonant tunneling structure by changing dynamic resistance, series inductance, and dynamic device capacitance is reported to model and analyze the resonant tunneling structure in small signal impedance. The high-frequency impedance model in quantum resonant structure is successfully constructed by combing a parallel R-C circuit along with a series *R-L* circuit with parasitic effect. The relationship between quantum impedance and frequency is obtained. Well-defined model corresponding to the quantum effect is implemented. The quantum inductance absolutely influenced the imaginary value of impedance. The positive resistance value of quantum resonant device is influenced by the quantum inductance, as the device current is expressed in thermal ionic effect but not resonant tunneling effect. The quantum inductance absolutely influenced the imaginary value of impedance, because the imaginary value of resonant impedance is the negative resistance effect. As the capacitance value is more than 0.01pF, the positive quantum impedance will be absolutely influenced by frequency. The quantum impedance is also absolutely influenced by frequency either in large capacitance or small capacitance.

* Corresponding author, E-mail:chchyang@mail.nkmu.edu.tw

1. INTRODUCTION

Since the invention of the double barrier resonant tunneling device (DBRTD, some different equivalent circuits have been proposed for the resonant device [1-3]. The basic compose of these models is parallel (lumped resistance-capacitance circuit with a series resistor and a series inductance used for the Esaki diode [4]. Impedance measurement is the most direct technique for thorny an equivalent circuit. The measured the impedance of RTD's is up to 12 GHz [1]. It showed that the experimental results can be represented by the traditional equivalent circuit. However, the capacitance, which is generally believed to be a function of voltage, was assumed a constant. Also, as was pointed out by some reports, the source of the inductance was not clear [1-4]. Further study was needed. The capacitance of the RTD is very small and at high frequencies the self-inductance which caused by very short connecting lines in a packaged device plays an important role [5-8].

2. RESEARCH MODEL AND METHOD

To estimate the packaging effects, on-wafer probing techniques to measure the small signal scattering parameters (S parameters of RTD's from 0 to 15 GHz at various series inductance or dynamic resistance or capacitance is used mathematical calculation software. It presented that the impedance is obtained over the whole frequency range which can be reproduced quite successfully by the equivalent circuit shown in Figure 1. Figure 1 reveals the quantum modeling circuit with parallel resistance-capacitance structure resembling the active quantum well region of the RTD. The series resistance (R_S and inductance (L_S represents the connecting lines, contact, and spreading resistance which are leaded from the double barrier structure. It is not directly related to the resonant tunneling, and we attribute that the phenomenon of the increased electron density within the quantum well width is due to the existence of a quasi bounded state. Figure 1 shows the equivalent circuit for quantum resonant device. Parallel *R-C* circuit represents the quantum well structure. The series *R-L* circuit accounts for the remaining structure.

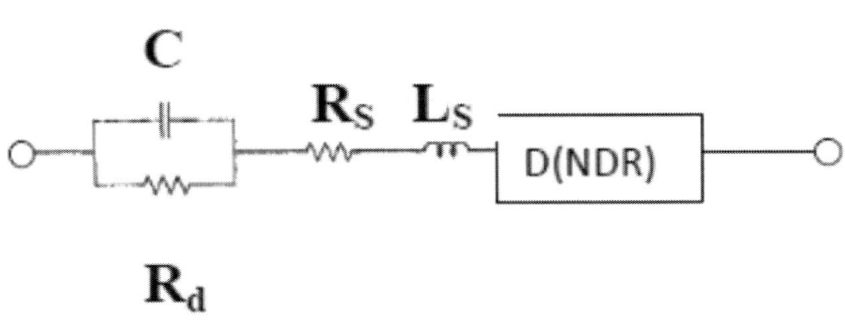

Figure 1. Equivalent circuit for quantum resonant device.

3. Results and Discussion

The quantum resonant tunneling device (RTD is a device with quantum effect that has attracted considerable attention during the past decade because of its strong negative differential resistance (NDR at room temperature and its high-speed switching and oscillation characteristics associated with the NDR effect. These structures may be used as an oscillator with operating frequencies as high as GHz and a mixer with operating frequency of THz. The remarkably NDR phenomenon of fast-acting RTDs make them with excellent devices in electronics application of ultra high frequency.

The impedance of the equivalent circuit quantum resonant phenomenon is expressed, as shown in Figure 1, in following equation:

$$Z = \frac{1}{\frac{1}{R_d + j\omega C} + R_S + j\omega L_S} \tag{1}$$

In Equation (1, the ω is the operation frequency, R_S is the series resistance, and L_S is the series inductance. R_d is the dynamic resistance in device and C is the dynamic device capacitance. R_d and C are directly related to the quantum well structure and are functions of bias. After simplifying the first equation, the second quantum formula is obtained as follow

$$Z = \left[\left(\frac{R_d}{1+\omega^2C^2R_d^2}\right) + R_S\right] + j[\omega L_S - (\frac{\omega C R_d^2}{1+\omega^2C^2R_d^2})] \tag{2}$$

Subsequently, we simulated the RTD's impedance equation on different variables by changing R_d, L_S and C. The R_d, L_S and C are dynamic resistance, series inductance, dynamic device capacitance, respectively. By changing R_d, we can obtain the relationship between frequency and impedance, as shown in Figure 2. The suitable choosing in fitting variables can achieve the practical quantum circuit in resonant tunneling function. We can see from Figure 2 that the calculated impedances agree very well with experimental results, indicating that the equivalent circuit used here is a good model in describing the RTD's high-frequency electrical properties.

The impedance in real part is not changed by changing the quantum inductance value of quantum resonant equivalent circuit, as shown in Figure 3. The results are fitted in the Equation (2. The positive resistance value of quantum resonant device is influenced by the quantum inductance, as the device current is expressed in thermal ionic effect but not resonant tunneling effect. In Figure 4, the impedance value of quantum resonant effect in imaginary part expresses the influence in variety of quantum inductance. The imaginary part value of impedance is smaller as the quantum inductance in quantum resonant device is smaller. The phenomenon is also presented in the Equation (2. The quantum inductance absolutely influenced the imaginary value of impedance, because the imaginary value of resonant impedance is the negative resistance effect.

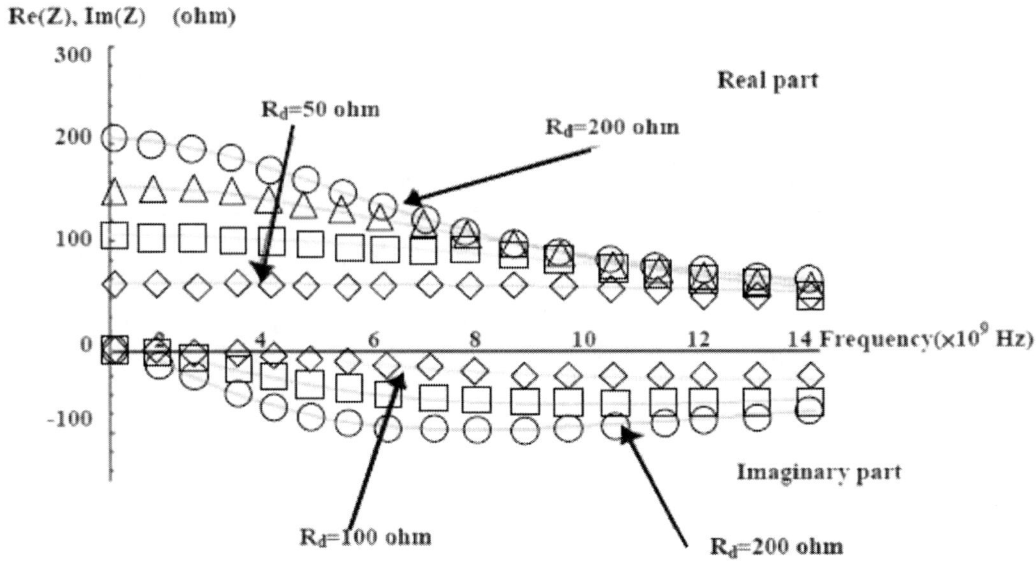

Figure 2. R_d is dynamic resistance, with range 0 to 200 ohm.

The influence of capacitance value in quantum resonant equivalent circuit is also obvious either in real part of quantum impedance or in imaginary part of quantum impedance. The capacitance revealed the device junction effect of quantum resonant device in the range from 0.01 pF to 0.1 pF. The fabrication technology of semiconductor device can eliminate the junction capacitance and diffusion capacitance of device, but it is still existed. The capacitance effect results in the resonant condition for the quantum device. The capacitance value is basically able to be controlled by precise growth control in device structure. As the capacitance is less, the real part of quantum impedance is more, which matches the express of Equation (2. The statement as above reveals the device must be small which results in the less capacitance value. As the capacitance value is less, the real part of quantum impedance will be influenced by frequency. As the capacitance value is more than 0.01 pF, the positive quantum impedance will be absolutely influenced by frequency, as shown in Figure 5. The positive quantum impedance decided the valley current density and the negative quantum impedance decided the peak current density of RTD. The capacitance value decided the quantum impedance whether is less than 0 or not. Figure 6 shows that the more capacitance value is, the less quantum impedance is. The quantum impedance is also absolutely influenced by frequency either in large capacitance or small capacitance.

The parasitic capacitance effect is verified in Figure 5 which the impedance of resonant tunneling structure remains constant at low capacitance value less than 0.01 pF up to gigahertz frequency. It reveals the true frequency response of the resonant tunneling structure including self resonant frequency and resonant frequency. The capacitance parasitic effect is originally from the depletion and accumulation charges of structure junction in resonant tunneling structure which stored in the quantum well with a bias voltage. The accumulation charge is resulted in parallel plate capacitance. The more scattering effect is in quantum well structure, the more parasitic capacitance value is. The real value of resonant tunneling impedance will become less as the parasitic capacitance is more at high frequency operation.

The result as above will result in the more dissipation power which the radio frequency transmittance will be worst. The depletion width is also a key factor in the calculation of parasitic capacitance effect. The depletion width of resonant tunneling structure is decided by the thickness of quantum layer and extrinsic doping concentration in the quantum layer. Therefore, the oscillating frequency of resonant tunneling structure is limited by the parasitic capacitance effect due to the formation of time constant and the traversal time respectively in transient phenomenon and small signal analysis.

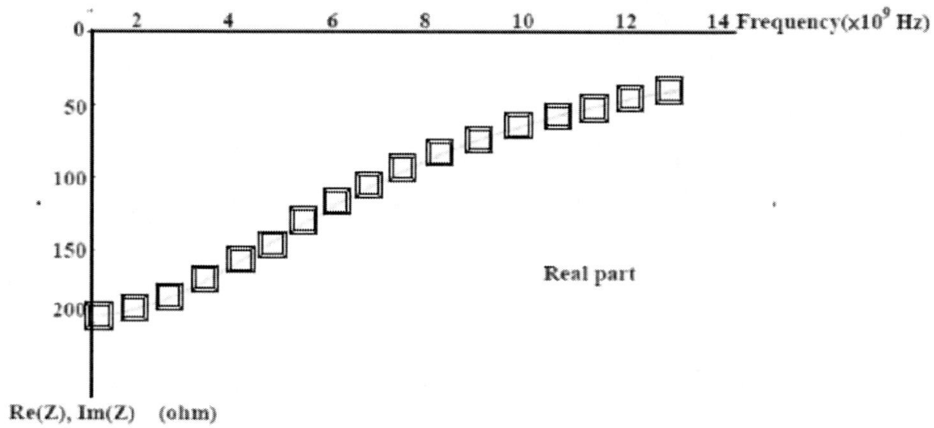

Figure 3. Variety of L_S with range from 0 to 50 pH.

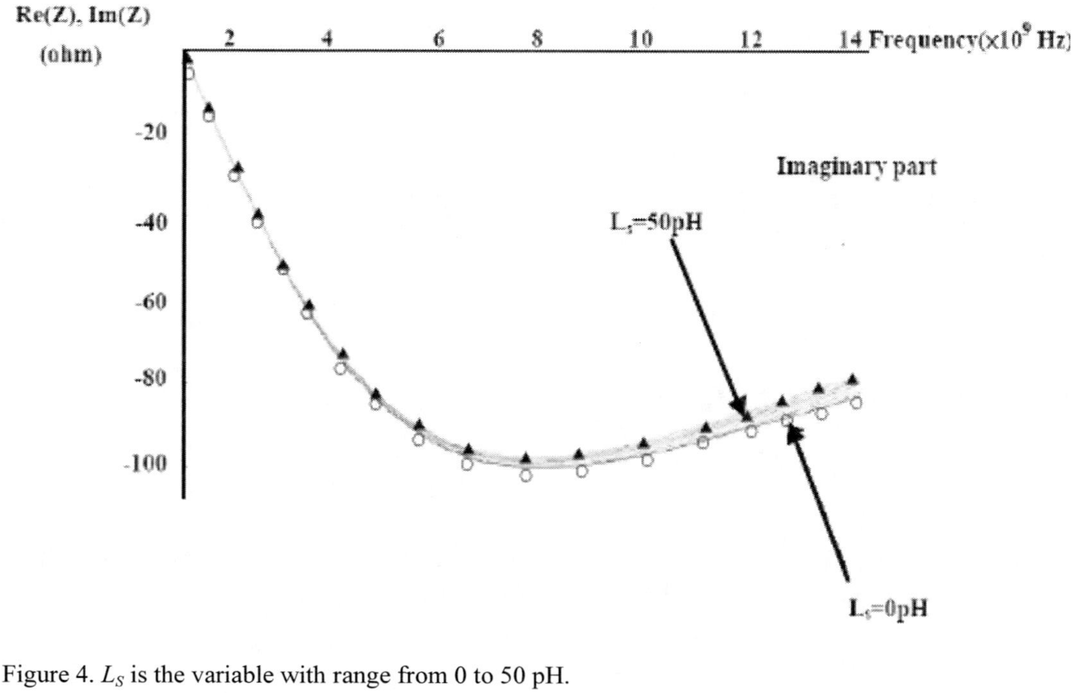

Figure 4. L_S is the variable with range from 0 to 50 pH.

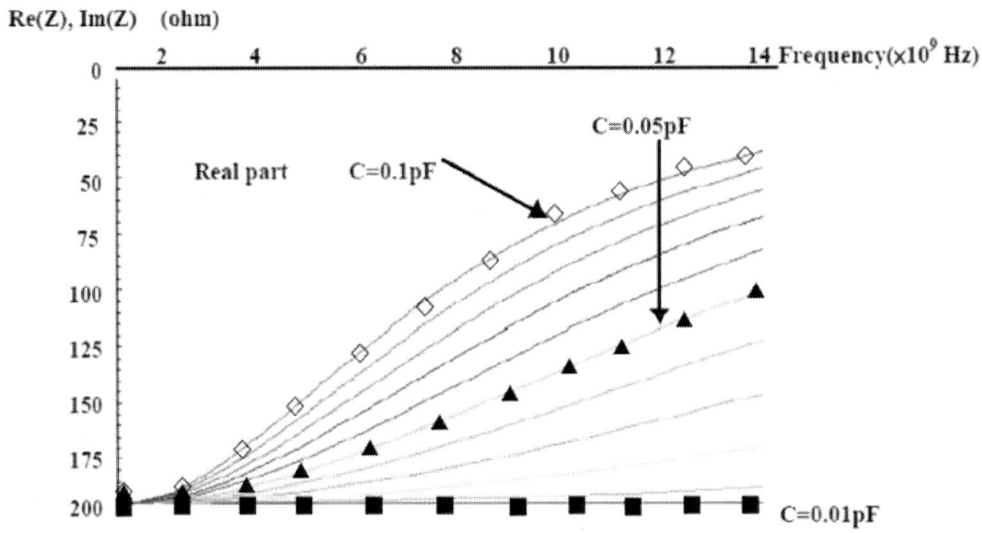

Figure 5. Real part of quantum impedance by capacitance variable from 0 pF to 0.1 pF.

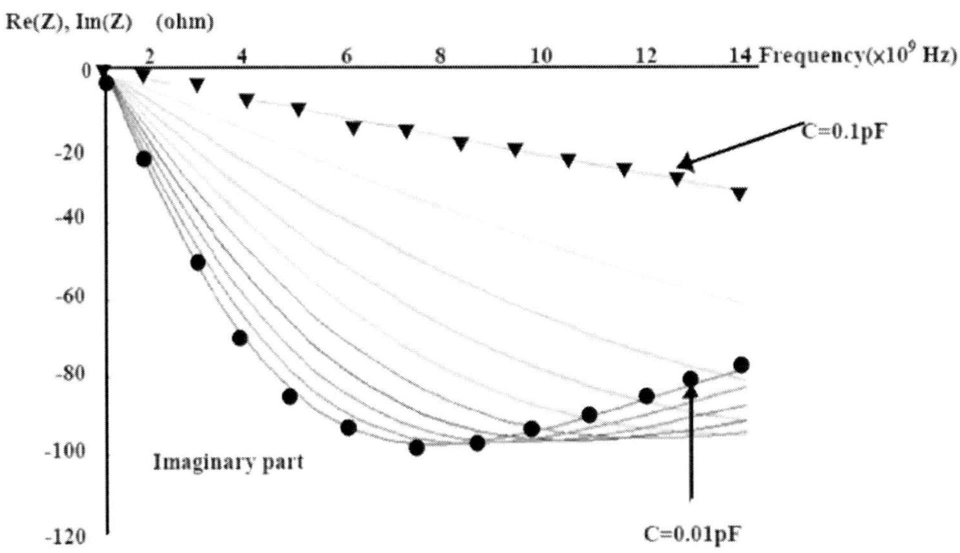

Figure 6. Imaginary part of quantum impedance by capacitance variable from 0 pF to 0.1pF.

The influence of parasitic capacitance effect in imaginary part resonant tunneling impedance is revealed in Figure 6. The small signal impedance as mentioned above determinates the imaginary impedance. The more parasitic capacitance is, the more oscillation frequency is. The high doping concentration in the resonant tunneling structure results in the more parasitic capacitance, which heavy doping is necessary in order to form the high quantum efficiency in quantum well. The parasitic capacitance effect always exists in any quantum well structure. In this study, the resonant frequency could be decided by the controlled parasitic capacitance. The imaginary part of resonant tunneling impedance determines the loss of radio frequency transmittance. Cutoff frequency is about 6 GHz at parasitic capacitance value of 0.01 pF. The more parasitic capacitance is, the more cutoff frequency is. As the parasitic capacitance is small, the scattering effect in the quantum well of resonant tunneling structure will be also less. The less scattering effect will raise peak current density of resonant tunneling structure for high power device, increase peak to valley current ratio of resonant tunneling structure for the operation of power amplify, and enhance the operation function with load line for multiple value logical element.

Conclusion

The equivalent circuit parameters of resonant tunneling diodes (RTD are extracted from numerical simulation results for quantum resonant devices. The RTD models used in this paper are quantum double barrier structures. In this report, the influence of frequency in impedance of resonant tunneling structure is completed by changing dynamic resistance, series inductance, and dynamic device capacitance. In high-frequency, the impedance can be successfully modeled by a parallel R-C circuit along with a series *R-L* circuit. The impedance-frequency relationship is simultaneously obtained. Well-defined modeling structure corresponding to the quantum effect is implemented. The increased capacitance after quantum resonant is revealed by the impedance-frequency curve, and it is attributed to the charge accumulation at the quasi bounded state within the quantum well structure.

Acknowledgement

This research was supported in part by the Russian-Taiwan Goal Oriented Researches on Interdisciplinary Collaborative Research Program (Grant No. 10-08-13300. The authors would also like to thank the National Science Council for financial support under the grant NSC 99-2923-E-022-001-MY3.

References

[1] Gering, J. M.; Crim, D. A.; Morgan, D. G.; Coleman, P. D.; Kopp, W.; Morkoc, H. *J. Appl. Phys.* 1987, *vol. 61*, 271-276.
[2] Nathanson, H. C.; Newell, W. E.; Wickson, R. A.; Davis, J. R. *IEEE Trans. on Electron Devices* 1967, *vol. 14*, 117-133.
[3] Brown, E. R.; Parker, C. D.; Sollner, T. C. L. G. *Appl. Phys. Lett.* 1989, *vol. 54*, 934-936.
[4] Chow, W. F.; *Principles of Tunnel Diode Circuits*; Wiley, New York, 1964; pp 1-360.
[5] Huynen, I. *IEEE Micro and Guided Wave Lett.* 1998, *vol. 8*, 72-74.
[6] Yeo, E. G.; Shi, L. P.; Zhao, R.; Lim, K. G.; Chong,T. C.; Adesida, I. *Appl. Phys. Lett.* 2010, *vol. 96*, 043506-3.

[7] Kashiwagi, M.; Takenaga, K.; Ichii, K.; Kitabayashi, T.; Tanigawa, S.; Shima, K.; Matsuo, S.; Fujimaki, M.; Himeno, K. *IEEE Journal of Quantum Electron.* 2011, *vol. 47*, 1136-1141.
[8] Schlenvogt, G. J.; Barnaby, H. J.; Rollins, J. D.; Wilkinson, J.; Morrison, S.; Tyler, L. *IEEE Trans. on Nuclear Sci.* 2011, *vol. 58*, 2863-2870.

In: Physics and Mechanics of New Materials ...
Editors: Ivan A. Parinov and Shun-Hsyung Chang
ISBN: 978-1-62618-535-7

Chapter 12

AN INTRODUCTORY STUDY ON ELECTROCHEMICAL PREVENTION OF BIOFOULING WITH SIMULTANEOUSLY POTENTIAL DRAG REDUCTION ON MODIFIED POROUS TiO_2/TI ELECTRODES

Wen-Ling Hong[a], Jr-Ping Wang[a], Kee-Rong Wu[*b], Chung-Wei Yeh[c], Jiing-Kae Wu[b] and Juinn-Bing Tan[d]

[a] Department of Naval Architecture and Ocean Engineering, National Kaohsiung Marine University, Kaohsiung, Taiwan

[b] Department of Marine Engineering, National Kaohsiung Marine University, Kaohsiung Taiwan

[c] Department of Information System, Kao Yuan University, Kaohsiung, Taiwan

[d] Department of Applied Foreign Languages, National Penghu University of Science and Technology, Penghu, Taiwan

ABSTRACT

This study investigated modified porous TiO_2/Ti electrodes that were activated electrochemically to evolve hydrogen and oxygen gases in seawater for potential applications in biofouling prevention and simultaneous drag reduction of seagoing ships. The porous TiO_2/Ti plates were first prepared on pure titanium plates using a micro-arc oxidation technique in the electrolyte solution containing Na_2WO_4 (15 $g \cdot L^{-1}$), $NaOH_3$ (2 $g \cdot L^{-1}$), and NaF (2 $g \cdot L^{-1}$) solutions. Pt or C thin layers were subsequently deposited on the porous TiO_2/Ti samples as modified porous Pt-TiO_2/Ti or C-TiO_2/Ti electrodes using a DC magnetron sputtering technique. Our first preliminary tests revealed that no attachment of marine organisms was found on the porous electrodes at an energy consumption rate of approximately $2.0 W \cdot m^{-2}$, after 15 days of field seawater test. By contrast, organisms, mainly hydroids and barnacles, attached and grew on the electrode and its fixture, which was applied without a DC bias (i.e., attachment of organisms can be

* Department of Marine Engineering, National Kaohsiung Marine University, 142 Haichuan Rd., Nanzih, Kaohsing 811, Taiwan, E-mail: krwu@mail.nkmu.edu.tw

prevented). Meanwhile, a higher DC bias powered the coupled Pt-TiO_2/Ti cathodes and C-TiO_2/Ti anodes, where hydrogen and oxygen microbubbles were electrolytically evolved in seawater. The modified porous electrodes continuously yielded a gas production rate of 33.4 $L{\cdot}h^{-1}m^{-2}$ at an energy consumption rate of approximately 48$W{\cdot}m^{-2}$. It is attributable to considerably rough and network-structured porous Pt-TiO_2/Ti and C-TiO_2/Ti electrodes with low resistivity, rendering more reactive sites for electrolytic reactions for gas evolution. Furthermore, the microbubbles could be distributed evenly over the submerged electrode surface, possibly providing a higher local void ratio close to the hull surface and potentially maximizing the drag reduction effect for seagoing ships. Along with its excellent tribological properties, the proposed electrodes can be potentially realized for electrochemical prevention of biofouling with simultaneously drag reduction of ship of which submerged surface is covered mostly with the electrode plates.

1. INTRODUCTION

Considerable efforts have been devoted to the reduction of energy cost in various applications since the first oil crisis in the early 1970s. However, worldwide primary energy consumption increased by 5.6% in 2010, the strongest growth since 1973; maritime transportation accounted for 1.8% of global energy consumption [1]. In general, the total resistance of a low-speed ship (approximately 15 nautical miles per hour) is mostly frictional resistance (90%) of submerged hull, and a small percentage from wave making resistance (5%), vortex flow loss (3%), and air resistance (2%). Decreasing frictional resistance of ships is one of the most effective approaches to reduce operating costs and CO_2 pollution. This has been a crucial issue in fluid dynamic research of ships in the past several decades. Several methods for reducing the frictional resistance were proposed, such as surface structure with nano-techniques and riblets for biomimetic drag reduction, artificial microblowing, microbubble injection, and antifriction coatings. Among these methods, antifriction coatings are one of the most convenient and economical, but least eco-friendly approaches; whereas microbubble injection is the most eco-friendly approach and provides excellent drag reduction [2].

It has been argued that slip may be present in several microfluidic systems; in particular, because of the presence of an adsorbed layer of nanobubbles [3, 4]. Because gas viscosity, μ_G, is much lower than liquid viscosity, μ_L, the air or vapor layer of thickness H leads to the apparent slip length, that is, $l = H\ (\mu_L/\mu_G - 1)$. Hence, a thin air film can reduce the frictional resistance of submerged surface substantially [4]. The pioneer work on drag reduction using microbubbles was first reported dated in 1973 [5]. In their work, hydrogen microbubbles generated by electrolysis produced up to 30% drag reduction over a body of revolution with less than 1% void fraction. Recently, particle tracking velocimetry was used to measure the velocity fields of dispersed microbubble phase in a turbulent boundary layer of a channel flow, in which an average size of 15 μm of the hydrogen and oxygen microbubbles were generated by electrolysis [6]. Drag reduction of up to 40% was obtained when the accumulation of microbubbles occurred in a critical zone within the buffer layer. They concluded that the crucial aspect in achieving high drag reductions is the accumulation of microbubbles in a critical zone in the buffer layer. That is, the local void ratio close to the wall is one of the essential parameters for drag reduction by microbubbles [7]. In addition, the electrochemistry reaction can yield numerous smaller-sized microbubbles close to the hull wall, which is beneficial to a higher local void ratio for drag reduction [8].

Alternatively, Madavan, et al. obtained a drag reduction of up to 80% by supplying compressed air through porous plates in 1985 [9], which was subsequently confirmed by Kato et al. by using an identical experimental apparatus [10]. However, microbubble injection through porous plates on a model-scale ship of six meters in length reported a drag reduction of less than 2% [11]. Furthermore, the bubbles injected into the seawater have a detrimental effect on the propeller efficiency of approximately 3%, and cause cavitations of the blades. The ship speed decreased upon microbubble injection under several experimental conditions in a training ship [12]. The possible reasons for such a considerable difference are as follows: first, the diameters of the bubbles generated by the porous plates tend to increase up to several millimeters because of the buoyant effect. The larger bubbles may adhere to the hull surface and increase the resistance to the hull. The area of the hull surface of the actual ship covered by the porous plates is too small, due to the limit number of air injectors installed. Thus, the microbubbles cannot cover most of the submerged hull surface. Finally, the microbubbles may overflow out of the boundary layer under the action of vertical velocity and increase microbubbles in the flow field around the hull, leading to an adverse effect on the drag reduction efficiency. In addition to its low drag reduction efficiency, the conventional microbubble injection device may incur several structural safety problems, including the risks of leakage of the hull and attachment of marine organisms on the air injectors.

Since the disadvantages of microbubble drag reduction using compressed air through porous plates, Wedin, et al. generated smaller sized microbubbles by electrolysis [13]. Because the nucleuses of microbubbles can be directly carried to the downstream of the model ship [8], the microbubbles can be distributed evenly over the submerged surface, providing a higher local void ratio close to the hull surface and maximizing the drag reduction effect. Recently, particle tracking velocimetry has been used to measure the velocity fields of dispersed hydrogen and oxygen microbubbles generated by electrolysis in a channel flow [14]. Drag reductions up to 40% has been found with an average bubble size of 15 μm, which mostly takes place in a critical zone within the buffer layer. It has been concluded that a combination of concentration and distribution of microbubbles in the boundary layer can achieve high drag reduction values [14]. One of advantages over conventional porous plates is that the only hull penetrations are those required for electrical leads to pass from the power supply to the electrodes by using the electrolytic scheme. The risks of hull leakage are minimized.

Seagoing ships and marine platforms are always subjected to environmental attacks from the surrounding ocean. Marine biological fouling, usually termed marine biofouling, as a result of attached organisms, such as slimes, hydroids, barnacles and mussels, to the ship hull and marine structures. It has been reported that more than 4000 marine organisms have been identified on fouled structures [15]. As a number of marine microorganisms adhere to the hull surface, the roughness of the hull surface and the overall weight of the hull may increases by up to 75% [16, 17]. Thus, biofouling can lead to a considerable increase in the cost of maritime transportation. For example, in a low-speed ship, the resistance to a ship may be increased by up to 80% and the fuel consumption may be increased by up to 86% under heavy biofouling [17]. Furthermore, the marine organisms inhabit any artificial structure or tool, such as set nets, anchor chains, and cooling pipelines of generators, thereby speeding up metal corrosion or increasing the flow resistance of water [15].

Marine biofouling is of crucial economic concern to all marine industries and consequently, the high costs relative to transport delays and hull repairs (estimated at US $150 billion per year) [16]. The shipping trade is particularly alert to the development of new antifouling strategies because of the stricter environmental effects of the compounds [18]. Antifouling (AF) coatings are necessary to avoid colonization of surfaces by biofoulers. Among various AF paints, self-polishing copolymer tributyltin (TBT) paint is commonly used because it is stable up to 60 months, but toxic to the ocean environment [19]. Thus, the International Maritime Organization (IMO) has forbidden the use of self-polishing copolymer TBT paint on hulls since January 1, 2008, to mitigate damage to the ocean environment [19].

Although conductive paint or conductive material is more effective and environment-friendly than tin-free antifouling paint, the cost is higher, and damage occurs easily. This method also consumes considerable electricity, and increases the possibility of corrosion of the hull. The chloride gas deliberated by electrolysis of seawater is highly hazardous to all marine organisms at a concentration higher than 0.5 $mg\cdot L^{-1}$ [20].

In 1998, Nakayama innovatively applied an alternating DC bias of 1.0 V and –0.6 V, slightly lower than the voltage for obtaining chlorine gas by electrolysis, on a film of titanium nitride (TiN), and observed effective inhibition of marine biofouling after 279 days of the experiment in seawater [21]. Since the formation of biofilms is crucial for the subsequent attachment and development of barnacles and mussels, the inhibiting mechanism is based on microelectric current on the TiN electrodes that destroys the marine microorganisms on which the marine organisms feed. Thus, parasitism of the marine microorganisms can be inhibited via the direct electron transfer reaction and OH radical generated through electrolysis of the electrodes to avoid adherence and growth of marine organisms, including barnacles, tube worms, and oysters [21, 22]. This antifouling method is eco-friendly and effective in a small scale of field test [21].

Micro-arc oxidation (MAO) was recently reported in related literature for preparing porous TiO_2/Ti layers coupled with various metal oxides, such as vanadia or tungsten oxide, on Ti substrates [23-26]. The main mechanism in MAO involves dielectric breakdown, which causes spark discharge and microarcing in the electrolyte, resulting in the formation of porous melt-quenched oxides and compounds on the anode substrate. Thus, porous TiO_2/Ti layer is constituent with species derived from the substrate and electrolytes [27]. Details of MAO process can be found in the relevant literatures [27-29]. One of the advantages of MAO process is the possibility of incorporating anionic and/or cationic ions into the TiO_2 layer on a Ti substrate by controlling the constituents of the electrolyte [23, 27-29]. The porous MAO-TiO_2 film with a thickness of 40 μm reportedly exhibited excellent tribological properties, demonstrating nanohardness and elastic modulus of 5.5 and 69.1 GPa, respectively, with an adhesion strength of approximately 40 MPa [30]. This approach has several potential applications, mostly in biocompatibility [25, 31-33], and wear protection [26]. However, only a few related studies have addressed the fabrication of porous TiO_2/Ti substrates using an MAO-analogous process on electrolytic applications [34, 35]. For instance, deposition of Pt on porous TiO_2/Ti substrates is economically possible for obtaining an inexpensive electrode that, electrochemically, behaves similarly to a pure Pt electrode; the resulting Pt-TiO_2/Ti electrodes constitute a particular type of dimensionally stable anodes (DSAs) for noble metal coated anodes [35]. In general, the nature of electrode material considerably influences the efficiency of the oxidation [36, 37]. Recently, Pt-TiO_2/Ti anodes prepared by the galvanostatic anodization of Ti sheets followed by galvanostatic deposition of overlying Pt have comparatively demonstrated the same electrocatalytic activity as bare Pt foils. This has been attributed to the increase in the active surface area of Pt deposits formed on the porous anode surface, compared with the bare metal [35].

2. EXPERIMENTAL PROCEDURE

2.1. Sample Preparation

Porous TiO_2/Ti plates were synthesized on Ti substrates ($25\times75\times1$ mm^3) using an in-house MAO system, in which the Ti substrate was applied as an anode, and a stainless steel container was used as a cathode. The surfaces of the Ti substrates were polished using silicon carbide paper, and ultrasonically cleaned in 100% acetone, 100% ethanol, and distilled water. The operating temperature of the reactor was maintained below 50 °C using a cooling water jacket. The electrolyte solution comprised Na_2WO_4 (15 $g\cdot L^{-1}$), $NaOH_3$ (2 $g\cdot L^{-1}$), and NaF (2 $g\cdot L^{-1}$) solutions. The Ti substrates were oxidized at a constant voltage of 300 V for 10 min. The as-oxidized TiO_2/Ti samples were rinsed with distilled water and subsequently dried in air at room temperature without post heat treatment. Pt and C thin layers with a thickness of approximately 20 nm were subsequently deposited on the porous TiO_2/Ti samples, denoted as Pt-TiO_2/Ti and C-TiO_2/Ti electrodes, by sputtering a Pt and graphite targets, respectively, using an in-house magnetron sputtering system (MIRDC, Taiwan).

2.2. Sample Characterization and Field Test

The resistivity of the samples was analyzed using a Hall-effect measurement (Ecopia HMS-3000). The crystal structures of the samples were analyzed using a high-resolution X-ray diffractometer (XRD, Rigaku ATX-E). The microstructures of the films were investigated using a scanning electron microscope (SEM, JEOL JSM-6700F). A high-resolution transmission electron microscope (HRTEM, Philips Tecnai 20, operated at 200 kV) was employed for microstructure characterization and composition analysis. TEM cross-sectional specimens of the samples were prepared using a focused ion beam (FEI Quant D 200) at a voltage of 30 kV.

An electrical contact with copper wire was made on the back of the electrodes using silver paste, and was subsequently electrically insulated by nonconductive epoxy resin. The electrolysis for hydrogen and oxygen evolution was performed using a standard two-cell system, which included a C-TiO_2/Ti anode and a Pt-TiO_2/Ti cathode in natural seawater. The hydrogen concentration was monitored by a gas chromatography (GC, HP4890) with flame ionization detection as its detector. The system was controlled by a potentiostat (CHI 610C). The DSA anodes were tested under the same conditions for electrolysis of hydrogen and oxygen evolution.

Preliminary field tests of biofouling on sample electrodes were performed at a depth of approximately 100 cm in seawater, in the inner harbor of Kaohsiung, Taiwan. A DC bias of 15 V was applied to the electrodes for an experimental period of 15 days; the applied DC bias was cyclically inversed every 30 min and maintained for 5 min, which was analogous to the scheme reported in [21]. Sample electrodes without the application of DC bias were also used for comparison. All samples were properly fixed to a submerged plastic container using cable ties and those electrical connected samples were tentatively sealed by epoxy resin.

3. RESULTS AND DISCUSSION

3.1. Sample Characterization and Field Test

Figure 1 shows the XRD patterns of TiO_2/Ti, Pt-TiO_2/Ti and C-TiO_2/Ti samples with a commercial DSA anode for comparison. The XRD patterns of the Pt-TiO_2/Ti, C-TiO_2/Ti and TiO_2/Ti reveal dominant anatase TiO_2 (101) phases along with a crystalline WO_3 (020) at $2\theta = 23.67^o$ and two strong Ti peaks at $2\theta = 38.4^o$ and $2\theta = 40.2^o$ of the unoxidized Ti substrate. For the C-TiO_2/Ti, there is an additional crystalline WO_3 (200) at $2\theta = 24.18^o$ and some minor peaks which might arise from the contents of electrolytes. These suggest that the WO_3 phase can be largely loaded onto the TiO_2 matrix [23, 38]. The relatively large amount of WO_3 was attributed to high absorption ability in cationic dye solution [38]. No noticeable peak related to Pt was detected for Pt-TiO_2/Ti sample because of the thin or hemisphere-shaped overlaid layer. But, a strong peak related to crystalline C (005) at $2\theta = 38.4^o$ was detected for C-TiO_2/Ti sample. Hence, the overlaid Pt and C layers could enhance the conductivity of the porous samples. Alternatively, the XRD pattern of DSA exhibited mostly IrO_2 (110) and (211) phases, and Ti with a slight anatase TiO_2 phase.

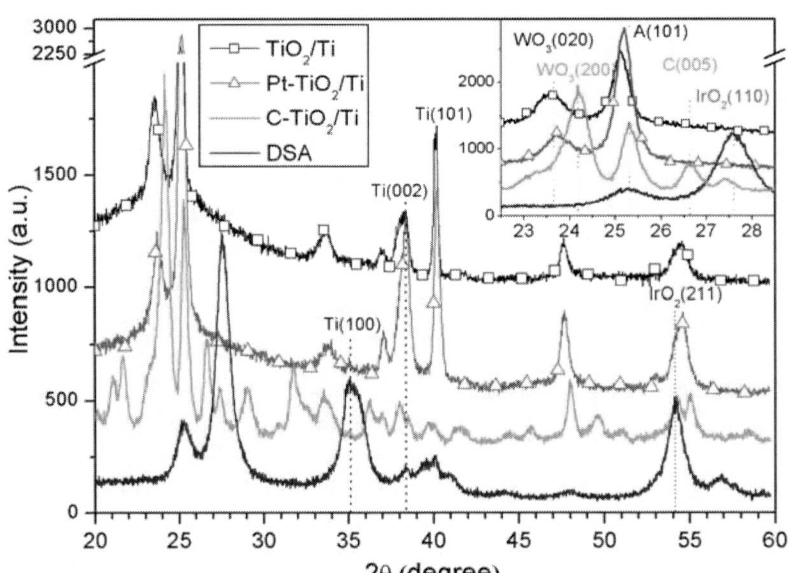

Figure 1. XRD patterns of TiO_2/Ti, Pt-TiO_2/Ti and C-TiO_2/Ti samples along with a commercial DSA anode for comparison, where A stands for anatase TiO_2.

Figure 2(a) shows the morphological SEM image of sample TiO_2/Ti, showing porous characteristic of MAO-prepared TiO_2 structure. Figure 2(b) presents a magnified image of the sample exhibiting a number of submicron pores on a large cavity, which is beneficial to a high surface area and more reactive sites of electrolysis. One can see that the Pt-TiO_2/Ti exhibits approximately the same porous characteristic as that of the sample TiO_2/Ti (see Figure 2(c)); the highly porous characteristic was maintained after platinization. Figure 3 shows a typical TEM image of Pt-TiO_2/Ti plate exhibiting a substantially rough and porous microstructure. However, various small-sized inner pores were randomly distributed with few open-ended pores, and few inner pores were not fully covered with Pt layer. The MAO-prepared TiO_2/Ti exhibited a BET (Brunauer–Emmett–Teller) surface area of 1.75 $m^2 \cdot g^{-1}$ [24], which is 11 times larger more than that of a typical plain TiO_2 film prepared by magnetron sputtering or sol-gel technique [39, 40]. Thus, considerably rough and network-structured porous TiO_2/Ti generally renders more reactive sites for electrical and chemical reactions in aqueous solution.

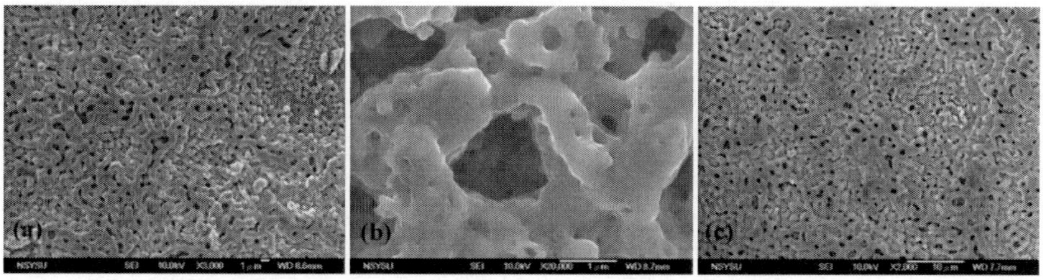

Figure 2. Morphological SEM images of (a) TiO_2/Ti, (b) magnified view of TiO_2/Ti, and (c) Pt-TiO_2/Ti plate.

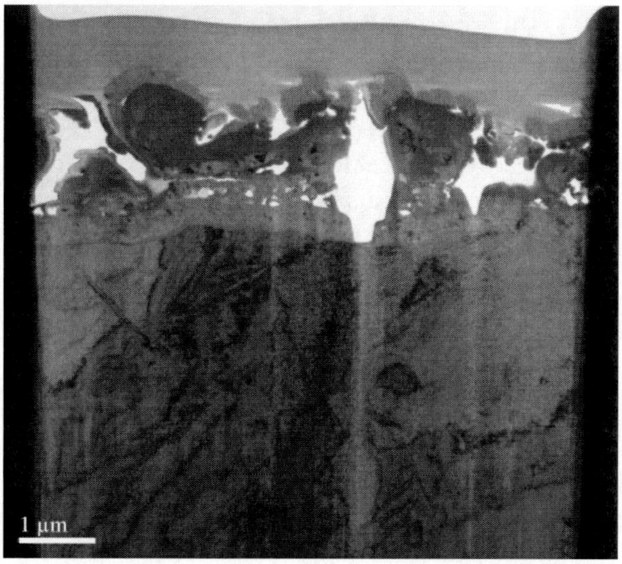

Figure 3. A typical TEM image of Pt-TiO_2/Ti plate where various small-sized inner pores are randomly distributed with few open-ended pores.

3.2. Electrolytic Properties

In general, TiO_2 semiconductor behaves similarly to an electrical insulator with a resistivity of 8×10^7 Ω·m in darkness [41]. A substantial decrease in resistivity was obtained by the MAO process and subsequent Pt or C plating, in which the resistivity is decreased from 5.8×10^4 Ω·m for the porous TiO_2/Ti to 2×10^{-2} Ω·m and 92 Ω·m for the Pt-TiO_2/Ti and C-TiO_2/Ti plates. A decrease in the resistivity for the porous TiO_2/Ti is related to unoxidized Ti and W elements in the TiO_2 layer prepared by MAO process [23, 38]. Low resistivity in the electrode indicates an easy path for the electron motions driven by an applied bias. This allows preparation of low-cost DSA for use as anodes, despite the presence of the underlying TiO_2 layer. Because low resistivity is one of the crucial factors of electrolytic electrodes for generating gases, the MAO process is an alternative method for fabricating porous TiO_2/Ti layer, facilitating effective transport of electrons during application of electric current, in addition to the excellent correction property of TiO_2/Ti layer in various application environments.

Figure 4 shows the cyclic voltammograms of sample electrodes obtained in 2.0 M $NaCO_3$ (pH ~11) with a scan rate of 5 mV·s^{-1}, whereas the inset displays the plain TiO_2/ITO electrode prepared by DC magnetron sputtering for comparison. The characteristics of MAO-prepared TiO_2/Ti and Pt-TiO_2/Ti plates differ considerably from that of the DSA; the oxygen evolution potential, which is characterized by a rapid increase in the current, is approximately 0.67 V versus SCE for the IrO_2-based DSA. However, the MAO-prepared TiO_2/Ti did not yield a substantial oxygen evolution potential up to a value higher than 1.5 V versus SCE. This is consistent with the results, in that DSAs are excellent electrodes for the oxygen evolution reaction (OER), rendering a lower energy consumption rate. This also implies that the Pt-TiO_2/Ti has high oxygen evolution overpotential, and further improvement is required to become a suitable OER electrode [35, 36]. The features of the MAO-prepared TiO_2/Ti and Pt-TiO_2/Ti plates closely resemble that of the plain TiO_2/ITO electrode prepared by DC magnetron sputtering; however it differs considerably from the magnitude of current densities in a more cathodic potential region (−0.8 V *vs.* SCE). This is possibly related to the unoxidized Ti element in the MAO-prepared TiO_2/Ti plate. The magnitude of current densities was further increased by platinization onto the TiO_2/Ti plate. As shown in Figure 4, the modified porous Pt-TiO_2/Ti exhibits a substantial deviation from the well-known curves obtained with pure Pt or DSA electrode under similar conditions [35]. The magnitude of current densities remains unchanged in the anodic potential region. This may be partially attributed to a partial coverage of the overlying Pt or C layer, which cannot be deposited onto the inner surfaces of micro-pores by using DC magnetron sputtering technique, as displayed in the TEM image in Figure 3. Oxides with excellent conductivity and chemical stability, such as IrO_2 or RuO_2, may be more suitable constituents in the MAO-prepared TiO_2/Ti electrode [36]. In addition, more efficient techniques, such as electrodeposition and galvanostatic deposition, must be considered in future studies [34, 35].

As shown in Figure 5(a), growing marine organisms were scarce on the electrode surface after 15 days of field seawater test at an applied bias of 15 V and 0.2 mA, i.e., an energy consumption rate of approximately 2.0 W·m^{-2}. The marine microorganisms are therefore inhibited via the direct electron transfer reaction and OH radical generated through electrolysis of the electrodes in seawater [21, 22]. Conversely, marine organisms of several hydroids and a number of small barnacles grew on the electrode and its fixture, which was tested without a DC bias, as shown in Figure 5(b). The attachment of microorganisms can be prevented by applying a cyclic DC bias on the Pt-TiO_2/Ti and C-TiO_2/Ti electrodes.

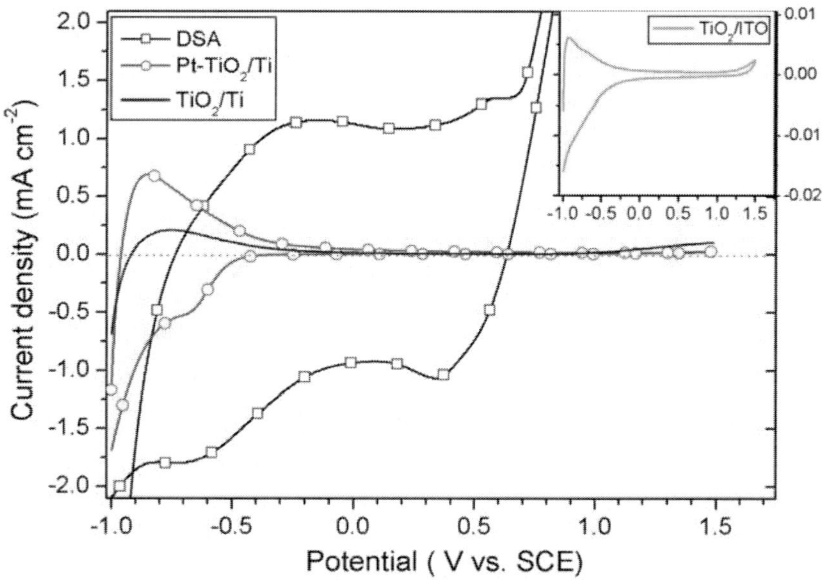

Figure 4. Cyclic voltammograms of the porous Pt-TiO_2/Ti and TiO_2/Ti plates and DSA electrode in 0.1 M $NaCO_3$, where the inset displays the plain TiO_2/ITO electrode prepared by DC magnetron sputtering.

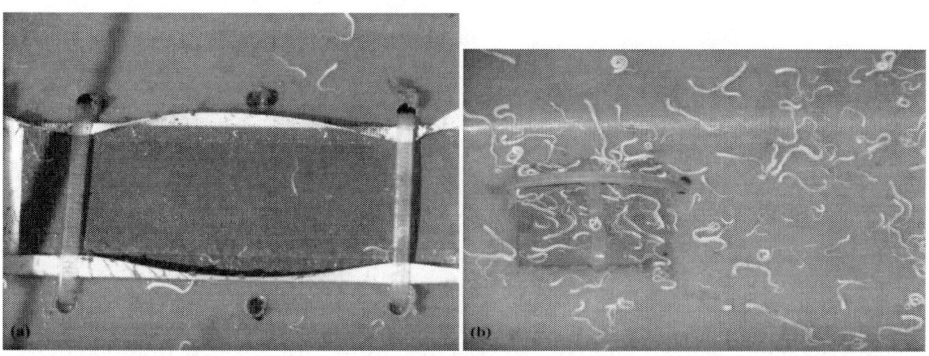

Figure 5. Photos of marine organisms grew on the modified porous electrodes, each was fixed by two cable ties, after 15 days of field seawater test: (a) at an applied DC power of 2.0 W·m^{-2} and (b) without applying DC power.

By applying a higher DC bias, oxygen and hydrogen were generated at C-TiO_2/Ti and Pt-TiO_2/Ti electrodes in seawater, respectively, as shown in Figure 6. The electrodes yielded a total gas (H_2 and O_2) production rate of 33.4 L·h^{-1}m^{-2} by electrolysis of seawater at an energy consumption rate of approximately 48 W·m^{-2}, compared to 185.5 L·h^{-1}m^{-2} at 50 W·m^{-2} of DSAs. This implies that MAO-prepared TiO_2/Ti film with a Pt or C top layer exhibits quite electrochemical property as electrolytic electrodes for gas evolution in seawater. One can optimistically expect that microbubbles could be involved evenly over the submerged electrode surface, possibly providing a higher local void ratio close to the hull surface for potential drag reduction for seagoing ships. The microbubble-induced drag reduction generated by MAO-prepared TiO_2/Ti film electrodes are currently under investigation in an in-house flow system using particle image velocimetry. Nevertheless, the energy gain along with durability of the proposed scheme should be further tested under various practical conditions. More efficient fabrication techniques must be applied to reduce resistivity of the electrodes to lower electrolytic energy consumption and increase the gas evolution rate in future studies [23, 35].

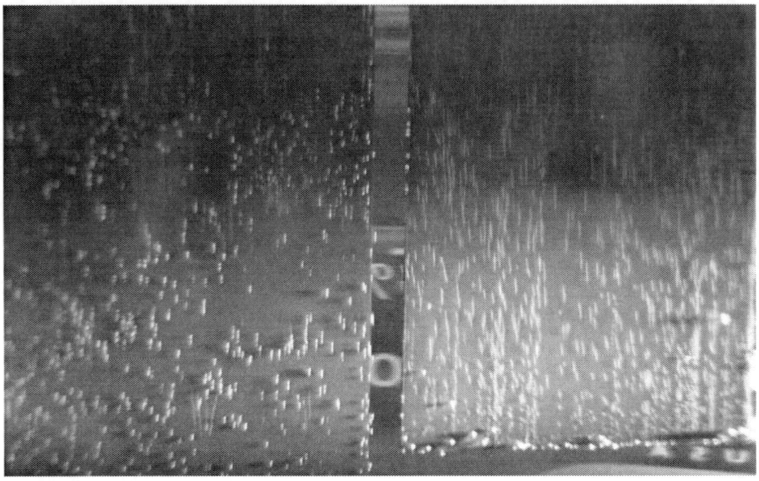

Figure 6: Photo of electrolytic evolution of O_2 and H_2 gases in C-TiO_2/Ti and Pt-TiO_2/Ti electrodes in seawater, respectively.

4. Conclusions

In this study, modified porous Pt-TiO_2/Ti and C-TiO_2/Ti electrodes were successfully obtained by micro-arc oxidation, and subsequently, magnetron sputtering technique. A low resistivity, which offers an easy path for the electron motions in the porous electrodes, was obtained by overlying a thin Pt or C layer. Considerably rough and network-structured porous Pt-TiO_2/Ti and C-TiO_2/Ti electrodes with low resistivity and tribological properties may provide more reactive sites for electrolytic reactions for gas evolution in practical applications. However, it yielded only one-sixth of the gas evolution rate of commercial DSA anodes. The MAO process, including reactive electrolytes and process parameters, along with overlying Pt or C layer must be justified accordingly. However, a submerged ship hull that includes a series of coupled anodic and cathodic electrodes of modified porous Pt-TiO_2/Ti and C-TiO_2/Ti plates powered cyclically by a DC bias can electrochemically generate microbubble-induced surfaces for potential biofouling prevention and simultaneous drag reduction of seagoing ships.

Acknowledgments

Free DSA samples given by Xinxiang Future Hydrochemistry Co., Ltd. (Henan, China) are highly appreciated. The authors wish to thank Prof. Chung-Hsuang Hung of National Kaohsiung First University of Science and Technology and Mr. Jui-Ching Sun of National

Kaohsiung Marine University for their kind assistance in electrolysis tests and sample preparations.

REFERENCES

[1] *BP Statistical Review of World Energy* 2011, 2011.
[2] Almeida, E.; Diamantino, T. C.; de Sousa, O. *Prog. Org. Coat.* 2007, vol. 59, 2.
[3] Nosonovsky, M.; Bhushan, B. *J. Phys. Condens. Matter.* 2008, vol. 20, 225009.
[4] Nosonovsky, M.; Bhushan, B. *Curr. Opin. Colloid. Interface. Sci.* 2009, vol. 14, 270.
[5] McCormick, M. E.; Bhattacharyya, R. *Nav. Eng. J.* 1973, vol. 85, 11.
[6] Ortiz-Villafuerte, J.; Hassan, Y. A. *J. Fluids Eng. - Trans. ASME.* 2006, vol. 128(3), 507.
[7] Yoshiaki Kodama; Akira Kakugawa; Takahito Takahashi; Hideki Kawashima, *Int. J. Heat Fluid Flow.* 2000, vol. 21, 582.
[8] Wu, S. J.; Hsu, C. H.; Lin, T. T. *Ocean Eng.* 2007, vol. 34, 83.
[9] Madavan, N. K.; Deutsch, S.; Merkle, C. L.; *J. Fluid.* 1985, vol. 156, 237.
[10] Kato, H.; Miyanaga, M.; Yamaguchi, H. *Cavitation and Gas-Liquid Flow in Fluid Machinery and Device ASME.* 1994, vol. 90, 185.
[11] Tsai, J. F.; Chen, C. C. *Int. Sch. Res. Netw. ISRN Mech. Eng.* 2011 (2011) 9 pages, doi:10.5402/2011/405701.
[12] Guin, M. M.; Kato, H.; Yamaguchi, H.; Maeda, M.; Miyanaga, M. *J. Mar. Sci. Technol.* 1996, vol. 1(5), 241.
[13] Wedin, R.; Davoust, L.; Cartellier, A.; Byrne, P. *Exp. Therm. Fluid Sci.* 2003, vol. 27, 685.
[14] Ortiz-Villafuerte, J.; Hassan, Y. A. *ASME J. Fluids Eng.* 2006, vol. 107, 370.
[15] Yebra, D. M.; Kiil, S.; Dam-Johansen, K. *Prog. Org. Coat.* 2004, vol. 50, 75.
[16] Schultz, M. P. *Biofouling.* 2007, vol. 23(5), 331.
[17] Chambers, L. D.; Stokes, K. R.; Walsh, F. C.; Wood, R. J. K. *Surf. Coat. Technol.* 2006, vol. 201, 3642.
[18] *Introduction in Advances in Marine Antifouling Coatings and Technologies,* C. Hellio, D. M. Yebra (Eds.), Woodshead Publishing, Cambridge, UK, 2009, Ch. 1.
[19] Maréchal, J. P.; Hellio, C. *Int. J. Mol. Sci.* 2009, vol. 10, 4623.
[20] Huang, J. R.; Lin, W. T.; Huang, R.; Wu, J. K. *J. Mar. Sci. Technol.* 2009, vol. 17(3), 173.
[21] Nakayama, T.; Wake, H.; Ozawa, K.; Nakamura, N.; Matsunaga, T. *Appl. Microbiol. Biotechnol.* 1998, vol. 50, 502.
[22] Matsunaga, T.; Nakayama, T.; Wake, H.; Ozawa, K.; Nakamura, N.; Murakami, N.; Takahashi, H.; Takimoto, T.; Kadoi, H. *US Patent* (2001) 6,197,168 B1.
[23] Bayati, M. R.; Golestani-Fard, F.; Moshfegh, A. Z. *Appl. Catal. A-Gen.* 2010, vol. 382, 322.
[24] Yeh, C. W.; Wu, K. R.; Hung, C. H.; Chang, H. C.; Hsu, C. *J. Int. J. Photoenergy* 2012 (2012) 9 pages, doi:10.1155/2012/285129.
[25] Deng, F.; Zhang, W.; Zhang, P.; Liu, C.; Ling, *J. Mater. Sci. Eng. C-Mater. Biol. Appl.* 2010, vol. 30, 141.

[26] Khan, R. H. U.; Yerokhin, A. L.; Matthews, A. *Philos. Mag*. 2008, vol. 88, 795.
[27] Li, J.; Wan, L.; Feng, J. *J. Mater. Processing Technol*. 2009, vol. 209, 762.
[28] Yerokhin, A. L.; Nie, X.; Leyland, A.; Matthews, A.; Dowey, S. *J. Surf. Coat. Technol.* 1999, vol. 122, 73.
[29] Snizhko, L. O.; Yerokhin, A. L.; Pilkington, A.; Gurevina, N. L.; Misnyankin, D. O.; Leyland, A.; Matthews, A. *Electrochim. Acta*. 2004, vol. 49, 2085.
[30] Wang, Y.; Lei, T.; Jiang, B.; Guo, L.*Appl. Surf. Sci.* 2004, vol. 233, 258.
[31] Wei, D.; Zhou, Y.; Jia, D.; Wang, Y. *Acta Biomater*. 2007, vol. 3, 817.
[32] Li, L. H.; Kong, Y. M.; Kim, H. W.; Kim, Y. W.; Kim, H. E.; Heo, S. J.; Koak, J. Y. *Biomaterials*. 2004, vol. 25, 2867.
[33] Chen, H. T.; Hsiao, C. H.; Long, H. Y.; Chung, C. J.; Tang, C. H.; Chen, K. C.; He, J. L. *Surf. Coat. Technol.* 2009, vol. 204, 1126.
[34] Devilliers, D.; Mahe, E. *Electrochim. Acta.* 2010, vol. 55, 8207.
[35] Gueneau de Mussy, J.-P. ; Macpherson, J. V. ; Delplancke, J.-L. *Electrochim. Acta.* 2003, vol. 48, 1131.
[36] Panizza, M.; Cerisola, G. *Appl. Catal. B-Environ*. 2007, vol. 75, 95.
[37] Carlesi Jara, C.; Fino, D.; Specchia, V.; Saracco, G.; Spinelli, P. *Appl. Catal. B-Environ*. 2007, vol. 70, 479.
[38] He, J.; Cai, Q. Z.; Ji, Y. G.; Luo, H. H.; Li, D. J.; Yu, B. *J. Alloys Compd.* 2009, vol. 482, 476.
[39] Wu, K. R.; Hung, C. H.; Tsai, M. H. *Appl. Catal. B-Environ.* 2009, vol. 92, 357.
[40] Wu, K. R.; Hung, C. H. *Appl. Surf. Sci*. 2009, vol. 256, 1595.
[41] Poznyak, S. K.; Talapin, D. V.; Kulak, A. I. *Thin Solid Films*. 2002, vol. 405, 35

In: Physics and Mechanics of New Materials ...
Editors: Ivan A. Parinov and Shun-Hsyung Chang
ISBN: 978-1-62618-535-7

Chapter 13

NOVEL OPTICAL MEANS FOR MEASUREMENT OF DISPLACEMENTS OF SURFACES OF THE CONTROL OBJECTS AT DIAGNOSTICS OF MATERIALS AND GOODS

I. P. Miroshnichenko*[1], I. A. Parinov[2], E. V. Rozhkov[2], V. P. Sizov[2] and V. A. Shevtsov[2]
[1]Don State Technical University, Rostov-on-Don, Russia
[2]Southern Federal University, Rostov-on-Don, Russia

ABSTRACT

The paper focuses on original technique and technological means intended to improve measurements and optical interference methods based on novel solutions and directed to solution of practical diagnostic problems of the technical state of materials and goods. The work presents the developed methods patented in Russia and awarded by medals of many international exhibitions.

1. INTRODUCTION

Now, one of the most perspective directions to solve the practical problems of diagnostics of the construction state and experimental investigations of damage processes in novel materials is connected with development and use of high precise meters for measurement of displacements of the control object surfaces. These meters are based on application of laser interferometers.

This technique allows one considerably improve precision and informativeness at analysis of elastic wave fields by ultra-sound defectoscopy method, diagnostics of state of different goods by acoustic-emission method, etc.

---

* Email: ipmir@rambler.ru

The main directions of improving laser interferometers, used for solution of above problems, are the following ones: (i) development of novel optical schemes for interferometers, (ii) creation of advanced technologies for information registration and processing based on the interference pictures, (iii) improvement of sources and receivers of the optical radiation, and (iv) extension of their possibilities allowing one to carry out continuous account of different influence factors with simultaneous fulfillment of corrections and introduction of the corrections in results of the measurements in real time scale.

The purpose of the paper focuses on presentation of original technique and technological means directed to improvement of measurements by using novel optical interference methods and oriented to solution of practical diagnostic problems of the technical state of materials and goods. The work presents the technical solutions patented in Russia and awarded by medals of many international exhibitions.

2. Novel Technical Solutions

It is well known [1, 2] an optical interference meter of displacements, based on two-way laser interferometer with combined branches providing a possibility of direct application for solution of above problems. The meter consists of optically coupled and consecutively situated: (i) source of coherent optical radiation, (ii) optical system, (iii) beam-splitter, (iv) reflector fastened to surface of a control object , and (v) screen on which an interference picture (presenting itself a set of rings with various intensities) is projected. The photo-detectors are also stated on the meter. The beam-splitter and reflector dispose one from other under angle α, and the photo-detectors situate into rings of interference picture.

The action principle of the displacement meter consists in the following. Radiation of the source of coherent optical radiation after passing optical scheme transforms in diverging light beam and catches on the beam-splitter, where its splitting is carried out, namely first part reflects from surface of the beam-splitter (reference beam), and second part reflects from the reflector (object beam), fastened to the surface of the control object. The interference picture, obtained in combination of the reference and object beams, presenting itself the set of rings 9 with various intensities, projects on the screen with the photo-detectors stated on it. At the displacement of the surface of the control object, the photo-detectors measure an intensity of optical field of the interference picture in place of their disposition, directly coupled with the pointed displacement.

The intensity change of the interference picture rings from minimal to maximal value occurs at displacement of the control object surface on the value of $\lambda/4$, where λ is the wave length of optical radiation (for example, for optical radiation of helium-neon gas laser of red color $\lambda = 0.63$ μm). This meter provides without-contact measurement of small displacements of control object surface and has high precise characteristics.

The authors have carried out numerical and experimental researches of functional characteristics of the displacement meter, that allowed one to formulate a set of actual directions for its improvement [3 – 6]. The papers [7 – 16] present numerical and experimental fundamentals of novel methods for displacement measurements of control object surfaces. They describe technical and technological solutions leading to increasing quality of the measurement results and to extension of functional possibilities of the meter considered, and also present theoretical and experimental confirmations for the results obtained by using test specimens.

The significant shortcomings of the device, presented in [1, 2], are connected with high labor-intensiveness of technical preparation of the measurements and low accuracy of the obtained experimental results. The high labor-intensiveness of technical preparation of the measurements is caused by that before conducting any kind of tests, and respectively of measurements, it is necessary to carry out individual statement and tuning each of optical elements in order to rich a combination of reference and object light beams and receiving interference picture in area of screen disposition. It increases significantly the labor-intensiveness of technical preparation of the tests. At the same time, the low accuracy of measurement results is caused by that the external sources of optical radiation (natural lightening, illuminating equipment, etc.) influence on optical elements of the device [1, 2] and photo-detector during conducting the measurements. In this case, intensiveness of these radiations and their changes during time demonstrate accidental character leading to accidental errors in the measurement results decreasing their accuracy.

Thus, it is obvious the great actuality of solution of the pointed problems, directed to decreasing labor-intensiveness of technical preparation of the measurements and increasing accuracy of the obtained experimental results.

With this aim, it has been developed an optical interference device measuring displacements which allows one to solve the stated problems. Essence of the proposed device is explained by the scheme presented in Figure 1. The device consists of base 1, bedplate 2 with central hole 3, fastened to the base 1 with help of device 4 for adjustment and fixation of the bedplate disposition, cylindrical hull 5, in cavity 6 of which source 7 of coherent optical radiation and optical system 8 are disposed, photo-detector 9 in the form of rectangular matrix of photo-detectors with central hole 10, which is to be co-axis with the central hole 3 of the plate 2, conical hull 11 with flanges 12 and 13 rigidly fastened to great and small bases of the cone, respectively, beam-splitter 14 is rigidly fixed to the flange 13, and reflector 15 rigidly fastened to surface 16 of control object 17.

The plate 2 has external 18 and internal 19 surfaces, by this the last one is directed to the surface 16 of the control object 17. The photo-detector 9 is rigidly fastened to the internal surface 19 of the plate 2. The cylindrical hull 5 disposes on external surface 18 of the plate 2 co-axially to the hole 3 of the plate 2 and the hole 10 of the device 9 by using fastening elements 20, ensuring adjustment and fixation of the device location. The conical hull 11 is stated on the internal surface 19 co-axially to the cylindrical hull 5, the hole 3 of the plate 2 and the hole 10 of the photo-detector 9 by using fastening elements 21, ensuring adjustment and fixation of the detector location. At the flange 13 of the conical hull 11, it is rigidly fastened by one edge an elastic light-proof membrane 22 in the form of ring and resting by contrary edge on the surface 16 of the control object 17.

The device works in the following scheme. Radiation of the source 7 of coherent optical radiation after passing the optical system 8 transforms in diverging light beam, passes through hole 3 of the plate 2 and the hole 10 of the photo-detector 9 and catches on the beam-splitter 14, where its splitting is carried out, namely first part reflects from surface of the beam-splitter 14 (reference beam), and second part reflects from the reflector 15 (object beam), fastened rigidly to the surface 16 of the control object 17. The interference picture, obtained in combination of the reference and object beams, presenting itself a set of rings with various intensities projects on the photo-detector device 9.

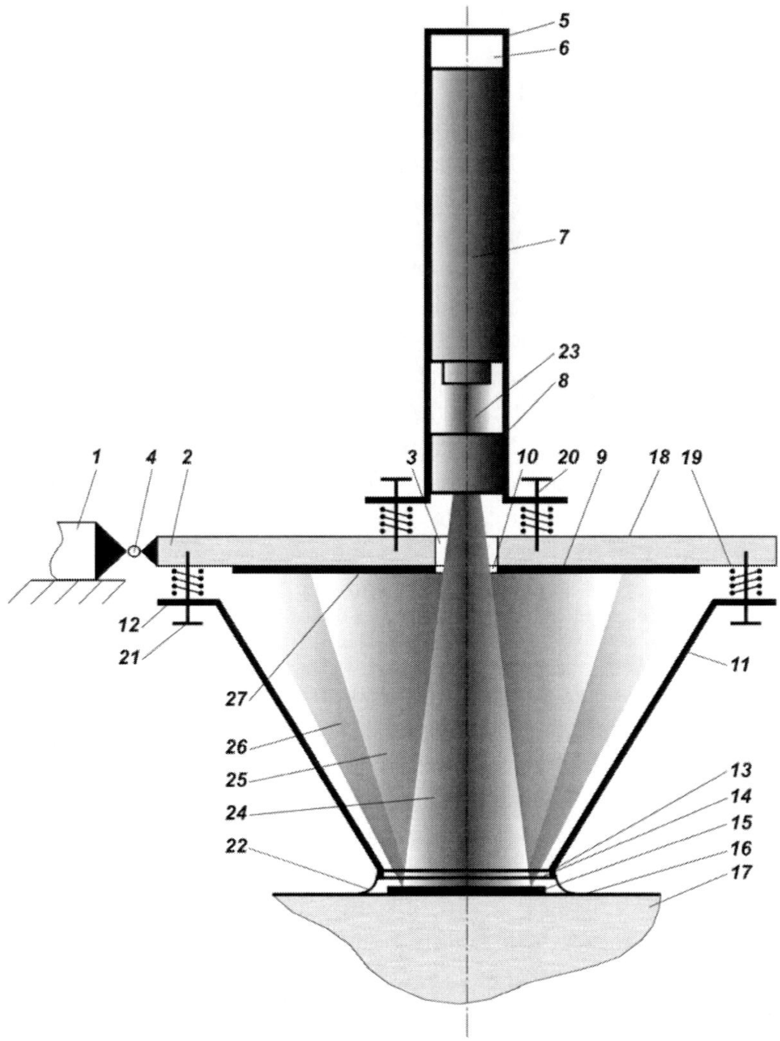

Figure 1. Scheme of the proposed optical interference device for measurement of displacements (the designations are explained into text).

At the displacement of the surface 16 of the control object 17, the photo-detector device 9 registers an intensity change of optical field of the interference picture, which corresponds to value of the displacement. The measurement results transfer from the photo-detector 9 to the device for registration and reflection of the measurement results.

In order to confirm experimentally the functional properties of the proposed technical solution, it has been manufactured its test sample, presented in Figures 2 – 5, by this the position numbers of the device elements correspond to the ones depicted in Figure 1; protective shutter 28 of the conic hull 11 is shown in open position, but the elastic light-proof membrane 22 is taken off. Figure 2 shows common picture of the proposed device, Figure 3 presents the common picture of the device with possibility to exam the photo-detector device 9, but Figures 4 and 5 depict a look from above and from below.

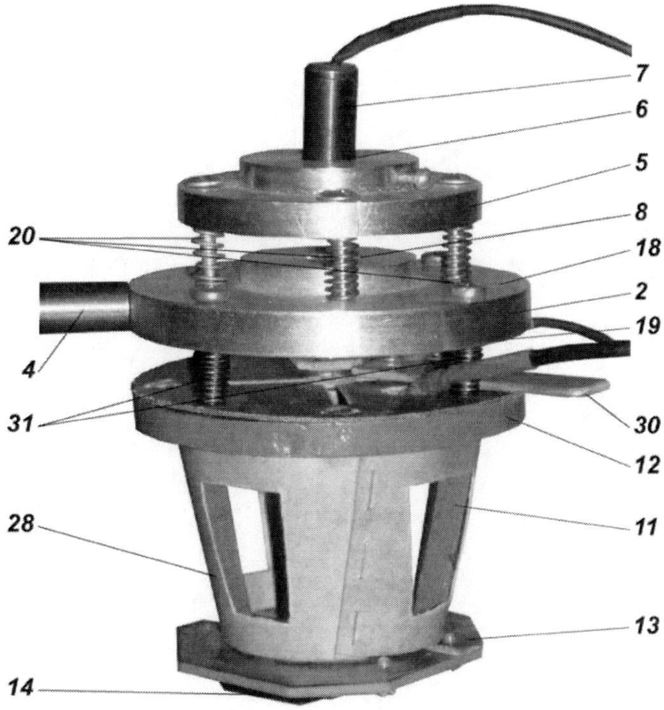

Figure 2. Common picture of the proposed device.

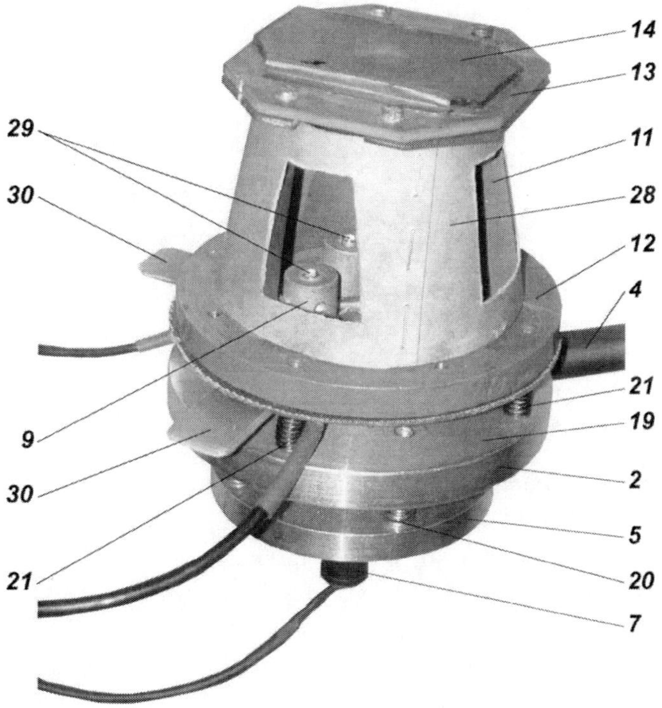

Figure 3. Common picture of the device with possibilities to exam the photo-detector device 9.

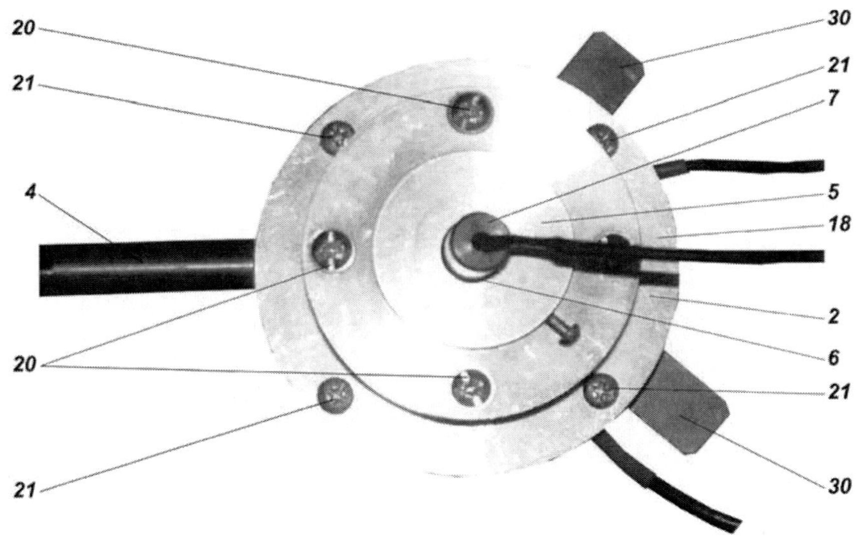

Figure 4. A look from above on the device.

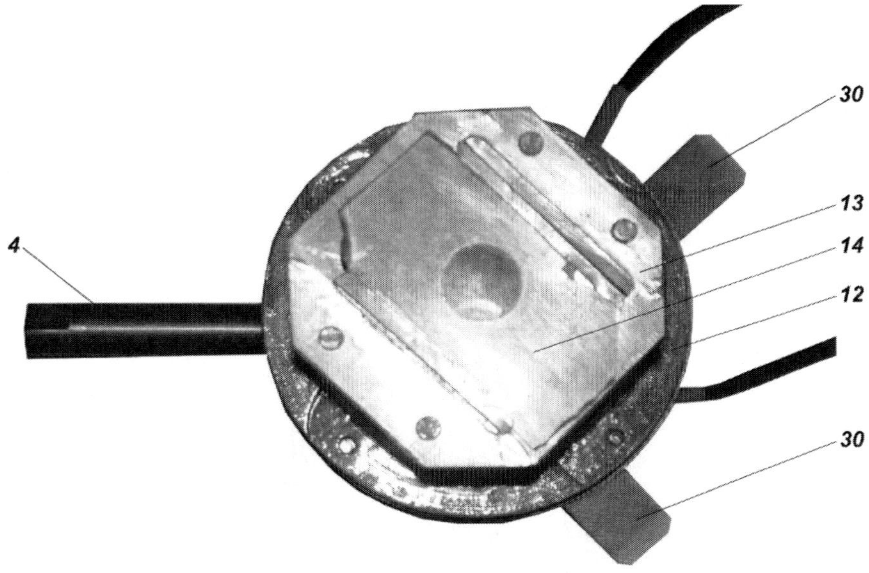

Figure 5. A look from below on the device.

In the test sample, the source 7 of coherent optic radiation presents itself the laser point module of the type KLM-B635-3-5 (semi-conductive laser) with the wave length $\lambda = 0.635$ μm and power of the laser radiation of 3 mW, the beam-splitter 14 presents itself a semi-transparent mirror, and as the photo-detector device 9 here were used two photo-diodes 29 of the type FD9-111E, each of which additionally was equipped by device 30 for regulation and fixation of a position into area of interference picture 27.

The analysis of the research results supported correctness of the technical solutions used in the device construction for attainment of required technical-economic effect and allowed one to point that the proposed device, keeping all positive qualities of the known meters, decreases labor-intensiveness of technical preparation of the measurements and improves accuracy of the measurement results.

Significant shortcoming of the device [1, 2] is also connected with low accuracy of the measurement results due to during measurement process arise external non-stationary and, as rule, accidental perturbations (vibrations of technological equipment, influences of different nature, etc.), which transfer on the controlled surface and lead to its additional displacements causing introduction of accidental errors in the measurement results that decreases their accuracy. Therefore, the solution of the problem of registration and compensation of the pointed perturbations during control is to be very actual especially for high-exact optical measurement means working in structure of moving diagnostic systems.

It has been developed perspective device for registration and compensation of external influencing loading. It has been realized into construction of known optic interference meter of small displacements.

In order to confirm principal possibility to realize the proposed device and fulfill its experimental verification, it has been developed and manufactured experimental device. The functional scheme of the device is presented in Figure 6.

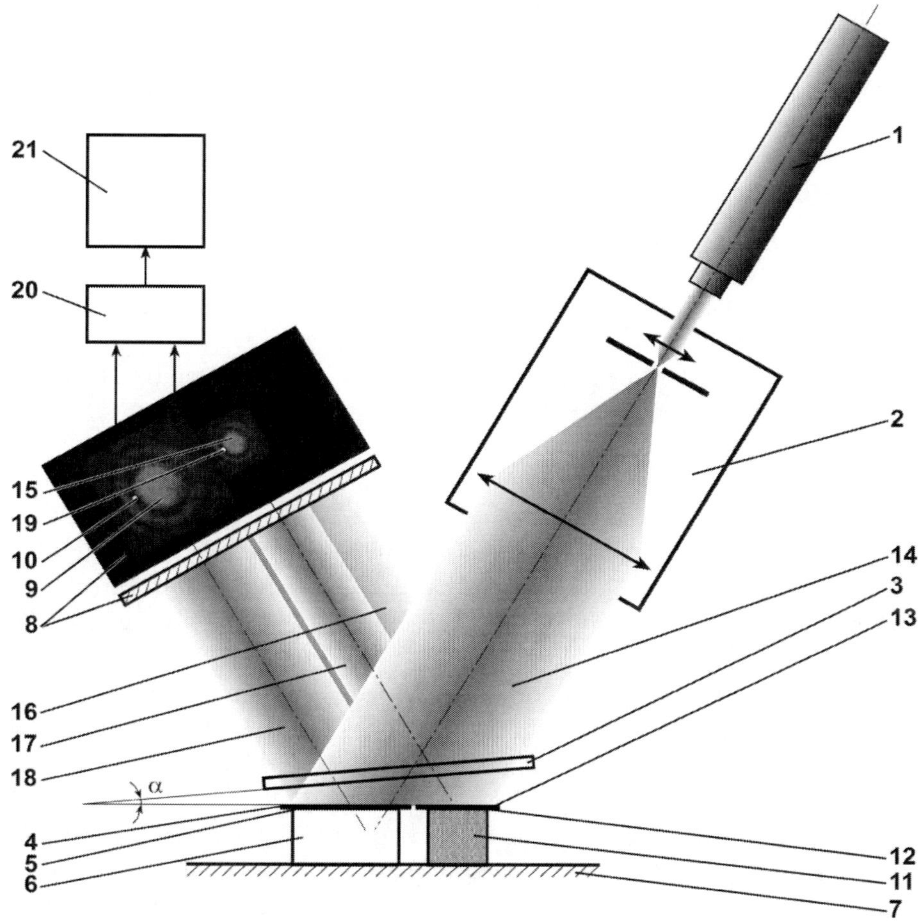

Figure 6. Scheme of device for registration and compensation of effects of the external destabilizing influences (the designations are explained into text).

The device consists of optically coupled and consecutively situated: (i) source 1 of coherent optical radiation (helium-neon laser of continuous type LNG-207B with wave length $\lambda = 0.63$ μm), (ii) optical system 2, (iii) beam-splitter 3, (iv) reflector 4, fastened to surface 5 of control object 6, rigidly fastened to supporting base 7, and (v) screen 8, disposed under angle α to the beam-splitter 3. Photo-detector device 10 is stated in the plane of the screen 8 in the given region of interference picture 9. Moreover, the device includes imitator 11 of control object with additional reflector 13, stated on its surface 12. The imitator 11 is rigidly fastened to the supporting base 7 in the area, irradiated by beam 14 of optical radiation, behind the optical system 2 and is provided by devices for adjustment of position of the additional reflector 13. The additional reflector 13 is disposed in one plane with the reflector 4, fastened to the surface 5 of the control object 6. In the screen 8, it is additionally projected interference picture 15, obtained at combination of reference beam 16 from the beam-splitter 3 and object beam 17 from the additional reflector 13. In the screen plane 8, in its region is similar to the area of the interference picture 9, obtained in combination of the reference beam 16 from the beam-splitter 3 and object beam 18 from the reflector 4, fastened to the surface 5 of the control object 6, it is stated additional photo-detector device 19. So, the photo-detector device 10 and additional photo-detector device 19 are stated, respectively in rings of first order of the interference pictures 9 and 15, as it is shown in Figure 6. Outputs of the photo-detector devices 10 and 19 are electrically coupled with inputs of differential amplifier 20. Output of the amplifier 20 is coupled with device 21 of registration, treatment and image of measurement results.

In the process of tests, the next variants have been modeled:

i. it has been initiated a harmonic influence, applied to the supporting base 7 at fixed (constant) displacement of the surface 5 of the control object 6;
ii. it has been initiated a harmonic influence, applied to the supporting base 7 at impulse displacement of the surface 5 of the control object 6;
iii. it has been initiated an impulse influence, applied to the supporting base 7 at fixed (constant) displacement of the surface 5 of the control object 6;
iv. it has been initiated an impulse influence, applied to the supporting base 7 at harmonic displacement of the surface 5 of the control object 6.

During testing, there were registered signals at outputs of the photo-detectors 10 and19, and also the differential amplifier 20. In these cases, the photo-detector devices 10 and 19 were stated in the screen plane 8 in the rings of first order of the interference pictures 9 and 15, respectively.

The test results are presented in the form of oscillograms in Figure 7, a-d, respectively for (i) – (iv) variants of the modeling. In Figure 8, curve 1 presents output signal of the device 10, curve 2 presents output signal of the device 19, and curve 3 presents output signal of the amplifier 20.

Analysis of the obtained results allows one to make conclusion on possibility of realization of the proposed device and solution of the stated problem on increasing accuracy of the results at measurements of small displacements of the control objects. The method realized in the proposed device could be extended to interference measurements, in whole. Moreover, its application allows one to increase accuracy of the test results up to 40%.

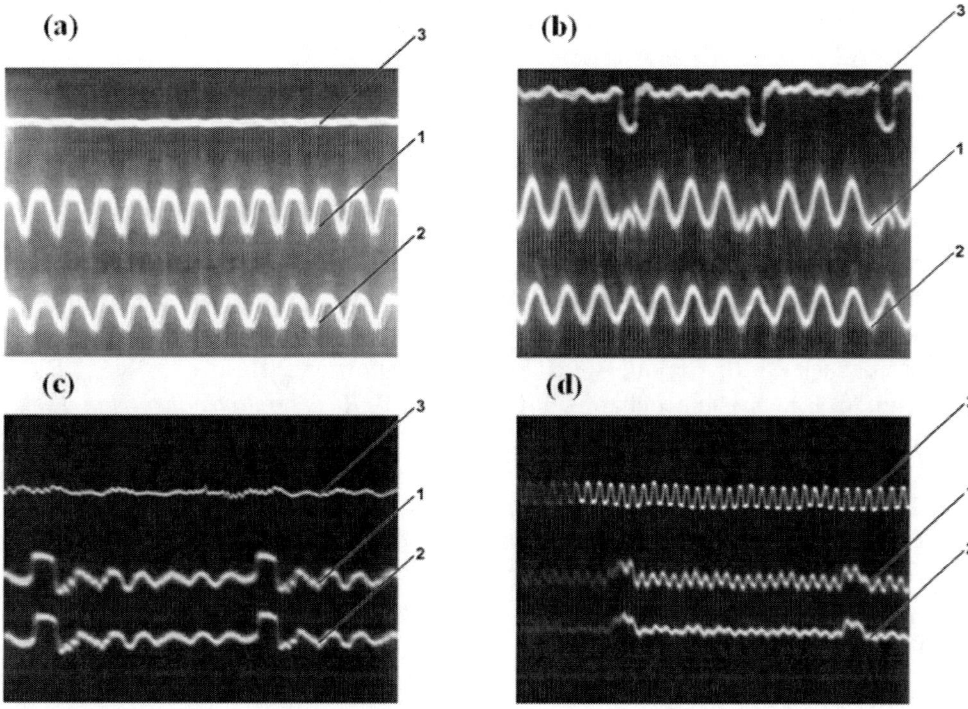

Figure 7. Test results, presented in the form of oscillograms (designations are presented into text).

The proposed measurement means have been successively applied in studies of the damage-initiation processes in high-temperature superconducting tapes and at development of technology of their qualitative control during manufacture [17, 18].

The proposed technical solutions have been patented in Russia [19, 20] and awarded by gold medals and special prizes on several Moscow international exhibitions, in particular "New Time" (2010) and "Archimedes" (2009-2011).

ACKNOWLEDGMENTS

This research was supported in part by the Russian Foundation for Basic Research (Grant No. 10-08-00136).

REFERENCES

[1] Matveev, A. N. *Optics*; Vysshaya Shkola: Moscow, 1985; pp. 1 – 351.
[2] Batrakov, A. S.; Butusov, M. M; Grechka, G. P. *Laser Measuring Systems*; Radio i Svyaz: Moscow, 1981; pp. 1 – 456.
[3] Miroshnichenko, I. P.; Serkin, A. G. *Measurement Techniques*. 2006. *vol. 49(5)*, 22 – 25.

[4] Miroshnichenko, I. P.; Serkin, A. G. *Russian Journal of Nondestructive Testing*. 2007. *vol. 43(4)*, 31 –38.

[5] Miroshnichenko, I. P.; Serkin, A. G.; Sizov, V. P. *Izv. VUZov. Priborostroenie*. 2007. *12*, 34 – 40.

[6] Miroshnichenko, I. P.; Serkin, A. G.; Sizov, V. P. *Journal of Optical Technology*. 2008, *vol. 75(7)*, 41 – 47.

[7] Miroshnichenko, I. P.; Serkin, A. G.; Sizov, V. P. *Measurement Techniques*. 2007. *vol. 50(1)*, 9 – 13.

[8] Miroshnichenko, I. P.; Serkin, A. G. *Russian Journal of Nondestructive Testing*. 2008. *vol. 44(5)*, 23 – 27.

[9] Miroshnichenko, I. P.; Serkin, A. G. *Control. Diagnostics*. 2011. *No. 7,* 46 – 50.

[10] Alekhin, V. E.; Miroshnichenko, I. P.; Sizov, V. P. *Russian Journal of Nondestructive Testing*. 2007. *vol. 43(2)*, 53 – 61.

[11] Alekhin, V. E.; Miroshnichenko, I. P.; Nesterov, V. A.; Sizov, V. P. *Russian Journal of Nondestructive Testing*. 2007. *vol. 43(9)*, 37 – 43.

[12] Alekhin, V. E.; Miroshnichenko, I. P.; Serkin, A. G. *Measurement Techniques*. 2008, *vol. 51(10)*, 26 – 30.

[13] Miroshnichenko, I. P. *Control. Diagnostics.* 2010. *No. 2*, 34 – 37.

[14] Miroshnichenko, I. P.; Serkin, A. G. *Datchiki i Sistemy.* 2008. *No. 3*, 28-30.

[15] Miroshnichenko, I. P.; Serkin, A. G.; Sizov, V. P. *Nanotekhnika*. 2008. *No. 4(16)*, 56 – 59.

[16] Miroshnichenko, I. P. *Control. Diagnostics.* 2010. *No. 1*, 45 – 49.

[17] Miroshnichenko, I. P.; Parinov, I. A.; Rozhkov, E. V.; Serkin, A. G. *Metallurgist*. 2006. *vol. 50(7 – 8)*, 408 – 410.

[18] Miroshnichenko, I. P.; Serkin, A. G. *Metallurgist*. 2010. *vol. 54(3 – 4),* 189 – 191.

[19] Miroshnichenko, I. P.; Parinov, I. A.; Rozhkov, E. V.; Serkin, A. G. *Optical Device for Measurement of Displacements*. Russian Patent No. 2407988, 12. 27.2010.

[20] Miroshnichenko, I. P.; Nesterov, V. A.; Serkin, A. G.; Sizov, V. P. Shevtsov, V. A. *Interference Meter of Small Displacements*. Russian Patent No. 2410642, 01. 27.2011.

III. Mechanics of New Materials

In: Physics and Mechanics of New Materials … ISBN: 978-1-62618-535-7
Editors: Ivan A. Parinov and Shun-Hsyung Chang

Chapter 14

INFLUENCE OF DISLOCATIONS AND DISCLINATIONS ON NONLINEAR DEFORMATION OF THIN-FILM MATERIALS

S. Derezin* and L. Zubov
Southern Federal University, Rostov-on-Don, Russia

ABSTRACT

Thin-film materials and carbon nanostructures have unique physical properties that are directly related to their exotic geometry. It was experimentally found that graphene, for example, is always corrugated and covered by ripples that may have intrinsic character (caused by topological defects like dislocations and disclinations) or be induced by a roughness of substrate. We propose a nonlinear elastic shell theory of Kirchhoff-Love type which takes into account both isolated and continuously distributed dislocations and disclinations. In contrast to the usual shell theory the strain compatibility conditions are replaced by the so-called incompatibility equations of the first and second order containing densities of defects. We present some exact solutions to a number of practically important problems, e.g. inflation of a fullerene containing screw dislocations, torsion, tension, inflation, and bending of a cylindrical nanotube with distributed dislocations of various types, buckling of a nonlinearly elastic flat membrane due to the given densities of dislocations and disclinations.

1. INTRODUCTION

Carbon nanostructures have attracted recently great attention. Carbon is known for its ability to assume extended two-dimensional sheet surfaces. The in-sheet bonding is extremely strong, the Young's modulus of graphene is one of the highest values known for any material. Therefore these sheets are rather stable both as isolated objects, and also when curved into cylindrical geometries (nanotubes) or quasispherical geometries (fullerenes).

* E-mail: svd@hrz.tu-chemnitz.de.

In [1–3] several shell theories were proposed for modeling of carbon nanotubes. It was found experimentally [4] that graphene is always corrugated and covered by ripples that may have intrinsic character or be induced by a roughness of substrate. Understanding the influence of defects on various slender bodies like thin films is also critical for the improvement of the reliability of many micro-electro-mechanical systems [5].

A natural way that leads to emergence of local curvature in the hexagonal graphene lattice is the substitution of some of the hexagons by pentagons (positive disclinations) or heptagons (negative disclinations).

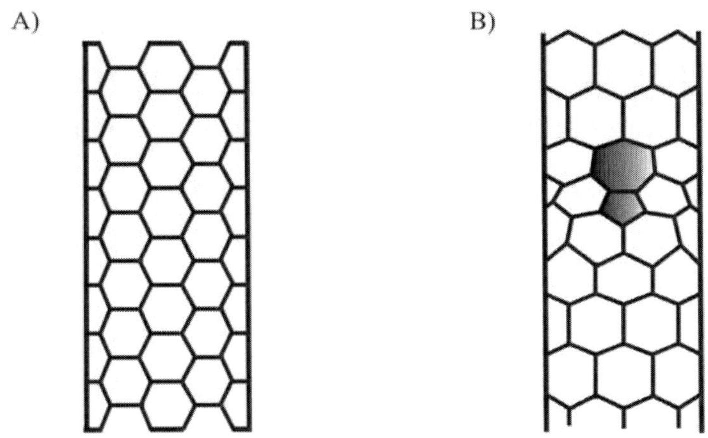

Figure 1. Different chiral nanotubes: A) An ideal armchair nanotube B) A zigzag nanotube with two opposite disclinations.

The curvature of the graphene surface in the presence of a disclination in the continuum limit has a delta-function singularity at the position of the defective ring. An edge dislocation (Figure 1) can be formed by joining a pentagon and a heptagon through a line which will represent the Burgers vector of the dislocation.

Although a dislocation phenomenon appeared from the discrete structure of the crystal lattice, a continuum theory proved to be very effective in the modeling of imperfections in thin films [6, 7]. It allows one, for example, to smooth out singularities from isolated defects that make the elastic strain energy density non-integrable. Between the film and the substrate, stresses usually arise from incompatibilities, or misfit, caused by different properties of materials.

2. Dislocations in Shells

2.1. Density of Dislocations in Shells and Incompatibility

Let σ be the elastic shell surface considered as a two-dimensional material continuum in the reference configuration. We denote the position vector of a point on σ as $\mathbf{r}$ and the Gaussian coordinates as q^{α} $(\alpha = 1, 2)$. We introduce the vectors $\mathbf{r}_{\alpha}$ of the principal basis and the vectors $\mathbf{r}^{\beta}$ of the reciprocal basis, as well as the gradient operator on σ:

$$\mathbf{r}_\alpha = \frac{\partial\, \mathbf{r}}{\partial\, q^\alpha}, \ \mathbf{r}^\beta \cdot \mathbf{r}_\alpha = \delta_\alpha^\beta, \ \mathbf{r}^\beta \cdot \mathbf{n} = 0, \ \mathrm{grad} = \mathbf{r}^\beta \frac{\partial}{\partial\, q^\beta}. \tag{1}$$

Here **n** is the unit normal vector to the surface σ. We designate the position vector of the deformed surface Σ referred to the same curvilinear coordinates q^α as **R** and the vectors of the principal basis, the reciprocal basis, and the unit normal to Σ as $\mathbf{R}_\alpha$, $\mathbf{R}^\beta$, and **N**, respectively. Assuming that the surface distortion tensor [8] $\mathbf{F} = \mathrm{grad}\,\mathbf{R} = \mathbf{r}^\alpha \otimes \mathbf{R}_\alpha$ is a single-valued and continuously differentiable function of coordinates q^α, we consider the problem of determining the field of shell displacements $\mathbf{u} = \mathbf{R} - \mathbf{r}$ via the field of the tensor **F** prescribed in the multiply connected region. The vector field $\mathbf{u}(\mathbf{r})$ is determined in this case, generally speaking, not uniquely, which indicates the presence of translational dislocations in the shell, each of which is characterized by the Burgers vector. Assuming that the number of dislocations on a limited part of the shell is high and following the method of [9], we make a transition from a discrete set of dislocations to their continuous distribution. Let us determine the dislocation density $\boldsymbol{\alpha}$ as a vector field, which integral over an arbitrary region on σ is equal to the total Burgers vector of all dislocations contained in this region. This definition results in the following equation with respect to the distortion:

$$\mathrm{div}\,(\mathbf{e} \cdot \mathbf{F}) = \boldsymbol{\alpha}, \ \ \mathbf{e} = -\,\mathbf{E} \times \mathbf{n}. \tag{2}$$

Here **E** is the three-dimensional identity tensor, div is the divergence operator on σ, and **e** is the permutation tensor on the surface [8]. When $\boldsymbol{\alpha} = 0$, equation (2) is known to be the compatibility equation of the first order [10] for the distortion tensor. If $\boldsymbol{\alpha} \neq 0$, there is no displacement field **u**; for this reason, (2) is called then the incompatibility equation.

We mention also here the alternative way of using the modified Gauss-Codazzi equations deduced in [10], which are the compatibility (incompatibility) conditions of the second order

$$\mathrm{div}(\overset{\frown}{\mathbf{K}} - \mathbf{e} \cdot \mathbf{h} \otimes \mathbf{n}) + \mathbf{h} \cdot \overset{\frown}{\mathbf{K}} \cdot \mathbf{e} = (\det \overset{\frown}{\mathbf{K}})\mathbf{n}, \tag{3}$$

$$\overset{\frown}{\mathbf{K}} = \mathbf{e} \cdot (\mathbf{P} \cdot \mathbf{U}^{-1} - \mathbf{b}) \cdot \mathbf{e}, \ \mathbf{h} = (\det \mathbf{U})^{-1} [\mathbf{U} \cdot \mathrm{div}(\mathbf{e} \cdot \mathbf{U}) - \boldsymbol{\alpha}_0],$$

$$\mathbf{U}^2 = \mathbf{F} \cdot \mathbf{F}^T, \ \mathbf{P} = -\mathrm{grad}\,\mathbf{N} \cdot \mathbf{F}^T, \ \mathbf{b} = -\mathrm{grad}\,\mathbf{n}, \ \boldsymbol{\alpha}_0 = \boldsymbol{\alpha} \cdot \mathbf{F}^T.$$

These equations are certainly very complicated, but allow to consider more general defects, for example, single disclinations.

2.1. Nonlinear Statics of Shells with Distributed Dislocations

Within the framework of the Kirchhoff-Love shell model, the equilibrium equations in terms of force stresses and moment stresses can be presented as [11]

$$\operatorname{div}[\mathbf{D} + (\operatorname{div}\mathbf{H}) \cdot (\operatorname{Grad}\mathbf{r})\mathbf{N}] + \mathbf{f} = 0\,, \ \operatorname{Grad}\mathbf{r} = \mathbf{R}^{\beta} \otimes \frac{\partial\, \mathbf{r}}{\partial\, q^{\beta}}\,, \tag{4}$$

where **D** is the force stress tensor, **H** is the moment stress tensor. Both tensors **D** and **H** are of Piola-Kirchhoff type and may be expressed through the distortion tensor **F** by means of the constitutive equations :

$$\mathbf{D} = \frac{\partial\, W}{\partial\, \mathbf{F}}\,, \ \mathbf{H} = \frac{\partial\, W}{\partial\, \mathbf{K}}\,, \ W = W(\mathbf{F} \cdot \mathbf{F}^T, \mathbf{K} \cdot \mathbf{F}^T) \tag{5}$$

and the geometrical relations:

$$\mathbf{K} = \operatorname{grad}\mathbf{N} = -(\operatorname{Grad}\mathbf{r}) \cdot (\operatorname{grad}\mathbf{F}) \cdot \mathbf{N}\,, \tag{6}$$

$$\operatorname{Grad}\mathbf{r} = J^{-2}\mathbf{F}^T \cdot [\mathbf{g}\ \operatorname{tr}(\mathbf{F} \cdot \mathbf{F}^T) - \mathbf{F} \cdot \mathbf{F}^T]\,, \ \mathbf{g} = \mathbf{r}^{\alpha} \otimes \mathbf{r}_{\alpha} = \mathbf{E} - \mathbf{n} \otimes \mathbf{n}\,, \tag{7}$$

$$J\mathbf{N} \otimes \mathbf{n} = \mathbf{F}^2 - \mathbf{F}\ \operatorname{tr}\mathbf{F} + \frac{1}{2}\mathbf{E}(\operatorname{tr}^2\mathbf{F} - \operatorname{tr}\mathbf{F}^2)\,, \ J^2 = \frac{1}{2}[\operatorname{tr}^2(\mathbf{F} \cdot \mathbf{F}^T) - \operatorname{tr}(\mathbf{F} \cdot \mathbf{F}^T)^2]\,. \tag{8}$$

In (4)–(8) **f** is the vector of reduced external load acting on the unit area of the surface σ, W is the specific (per unit area of the surface σ) shell deformation energy, Grad **r** is the inverse distortion tensor, **g** is the first fundamental tensor of the surface σ, and $(J - 1)$ is the local relative variation of the surface area under deformation. Making use of (5)–(8), all values in the square brackets in (4) are expressed through the distortion tensor **F**. Adding incompatibility equation (2) to (4) we obtain a closed system of equations with respect to the distortion tensor which can be solved.

In what follows we will use the proper orthogonal tensor field **A** of rotations of elementary surface areas expressed through the distortion field by means of the formula [8]:

$$\mathbf{A} = (\mathbf{F} \cdot \mathbf{F}^T)^{-1/2} \cdot \mathbf{F} + \mathbf{n} \otimes \mathbf{N}\,. \tag{9}$$

If the shell thickness is quite small, it is possible to neglect its flexural rigidity and to use the nonlinear membrane (momentless) theory of shells. In the case of the membrane theory, the specific energy W is independent of the tensor **K**, and the moment tensor **H** is zero, while equilibrium equation (4) becomes

$$\operatorname{div}\mathbf{D} + \mathbf{f} = 0, \ \mathbf{D} = \frac{d\,W(\mathbf{F} \cdot \mathbf{F}^T)}{d\,\mathbf{F}}\,. \tag{10}$$

Additionally, if the shell material is isotropic, the specific energy W is a function of the deformation invariants J and $\operatorname{tr}(\mathbf{F} \cdot \mathbf{F}^T)$. In particular, the constitutive equations of the membrane made of a neo-Hookean [12] material can be written as [13]

$$W = \frac{1}{2}\mu h[\mathrm{tr}(\mathbf{F}\cdot\mathbf{F}^T) + J^{-2} - 3], \tag{11}$$

$$\mathbf{D} = \mu h J^{-4}\{[J^4 - \mathrm{tr}(\mathbf{F}\cdot\mathbf{F}^T)]\mathbf{F} + \mathbf{F}\cdot\mathbf{F}^T\cdot\mathbf{F}\}, \tag{12}$$

where h is the shell thickness, and μ is the shear modulus.

Let us assume that the shell boundary $\partial\sigma$ consists of two connected parts: $\partial\sigma = \gamma' \cup \gamma''$. The most typical boundary conditions of the membrane theory are the distortion conditions on γ' and the force conditions on γ'', respectively:

$$\mathbf{t}\cdot\mathbf{F} = \mathbf{V}(s),\ \mathbf{m}\cdot\mathbf{D} = \boldsymbol{\tau}(s). \tag{13}$$

Here $\mathbf{m}$ is the unit vector of the normal to the contour $\partial\sigma$ ($\mathbf{m}\cdot\mathbf{n} = 0$), $\mathbf{t}$ is the unit vector tangent to $\partial\sigma$, s is the current arch length on $\partial\sigma$, and $V(s)$ and $\boldsymbol{\tau}(s)$ are given vector functions.

2.2. Inflation of a Fullerene with Screw Dislocations

Let a closed spherical shell of radius r_0 be loaded from inside with a uniform hydrostatic pressure of intensity p. The dislocation density vector is taken to be $\boldsymbol{\alpha} = a_0\mathbf{n}$, $a_0 = \mathrm{const}$, which corresponds to the uniform distribution of screw dislocations. Taking into account the symmetry of the problem, the solution of (2) and (4) will be sought in the following form, where F_0 and H_0 are assumed to be constant

$$\mathbf{F} = F_0\mathbf{g} + H_0\mathbf{e}. \tag{14}$$

In virtue of incompatibility equation (2), the value of H_0 is expressed via the screw dislocation density as $2H_0 = a_0 r_0$, and the value of F_0 can be written as $F_0 = \sqrt{J - H_0^2}$ from (8). Using (5)–(7), it is possible to verify that, in this case, the tensors $\mathbf{K}$ and Grad $\mathbf{r}$ and also the force tensor $\mathbf{D}$ and the moment tensor $\mathbf{H}$ have representations similar to (14) for an isotropic shell. The equilibrium equation (4) is reduced now to a single scalar algebraic equation involving the pressure p, the relative variation of the area ($J - 1$), and the dislocation density a_0. In particular, for the spherical membrane of the neo-Hookean material, this relationship has the form:

$$p = 2\mu h r_0^{-1} J^{-4}(J^3 - 1)\sqrt{J - \frac{1}{4}r_0^2 a_0^2}\,. \tag{15}$$

The last formula shows that the presence of the distributed dislocations reduces the resistance of the closed elastic sphere to the action of the applied load.

2.3. Torsion, Tension, and Inflation of a Nanotube with Screw Dislocations

Let x_1, x_2, and x_3 be Cartesian coordinates in space, and let $\mathbf{i}_1$, $\mathbf{i}_2$, and $\mathbf{i}_3$ be the corresponding basis vectors. We define the equation of the cylindrical shell surface as $x_1 = r_0 \cos\varphi$, $x_2 = r_0 \sin\varphi$, $x_3 = z$ and assume $q^1 = \varphi$ and $q^2 = z$ as the Gaussian coordinates (Figure 2).

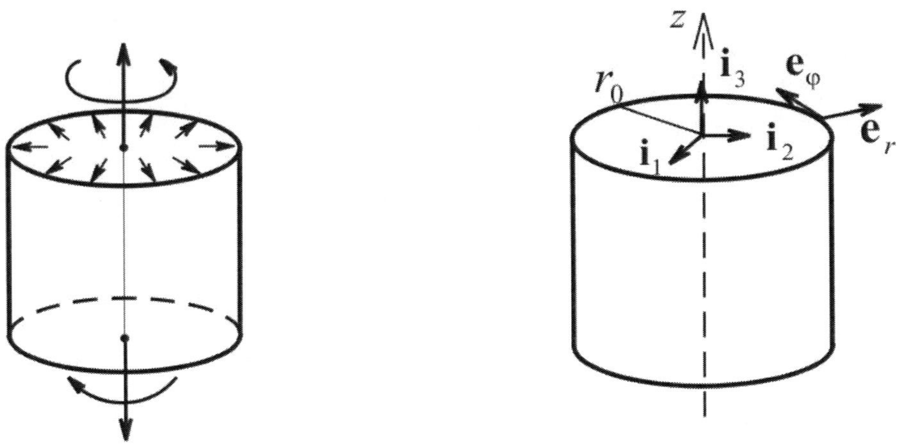

Figure 2. Deformation of a cylindrical nanotube.

We introduce the unit vectors (ψ = const):

$$\mathbf{e}_r = \mathbf{i}_1 \cos\varphi + \mathbf{i}_2 \sin\varphi,\ \mathbf{e}_\varphi = -\mathbf{i}_1 \sin\varphi + \mathbf{i}_2 \cos\varphi, \tag{16}$$
$$\mathbf{e}_R = \mathbf{e}_r \cos(\psi z) + \mathbf{e}_\varphi \sin(\psi z),\ \mathbf{e}_\Phi = -\mathbf{e}_r \sin(\psi z) + \mathbf{e}_\varphi \cos(\psi z)$$

and set the distortion field describing the torsion of the cylindrical tube with the twist angle ψ, which is accompanied by the inflation and axial tension–compression in such a form:

$$\mathbf{F} = F_{11}\mathbf{e}_\varphi \otimes \mathbf{e}_\Phi + F_{21}\mathbf{i}_3 \otimes \mathbf{e}_\Phi + F_{22}\mathbf{i}_3 \otimes \mathbf{i}_3,\ F_{\alpha\beta} = \text{const}. \tag{17}$$

Let us assume that the screw dislocations are distributed with the density $\boldsymbol{\alpha} = a_0 \mathbf{e}_R$, a_0 = const, and the shell is loaded by an internal uniform hydrostatic pressure p so that the vector $\mathbf{f}$ has the form $\mathbf{f} = pJ\mathbf{e}_R$.

If the material of the momentless shell is isotropic, the Piola force tensor has the following representation according to (17):

$$\mathbf{D} = D_{11}\mathbf{e}_\varphi \otimes \mathbf{e}_\Phi + D_{12}\mathbf{e}_\varphi \otimes \mathbf{i}_3 + D_{21}\mathbf{i}_3 \otimes \mathbf{e}_\Phi + D_{22}\mathbf{i}_3 \otimes \mathbf{i}_3,\ D_{\alpha\beta} = \text{const}. \tag{18}$$

Due to (17) and (18), incompatibility equation (2) and equilibrium equation (10) are reduced to the algebraic relations:

$$r_0\psi F_{11} - F_{21} = r_0 a_0\,,\; D_{11} + r_0\psi D_{21} = r_0 F_{11} F_{22} p\,. \tag{19}$$

Here it is assumed that the stresses $D_{\alpha\beta}$ are expressed through the distortion components by means of the constitutive equations. Regarding the axial force P and the twisting moment M acting in the tube cross section $z = \text{const}$ as given functions, we obtain in addition to (19) two other relations for determination of the constants F_{11}, F_{21}, F_{22}, ψ:

$$P = 2\pi\, r_0 D_{22}\,, M = 2\pi\, r_0^2 F_{11} D_{21}\,. \tag{20}$$

According to (19), an isometric deformation is realized in the shell without external loads, when $F_{11} = 1$, $F_{22} = 1$, $F_{21} = 0$, and $D_{\alpha\beta} = 0$, while the distortion tensor is expressed through a proper orthogonal tensor $\mathbf{A}_1$ in such a way:

$$\mathbf{F} = \mathbf{g}\cdot\mathbf{A}_1,\; \mathbf{A}_1 = \mathbf{e}_\varphi \otimes \mathbf{e}_\Phi + \mathbf{i}_3 \otimes \mathbf{i}_3 + \mathbf{e}_r \otimes \mathbf{e}_R\,. \tag{21}$$

For the isometric deformation, it follows from (19) that $\psi = a_0$. By the fact that the stresses are zero, it means that the tube twists with no resistance to the torsion, and the twist angle is equal to the screw dislocation density.

2.4. Nonlinear Bending of a Nanotube with Edge Dislocations

We introduce the unit vectors ($\omega = \text{const}$) (see Figure 3)

$$\mathbf{e}_2 = \mathbf{i}_2 \cos(\omega z) + \mathbf{i}_3 \sin(\omega z)\,,\; \mathbf{e}_3 = -\mathbf{i}_2 \sin(\omega z) + \mathbf{i}_3 \cos(\omega z) \tag{22}$$

and suppose that edge dislocations are distributed with the density $\boldsymbol{\alpha} = b(\varphi)\mathbf{e}_3$ in the cylindrical tube loaded from inside by the internal hydrostatic pressure p. We seek the solution for the membrane shell in a form enabling us to reduce the initial two-dimensional problem to the one-dimensional one:

$$\mathbf{F} = F_{11}(\varphi)\mathbf{e}_\varphi \otimes \mathbf{i}_1 + F_{12}(\varphi)\mathbf{e}_\varphi \otimes \mathbf{e}_2 + F_{23}(\varphi)\mathbf{i}_3 \otimes \mathbf{e}_3\,. \tag{23}$$

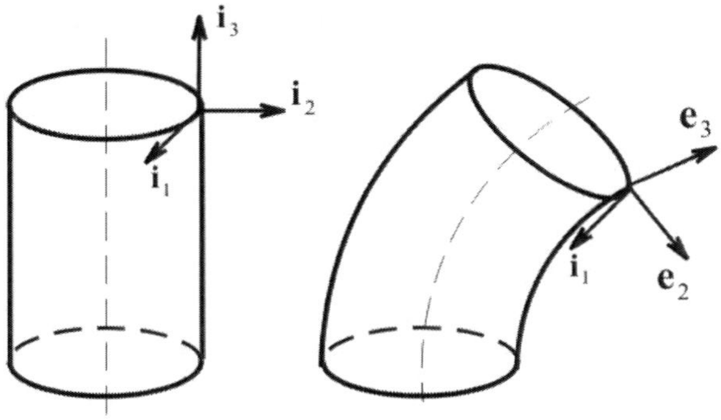

Figure 3. Nonlinear bending of a cylindrical nanotube.

Representation (23) describes pure bending of the tube, when the shell cross sections rotate with the angle ωz around the basis vector $\mathbf{i}_1$ and deform in its plane as well. For an isotropic momentless shell, the Piola force tensor has the following representation according to (23):

$$\mathbf{D} = D_{11}(\varphi)\mathbf{e}_\varphi \otimes \mathbf{i}_1 + D_{12}(\varphi)\mathbf{e}_\varphi \otimes \mathbf{e}_2 + D_{23}(\varphi)\mathbf{i}_3 \otimes \mathbf{e}_3 . \tag{24}$$

The substitution of (22)–(24) into incompatibility equation (2) and equilibrium equation (10) results in a system of nonlinear ordinary differential equations with respect to $F_{11}(\varphi)$, $F_{12}(\varphi)$, and $F_{23}(\varphi)$:

$$\frac{dF_{23}}{d\varphi} - r_0\omega F_{12} = r_0 b(\varphi), \quad \frac{dD_{11}}{d\varphi} + r_0 F_{23} F_{12} p = 0, \quad \frac{dD_{11}}{d\varphi} - r_0\omega D_{23} - r_0 F_{23} F_{11} p = 0 . \tag{25}$$

For determination of the constant ω, it is necessary to couple (25) with the relation for the given bending moment G acting on the end faces of the cylindrical shell. If there are no external loads ($p = G = 0$), and $a(\varphi) = b_0 \cos\varphi$, $b_0 = \text{const}$, the system (25) has the following solution in the form of the isometric deformation allowing the stresses in the momentless shell be zero:

$$\mathbf{F} = -\sin\varphi\ \mathbf{e}_\varphi \otimes \mathbf{i}_1 + \cos\varphi\ \mathbf{e}_\varphi \otimes \mathbf{e}_2 + \mathbf{i}_3 \otimes \mathbf{e}_3 = \mathbf{g} \cdot \mathbf{A}_2 , \tag{26}$$

$$\mathbf{A}_2 = \mathbf{A}_2^{-T} = \mathbf{i}_1 \otimes \mathbf{i}_1 + \mathbf{i}_2 \otimes \mathbf{e}_2 + \mathbf{i}_3 \otimes \mathbf{e}_3 , \quad \omega = -b_0 .$$

This solution means that the nanotube with distributed edge dislocations can bend without occurrence of stresses.

2.4. Buckling of a Flat Membrane with Edge Dislocations

Let us introduce the polar coordinates r and φ in the plane of a circular membrane of radius r_0 and assume that the edge dislocations are distributed in the membrane with the density $\boldsymbol{\alpha} = \gamma(r)\mathbf{e}_\varphi$. The distortion tensor will be sought in the form that corresponds to the axisymmetric bending of the membrane

$$\mathbf{F} = F_1(r)\mathbf{e}_r \otimes \mathbf{e}_r + F_2(r)\mathbf{e}_\varphi \otimes \mathbf{e}_\varphi + F_3(r)\mathbf{e}_r \otimes \mathbf{i}_3, \tag{27}$$

where the vectors $\mathbf{e}_r$ and $\mathbf{e}_\varphi$ are defined as in (16). In the case of an isotropic material, the Piola force tensor has a representation similar to (27):

$$\mathbf{D} = D_1(r)\mathbf{e}_r \otimes \mathbf{e}_r + D_2(r)\mathbf{e}_\varphi \otimes \mathbf{e}_\varphi + D_3(r)\mathbf{e}_r \otimes \mathbf{i}_3. \tag{28}$$

Consider the hydrostatic pressure $\mathbf{f} = pJ\mathbf{N}$ acts on the membrane surface, we transform then the system of equations (2) and (10) taking into account (27) and (28) in the following form:

$$\frac{dF_2}{dr} + \frac{F_2 - F_1}{r} = \gamma(r), \quad \frac{dD_1}{dr} + \frac{D_1 - D_2}{r} - F_2F_3p = 0, \quad \frac{dD_3}{dr} + \frac{D_3}{r} + F_1F_2p = 0. \tag{29}$$

It is necessary to supplement the system (29) of ordinary differential equations with the boundary condition $D_1(r_0) = \tau_0$, where τ_0 is the intensity of the radial edge load. In the absence of external loads ($p = \tau_0 = 0$), the system (29) has a solution, where $F_3 = D_3 = 0$. This corresponds to the plane stress state of the membrane caused by the distributed dislocations. At the same time, when $p = \tau_0 = 0$, (29) has also another solution:

$$\begin{aligned} &F_1 = \cos\chi(r), \; F_2 = 1, \; F_3 = -\sin\chi(r), \\ &\cos\chi(r) = 1 - r\gamma(r), \; D_1 = D_2 = D_3 = 0. \end{aligned} \tag{30}$$

Solution (30) describes the membrane buckling which results in the release of residual stresses caused by dislocations. Although there is no shell displacement field with the distributed dislocations in the general case, in this special case, when $\boldsymbol{\alpha} \cdot \mathbf{i}_3 = 0$, it is possible to find the normal membrane deflection $w = \mathbf{R} \cdot \mathbf{i}_3$ from the relation $\mathbf{F} \cdot \mathbf{i}_3 = \operatorname{grad}\mathbf{R} \cdot \mathbf{i}_3$. Using (30) and taking that $w(r_0) = 0$, we get

$$w(r) = \pm \int_r^{r_0} \sqrt{2\rho\gamma(\rho) - \rho^2\gamma^2(\rho)}\,d\rho. \tag{31}$$

3. DISCLINATIONS

3.1. Disclinations in Shells

Disclinations are the most important defects in carbon structures because of their hexagonal lattice structure. The values **F** of the distortion tensor on the opposite borders of the cut which makes a (m–1)-connected domain from a m-connected one are related by the jump [11]

$$\mathbf{F}_+ = \mathbf{F}_- \cdot \mathbf{Q}, \tag{32}$$

where

$$\mathbf{Q} = \mathbf{F}^{-1} \cdot \overset{\frown}{\oint} (\mathbf{E} + d\mathbf{r} \cdot \mathbf{\Pi}) \cdot \mathbf{F} \tag{33}$$

Here **Q** is the disclination charge (a proper orthogonal tensor) expressed via the path-ordered exponential, **Π** is the so-called affine deformation tensor [12].

The transition to continuous distribution of disclinations is hampered by the fact of noncommutativity of the finite rotation tensors **Q** for different cuts. In the framework of nonlinear plane strain, however, the finite rotations commute, and the transition was performed in [9]. Nevertheless, the possible application of the noncommutative Stokes' theorem for product-type integrals (33) invented by Schlesinger [14] may help to resolve this problem for general surfaces.

3.2. Buckling of a Flat Membrane with Distributed Disclinations

In this section we consider a buckling problem for an elastic plate containing continuously distributed wedge disclinations. In contrast to the Föppl-von Kármán model [15, 16] the tangential strains are not supposed to be small. In the case of very thin plate (membrane) that doesn't resist bending we show that there exists along with the plane stress state solution also the bent form which corresponds to the relaxed state of the membrane.

The starting point is the general incompatibility equation [9] of the plane nonlinear elasticity theory:

$$\mathrm{div}[(\det\mathbf{U})^{-1}\mathbf{e} \cdot \mathbf{U} \cdot \mathrm{div}(\mathbf{e} \cdot \mathbf{U})] = \mathrm{div}[(\det\mathbf{U})^{-1}\mathbf{e} \cdot \mathbf{G}^* \cdot \boldsymbol{\alpha}_0] + \beta. \tag{34}$$

Here $\mathbf{G}^* = \mathbf{U}^2$ is the Cauchy-Green strain measure, $\mathbf{e} = -\mathbf{E} \times \mathbf{i}_3$ is the 2D permutation tensor, **E** is the identity tensor, $\boldsymbol{\alpha}_0$ is the density of edge dislocations, β is the density of wedge disclinations.

The buckling process here means relaxation of stresses by the escape in the third dimension (see Figure 4). Above w is used again to denote the transverse deflection of the plate. Then

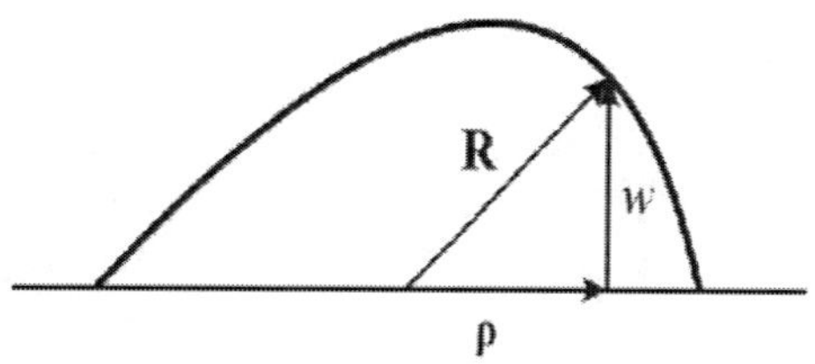

Figure 4. Buckling of a membrane with distributed disclinations.

$$\mathbf{G} = \nabla\mathbf{R} \cdot \nabla\mathbf{R}^T = \nabla\boldsymbol{\rho} \cdot \nabla\boldsymbol{\rho}^T + \nabla w \otimes \nabla w = \mathbf{G}^* + \nabla w \otimes \nabla w. \tag{35}$$

Substituting (35) into (34) and assuming $\boldsymbol{\alpha}_0 = 0$ we have in the membrane limit ($G = E$)

$$[w, w] = [1 - (\nabla w)^2]^{3/2} \beta, \; [w, w] = \frac{\partial^2 w}{\partial x_1^2} \frac{\partial^2 w}{\partial x_2^2} - \left(\frac{\partial^2 w}{\partial x_1 \partial x_2} \right)^2, \tag{36}$$

where $[w, w]$ is the Monge-Ampère operator. In the case of the Föppl-von Kármán theory [15, 16] instead of (36) one should use $[w, w] = \beta$. This type of equation according to the general theory [17] gives no direct way of taking into account negative β.

Let the domain occupied by the plate, the distribution of disclinations β and the transverse deflection w be axially-symmetric. Then equation (36) admits exact integration. Moreover we assume the usual zero-slope condition [18] in the center of the membrane: $w'(r)\big|_{r=0} = 0$.

Under such conditions for constant positive β we get

$$w(r) = \frac{1}{\beta}\sqrt{\beta^2 r^2 + 4\beta} - \frac{1}{\sqrt{2\beta}} \log \left| \frac{\sqrt{2\beta} + \sqrt{\beta^2 r^2 + 4\beta}}{\sqrt{2\beta} - \sqrt{\beta^2 r^2 + 4\beta}} \right| - C, \tag{37}$$

whereas for constant negative β we have

$$w(r) = \frac{1}{\beta}\sqrt{\beta^2 r^2 + 4\beta} + \frac{1}{\sqrt{-2\beta}} \arctan \frac{\sqrt{\beta^2 r^2 + 4\beta}}{\sqrt{-2\beta}} - C, \tag{30}$$

the constant C in both cases is provided by vanishing $w(r)$ on the outer radius $r = r_0$ of the plate.

For negative β the solution exists only in some part of the circular disk, where $r \geq \sqrt{-4/\beta}$ (Figure 5).

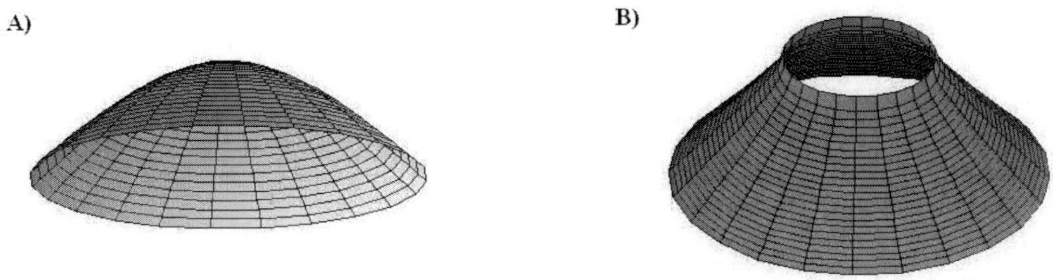

Figure 5. Buckling of a membrane with A) Positive disclination density; B) Negative disclination density.

4. Conclusions

The structural flexibility of graphene is reflected in its electronic properties. In fact, graphene is unique in the sense that it shares properties of semi-metallic soft membranes, while all previously studied membranes were actually insulators [19]. The coupling between geometry and electron propagation is unique to various graphene structures, and results in additional scattering and resistivity due to ripples [4, 20].

The graphene ripples have the ability to trigger large deformations, this makes it ideal for flexible electronic applications [5]. It was demonstrated [21] that both graphene ribbon and nanographene film ripples may be used for strain sensors, and their resistance changes upon different strains were studied. This simple and controllable process of buckled graphene provides a feasible fabrication for graphene flexible electronic devices and strain sensors due to its novel mechanical and electrical properties [22].

Acknowledgements

The work was supported by the Russian Foundation for Basic Research (project no 12-01-00038), and the Federal target programme "Research and Pedagogical Cadre for Innovative Russia" for 2009-2013 years (government contract P596).

References

[1] Kolesnikova, A. L.; Romanov, A. E. *Phys. Sol. State* 1998, *vol. 40(6)*, 1075–1077.

[2] Romanov, A. E.; Sheinerman, A. G. *Phys. Sol. State* 2000, *vol.42 (8)*, 1569–1574.

[3] Favata, A.; Podio-Guidugli, P. In: *Shell-Like Structures – Non-classical Theories and Applications.* H. Altenbach, V. A. Eremeyev (Eds.). Berlin: Springer, 2011. pp. 561–570.

[4] Vozmediano, M. A. H.; Katsnelson, M. I.; Guinea, F. *Physics Reports* 2010, *vol. 496*, 109–148.

[5] Khang, D.-Y.; Rogers, J. A.; Lee H. H. *Adv. Funct. Mater.* 2008, *vol. 18*, 1–11.

[6] Freund, L. B.; Suresh, S. *Thin Film Materials: Stress, Defect Formation and Surface Evolution*. Cambridge: Cambridge University Press, 2003.
[7] Chen, S.; Chrzan, D. C. *Phys. Rev. B* 2011, *vol. 84*, 214103.
[8] Zubov, L. M. *Methods of Nonlinear Elasticity in Shell Theory (in Russian)*. Rostov-on-Don: Rostov University Press, 1982.
[9] Derezin, S.; Zubov, L. *Z. Angew. Math. Mech.* 2011, *vol. 91(6)*, 433–442.
[10] Derezin, S. In: *Shell-Like Structures – Non-classical Theories and Applications*. H. Altenbach, V. A. Eremeyev (Eds.). Berlin: Springer, 2011. pp. 531–548.
[11] Zubov, L. M. *Nonlinear Theory of Dislocations and Disclinations in Elastic Bodies*. Berlin: Springer, 1997.
[12] Lurie, A. I. *Nonlinear Theory of Elasticity*. Amsterdam: North-Holland, 1990.
[13] Kolesnikov, A. M.; Zubov, L. M. *Z. Angew. Math. Mech.* 2009, *vol. 89(6)*, 288–305.
[14] Schlesinger, L. *Math. Ann.* 1928, *B. 99*, 413–434.
[15] Seung, H. S.; Nelson, D. R. *Phys. Rev. A* 1988, *vol. 38(2)*, 1005–1018.
[16] Zubov, L. M. *Dokl. Phys.* 2007, *vol. 52(1)*, 67–70.
[17] Pogorelov, A. V. *Multidimensional Monge-Ampère Equation*. Cambridge: Cambridge Scientific Publishers, 2008.
[18] Timoshenko, S. P.; Woinowsky-Krieger, S. *Theory of Plates and Shells*. New York: McGraw Hill, 1991.
[19] Abedpour, N.; Asgari, R.; Guinea, F. *Phys. Rev. B* 2011, *vol. 84*, 115437.
[20] Castro Neto, A. H.; Guinea, F.; Peres, N. M. R.; Novoselov, K. S.; Geim, A. K. *Rev. Mod. Phys.* 2009, *vol. 81(1)*, 109–162.
[21] Pereira, V. M.; Castro Neto, A. H.; Liang, H. Y.; Mahadevan, L. *Phys. Rev. Lett.* 2010, *vol. 105*, 156603.
[22] Wang, Y.;Yang, R.; Shi, Z.; Zhang, L.; Shi, D.; Wang, E.; Zhang, G. *ACS Nano* 2011, *vol. 5*, 3645–3650.

In: Physics and Mechanics of New Materials ...
Editors: Ivan A. Parinov and Shun-Hsyung Chang ISBN: 978-1-62618-535-7

Chapter 15

PHYSICAL, MECHANICAL AND STATISTICAL ASPECTS OF ACOUSTIC EMISSION DIAGNOSTICS

S. I. Builo*
Vorovich Mechanics and Applied Mathematics Research Institute,
Southern Federal University, Rostov-on-Don, Russia

ABSTRACT

The relationships between the sequence of deformations and fracture stages in solids and the registered parameters of an accompanying acoustic emission (AE) are analyzed. A physical-mechanical model of AE from a dislocation pile-up during its passage through a boundary is considered. The possibilities and limits of existing methods of AE diagnostics are described. It is shown, that as a result of multimode propagation, reflection and transformation of the wave mode, damping of high-frequency components, and resonance properties of the receiving equipment, the detected AE signals differ greatly from emitted elastic AE pulses. To solve this problem, we developed physical, mechanical and statistical methods for reconstructing the true initial AE-event stream parameters inside the material and the process of flaw storage using the recorded AE signals.

The method of identifying the fracture stages, based on the positions of the break points in the AE intensity (or total number of AE events) curve, estimating the concentration of defects and their multiplication constants, is shown. The application of these results can significantly (by at least 2-fold) increase the reliability of the AE diagnostics. Our methods and algorithms are quite versatile, so that may be useful in solving problems of improving the accuracy and reliability of the results from the use of the AE phenomena in studies of the dynamics of various structural changes in solids. The results have been used in the diagnostics of the strength of the Tsar-bell in the Moscow Kremlin, the thermal protection tiles of the Russian space shuttle "Buran", titanium turbine blades of aircraft engines and in the diagnostics of other products and objects of critical applications.

* Vorovich Mechanics and Applied Mathematics Research Institute, Southern Federal University, 200/1, Stachki Ave, Rostov-on-Don, 344090, Russia, E-mail: bsi@math.sfedu.ru

1. Introduction

The diagnostics and modeling of the early stages of a solid fracture encounter considerable difficulties. Classic methods of continuum mechanics and fracture mechanics can be applied if the characteristic size of a single defect (crack) is at least of 1 mm. This means that the presence of a single crack has a considerable effect on the strain-stress state in the vicinity of a crack. In view of this, a new approach based on the study of the relationship between acoustic emission (AE) parameters and the kinetics of flaw accumulation in solids has actively been developed for over 40 years, along with conventional approaches [1-19].

Experimental data on the fracture physics and mechanics of materials show that deformation and destruction processes realize at the micro-level in the form of discrete collective dynamic processes. The discrete character of these processes is likewise confirmed by the data of AE tests, which testify that, in each collective event of structural damage (motion of a dislocation avalanche, merging of microcracks, etc.), a primary elastic pulse is generated, the emission process of which involves, so-called AE event. An AE event is considered to mean the process of emission of a primary elastic pulse during a collective discrete event of structural transformation (or damage) in a solid.

Hence, measuring intensity of the stream of AE events, $\dot{N}_a$ (the number per unit time), or their total number, N_a, in principle makes it possible to perform quantitative study of the damage-accumulation kinetics and diagnostics of early damage stages.

As a result of multimode propagation, reflection, and transformation of the wave mode, damping of high-frequency components, and resonance properties of the receiving equipment, the detected AE signals differ greatly from emitted elastic AE pulses. For example, we have revealed that, in the case of a high-intensity emitted AE-event stream, both false minima in the recorded activity and false maxima in the amplitude distribution of recorded AE pulses may appear. To solve this problem, we are developing physical, mechanical and statistical methods for reconstructing the true initial AE-event stream parameters inside the material and the process of flaw storage using the recorded AE signals.

2. Physical and Mechanical Nature of Acoustic Emission

The AE method is consists in analyzing the parameters of the ultrasonic radiation, that accompanying any change or damage in the structure of metals, their alloys, and composite materials. The physical and mechanical nature of the dislocation AE event is shown in Figure 1.

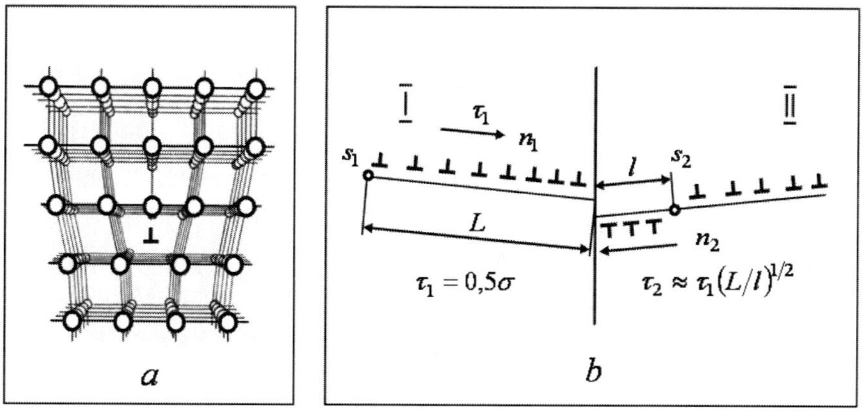

Figure 1. Physical and mechanical nature of an acoustic emission event: (a) edge dislocation (linear defect) in a simple cubic lattice, (b) the formation of dislocation pile-up at the boundary under mechanical loading.

For simplicity, we consider accumulation of edge dislocations, each of which can be represented by the presence an extra atomic half-plane in crystal (see Figure 1a).

Parameters of event, we estimate on example of AE radiation under mechanical loading of the material. It is known, that during deformation of materials with applied stress σ, moving dislocations are gathered near boundaries, that lead to the appearance of tangent stress $\tau_2 n_2$, which exceeds the stress activating the source (see Figure 1b):

$$\tau_2 n_2 \approx n_2 \tau_1 (L/l)^{1/2} = 0{,}5 n_2 \sigma (L/l)^{1/2}; \quad (L/l) >> 1.$$

where τ_1 is the shear stress in a glide line of length L of the first dislocation source s_1; τ_2, n_2, l are the shear stress, the number of dislocations, and the length of the glide line of second dislocation source s_2, respectively

Assuming that each dislocation that crosses the boundary at moment t_i radiates a short exponential pulse of amplitude x_{oi} and taking into account that their superposition yields a detectable pulse of amplitude X, we have obtained the following form of the normalized amplitude of the total AE pulse [2-7]:

$$A(\lambda) = \frac{X}{n x_{oi}} = \frac{1}{n} \sum_{k=0}^{n-1} \exp\left(-\frac{\lambda k}{n} \right); \; \lambda^{-1} \sim \bar{v}; \; \dot{\varepsilon} = M \rho_0 b \bar{v},$$

$$A(\dot{\varepsilon}) \approx \int_0^{n-1} \exp\left(-\frac{Ck}{n\dot{\varepsilon}} \right) dk \approx \frac{\dot{\varepsilon}}{\sqrt{\dot{\varepsilon}_{\min} \dot{\varepsilon}_{\max}}} \left[1 - \exp\left(-\frac{\sqrt{\dot{\varepsilon}_{\min} \dot{\varepsilon}_{\max}}}{\dot{\varepsilon}} \right) \right]. \tag{1}$$

where n is the number of dislocations in the pile-up, $\lambda = t_r / \tau_a$ is the ratio of the time $t_r = n(t_i - t_{i-1})$ of crossing the border to the stress relaxation time constant τ_a, $\bar{v}$ is the average velocity of dislocations, $\dot{\varepsilon}$ is the strain rate, M is the orientation factor, ρ_o is the density of mobile dislocations, b is the modulus of Burgers vector; $\dot{\varepsilon}_{\min}$ and $\dot{\varepsilon}_{\max}$ are the minimum and maximum strain rate in the experiment, respectively.

The application of this model of AE event is shown in Figure 2 as an example of microcrack nucleation on Cottrell dislocation mechanism. Where τ is the after-sounding time constant of the transducer and article; u_0 is the maximum (peak) of the AE pulse, and u_t is the discrimination threshold.

Figure 3 shows plots of relationships (1) and the results of experimental measurements of the average amplitude m_u at several deformation rates for specimens of steel 07X16H6. The theoretical curves (1) and experimental results agree well.

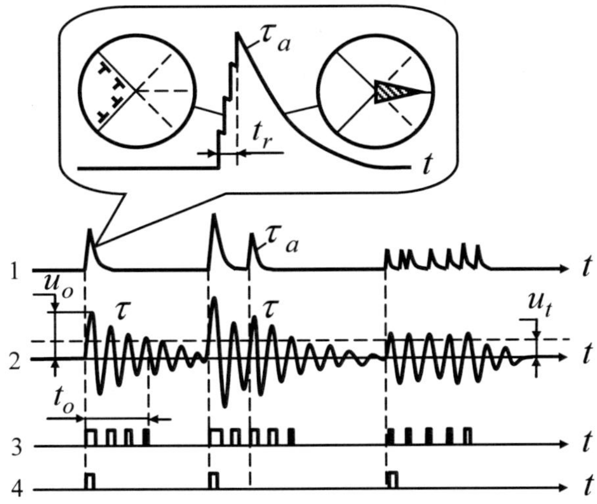

Figure 2. Acoustic emission event during microcrack nucleation on Cottrell dislocation mechanism and its detection by the receiving devices: 1 - stream of emitted elastic AE pulses, 2 - stream of detected AE radio signals, 3 - stream of AE oscillations (N), 4 - stream of envelopes of the detected AE pulses N_{Σ} .

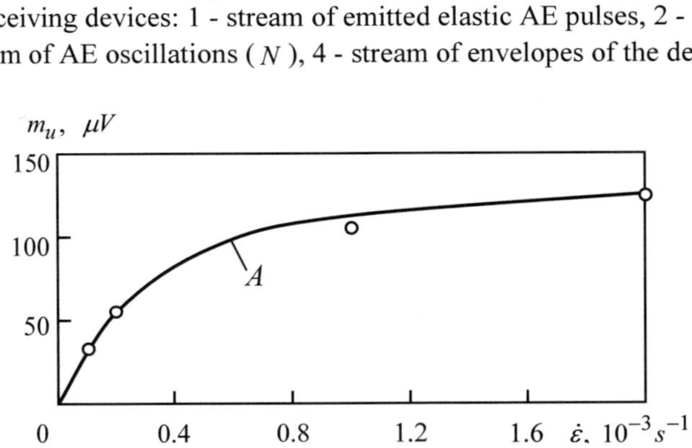

Figure 3. Average AE amplitude as a function of the deformation rate: the material is steel 07X16H6 (Fe; 0.07 %C; 16 %Cr; 6 %Ni).

These obtained results confirm the validity of our proposed physical-mechanical model of the phenomenon of acoustic emission.

3. The Distribution Density of Time Intervals between Events and Poisson Stream as a Model for Acoustic Emission

During determination of the parameters of an AE event stream, we founded (see, for example, Figure 4) that, in most cases, the distribution density of time intervals Δt between AE events, $w(\Delta t)$, along the entire loading curve is described by the exponential dependence [2, 3, 5 ,6, 8-10]:

$$w(\Delta t) = \dot{N}_a \exp(-\dot{N}_a \Delta t). \tag{2}$$

where $\dot{N}_a$ is the intensity (number per time unit) of events in the stream.

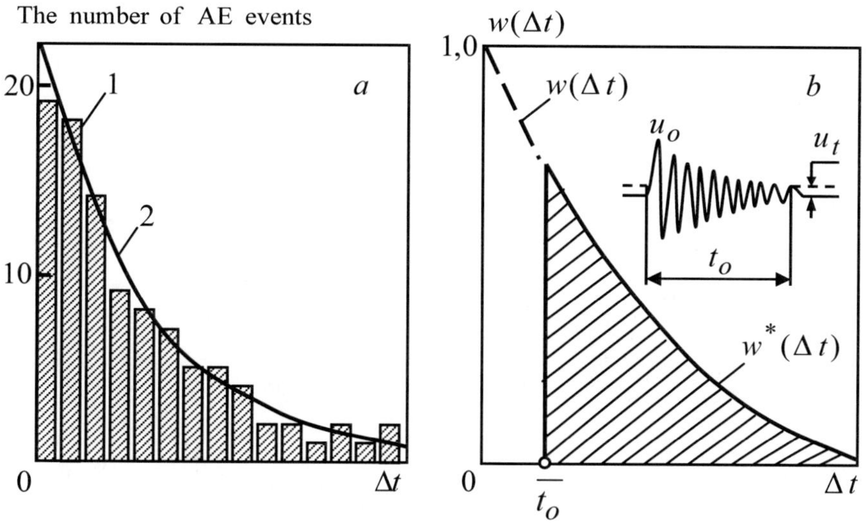

Figure 4. Experimental histograms of the distribution of the time intervals between AE events (a), and distribution density of intervals between events of a Poisson flow in the presence of a dead time (b). The material is steel 95X18 (Fe; 1.0 %C; 18 %Cr), a sample size of 104 events, strain = 2.16 %, number of analyzer channels is equal to 15, and channel width = 1.92 ms: (1) experimental values, and (2) approximating Poisson stream.

Equation (2) corresponds to the distribution density of time intervals between events of the Poisson stream [2, 5, 6, 9]. In such a stream, the probability $P_r(\Delta t)$ of the appearance of r events within a time interval Δt obeys the Poisson distribution:

$$P_r(\Delta t) = [(\dot{N}_a \Delta t)^r / r!] \exp(-\dot{N}_a \Delta t). \tag{3}$$

The first experimental confirmation of the Poisson character of the AE phenomenon was obtained in our investigations in the end of 1970s.

If experimental data are absent, the hypothesis of the Poisson stream of events may be applied *a priori* in many cases on the basis of the limiting theorem of streams, which asserts that the resulting stream is of Poisson character if it consists of a superposition of any independent low intensity streams. These conditions are usually met during recording of AE signals [2, 5, 6], because the recorded AE stream is almost always produced by several microsources, which are most frequently independent and distributed over a certain local microvolume.

4. Method of Reconstructing AE Event Stream Grom Recorded AE Signals

The essence of the reconstruction method we develop can be briefly described as follows. Proceeding from the physical nature of AE, the kinetic concept of strength that utilizes a statistical Poisson's model of an event stream is used to determine a specific form of the distribution function (or density) for a given parameter in the signals that have not yet overlapped. This distribution is then extrapolated to the region where the recorded parameters have been strongly distorted during signal propagation and receipt. In this way, it becomes possible to evaluate the true distributions of AE parameters. Subsequently, the desired characteristics of AE sources (flaws) are calculated on the basis of the information on the reconstructed distribution functions [2, 5, 6, 8-10].

If an AE pulse at the output of the receiving channel is described by an exponentially decaying radio-pulse, the duration t_0 of its envelope at the discrimination level u_t is determined by the expression (see Figure 4b):

$$t_0 = \tau \ln(u_0 / u_t), \tag{4}$$

where τ is the after-sounding time constant of the transducer and article, u_0 is the maximum (peak) of the AE pulse, and u_t is the discrimination threshold.

The duration t_0 can be regarded as the system's "dead time" [2, 5, 6], because, if the time interval between AE events is shorter than t_0, these events are detected as a single pulse. With allowance for (2), the truncated distribution density $w^*(\Delta t)$ of time intervals Δt provided that a part of pulses overlaps has the form (see Figure 4b):

$$w^*(\Delta t) = 0 \text{ for } \Delta t < \overline{t_0}; \; w^*(\Delta t) = w(\Delta t) \text{ for } \Delta t \geq \overline{t_0}. \tag{5}$$

where $\overline{t_0}$ is the dead time average duration.

To quantitatively describe the degree of overlap of AE pulses, an overlap parameter K is introduced [5, 6, 8, 9]:

$$K = \dot{N}_a \overline{t_0} \approx \Sigma t_0 / (T - \Sigma t_0), \tag{6}$$

where Σt_0 is the total duration of AE pulses (total dead time) at the output of the detecting channel within the interval T of intensity measurements (for example, $T = 1$ s).

The activity $\dot{N}_\Sigma$ of detected AE pulses is equal to the product of the actual AE event stream intensity and the area under curve (5):

$$\dot{N}_\Sigma = \dot{N}_a \int_{\overline{t_0}}^{\infty} \exp(-\dot{N}_a \Delta t) d\Delta t = \dot{N}_a \exp(-\dot{N}_a \overline{t_0}) = \dot{N}_a \exp(-K). \tag{7}$$

Analysis of (7) shows that, during calculation of detected AE pulses, $\dot{N}_\Sigma = \dot{N}_a$ only at $K << 1$. For $K = 1$, the detected activity has an extreme: $\dot{N}_\Sigma^{\max} = 0.37 \dot{N}_a$, and at $K > 1$, the detection method leads to dips in the detected AE activity $\dot{N}_\Sigma$, because $\lim \dot{N}_\Sigma = 0$ for $\dot{N}_a \to \infty$ ($K >> 1$).

Solving (7) with respect to $\dot{N}_a$ and taking into account (6), we obtain fundamental equation for reconstructing the true intensity of the AE event stream from the detected AE signals:

$$\dot{N}_a = \dot{N}_\Sigma \exp[\Sigma t_0 / (T - \Sigma t_0)]. \tag{8}$$

For devices with constant or no prolonged type of “dead time” (the duration of no prolonged “dead time” is independent of the pulses arriving during the dead time) we obtain the following expressions for reconstructing the intensity of the AE event stream from the stream intensity of registered AE pulses $\dot{N}_r$ [2, 5, 6, 8]:

$$\dot{N}_a = \dot{N}_r / (1 - \dot{N}_r \overline{t_0}) = N_r / (T - \Sigma t_0). \tag{9}$$

where N_r is the number of AE pulses registered within a time T.

The obtained results also allowed us to reconstruct $\dot{N}_a$ from the AE count rate $\dot{N}$ [5, 6, 9]:

$$\dot{N}_a = \dot{N} / (f - \dot{N}) \tau L(m, D), \tag{10}$$

where f is the filling frequency of AE signals (which is approximately equal to the transducer resonance frequency), $L(m,D)=(m-1)^{-1}-(D^{m-1}-1)^{-1}\ln D$, m is the power factor of the distribution density of peak amplitude AE u_0 [3, 9], $D=u_{0\max}/u_t$ is the dynamic range of AE signals and $u_{0\max}$ is the maximum amplitude corresponds to the upper boundary of the dynamic range of AE signals.

Very often $D>>1$, $m\approx 2$, and we can obtain very simple equation [5, 6, 9]:

$$\dot{N}_a \approx \dot{N}/(f-\dot{N})\tau. \tag{11}$$

Our reconstruction method uses the Poisson stream model and appreciably increases the accuracy and reliability of the results of AE diagnostics of stages of flaw formation processes.

5. Method for Identifying Stages of Destruction by Location of Critical Points of the Reconstructed Stream of AE Events

The reconstruction of the parameters of an AE-event stream makes it possible to obtain the true form of the dependences of AE parameters along the loading curve and to substantially increase the reliability of AE diagnostics results. For example, we discovered, that in the most general case, the intensity of the reconstructed (i.e., emitted inside a material) AE event stream is accompanied by a local peak in the yield zone, a much lower local peak in the strengthening region, a lower inflection point corresponding to the onset of dispersed accumulation of microscopic discontinuities, and an upper inflection point corresponding to a transition from the dispersed to localized flaw accumulation [5, 6, 9].

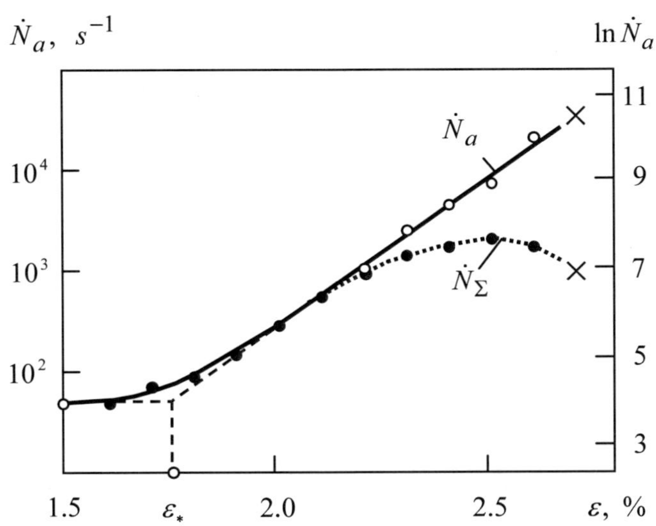

Figure 5. Dependence of the reconstructed intensity of the AE event stream on a specimen deformation with the lower break point of the AE event stream. The material is steel 95X18 (Fe; 1.0 %C; 18 %Cr), the strain rate is $2\times10^{-4}s^{-1}$, the crosses indicate the moment of fracture.

The intensities of flaw accumulation derived from reconstructed AE event intensities are in agreement with estimates based on statistical models of the rupture strength of solids [5, 6, 12, 13]. This allows us to determine kinetic parameters of flaw accumulation in materials with nonlinear elastic and strength characteristics on the base of restored AE events on formula (8) for detected AE signals. We developed, that the concentration of flaw generation in a material C can be described by the following equation [6, 12]:

$$C = C_o \exp[\gamma(\varepsilon - \varepsilon_*)],\ C = N_f / V, \tag{12}$$

where C_0 is the initial (at point $\varepsilon = \varepsilon_*$) density of flaws, γ is the multiplication constant, ε_* is the strain at the initial point of damage accumulation, ε is the current strain, N_f is the total number of flaws, and V is the material volume. All the parameters of the process in Equation (12) can be estimated using data of AE tests (see Figure 5).

If each event of fracture in the material generates an AE pulse, the process described by (12) should be characterized by the following intensity of emitted AE pulses as a function of the strain in the material [6, 12]:

$$\dot{N}_a = VC_o\gamma\dot{\varepsilon} \exp[\gamma(\varepsilon - \varepsilon_*)]. \tag{13}$$

Equation (13) describes the exponential function that starts with a certain threshold strain ε_* and characterizes exponential multiplication of AE events, which is, indeed, often observed in practical tests. The specific value of threshold strain ε_* is derived from the position of the upper turning point on the experimental curve, $\log \dot{N}_a = \log \dot{N}_a(\varepsilon)$. The multiplication constant γ equals the tangent of slope of this curve, and c_0 is derived from (13) at $\varepsilon = \varepsilon_*$, and the value of $\dot{N}_a$ corresponds to this point.

Thus, after deriving from AE data the specific values of c_0, γ and ε_*, one can calculate by formula (12) the real concentration of flaws at any moment of the loading process, including the moment of fracture.

For example, for the specimen in Figure 5, we obtain the following results: $\varepsilon_* = 1.7\times10^{-2}$, $\gamma = 6.22\times10^{2}$, $C_0 = 1.9\times10^{2}\,\text{cm}^{-3}$. From these, considering (12), we obtain the following expression for the concentration C of flaws in the given specimen at any point along the stress-strain curve, including the moment of fracture:

$$C = 190\exp(622\varepsilon - 10.57),\ \text{at}\ \varepsilon \geq 1.7\times10^{-2}. \tag{14}$$

One can see that the plot of $\log \dot{N}_a$ often consists of three linear sections, two of which refer to exponential functions. This kind of dependence agrees well with the S. Zhurkov's kinetic concept of strength [5, 6].

The results of our recent experiments show that the reconstructed intensity and the total number of AE events of different materials often have not only one, but two critical points and S-shaped functions observed along the stress-strain curve (see, for example, Figure 6).

Analysis of the obtained results allows us to state that at $\varepsilon < \varepsilon_*$ (below the lower break point ε_*), accumulation of AE events occurs due to collective effects produced during the plastic deformation of the material without a noticeable accumulation of defects. This conclusion follows from calculations described below for initial concentration c_0 of flaws (micro defects) obtained from AE data presented in Figure 5 and 6a. This data demonstrates that at this stage the initial concentration of flaws at the lower break point of the curve is in many times smaller than the total number of AE events accumulated before this point.

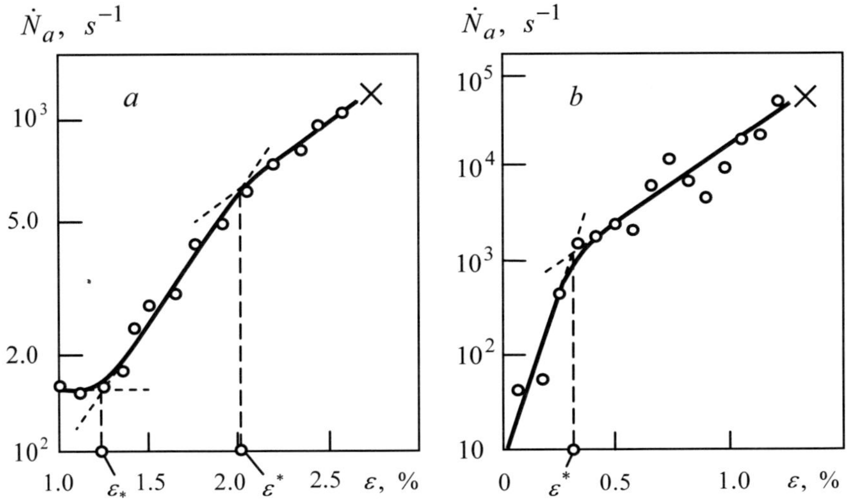

Figure 6. Dependence of the reconstructed intensity of the AE event stream on a specimen's; the material is: a - composite based on laminated fiberglass, b – coal; the crosses indicate the moment of fracture.

The region of deformation at $\varepsilon_* \le \varepsilon < \varepsilon^*$ corresponds to the stage of stochastic (chaotic over volume and time) nucleation and exponential accumulation of microcracks. The chaotic character of defect accumulation in this early destruction stage is confirmed experimentally by the Poisson distribution of the accompanying AE stream.

The upper break point in the curve of the stream intensity of AE events for strain values $\varepsilon \ge \varepsilon^*$ is determined by us as the beginning some localization of defect accumulation before the generation of a macrocrack.

Based on the results of AE studies, we suggest the following relationship for the number of flaws (and the corresponding total number of AE events) along the stress-strain curve with two critical points [5, 6, 13]:

$$N_f = VC_0 \exp[\gamma_1(\varepsilon - \varepsilon_*)], \text{ at } \varepsilon_* \le \varepsilon < \varepsilon^*; \qquad (15)$$

$$N_f = VC_0 \exp[\gamma_1(\varepsilon^* - \varepsilon_*)] - N^* \{1 - \exp[\gamma_2(\varepsilon - \varepsilon^*)]\}, \text{ at } \varepsilon \geq \varepsilon^*,$$

where N_f is the total number of flaws (microdefects) in a specimen (also see destinations in equations (12) and (13), γ_1 and γ_2 are the multiplication constants, ε_* is the strain of transition to the stage of exponential multiplication of flaws, N^* is a parameter that describes the number of growing flaws at point $\varepsilon = \varepsilon^*$, and ε^* is the strain corresponding to the start of macrodefect localization.

All the necessary parameters for (15) can be found from data of AE tests. Taking into account that each AE event corresponds to a flaw generation event, and after differentiating (15), we obtain:

$$\dot{N}_a = VC_0 \gamma_1 \dot{\varepsilon} \exp[\gamma_1(\varepsilon - \varepsilon_*)] \text{ , at } \varepsilon_* \leq \varepsilon < \varepsilon^*;$$

$$\dot{N}_a = N^* \gamma_2 \dot{\varepsilon} \exp[\gamma_2(\varepsilon - \varepsilon^*)], \text{ at } \varepsilon \geq \varepsilon^*. \tag{16}$$

From relationship (16), it is easy to estimate concrete values of C_0, N^*, γ_1, γ_2, ε_*, ε^* and calculate by Equation (15) the real concentration of flaws at any moment of the loading process, including the moment of fracture [13]. For example, by substituting concrete values of fracture deformation and other calculated parameters on the base of AE data for specimen in Figure 5 into (15) or (12), we obtain the critical concentration of flaws $C_{cr} = 1.02 \times 10^5 \text{cm}^{-3}$ at the moment of fracture. Process of fracture takes place, when $C \geq C_{cr}$.

For many specimens that we used, the results of our AE investigations give the following estimate for the critical flaw concentration: $C_{cr} \approx 10^{-1} n^*$, where n^* is the total number of structural elements (e.g., grains) in a unit volume of the material. Therefore one can use the concentration criterion of destruction and quantitatively estimate the critical concentration of flaws (defects) on the base of AE data analysis.

6. Summary and Conclusions

The developed methods are based on physical, mechanical and stochastic nature of AE. Our algorithms provide a significant increase in the accuracy of estimating the parameters of processes of structural changes and damage accumulation in materials according to the AE tests. Their use provides valuable new information that is difficult and often impossible (especially in the dynamics) obtained experimentally by any other physical or mechanical methods of investigation.

The application of these results can significantly (by at least 2-fold) increase the reliability of the AE method, which opens up new possibilities in solving the problems of diagnostics of strength, fracture and thermal destruction of materials. The proposed in the paper the approaches, methods, and the resulting algorithms are quite versatile, so that may be useful in solving problems of improving the accuracy and reliability of the results from the use of the AE effect in studies of the dynamics of various structural changes in solids. It can be plastic and yield deformations, destruction, phase transitions, martensitic transformations and other phenomena.

The results have been used in the diagnostics of the strength of the Tsar-bell in the Moscow Kremlin [15], the thermal protection tiles of the first Russian space shuttle "Buran" [16], titanium turbine blades of aircraft engines, the development of AE thermal destruction [17, 18], as well as in the diagnostics of other products and objects [19] of critical applications.

ACKNOWLEDGMENTS

This research was carried out with partial financial support from Russian Foundation for Basic Research (project codes 06-08-01039, 10-08-13300, 11-08-13152, and 12-08-01190).

REFERENCES

[1] *Acoustic Emission;* ASTM Special Technical Publication 505: Baltimore, 1972; pp 1-337
[2] Builo, S. I. *Investigation of Acoustic Emission during Mechanical Deformation. PhD Dissertation*; Rostov-on-Don, Rostov State University, 1983; pp. 1–191, [in Russian]
[3] Tripalin, A. S.; Builo, S. I. *Acoustic Emission. Physical and Mechanical Aspects*, Rostov-on-Don: Rostov State University, 1986, pp 1-160 [in Russian]
[4] Builo S. I.; Tripalin A. S. *Strength of Materials*, 1986*, vol. 18(2)*, 250-254
[5] Builo, S. I. *Physicomechanical and Statistical Aspects of Increasing the Reliability of Results of Acoustic-Emission Testing and Diagnostics,* Rostov-on-Don: Southern Federal University, 2008, pp 1-192 [in Russian]
[6] Builo, S. I. *Physicomechanical and Statistical Aspects of Acoustic Emission Predestructive State Diagnostics. Doctor of Sciences in Physics and Mathematics Dissertation;* Rostov-on-Don, Southern Federal University, 2009, pp 1-279 [in Russian]
[7] Builo,. S. I. *Russian Journal of Nondestructive Testing*, 2007, *vol. 43(3)*, 188–195
[8] Builo, S. I. *Russian Journal of Nondestructive Testing*, 1995, *vol. 31(7)*, 492–503
[9] Builo, S. I. *Russian Journal of Nondestructive Testing*, 2008, *vol. 44(8)*, 517–526
[10] Builo, S. I. *Russian Journal of Nondestructive Testing,* 1999, *vol. 35(4)*, 267-273
[11] Builo, S. I. *Russian Journal of Nondestructive Testing,* 1996, *vol. 32(5)*, 353-360
[12] Builo, S. I. *Russian Journal of Nondestructive Testing,* 1997, *vol. 33(7)*, 512-516
[13] Builo, S. I. *Russian Journal of Nondestructive Testing*, *vol. 40(8)*, 2004, 552–560
[14] Builo, S. I.; Kozinkina, A. I. *Physic Solid State*, 1996, *vol. 38(11)*, 1844–1845
[15] Builo, S. I. *Russian Journal of Nondestructive Testing*, 2004, *vol. 40(7)*, 498–501
[16] Builo, S. I. *Russian Journal of Nondestructive Testing*, *vol. 39(10)*, 2003, 807–810
[17] Builo, S. I.; Belozerov, V. V.; Prus, Yu. V. *Russian Journal of Nondestructive Testing*, 2008, *vol. 44(3)*, 212–214

[18] Belozerov, V. V.; Bosyj, S. I.; Builo, S. I.; Prus, J. V. *Abstracts of Russian-Taiwanese Symposium "Physics and Mechanics of New Materials and Their Applications (PMNM-2012)"*, Rostov-on-Don, Russia, June 4-6, 2012, pp 12-13

[19] Builo S. I. *Abstracts of Russian-Taiwanese Symposium "Physics and Mechanics of New Materials and Their Applications (PMNM-2012)"*, Rostov-on-Don, Russia, June 4-6, 2012, pp 15-16

In: Physics and Mechanics of New Materials ...
Editors: Ivan A. Parinov and Shun-Hsyung Chang

ISBN: 978-1-62618-535-7

Chapter 16

MULTISCALE COMPUTER SIMULATION OF PIEZOELECTRIC DEVICES WITH ELEMENTS FROM POROUS PIEZOCERAMICS

A. V. Nasedkin[1,*] ***and M. S. Shevtsova***[2,†]

[1]Department of Mathematics, Mechanics and Computer Sciences, Southern Federal University, Rostov-on-Don, Russia
[2]Southern Scientific Center of Russian Academy of Sciences, Rostov-on-Don, Russia

ABSTRACT

Due to increased attention of engineers and scientists to use of porous piezoelectric materials for acoustic applications we present the developed method for prediction of the material constants for porous piezoelectric ceramics and some results of the comparative numerical analysis of some acoustic piezoelectric based devices made of dense and porous ceramics.

At microlevel, to calculate the effective moduli of 3-0 and 3-3 porous piezocomposites, we use an approach based on the effective moduli methods. The effective moduli for different types of porous piezoceramics were calculated by the finite element (FE) modeling of representative volumes [1, 2]. A comparison of these basic characteristics for porous piezocomposites with the experimental data, obtained in Research Institute of Physics, (Southern Federal University), was presented and discussed.

On the macrolevel we investigate 1-3 piezocomposite, hydro-acoustic projectors made of solid ceramics with intermediate matching non-piezoelectric layers and piezoelectric emitter from porous piezoceramics without such layers. From the FE solutions of static and dynamic problems we conclude a good efficiency of high porous piezoceramics for quasistatic and low-frequency applications in piezoelectric energy harvesters as its piezo-sensitivity rapidly increases at the porosity growth.

* Southern Federal University. 8a, Milchakova Street, 344090 Rostov-on-Don, Russia. nasedkin@math.rsu.ru.
† 41, Tchekhov Street, 344006 Rostov-on-Don, Russia, mariamarcs@bk.ru.

1. INTRODUCTION

Nowadays piezoceramic materials are widely used as active parts in ultrasonic transducers, hydrophones, medical diagnostic equipment, electronic, optoelectronic applications and other piezoelectric devices [3-5]. Sonar transducers transmit and receive acoustic energy for the detection of underwater objects. Ultrasound imaging transducers generate an image of a part of the body, based on the echo signals. Hydrophone devices detect underwater noises and convert mechanical vibration into electrical signal. The main advantages of piezoelectric ceramics are the efficient conversion of electrical and mechanical energy, low mechanical and dielectrical losses for achievement of high sensitivity. On the other hand the main disadvantage of using the dense ceramics is its high acoustical impedance that makes it necessary to add at least two matching layers for the sufficient acoustical agreement of piezoelectric material with the acoustic medium. Compared to the dense ceramics the porous piezoelectric materials have ultra-high volume piezosensitivity, broadened operating frequency range and lower impedance, giving a better acoustic agreement between piezoelectric and acoustic media. The other advantage of porous ceramics is that its thickness piezoelectric module, responsible for the efficiency of electrical and mechanical energy conversion at the thickness vibrations of the transversaly-polarized ceramics, remains almost the same as for the dense ceramics. All this determines the relevance of research and use of the porous piezoceramics.

The microstructural characterization, analysis and classification of piezoelectric composites was initiated by Newnham's connectivity theory [6]. In compliance with this theory porous piezoceramics can be classified as two phase composite. Depending on whether their pores are unconnected or connected, such piezoceramics are said to have 3-0 or 3-3 connectivity. Here, the first number denotes three-dimensional interconnection of the ceramic material and the second number marks the dimensions in which the pore volume is interconnected. As is well known, the 3-0 and 3-3 piezocomposites in the form of porous Lead Zirconate Titanate (PZT) materials show considerably improved transducer characteristics [10-14, 17-19, 23].

Porous composite piezoceramics having pore or inclusion sizes less than 100 μm may be accepted as a quasi homogeneous medium with some effective moduli for most applications. At present, there are many publications in which the effective properties of 3-0, 3-3 piezocomposite media have been analyzed [7-14]. The approximate formulas of engineering character for piezocomposites with various types of connectivity have been obtained. However, the strong mathematical approaches of mechanics of composites were used only in few of papers [15-16, 20-21].

In the first part of our work we present the developed methodology to predict the effective moduli using the finite element technique in accordance with [22-25]. New aspects of this paper relate to the models of the representative volumes and with taking into account the inhomogeneous polarization. In the second part of our work we investigate 1-3 piezocomposite hydro-acoustic projectors made of solid ceramics with elastic matching layers and piezoelectric emitter from porous piezoceramics without such additional layers. To conclude a good efficiency of high porous piezoceramics for quasistatic and low-frequency applications we perform harmonic and transient analysis for 1-D FE models of the projectors with two, one and without matching layers.

2. MATHEMATICAL ASPECTS OF THE EFFECTIVE MODULI METHOD

Let Ω be a region occupied by a heterogeneous body with piezoelectric properties, $\Gamma = \partial\Omega$ is the boundary of the region, $\mathbf{n}(\mathbf{x})$ is the vector of the external unit normal to Γ ($\mathbf{x} \in \Gamma$), $\mathbf{u}(\mathbf{x})$ is the displacement vector-function, $\varphi(\mathbf{x})$ is the electric potential function. We denote by $\boldsymbol{\varepsilon}$ the deformation tensor, $\boldsymbol{\sigma}$ - the stress tensor, $\mathbf{E}$ - the electric field vector, $\mathbf{D}$ - the electric displacement vector. The fields $\boldsymbol{\varepsilon}$ and $\mathbf{E}$ are determined from the functions $\mathbf{u}$ and φ by the following form:

$$\boldsymbol{\varepsilon} = \frac{1}{2}(\nabla\mathbf{u} + \nabla\mathbf{u}^*), \qquad \mathbf{E} = -\nabla\varphi, \tag{1}$$

where superscript "*" denotes transposition operation.

On the region Γ we will consider the mechanical stress vector $\mathbf{p}$ and surface density of electric charges q

$$\mathbf{p} = \mathbf{n} \cdot \boldsymbol{\sigma}, \tag{2}$$

$$q = -\mathbf{n} \cdot \mathbf{D}. \tag{3}$$

As usual, we will denote the volume-averaged quantities in the broken brackets

$$\langle ... \rangle = \frac{1}{\Omega} \int_\Omega (...) \, d\Omega. \tag{4}$$

Following a substantiation of a method of effective moduli for elastic media, for piezoelectric body we will formulate the auxiliary statements. These statements are proved under similar techniques, as for an elastic body [26]. The following states are found on the basis of effective moduli theory.

Lemma 1. These representations take place for volume averaging field characteristic by means of appropriate values on boundary Γ

$$\text{(a)} \quad \boldsymbol{\varepsilon} = \frac{1}{2}(\nabla\mathbf{u} + \nabla\mathbf{u}^*) \quad \Rightarrow \quad \langle \boldsymbol{\varepsilon} \rangle = \frac{1}{2\Omega} \int_\Gamma (\mathbf{un}^* + \mathbf{nu}^*) \, d\Gamma, \tag{5}$$

$$\text{(b)} \quad \mathbf{E} = -\nabla\varphi \quad \Rightarrow \quad \langle \mathbf{E} \rangle = -\frac{1}{\Omega} \int_\Gamma \mathbf{n}\varphi \, d\Gamma, \tag{6}$$

$$\text{(c)} \quad \mathbf{p} = \mathbf{n} \cdot \boldsymbol{\sigma}; \quad \nabla \cdot \boldsymbol{\sigma} = 0 \quad \Rightarrow \quad \langle \boldsymbol{\sigma} \rangle = \frac{1}{2\Omega} \int_\Gamma (\mathbf{px}^* + \mathbf{xp}^*) \, d\Gamma, \tag{7}$$

(d) $q = -\mathbf{n}\cdot\mathbf{D}$; $\nabla\cdot\mathbf{D} = 0 \quad \Rightarrow \quad \langle\mathbf{D}\rangle = -\frac{1}{\Omega}\int_\Gamma \mathbf{x}q\,d\Gamma$. (8)

Lemma 2.

(a) Let $\mathbf{u} = \mathbf{x}\cdot\boldsymbol{\varepsilon}_0|_\Gamma$ for $\mathbf{x}\in\Gamma$, where $\boldsymbol{\varepsilon}_0 = \boldsymbol{\varepsilon}_0{}^* = \text{const}$. Then we have $\langle\boldsymbol{\varepsilon}\rangle = \boldsymbol{\varepsilon}_0$.

(b) Let $\varphi = -\mathbf{x}\cdot\mathbf{E}_0|_\Gamma$ for $\mathbf{x}\in\Gamma$, where $\mathbf{E}_0 = \text{const}$. Then we have $\langle\mathbf{E}\rangle = \mathbf{E}_0$.

(c) Let $\mathbf{p} = \mathbf{n}\cdot\boldsymbol{\sigma}_0|_\Gamma$ for $\mathbf{x}\in\Gamma$, where $\mathbf{p}$ is the stress vector from (2) and $\boldsymbol{\sigma}_0 = \text{const}$. Then we have $\langle\boldsymbol{\sigma}\rangle = \boldsymbol{\sigma}_0$.

(d) Let $q = -\mathbf{n}\cdot\mathbf{D}_0|_\Gamma$ for $\mathbf{x}\in\Gamma$, where q is the surface density of electric charges from (3) and $\mathbf{D}_0 = \text{const}$. Then we have $\langle\mathbf{D}\rangle = \mathbf{D}_0$.

Lemma 3.

(a) Let $\mathbf{u} = \mathbf{x}\cdot\boldsymbol{\varepsilon}_0|_\Gamma$ for $\mathbf{x}\in\Gamma$, where $\boldsymbol{\varepsilon}_0 = \boldsymbol{\varepsilon}_0{}^* = \text{const}$, and the equilibrium equation $\nabla\cdot\boldsymbol{\sigma} = 0$ takes place. Then we have $\langle\boldsymbol{\varepsilon}\cdot\cdot\boldsymbol{\sigma}\rangle/2 = \langle\boldsymbol{\varepsilon}\rangle\cdot\cdot\langle\boldsymbol{\sigma}\rangle/2$.

(b) Let $\varphi = -\mathbf{x}\cdot\mathbf{E}_0|_\Gamma$ for $\mathbf{x}\in\Gamma$, where $\mathbf{E}_0 = \text{const}$, and the equation of quasielectrostatic $\nabla\cdot\mathbf{D} = 0$ takes place. Then we have $\langle\mathbf{D}\cdot\mathbf{E}\rangle/2 = \langle\mathbf{D}\rangle\cdot\langle\mathbf{E}\rangle/2$.

(c) Let $\mathbf{p} = \mathbf{n}\cdot\boldsymbol{\sigma}_0|_\Gamma$ for $\mathbf{x}\in\Gamma$, where $\boldsymbol{\sigma}_0 = \text{const}$, $\mathbf{p}$ is the stress vector from (2), and the equilibrium equation $\nabla\cdot\boldsymbol{\sigma} = 0$ takes place. Then we have $\langle\boldsymbol{\varepsilon}\cdot\cdot\boldsymbol{\sigma}\rangle/2 = \langle\boldsymbol{\varepsilon}\rangle\cdot\cdot\langle\boldsymbol{\sigma}\rangle/2$.

(d) Let $q = -\mathbf{n}\cdot\mathbf{D}_0|_\Gamma$ for $\mathbf{x}\in\Gamma$, where $\mathbf{D}_0 = \text{const}$, q is the surface density from (3), and the equation of $\nabla\cdot\mathbf{D} = 0$ takes place. Then we have $\langle\mathbf{D}\cdot\mathbf{E}\rangle/2 = \langle\mathbf{D}\rangle\cdot\langle\mathbf{E}\rangle/2$.

In accordance with four equivalent fundamental forms of constitutive equations [22] we will introduce the moduli of piezoelectric medium $\mathbf{c}^E$, $\mathbf{e}$, $\boldsymbol{\varepsilon}^S$ and so on:

– in independent variables $\boldsymbol{\varepsilon}$, $\mathbf{E}$

$$\boldsymbol{\sigma} = \mathbf{c}^E\cdot\cdot\boldsymbol{\varepsilon} - \mathbf{e}^*\cdot\mathbf{E}, \qquad \mathbf{D} = \mathbf{e}\cdot\cdot\boldsymbol{\varepsilon} + \boldsymbol{\varepsilon}^S\cdot\mathbf{E}, \tag{9}$$

– in independent variables $\boldsymbol{\sigma}$, $\mathbf{E}$

$$\boldsymbol{\varepsilon} = \mathbf{s}^E\cdot\cdot\boldsymbol{\sigma} + \mathbf{d}^*\cdot\mathbf{E}, \qquad \mathbf{D} = \mathbf{d}\cdot\cdot\boldsymbol{\sigma} + \boldsymbol{\varepsilon}^T\cdot\mathbf{E}, \tag{10}$$

– in independent variables $\boldsymbol{\varepsilon}$, $\mathbf{D}$

$$\boldsymbol{\sigma} = \mathbf{c}^D\cdot\cdot\boldsymbol{\varepsilon} - \mathbf{h}^*\cdot\mathbf{D}, \qquad \mathbf{E} = -\mathbf{h}\cdot\cdot\boldsymbol{\varepsilon} + \boldsymbol{\beta}^S\cdot\mathbf{D}, \tag{11}$$

– in independent variables $\boldsymbol{\sigma}$, $\mathbf{D}$

$$\boldsymbol{\varepsilon} = \mathbf{s}^D\cdot\cdot\boldsymbol{\sigma} + \mathbf{g}^*\cdot\mathbf{D}, \qquad \mathbf{E} = -\mathbf{g}\cdot\cdot\boldsymbol{\sigma} + \boldsymbol{\beta}^T\cdot\mathbf{D}. \tag{12}$$

Here $\mathbf{c}^E$, $\mathbf{c}^D$ are the fourth rank tensors of elastic stiffness moduli at constant electric field and at constant electric displacement, respectively; $\mathbf{e}$, $\mathbf{d}$, $\mathbf{h}$, $\mathbf{g}$ are the third rank tensors of piezoelectric moduli (stress coefficients, charge coefficients, strain coefficients, voltage coefficient, respectively); $\boldsymbol{\varepsilon}^S$, $\boldsymbol{\varepsilon}^T$ are the second rank tensors of dielectric permittivity moduli at constant mechanical stress and at constant mechanical strain, respectively; $\mathbf{s}^E$, $\mathbf{s}^D$ are the fourth rank tensors of elastic compliance moduli at constant electric field and at constant electric displacement, respectively; $\boldsymbol{\beta}^S$, $\boldsymbol{\beta}^T$ are the second rank tensors of dielectric impermittivity at constant mechanical stress and at constant mechanical strain, respectively. These different material constants are connected by relations: $\mathbf{s}^E=(\mathbf{c}^E)^{-1}$, $\mathbf{s}^D=(\mathbf{c}^D)^{-1}$, $\boldsymbol{\beta}^S=(\boldsymbol{\varepsilon}^S)^{-1}$, $\boldsymbol{\beta}^T=(\boldsymbol{\varepsilon}^T)^{-1}$, $\mathbf{c}^D=\mathbf{c}^E+\mathbf{e}^*\cdot\mathbf{h}$, $\mathbf{s}^D=\mathbf{s}^E-\mathbf{d}^*\cdot\mathbf{g}$, $\boldsymbol{\varepsilon}^T=\boldsymbol{\varepsilon}^S+\mathbf{d}\cdot\cdot\mathbf{e}^*$, $\boldsymbol{\beta}^T=\boldsymbol{\beta}^S-\mathbf{g}\cdot\cdot\mathbf{h}^*$, $\mathbf{e}=\mathbf{d}\cdot\cdot\mathbf{c}^E$, $\mathbf{d}=\boldsymbol{\varepsilon}^T\cdot\mathbf{g}$, $\mathbf{g}=\boldsymbol{\beta}^T\cdot\mathbf{d}$, $\mathbf{h}=\mathbf{g}\cdot\cdot\mathbf{c}^D$.

For inhomogeneous body these moduli will be functions of coordinates $\mathbf{x}$: $\mathbf{c}^E=\mathbf{c}^E(\mathbf{x})$; $\mathbf{e}=\mathbf{e}(\mathbf{x})$; $\boldsymbol{\varepsilon}^S=\boldsymbol{\varepsilon}^S(\mathbf{x})$, etc.

Let Ω be the representative volume of heterogeneous piezoelectric materials. We will determine the effective moduli $\mathbf{c}^{Eeff}$, $\mathbf{e}^{eff}$, $\boldsymbol{\varepsilon}^{Seff}$ by the following technique [22-25].

We consider the static piezoelectric problem for representative volume Ω

$$\nabla\cdot\boldsymbol{\sigma}=0,\qquad \nabla\cdot\mathbf{D}=0,\qquad \mathbf{x}\in\Omega; \tag{13}$$

$$\mathbf{u}=\mathbf{x}\cdot\boldsymbol{\varepsilon}_0,\qquad \varphi=-\mathbf{x}\cdot\mathbf{E}_0,\ \ \mathbf{x}\in\Gamma=\partial\Omega \tag{14}$$

We call the problem (1), (9), (13), (14) the $u\varphi$ - problem and denote the solution of this problem by $\mathbf{u}^{u\varphi}$, $\varphi^{u\varphi}$. From the obtained solution and (1), (9) we find $\boldsymbol{\varepsilon}^{u\varphi}$, $\mathbf{E}^{u\varphi}$, $\boldsymbol{\sigma}^{u\varphi}$ and $\mathbf{D}^{u\varphi}$, where $\boldsymbol{\varepsilon}^{u\varphi}=\boldsymbol{\varepsilon}(\mathbf{u}^{u\varphi})$ and so on.

Due to linearity of $u\varphi$ - problem (13), (14) there exist the tensor functions $\mathbf{A}^{u\varphi}$, $\mathbf{B}^{u\varphi}$, $\mathbf{F}^{u\varphi}$ and $\mathbf{G}^{u\varphi}$ depending on the spatial coordinates. By using these functions we join $\boldsymbol{\varepsilon}^{u\varphi}$ and $\mathbf{E}^{u\varphi}$ with $\boldsymbol{\varepsilon}_0$ and $\mathbf{E}_0$ from (14)

$$\boldsymbol{\varepsilon}^{u\varphi}=\mathbf{A}^{u\varphi}\cdot\cdot\boldsymbol{\varepsilon}_0+\mathbf{B}^{u\varphi}\cdot\mathbf{E}_0;\ \ \mathbf{E}^{u\varphi}=\mathbf{F}^{u\varphi}\cdot\cdot\boldsymbol{\varepsilon}_0+\mathbf{G}^{u\varphi}\cdot\mathbf{E}_0 \tag{15}$$

By substituting (15) in the constitutive equations (9), we obtain the relations $\boldsymbol{\sigma}^{u\varphi}$, $\mathbf{D}^{u\varphi}$ with $\boldsymbol{\varepsilon}_0$ and $\mathbf{E}_0$

$$\boldsymbol{\sigma}^{u\varphi}=(\mathbf{c}^E\cdot\cdot\mathbf{A}^{u\varphi}-\mathbf{e}^*\cdot\mathbf{F}^{u\varphi})\cdot\cdot\boldsymbol{\varepsilon}_0+(\mathbf{c}^E\cdot\cdot\mathbf{B}^{u\varphi}-\mathbf{e}^*\cdot\mathbf{G}^{u\varphi})\cdot\mathbf{E}_0 \tag{16}$$

$$\mathbf{D}^{u\varphi}=(\mathbf{e}\cdot\cdot\mathbf{A}^{u\varphi}+\boldsymbol{\varepsilon}^S\cdot\mathbf{F}^{u\varphi})\cdot\cdot\boldsymbol{\varepsilon}_0+(\mathbf{e}\cdot\cdot\mathbf{B}^{u\varphi}+\boldsymbol{\varepsilon}^S\cdot\mathbf{G}^{u\varphi})\cdot\mathbf{E}_0$$

We note from Lemma 2, that for $u\varphi$ - problem $\langle\boldsymbol{\varepsilon}^{u\varphi}\rangle=\boldsymbol{\varepsilon}_0$ and $\langle\mathbf{E}^{u\varphi}\rangle=\mathbf{E}_0$.

We supply in conformity to the initial heterogeneous medium some "equivalent" homogeneous medium with effective moduli $\mathbf{c}^{Eeff}$, $\mathbf{e}^{eff}$ and $\boldsymbol{\varepsilon}^{Seff}$. The constitutive equations for "equivalent" medium, analogous (9), are in the forms

$$\boldsymbol{\sigma}_0 = \mathbf{c}^{Eeff} \cdot\cdot \boldsymbol{\varepsilon}_0 - \mathbf{e}^{eff} * \cdot \mathbf{E}_0 , \ \mathbf{D}_0 = \mathbf{e}^{eff} \cdot\cdot \boldsymbol{\varepsilon}_0 + \boldsymbol{\varepsilon}^{Seff} \cdot \mathbf{E}_0 . \tag{17}$$

For $u\varphi$ - problem we accept the following equations such as relations for definition of effective moduli from (17)

$$\langle \boldsymbol{\sigma}^{u\varphi} \rangle = \boldsymbol{\sigma}_0 ; \ \langle \mathbf{D}^{u\varphi} \rangle = \mathbf{D}_0 . \tag{18}$$

The moduli, found from these conditions, are marked with superscripts "$u\varphi$". As a result from (16)–(18) we obtain the formulae for $u\varphi$ - moduli

$$\mathbf{c}^{Eeff,u\varphi} = \langle \mathbf{c}^E \cdot\cdot \mathbf{A}^{u\varphi} - \mathbf{e} * \cdot \mathbf{F}^{u\varphi} \rangle ; \qquad \boldsymbol{\varepsilon}^{Seff,u\varphi} = \langle \mathbf{e} \cdot\cdot \mathbf{B}^{u\varphi} + \boldsymbol{\varepsilon}^S \cdot \mathbf{G}^{u\varphi} \rangle ,$$

$$\mathbf{e}^{Eeff,u\varphi} = \langle \mathbf{e} \cdot\cdot \mathbf{A}^{u\varphi} + \boldsymbol{\varepsilon}^S \cdot \mathbf{F}^{u\varphi} \rangle = -\langle (\mathbf{c}^E \cdot\cdot \mathbf{B}^{u\varphi} - \mathbf{e} * \cdot \mathbf{G}^{u\varphi}) * \rangle . \tag{19}$$

Note, that with Lemma 3 the average energies are equal for heterogeneous and for "equivalent" homogeneous piezoelectric media

$$\langle \boldsymbol{\sigma}^{u\varphi} \cdot\cdot \boldsymbol{\varepsilon}^{u\varphi} + \mathbf{D}^{u\varphi} \cdot \mathbf{E}^{u\varphi} \rangle / 2 = (\boldsymbol{\sigma}_0 \cdot\cdot \boldsymbol{\varepsilon}_0 + \mathbf{D}_0 \cdot \mathbf{E}_0) / 2 . \tag{20}$$

This variant of equivalent moduli method for piezoelectric media follows from [22, 23]. However, for piezoelectric media it is possible to suggest the other ways of introduction of effective moduli, considering the problems with other mechanical and electric boundary conditions from Lemma 2 [24].

From the solutions of the problems (13), (14) it is possible to define all effective moduli of porous piezocomposites. Piezoceramic, homogeneously polarized along the x_3 axis, is anisotropic material of crystallographic class 6mm and have 10 independent moduli (5 elastic, 3 piezoelectric and 2 dielectric moduli). Thus, for all piezoceramic moduli determination it is enough to consider five various $u\varphi$ - problems for calculation of the moduli in two-index notation ($i, j = 1,2,3$; $\alpha = 1,2,...,6$):

I-$u\varphi$. $\boldsymbol{\varepsilon}_0 = \varepsilon_0 \mathbf{e}_1 \mathbf{e}_1$, ($\varepsilon_{011} = \varepsilon_0$), $\qquad \mathbf{E}_0 = 0 \ \Rightarrow$

$$c_{11}^{E\,\mathrm{eff},u\varphi} = \langle \sigma_{11}^{u\varphi} \rangle / \varepsilon_0 ; \ c_{12}^{E\,\mathrm{eff},u\varphi} = \langle \sigma_{22}^{u\varphi} \rangle / \varepsilon_0 ; \ c_{13}^{E\,\mathrm{eff},u\varphi} = \langle \sigma_{33}^{u\varphi} \rangle / \varepsilon_0 ; \ e_{31}^{\mathrm{eff},u\varphi} = \langle D_3^{u\varphi} \rangle / \varepsilon_0 \tag{21}$$

II-$u\varphi$. $\boldsymbol{\varepsilon}_0 = \varepsilon_0 \mathbf{e}_3 \mathbf{e}_3$, ($\varepsilon_{033} = \varepsilon_0$), $\qquad \mathbf{E}_0 = 0 \ \Rightarrow$

$$c_{33}^{E\,\mathrm{eff},u\varphi} = \langle \sigma_{33}^{u\varphi} \rangle / \varepsilon_0 ; \ c_{13}^{E\,\mathrm{eff},u\varphi} = \langle \sigma_{11}^{u\varphi} \rangle / \varepsilon_0 = \langle \sigma_{22}^{u\varphi} \rangle / \varepsilon_0 ; \ e_{33}^{\mathrm{eff},u\varphi} = \langle D_3^{u\varphi} \rangle / \varepsilon_0 \tag{22}$$

III-$u\varphi$. $\boldsymbol{\varepsilon}_0 = \varepsilon_0(\mathbf{e}_2\mathbf{e}_3 + \mathbf{e}_3\mathbf{e}_2)$, ($\varepsilon_{023} = \varepsilon_{032} = \varepsilon_0$), $\mathbf{E}_0 = 0 \Rightarrow$

$$c_{44}^{E\,\mathrm{eff},u\varphi} = \langle \sigma_{23}^{u\varphi} \rangle /(2\varepsilon_0);\; e_{15}^{\mathrm{eff},u\varphi} = \langle D_2^{u\varphi} \rangle /(2\varepsilon_0) \quad (23)$$

IV-$u\varphi$. $\boldsymbol{\varepsilon}_0 = 0$, $\mathbf{E}_0 = E_0\mathbf{e}_1$ ($E_{01} = E_0$) $\Rightarrow$

$$e_{15}^{\mathrm{eff},u\varphi} = -\langle \sigma_{13}^{u\varphi} \rangle / E_0;\; \varepsilon_{11}^{S\,\mathrm{eff},u\varphi} = \langle D_1^{u\varphi} \rangle / E_0 \quad (24)$$

V-$u\varphi$. $\boldsymbol{\varepsilon}_0 = 0$, $\mathbf{E}_0 = E_0\mathbf{e}_3$ ($E_{03} = E_0$) $\Rightarrow$

$$e_{31}^{\mathrm{eff},u\varphi} = -\langle \sigma_{11}^{u\varphi} \rangle / E_0 = -\langle \sigma_{22}^{u\varphi} \rangle / E_0;\; e_{33}^{\mathrm{eff},u\varphi} = -\langle \sigma_{33}^{u\varphi} \rangle / E_0;\; \varepsilon_{33}^{S\,\mathrm{eff},u\varphi} = \langle D_3^{u\varphi} \rangle / E_0 \quad (25)$$

In (21)–(25), we use the vector-matrix forms for moduli [27]: $\mathbf{c}^E$ is the 6 × 6 matrix (pseudomatrix) of elastic moduli, $c_{\alpha\beta}^E = c_{ijkl}^E$; $\alpha, \beta = 1,2,...,6$; $i, j, k, l = 1,2,3$ with correspondence law $\alpha \leftrightarrow (ij)$, $\beta \leftrightarrow (kl)$, $1 \leftrightarrow (11)$, $2 \leftrightarrow (22)$, $3 \leftrightarrow (33)$, $4 \leftrightarrow (23) = (32)$, $5 \leftrightarrow (13) = (31)$, $6 \leftrightarrow (12) = (21)$; $\mathbf{e}$ is the 3 × 6 matrix (pseudomatrix) of piezoelectric moduli ($e_{i\beta} = e_{ikl}$).

Note that from the first $u\varphi$ - problem from (21) it is possible to find the following essentially non-zero effective moduli: $c_{11}^{Eeff,u\varphi}$, $c_{12}^{Eeff,u\varphi}$, $c_{13}^{Eeff,u\varphi}$ and $e_{31}^{eff,u\varphi}$. Analogously, from the second $u\varphi$ - problem from (22) we can determine the effective moduli $c_{13}^{Eeff,u\varphi}$, $c_{33}^{Eeff,u\varphi}$, $e_{33}^{eff,u\varphi}$; from the third $u\varphi$ - problem from (23) we can find $c_{44}^{Eeff,u\varphi}$, $e_{15}^{eff,u\varphi} = e_{24}^{eff,u\varphi}$; from the fourth $u\varphi$ - problem from (24) we can find $e_{15}^{eff,u\varphi}$, $\varepsilon_{11}^{Seff,u\varphi}$; and from the fifth $u\varphi$ - problem from (25) we can find $e_{31}^{eff,u\varphi}$, $e_{33}^{eff,u\varphi}$ and $\varepsilon_{33}^{Seff,u\varphi}$. As a result we shall have a full set of effective moduli of porous piezoceramics.

3. Hypothesis of Inhomogeneous Polarization and Models of Representative Volumes

In order to take into account the inhomogeneous polarization, we previously solved the electrostatic problem for the dielectric composite. In this case on first stage the process of partial polarization of porous ceramics in z - axis direction is modelled. For this first a quasielectrostatic problem for inhomogeneous dielectric is solved in region Ω:

$$\nabla \cdot \mathbf{D} = 0,\; \mathbf{D} = \varepsilon_* \mathbf{E},\; \mathbf{E} = -\nabla\varphi,\; \mathbf{x} \in \Omega, \quad (26)$$

$$\varphi = V_j,\quad \mathbf{x} \in \Gamma_{\varphi j}, \quad (27)$$

$$\mathbf{n} \cdot \mathbf{D} = 0,\; \mathbf{x} \in \Gamma_q, \quad (28)$$

where $\Gamma = \Gamma_{\varphi 1} \cup \Gamma_{\varphi 2} \cup \Gamma_q$; $\Gamma_{\varphi 1}$, $\Gamma_{\varphi 2}$ are the electrodes $z = L$ and $z = 0$ respectively; $V_1 = -E_p L$, $V_2 = 0$, E_p is a preliminary value for polarization field.

Here on the faces of a representative volume the potential difference was applied, and then the electric field was determined. Such problems have been modeling the piezoceramics polarization process. Further, according to the obtained electric field, the polarization vector directions and values for piezoceramic matrix of the multiple-phase piezocomposite have been finding. After that each finite element of a representative volume was associated with a new coordinate system so that the third axis was chosen codirectional to the element's polarization vector. As a result we have a representative volume for the piezocomposite with inhomogeneous properties of piezoceramic matrix.

In order to consider the structure of real ceramics we implemented several methods for generating a representative volume of the two-phase cubic piezocomposite: random method, method of the initial concentration, method of longitudinal plies, method of transversal plies, the DLA (diffusion limited aggregation) Witten-Sander method, and the DLA "growth from the plane" method. All these techniques are described in details in the work [2]. Several examples of the composite cube sections generated by these methods are shown below.

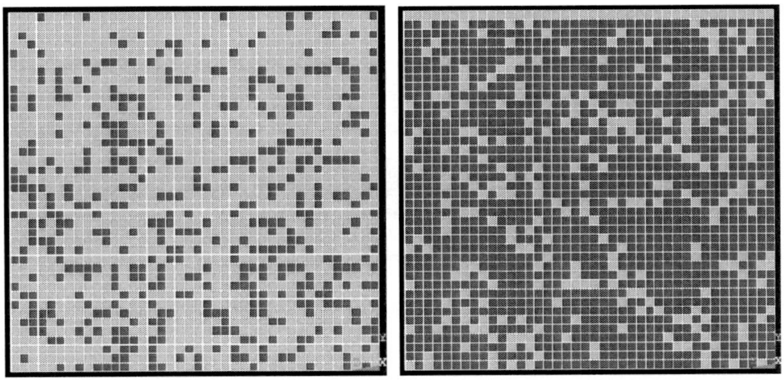

Figure 1. Central sections of the composite cube, generated by the initial concentration method (at 20% and 80% of the filler concentration, respectively).

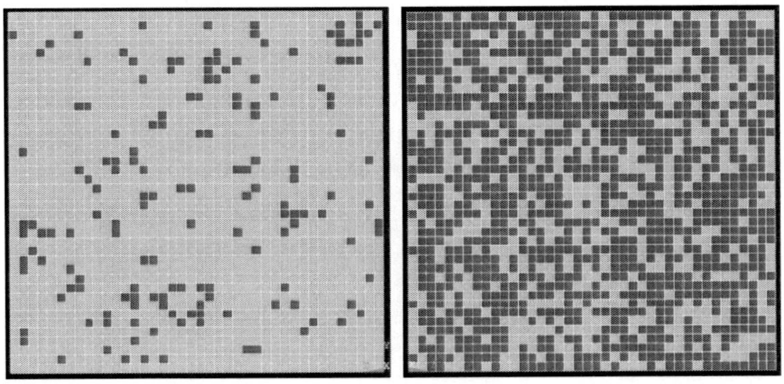

Figure 2. Central section of the composite cube, generated by the DLA method (at 20% and 80% of the filler concentration, respectively).

4. NUMERICAL RESULTS FOR THE EFFECTIVE MODULI DETERMINATION

Let us consider the dependencies of piezoelectric moduli relations for porous ceramics $r(d_{3j}) = d_{3j}^{eff} / d_{3j}$ ($j = 1, 3$) from porosity. Solving the $u\varphi$ -problem, we examined different models, generated by the random method, method of the initial concentration and by the DLA method. Figures 3 and 4 demonstrate the results obtained at homogeneous and inhomogeneous polarization, respectively. Assuming the hypothesis of inhomogeneous polarization and using the method of the initial concentration we obtained values of d_{33} that are almost not decreasing with porosity growth and very close to the experimental results from [31, 32] (Figure 4a). The calculated values of d_{33} for porous ceramics at homogeneous polarization are in less conformity with the experiments (Figure 3b). Comparing Figures 3b and 4b one can see that the piezoelectric modulus d_{31} is decreasing with the porosity growth in both cases, it is also obvious that it is decreasing more at inhomogeneous polarization, which is in better agreement with experimental data, especially for the DLA method.

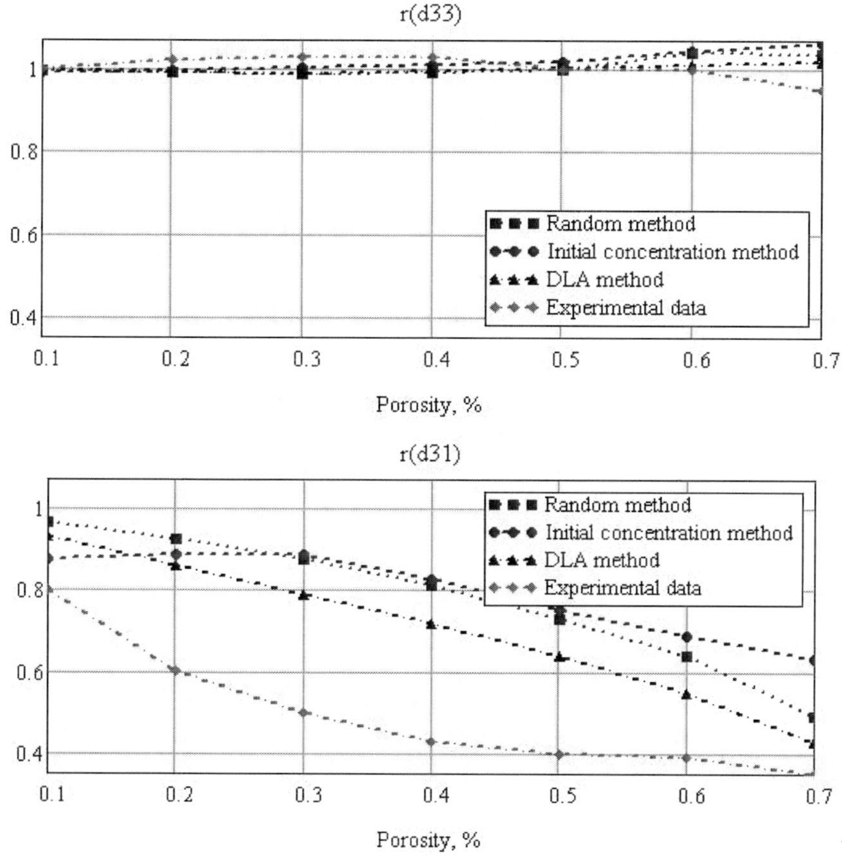

a) $r(d_{33})$

b) r(d_{31})

Figure 3. Dependencies of effective moduli on porosity at homogeneous polarization.

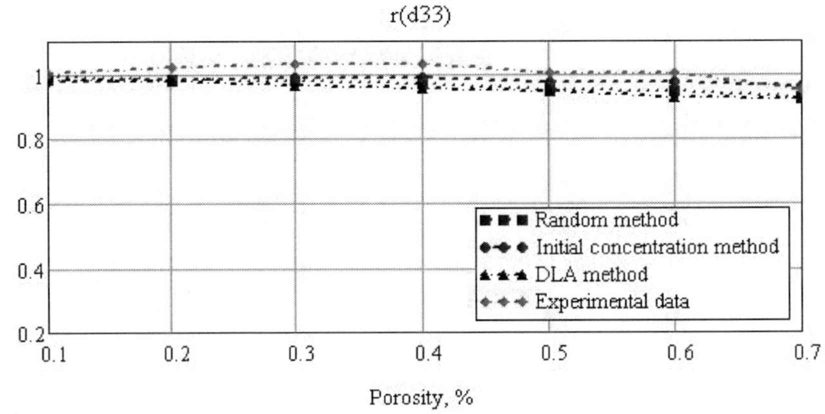

a) $r(d_{33})$

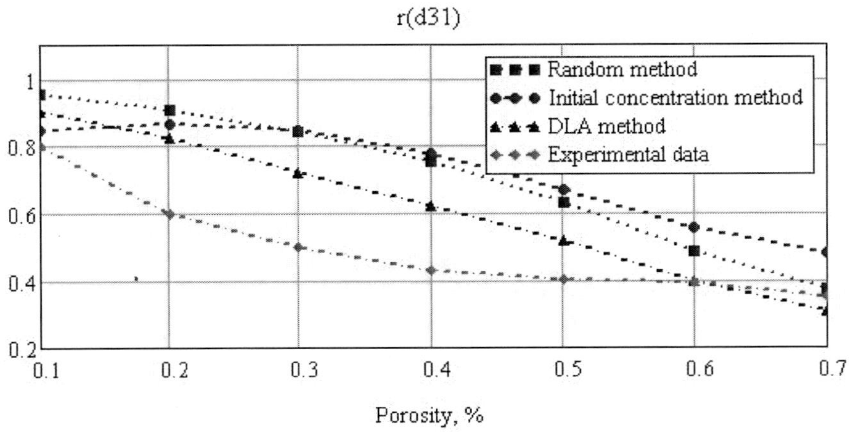

b) $r(d_{31})$

Figure 4. Dependencies of effective moduli on porosity at inhomogeneous polarization

Figures 5-7 demonstrate some simple dependencies of material constants on the porosity.

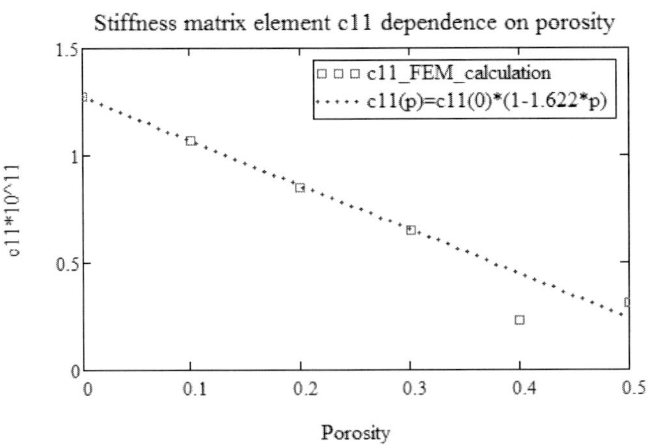

a) C_{11}

Figure 5. Continued on next page.

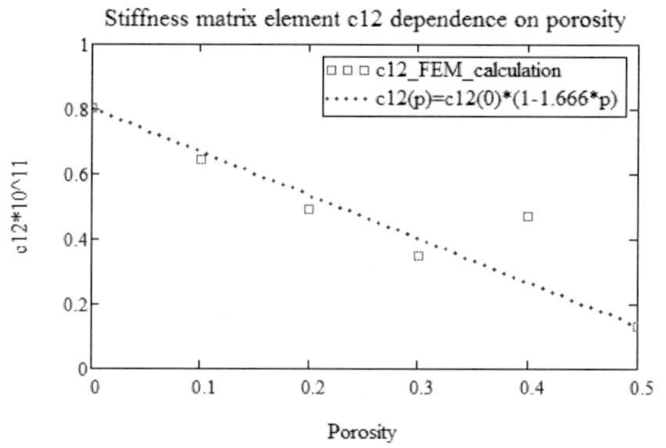

b) C_{12}

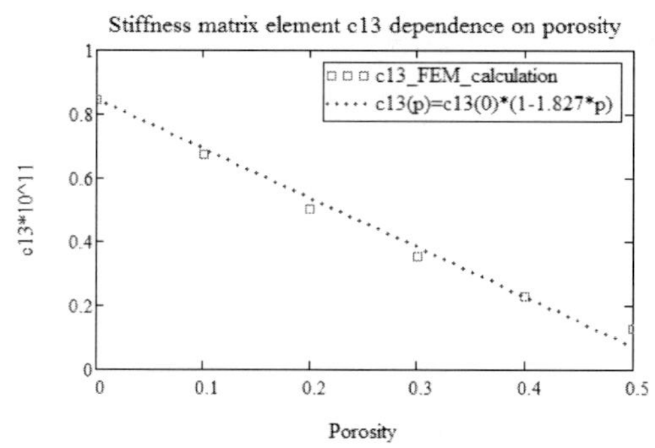

c) C_{13}

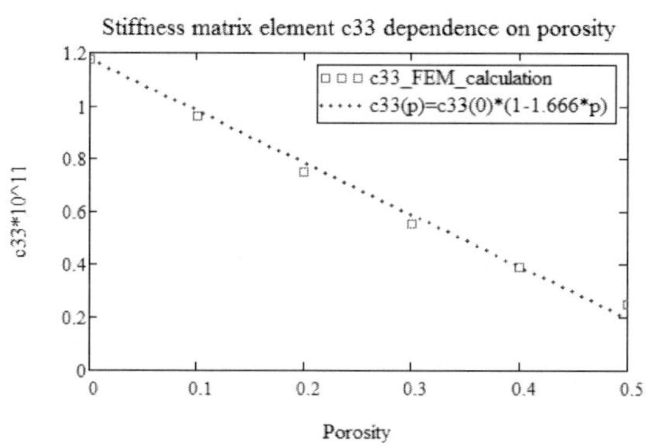

d) C_{33}

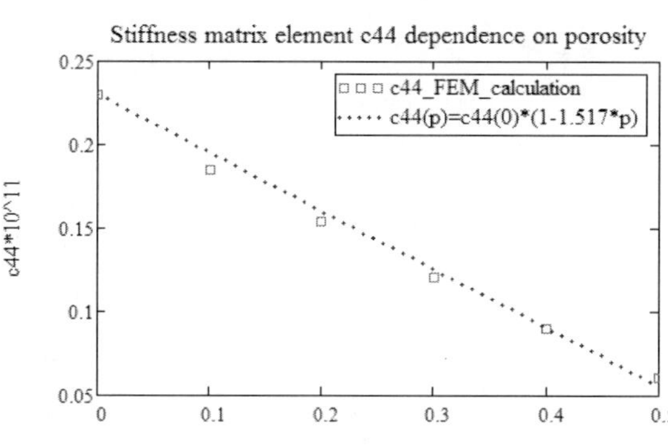

e) C_{44}

Figure 5. Dependencies of effective elastic moduli on porosity.

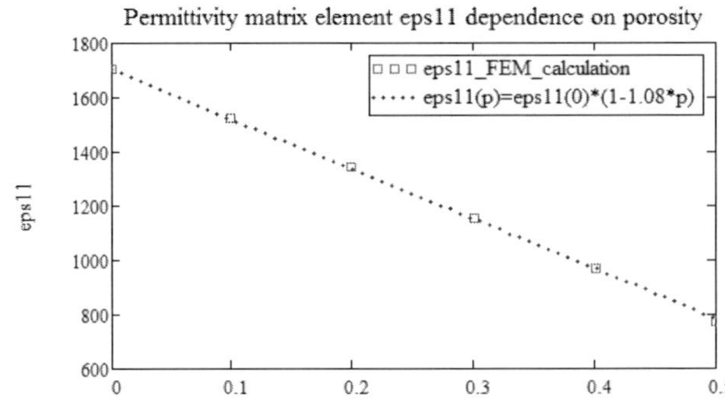

a) ε_{11}^{Seff}

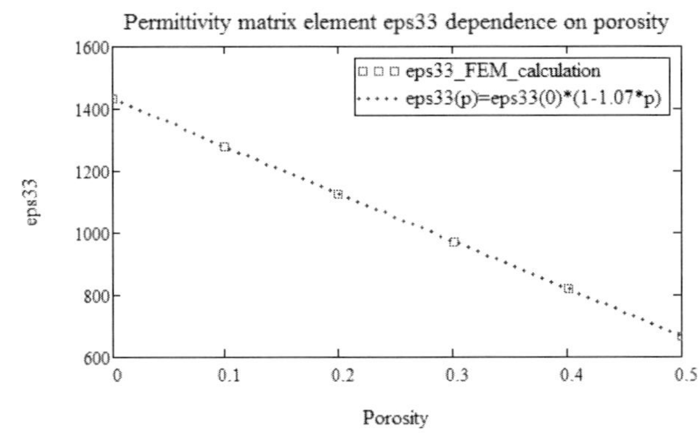

b) ε_{33}^{Seff}

Figure 6. Dependencies of effective dielectric permittivity moduli on porosity.

These results allow us to obtain simple empirical material constants dependencies on the porosity, thus reducing the number of design variables from 10 to 1 which is porosity. The resulting linear relationships are characterized by the different slope of the moduli dependencies on porosity, that ensures the possibility of finding an optimal porosity at various structural and technological constraints on the size and shape of the piezoelectric transducer, matching layers and the substrate.

5. Example of Piezoelectric Hydroacoustic 1-D Projector

FE complex ANSYS offers powerful tools for the analysis of piezoelectric devices. Let us consider the one-dimensional x-axis oriented continuum model of piezoelectric medium that is filling the region $x \in \lfloor x_k, x_m \rfloor$, $x_m - x_k = h_j$. We denote the displacement along the x axis by u, and the electric potential by φ. In a one-dimensional model with characteristic density ρ, elastic modulus c, dielectric constant ε and the piezoelectric modulus e, the stress σ and the electric displacement D are expressed in the following way:

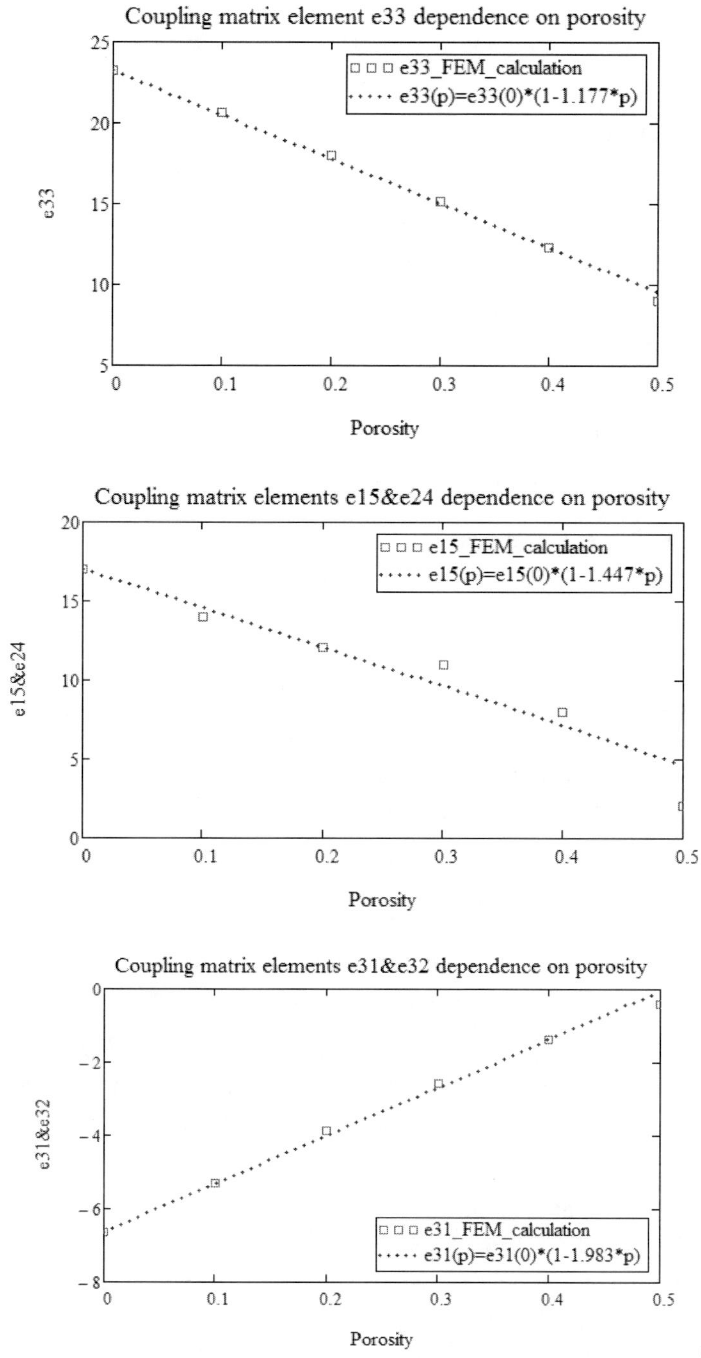

a) e_{33}^{eff}

b) e_{15}^{eff}

c) e_{31}^{eff}

Figure 7. Dependencies of effective piezoelectric moduli on porosity.

$$\sigma = cu' + e\varphi' , \; D = eu' - \varepsilon\varphi' , \tag{29}$$

and field equations have the form:

$$\sigma' = \rho \ddot{u}\,,\; D' = 0\,. \tag{30}$$

From (29) - (30) with a potential difference $V = \varphi(x_k, t) - \varphi(x_m, t)$ the equations for the mechanical functions can be obtained as

$$\bar{c}u'' = \rho \ddot{u}\,,\; \bar{c} = c + e^2/\varepsilon\,, \tag{31}$$

$$\sigma = \bar{c}u' + \frac{e^2}{\varepsilon h_j}\big(u(x_k, t) - u(x_m, t)\big) - \frac{e}{h_j}V\,. \tag{32}$$

According to (31)-(32), the behavior of piezoelectric medium with electrodes, powered by a potential difference V, can be simulated by an elastic medium with density ρ and of elastic modulus $\bar{c}$ with the external forces $F_k = -\frac{eA}{h_j}$ and $F_m = \frac{eA}{h_j}V$ at the boundary points $x = x_k$ and $x = x_m$, respectively (A is the cross-sectional area), and the medium connects points x_k and x_m by the spring with negative stiffness $-\frac{e^2 A}{\varepsilon h_j}V$.

In ANSYS the finite element LINK1 can be used to simulate the x-dimensional elastic problem, and the COMBIN14 element to simulate the spring. Acoustic medium in the 1-D case can be simulated by the elastic medium with the "density" ρ_a/v_a^2 (v_a is the the speed of sound) and "elastic modulus" ρ_a. The interaction of the elastic and acoustic medium can be accounted by creating two nodes at a contact point with the same coordinates and with different consecutive numbers. Then, for these nodes it is enough to create an element damping matrix $\mathbf{C^e}$ with the components: $C_{11}^e = C_{22}^e = 0$, $C_{12}^e = \rho_a$ and $C_{21}^e = -\rho_a$. The impedance boundary condition for the acoustic medium can be introduced into a one-dimensional FE model by setting the diagonal component of the damping matrix for the corresponding node ρ_a^2/Z_a, where Z_a is the impedance of the acoustic medium. The explicit values of the matrix $\mathbf{C}$ components in ANSYS can be set using the MATRIX27 finite element. Thus, a one-dimensional model of the acoustic, elastic and piezoelectric media interaction can be created in ANSYS using only the elastic elements MATRIX27, COMBIN14 and LINK1.

As an example we consider the 1-D model of a cylindrical piezoelectric generator assigned for the excitation of the shock acoustic pulses of high power and short duration in water. We investigate (i) the three-layer generator consisting of a piezoelectric element made of dense piezoceramic and two elastic matching layers, (ii) the generator made of porous piezoceramics with one matching elastic layer, and finally (iii) the generator which consists only of one porous piezoceramic layer without matching layers.

In all cases the projector consists of discs with the same radius $R = 15 \cdot 10^{-3}$ m. In the first example piezoelectric disc is made of dense ceramics PZT-5H with the thickness $h_1 = 3.8 \cdot 10^{-3}$ m, and polarized over the thickness. Piezoceramics is characterized by the density $\rho = 7.5 \cdot 10^3$ kg/m^3, elastic modulus $c = c_{33}^E = 11.7 \cdot 10^{10}$ Pa, piezoelectric modulus $e = e_{33} = 23.2$ C/m^2 and dielectric permittivity $\varepsilon = 12.7 \times 10^{-9}$ F/m. Electrodes are applied at the ends of the piezoelement. For the impedance matching of the piezoelectric ceramics with water the first disc is glued to the two elastic discs with lower impedances. The second disc is made of glass, its thickness is equal to $h_2 = 2.3 \cdot 10^{-3}$ m, the density $\rho_2 = 2.42 \cdot 10^3$ kg/m^3, the velocity of longitudinal waves $v_2 = 5.4 \cdot 10^3$ m / s. The third disc is made of ebonite, it has a thickness equal to $h_3 = 1 \cdot 10^{-3}$ m, the density $\rho_3 = 1.2 \cdot 10^3$ kg/m^3 and velocity of longitudinal waves $v_3 = 5.4 \cdot 10^3$ m/s. For the water we assumed the next constants: $\rho_a = 1 \cdot 10^3$ kg/m^3, $v_a = 1.5 \cdot 10^3$ m/s. The figures of merit were assumed $Q_1 = 65$, $Q_2 = 1200$ and $Q_3 = 60$, respectively. Operating projector generates impulses into water through the outside layer. The system is excited by the potential applied to the conductive surfaces of the piezoelectric disc.

In the case of two-layer piezoelectric projector, the piezoelement of porous ceramics is characterized by the following effective moduli: elastic modulus $c = c_{33}^E = 5.78 \cdot 10^{10}$ Pa, piezoelectric modulus $e = e_{33} = 15.45$ C/m^2 and dielectric permittivity $\varepsilon = 8.655 \times 10^{-9}$ F/m at 30% of porosity for piezoelectric element, and by the effective moduli: $c = c_{33}^E = 1.77 \cdot 10^{10}$ Pa, $e = 5.76$ C/m^2 and $\varepsilon = 4.486 \times 10^{-9}$ F/m at 60% of porosity. At that the thickness of the piezoelectric elements in the case of 30% and 60% to be equal to $h_1 = 2.8 \cdot 10^{-3}$ m and $h_1 = 1.54 \cdot 10^{-3}$ m, respectively, and the figures of merit $Q_1 = 60$ and $Q_1 = 55$, respectively. The thickness of the glass matching layer was assumed to be $h_2 = 2.3 \cdot 10^{-3}$ m.

The last considered example is a single-layer piezoelectric projector made of porous ceramics. In this case the piezoelement is characterized by the same effective moduli as in the previous example. Thickness of the piezoelectric layer, and the figures of merit at 30% and 60% of porosity were taken the same as in previous example of the two-layer piezoelectric projector.

In the one-dimensional theory only the thickness movement of piezoelectric projector are taken into account. The thickness of the piezoelectric layer is chosen to be a half the length of the longitudinal ferro-hard wave $h_1 = \lambda_1 / 2 = \sqrt{c/\rho} / (2 f_t)$ at frequency of $f_t = 500$ kHz. The thickness of the jth elastic layers were chosen equal to one quarter of the longitudinal waves $h_j = \lambda_j / 4 = v_j / (4 f_t)$ at the same frequency.

In considered examples the piezoceramic layer was divided into 20 equal finite elements, the elastic layer - into 10 elements and the acoustic medium with the length of $h_a = 6\cdot10^{-3}$ m - into 60 finite elements.

Figures 8-9 show the results of transient analysis, namely the dependence of the wave pulses from time at a fixed point of the acoustic medium. As can be seen from these figures when using porous piezoelectric ceramics the first amplitude pressure pulses and their shape remain almost the same. It is obvious that monolithic emitters without matching layers or with reduced number of layers have technological advantages. Thus, from the results of FE analysis for the acoustic electroeleasticity tasks it can be concluded that the high-porous piezoelectric ceramics is an effective material for different sonar applications.

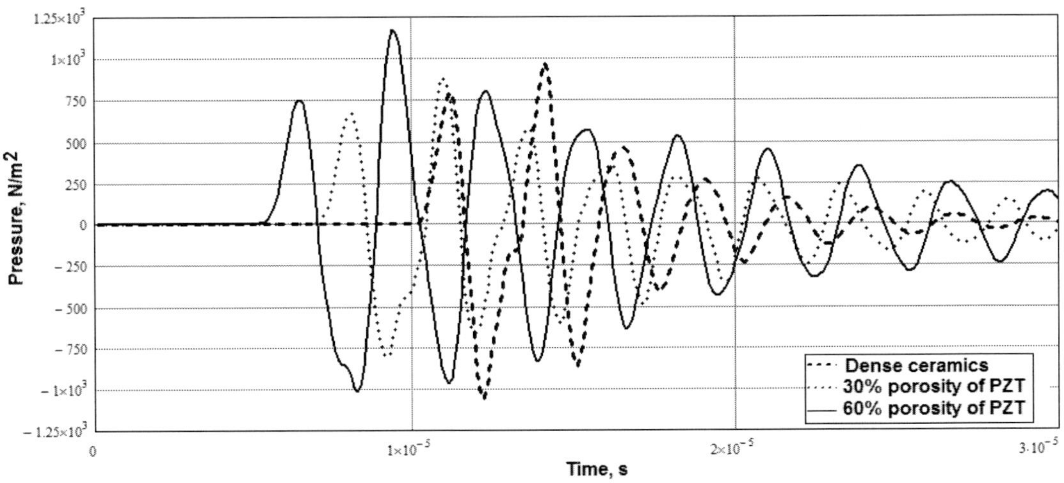

Figure 8. Pressure at fixed point, porous PZT with 1 matching layer (The DLA method).

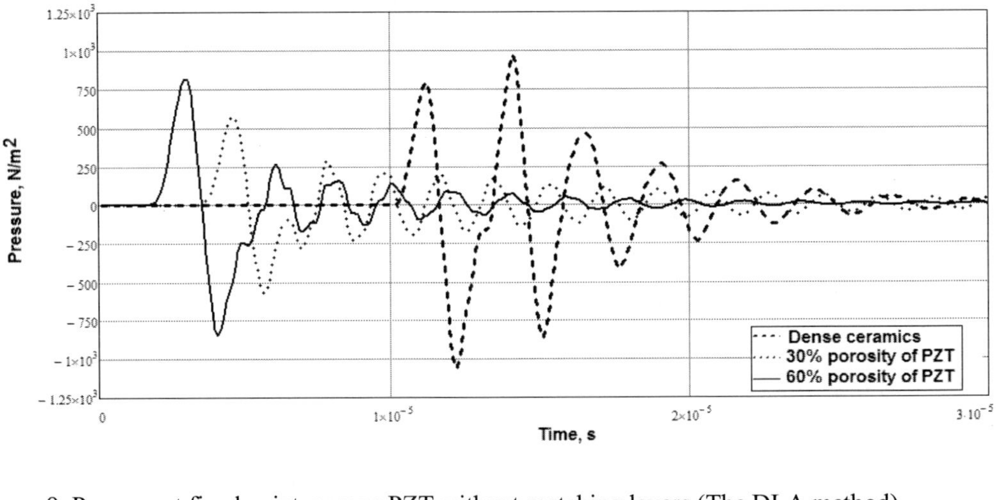

Figure 9. Pressure at fixed point, porous PZT without matching layers (The DLA method).

CONCLUSION

In our investigation, we developed and considered different FE models of porous piezoceramics based on the effective moduli method. The results of FEM modeling showed that the representative volume model with the diffusion limited aggregation (DLA) method gives better results compared to other models. The efficiency of proposed model and FE approximations were illustrated on the example of 1-3 piezocomposite projectors made of solid ceramics with elastic matching layers and piezoelectric emitter from porous piezoceramics without such layers. From the FE solutions of static and dynamic problems we conclude a good efficiency of high porous piezoceramics for quasistatic and low-frequency applications in piezoelectric energy harvesters, as its piezo-sensitivity rapidly increases at the porosity growth. We also verify the preference of porous ceramics for different hydro-acoustic ultrasonic applications.

ACKNOWLEDGEMENTS

This work is partially supported by the Russian Foundation for the Basic Research and by National Science Council of Taiwan (Grants Nos. 12-01-00829 and 10-08-13300-RT_omi)).

REFERENCES

[1] Nasedkin, A. V.; Shevtsova, M. S. In: *Ferroelectrics and Superconductors: Properties and Applications.* Ed. I. A. Parinov. Nova Science Publishers, N.-Y., 2011. Chapter 7. p. 231-254.

[2] Domashenkina, T. V.; Nasedkin, A. V.; Remizov, V. V.; Shevtsova, M. S. *Proc. 7th GRACM Int. Congress on Comput. Mechanics, Athens, Greece*, 2011. CD. Paper 141. 10 p.

[3] Willoughby, A. F. W. *Philosophical Transactions of the Royal Society.* 2006, *vol. 364*, 175-187.

[4] Boumchedda, K.; Hamadi, A.; Fantozzi, G. *Journal of the European Ceramic Society.* 2007, *vol. 27*, 4169–4171.

[5] Wersing, W. *Proceedings of the Sixth International Symposium on Applications of Ferroelectrics, IEEE, Piscataway*, 1986, pp. 212-223.

[6] Newnham, R. E.; Skinner, D. P.; Cross, L. E. *Vater. Res. Bull.*, 1978, *vol. 13*, 525-536.

[7] Rybianets, A. N., Nasedkin, A. V.; Turik, A. V. *Integrated Ferroelectrics.* 2004, *vol. 63*, 179-182.

[8] Wersing, W.; Lubitz, K.; Moliaupt, J. *Ferroelectrics.* 1986, *vol. 68*, 77-97.

[9] Banno, H. *Jpn. J. Appl. Phys.* 1993, *vol. 32*, 4214-4217.

[10] Rittenmyer, K.; Shrout, T.; Schulze, W.A.; Newnham, R.E. *Ferroelectrics.* 1982, *vol. 4*, 189-195.

[11] Perry, A.; Bowen, C. R.; Mahon, S. W. *Scripta Materialia.* 1999, *vol. 41*, 1001–1007.

[12] Bowen, C. R.; Perry, A.; Kara, H.; Mahon, S. W. *J. Eur. Ceram. Soc.* 2001, *vol. 21*, 1463-1467.

[13] Topolov, V. Yu.; Bowen, C. R. *Electromechanical Properties in Composites Based on Ferroelectrics*. Springer: London, 2009.

[14] Levassort, F.; Lethiecq, M.; Sesmare, R.; Tran-Huu-Hue, L. P. *IEEE Trans. Ultras. Ferr. Freq. Control*. 1999, *vol. 46*, 1028-1033.

[15] Dunn, H.; Taya, M. *J. Am. Ceram. Soc*. 1993, *vol. 76*, 1697-1706.

[16] Silva, E. C. N.; Fonseca, J. S. O.; Kikuchi, N. *Comput. Meth. Appl. Mech. Eng*. 1998, *vol. 159*, 49-77.

[17] Sokolkin, Yu. V.; Pan'kov, A. A. *Electroelasticity of Piezocomposites with Irregular Structure*. Fizmatlit: Moscow, 2003. (in Russian)

[18] Pan'kov, A. A. *Statistical Mechanics of Piezocomposites*. Perm State University Press: Perm, 2009. (in Russian)

[19] Ramesh, R.; Kara, H.; Bowen, C. R. *Comput. Materials Sci*. 2004, *vol. 30*, 397–403.

[20] Ramesh, R.; Kara, H.; Bowen, C. R. *Ultrasonics*. 2005, *vol. 43*, 173–181.

[21] Soloviev, A. N.; Vernigora, G. D. In: *Piezoceramic Materials and Devices*. Ed. I. A. Parinov. Nova Science Publishers, N.-Y., 2010, pp 217-240.

[22] Khoroshun, L. P.; Maslov, B. P.; Leshchenko, P. V. *Prediction of Effective Properties of Piezoactive Composite Materials*. Naukova Dumka: Kiev, 1989. (in Russian)

[23] Getman, I., Lopatin, S. *Ferroelectrics*. 1996, *vol. 186*, 301-304.

[24] Nasedkin, A. V. *Modern Problems of Continuum Mechanics, Proc. VII Int. Conf., Rostov on Don, 22-24 Oct. 2001*. Novaja Kniga: Rostov-on-Don, 2002, *vol. 1*, pp 182-188. (in Russian)

[25] Nasedkin, A.; Rybjanets, A.; Kushkuley, L.; Eshel, Y.; Tasker, R. *Proc. 2005 IEEE Ultrason. Symp., Rotterdam, Sept. 18 -21, 2005*, pp 1648-1651.

[26] Pobedria, B. E. *Mechanics of Composite Materials*. Moscow State University Press: Moscow, 1984. (in Russian)

[27] Berlincourt, D. A.; Curran, D. R.; Jaffe, H. *Piezoelectric and Piezomagnetic Materials and Their Application in Transducers.* Part A, In: W. P. Mason (Ed.), *Physical Acoustics*, vol. 1, Academic Press: New York, 1964.

[28] Kaltenbacher M. *Numerical Simulation of Mechatronic Sensors and Actuators*. Springer, Berlin - Heidelberg - New York, 2004.

[29] *ANSYS. Theory Ref.* Rel.11.0. Ed. P. Kothnke. ANSYS Inc. Houston, 2007.

[30] Nasedkin, A. V.; Rybjanets, A. N. *Russian Izvestiya, North-Caucas. Region. Ser. Techn Sciences, Spec. Issue*. 2004, pp 91-95. (in Russian)

[31] Dantziger, A. J.; et al. *Multicomponent Systems of Ferroelectric Solid Solutions: Physics, Crystallochemistry, Technology. Design Aspects of Piezoelectric Materials.* Rostov-on-Don: Rostov State University Press, *Vol. 1-2*, 2001. (in Russian)

[32] Rybyanets, A. N. In: *Piezoceramic Materials and Devices*. Ed. I. A. Parinov. Nova Science Publishers, N.-Y., 2010, pp 113-176.

In: Physics and Mechanics of New Materials ... ISBN: 978-1-62618-535-7
Editors: Ivan A. Parinov and Shun-Hsyung Chang

Chapter 17

USE OF GENERALIZED METHOD OF SCALARIZATION OF THE DYNAMIC ELASTIC FIELDS IN TRANSVERSE-ISOTROPIC MEDIA

***V. P. Sizov*[1] *and I. P. Miroshnichenko*[2,*]**
[1]Southern Federal University, Rostov-on-Don, Russia,
[2]Don State Technical University, Rostov-on-Don, Russia

ABSTRACT

The scalarization method, before developed for the boundary-value problems of dynamic theory of elasticity is generalized to the case of arbitrary disposition of axis of the material symmetry relatively the boundary between layers of transverse-isotropic composites.

1. INTRODUCTION

At the study of tensor elastic fields of displacements, stresses and strains into layered constructions from different materials (including transverse-isotropic composites), it has been successively applied an effective scalarization method of tensor fields. In this method, above-mentioned fields could be expressed via scalar potentials, corresponding quasi-longitudinal, quasi-transverse and pure-transverse waves [1].

This scalarization is possible, if the fields considered are tensor ones relatively subgroup of general coordinate transformations, when local affine basis has single invariant vector, coinciding with axis of material symmetry. The case, when this vector coincided with normal to boundary between layers of composite (or construction) has been considered in [2].

However, with practical viewpoint, other cases of mutual disposition of axis of the material symmetry and the boundary between layers are also interesting.

* Email: ipmir@rambler.ru

The aim of this paper consists in further development and generalization of the scalarization method in the boundary-value problems of the dynamic theory of elasticity to the case of arbitrary disposition of axis of the material symmetry relatively the boundary between layers of transverse-isotropic composites.

2. Generalized Method of Scalarization

Originally, the components of displacement and stress tensors are defined in coordinate system with admissible frame relatively of the frame. Then, by knowing all components of the tensor fields in the reference coordinate system, the normal and tangential components of the displacements and stresses maybe found in local coordinate system, connected with the boundary between layers of the composite. These components in the local coordinate system are used below to satisfy boundary conditions of the problem.

We present above approach by using the example of solution of the problems for plain-layered constructions. With this purpose, it is considered the case, when principal axis of symmetry for the transverse-isotropic material disposes under angle α relatively of normal to surface of plane boundary between layers (see Figure 1). In this case, we assume that the field considered is independent of coordinate $\overline{y}$ (perpendicular to the figure plane).

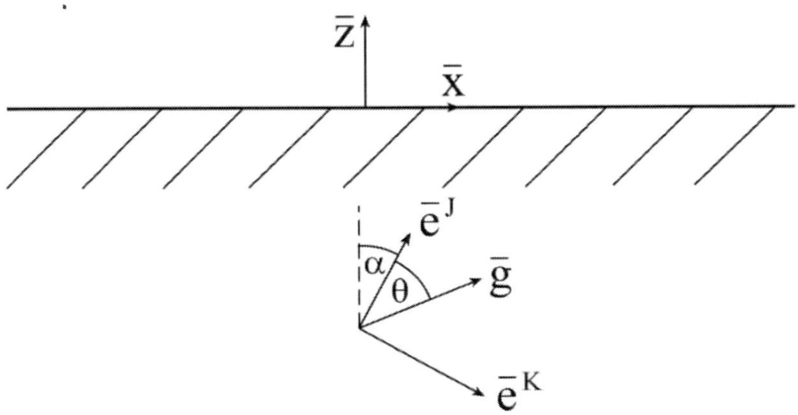

Figure 1. The statement of the problem for plain-layered constructions.

In accordance with [1] we have

$$U_i = \left(\overset{(L)}{D_1} \nabla_i + \delta_i^J \overset{(L)}{D_2} \nabla_J \right) \varphi + \left(\overset{(T)}{D_1} \frac{1}{\overset{(T)}{g}} \nabla_i \nabla_J + \overset{(T)}{g} \delta_i^J \right) w + \sqrt{g} (\delta_i^K \nabla^N - \delta_i^N \nabla^K) v;$$

$$\sigma_{ij} = (\overset{(L)}{d_1} g_{ij} + \overset{(L)}{d_2} \delta_i^J \delta_j^J + \overset{(L)}{d_3} \delta_{(i}^J \nabla_{j)} \nabla_J + \overset{(L)}{d_4} \nabla_i \nabla_j)\varphi +$$

$$+ (\overset{(T)}{d_1} g_{ij} \nabla_J + \overset{(T)}{d_2} \delta_i^J \delta_j^J \nabla_J + \overset{(T)}{d_3} \delta_{(i}^J \nabla_{j)} + \overset{(T)}{d_4} \nabla_i \nabla_J \nabla_j) w +$$

$$+2\sqrt{g}[a_2(\nabla_{(i}\delta_{j)}^{K}\nabla^{N}-\nabla_{(i}\delta_{j)}^{N}\nabla^{K})+a_4(\delta_{(i}^{J}\delta_{j)}^{K}\nabla^{N}-\delta_{(i}^{J}\delta_{j)}^{N}\nabla^{K})\nabla_J]v, \quad (1)$$

and components of tensor fields of displacements U_i and stresses σ_{ij} in the coordinate system, connected with the material anisotropy $\{x^J, x^K\}$ maybe written in the following forms:

$$U_J = i\overset{(L)}{\xi}(\overset{(L)}{D_1}+\overset{(L)}{D_2})\varphi+(-\overset{(T)}{D_1}\frac{\overset{(T)}{\xi^2}}{\overset{(T)}{g}}+\overset{(T)}{g})w;\ U_K = i\overset{(L)}{\beta}\overset{(L)}{D_1}\varphi-\overset{(T)}{D_1}\frac{\overset{(T)}{\xi}\overset{(T)}{\beta}}{\overset{(T)}{g}}w;$$

$$\sigma_{JJ}=[\overset{(L)}{d_1}+\overset{(L)}{d_2}-(\overset{(L)}{d_3}+\overset{(L)}{d_4})\overset{(L)}{\xi^2}]\varphi+i\overset{(T)}{\xi}(\overset{(T)}{d_1}+\overset{(T)}{d_2}+\overset{(T)}{d_3}-\overset{(T)}{d_4}\overset{(T)}{\xi^2})w;$$

$$\sigma_{JK}=-(\frac{1}{2}\overset{(L)}{d_3}\overset{(L)}{\xi}\overset{(L)}{\beta}+\overset{(L)}{d_4}\overset{(L)}{\xi}\overset{(L)}{\beta})\varphi+i\overset{(T)}{\beta}(\frac{1}{2}\overset{(T)}{d_3}-\overset{(T)}{d_4}\overset{(T)}{\xi^2})w;$$

$$\sigma_{KK}=(\overset{(L)}{d_1}+\overset{(L)}{d_4}\overset{(L)}{\beta^2})\varphi+i\overset{(T)}{\xi}(\overset{(T)}{d_1}-\overset{(T)}{d_4}\overset{(T)}{\beta^2})w. \quad (2)$$

Here, potentials of quasi-longitudinal φ and quasi-transverse w waves should satisfy a wave equation with corresponding wave number g ($\overset{(L)}{g}$ or $\overset{(T)}{g}$) and have the forms:

$$\varphi=\varphi_0 e^{i\overset{(L)}{\xi}x^J}e^{i\overset{(L)}{\beta}x^K};\ w=w_0 e^{i\overset{(T)}{\xi}x^J}e^{i\overset{(T)}{\beta}x^K}, \quad (3)$$

where

$$\xi^2+\beta^2=g^2;\qquad (\Delta+g^2)\begin{matrix}\varphi\\ w\end{matrix}=0. \quad (4)$$

The parameters D and d, which are presented in relationships (2), have been found in [2] as

$$\overset{(L)}{D_1}=\frac{\overset{(L)}{g^2}(C_{13}+C_{44})}{\omega^2\rho-\overset{(L)}{h^2}C_{44}-(\overset{(L)}{g^2}-\overset{(L)}{h^2})(C_{11}-C_{13}-C_{44})};$$

$$\overset{(L)}{D}_2 = \frac{\overset{(L)}{g}{}^2}{\overset{(L)}{h}{}^2} \frac{\omega^2 \rho - \overset{(L)}{h}{}^2 (2C_{44} + C_{13}) - (\overset{(L)}{g}{}^2 - \overset{(L)}{h}{}^2) C_{11}}{\omega^2 \rho - \overset{(L)}{h}{}^2 C_{44} - (\overset{(L)}{g}{}^2 - \overset{(L)}{h}{}^2)(C_{11} - C_{13} - C_{44})} = \frac{\overset{(L)}{g}{}^2}{\overset{(L)}{g}{}^2 - \overset{(L)}{h}{}^2} \overset{(L)}{D}_1 ;$$

$$\overset{(T)}{D}_1 = \frac{-\overset{(T)}{g}{}^2 (C_{13} + C_{44})}{\omega^2 \rho - \overset{(T)}{h}{}^2 (2C_{44} + C_{13}) - (\overset{(T)}{g}{}^2 - \overset{(T)}{h}{}^2) C_{11}} ; \tag{5}$$

$$\overset{(L)}{d}_1 = -\overset{(L)}{g}{}^2 a_1 - \overset{(L)}{h}{}^2 a_3 (\overset{(L)}{D}_1 + \overset{(L)}{D}_2) ; \; \overset{(L)}{d}_2 = -\overset{(L)}{g}{}^2 a_3 - \overset{(L)}{h}{}^2 (2a_4 \overset{(L)}{D}_2 + a_5 \overset{(L)}{D}_2 + a_5 \overset{(L)}{D}_1) ;$$

$$\overset{(L)}{d}_3 = 2a_2 \overset{(L)}{D}_2 + 2a_4 (2\overset{(L)}{D}_1 + \overset{(L)}{D}_2) ; \; \overset{(L)}{d}_4 = 2a_2 \overset{(L)}{D}_1 ;$$

$$\overset{(T)}{d}_1 = a_1 \overset{(T)}{g} (1 - \overset{(T)}{D}_1) + a_3 \left(\overset{(T)}{g} - \frac{\overset{(T)}{h}{}^2}{\overset{(T)}{g}} \overset{(T)}{D}_1 \right) ;$$

$$\overset{(T)}{d}_2 = a_3 \overset{(T)}{g} (1 - \overset{(T)}{D}_1) + 2a_4 \overset{(T)}{g} + a_5 \left(\overset{(T)}{g} - \frac{\overset{(T)}{h}{}^2}{\overset{(T)}{g}} \overset{(T)}{D}_1 \right) ;$$

$$\overset{(T)}{d}_3 = 2a_2 \overset{(T)}{g} + 2a_4 \left(\overset{(T)}{g} - 2\overset{(T)}{D}_1 \frac{\overset{(T)}{h}{}^2}{\overset{(T)}{g}} \right) ; \overset{(T)}{d}_4 = 2a_2 \overset{(T)}{D}_1 \frac{1}{\overset{(T)}{g}} ; \tag{6}$$

here

$$a_1 = C_{11} - 2a_2 ; a_2 = \frac{1}{2}(C_{11} - C_{12}) ;$$

$$a_3 = -\frac{1}{2}(C_{11} + C_{12}) + C_{13} + C_{44} - a_4 ;$$

$$a_4 = -\frac{1}{2}(C_{11} - C_{12}) + C_{44} ; a_5 = C_{11} + C_{33} - 2(C_{13} + 2C_{44}) , \tag{7}$$

where C_{ij} are the elastic moduli of the material.

For other parameters we have the next expressions:

$$h = \frac{\cos\theta}{\cos(\alpha+\theta)} g_z ; \beta = \frac{\sin\theta}{\cos(\alpha+\theta)} g_z ; g = \frac{g_z}{\cos(\alpha+\theta)} .$$

The components of displacements $\overline{U}_z$, $\overline{U}_x$ and stresses $\overline{\sigma}_{zz}$, $\overline{\sigma}_{zx}$, presented in the boundary conditions of dynamic problems of the elasticity theory, are written in the local coordinate system { $\overline{z}$, $\overline{x}$ }, connected with the boundary (see Figure 1).

The coordinate systems { $\overline{z}$, $\overline{x}$ } and { x^J, x^K } are mutually-connected by the relations:

$$\overline{z} = -x^K \sin\alpha + x^J \cos\alpha \; ; \; \overline{x} = x^K \cos\alpha + x^J \sin\alpha \tag{8}$$

or

$$x^K = \overline{x}\cos\alpha - \overline{z}\sin\alpha \; ; \; x^J = \overline{x}\sin\alpha + \overline{z}\cos\alpha . \tag{9}$$

We shall use relations (8) and (9) and the next expressions [2]:

$$\overline{U}_i = \frac{\partial x^m}{\partial \overline{x}^i} U_m ; \; \overline{\sigma}_{ij} = \frac{\partial x^m}{\partial \overline{x}^i} \frac{\partial x^p}{\partial \overline{x}^j} \sigma_{mp} , \tag{10}$$

where the parameters with upper line relate to the coordinate system { $\overline{z}$, $\overline{x}$ } (we shall call them "new" coordinates), and the parameters without the line relate to the coordinate system { x^J, x^K } ("old" coordinates). We write the components of the fields of displacements $\overline{U}_z$, $\overline{U}_x$ and stresses $\overline{\sigma}_{zz}$, $\overline{\sigma}_{zx}$, presented in the boundary conditions, via components from relations (2):

$$\overline{U}_z = \cos\alpha U_J - \sin\alpha U_K ; \; \overline{U}_x = \sin\alpha U_J + \cos\alpha U_K ; \tag{11}$$

$$\overline{\sigma}_{zz} = \cos^2\alpha\sigma_{JJ} - \sin 2\alpha\sigma_{JK} + \sin^2\alpha\sigma_{KK} ;$$

$$\overline{\sigma}_{zx} = \frac{1}{2}\sin 2\alpha(\sigma_{JJ} - \sigma_{KK}) + (1 - 2\sin^2\alpha)\sigma_{JK} . \tag{12}$$

The potential functions (3) in "new" coordinates may be represented as

$$\overline{\varphi} = \varphi_0 e^{i \overset{(L)}{\xi} (\overline{x}\sin\alpha + \overline{z}\cos\alpha)} e^{i \overset{(L)}{\beta} (\overline{x}\cos\alpha - \overline{z}\sin\alpha)} ;$$

$$\overline{w} = w_0 e^{i \overset{(T)}{\xi} (\overline{x}\sin\alpha + \overline{z}\cos\alpha)} e^{i \overset{(T)}{\beta} (\overline{x}\cos\alpha - \overline{z}\sin\alpha)} , \tag{13}$$

but the displacements (11) are written as

$$\overline{U}_z = \cos\alpha[i \overset{(L)}{\xi}(\overset{(L)}{D_1} + \overset{(L)}{D_2})\overline{\varphi} + (-\overset{(T)}{D_1}\frac{\overset{(T)}{\xi^2}}{\overset{(T)}{g}} + \overset{(T)}{g})\overline{w}] - \sin\alpha[i \overset{(L)}{\xi}\overset{(L)}{D_1}\overline{\varphi} + (-\overset{(T)}{D_1}\frac{\overset{(T)}{\xi^2}}{\overset{(T)}{g}} + \overset{(T)}{g})\overline{w}];$$

$$\overline{U}_x = \sin\alpha[i(\overset{(L)}{D_1} + \overset{(L)}{D_2})\overline{\varphi} + (-\overset{(T)}{D_1}\frac{\overset{(T)}{\xi^2}}{\overset{(T)}{g}} + \overset{(T)}{g})\overline{w}] + \cos\alpha[i \overset{(L)}{\beta}\overset{(L)}{D_1}\overline{\varphi} - \overset{(L)}{D_1}\frac{\overset{(T)}{\xi}\overset{(T)}{\beta}}{\overset{(T)}{g}}\overline{w}]. \quad (14)$$

Here the wave numbers ξ, β, g are defined relatively "old" frame (see Figure 2).

The wave numbers ξ, β, being by the projections of the vector $\overline{g}$ on the "old" frame { x^J, x^K } and wave numbers g_x, g_z, being by the projections of the vector $\overline{g}$ on the "new" frame { $\overline{z}$, $\overline{x}$ } are mutually-connected by the following relationships:

$$\xi = \frac{\cos\theta}{\sin(\alpha+\theta)}g_x = \frac{\cos\theta}{\cos(\alpha+\theta)}g_z; \beta = \frac{\sin\theta}{\sin(\alpha+\theta)}g_x = \frac{\sin\theta}{\cos(\alpha+\theta)}g_z;$$

$$g = \frac{g_x}{\sin(\alpha+\theta)} = \frac{g_z}{\cos(\alpha+\theta)}. \quad (15)$$

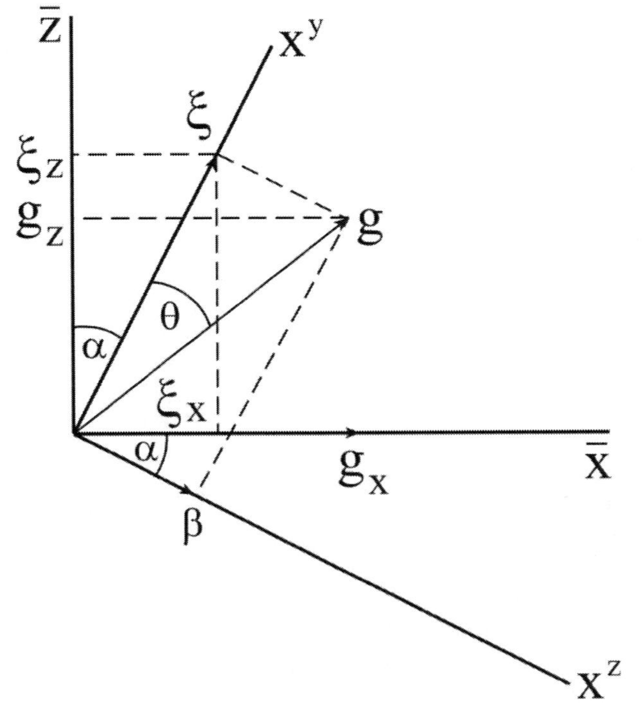

Figure 2. To definition of the wave numbers ξ, β, g.

By substituting (15) into (14), we obtain expressions for the displacements in the "new" coordinates as

$$\overline{U}_z = i\overset{(L)}{g}_z(\overset{(L)}{D}_1+\frac{\cos\theta\cos\alpha}{\cos(\alpha+\theta)}\overset{(L)}{D}_2)\overline{\varphi}+\overset{(T)}{g}_z(-\cos\theta\overset{(T)}{D}_1+\frac{\cos\alpha}{\cos(\alpha+\theta)})\overline{w}; \tag{16}$$

$$\overline{U}_x = i\overset{(L)}{g}_z[\frac{\cos\theta\sin\alpha}{\cos(\alpha+\theta)}(\overset{(L)}{D}_1+\overset{(L)}{D}_2)+\frac{\sin\theta\cos\alpha}{\cos(\alpha+\theta)}\overset{(L)}{D}_1]\overline{\varphi}+\overset{(T)}{g}_z[-\frac{\sin(\alpha+\theta)\cos\theta}{\cos(\alpha+\theta)}\overset{(T)}{D}_1+\frac{\sin\alpha}{\cos(\alpha+\theta)}]\overline{w},$$

where in accordance with (13) and (15), we obtain

$$\overline{\varphi}=\varphi_0 e^{i\overset{(L)}{g}_x\overline{x}}e^{i\overset{(L)}{g}_z\overline{z}};\ \overline{w}=w_0 e^{i\overset{(T)}{g}_x\overline{x}}e^{i\overset{(T)}{g}_z\overline{z}}. \tag{17}$$

Analogously, by using expressions (2), (14) and (15), we obtain the relations for the components of the stresses, presented in the boundary conditions of the problems, from (12) as

$$\overline{\sigma}_{zz}=\{\cos^2\alpha[\overset{(L)}{d}_1+\overset{(L)}{d}_2-(\overset{(L)}{d}_3+\overset{(L)}{d}_4)\frac{\cos^2\theta}{\cos^2(\alpha+\theta)}\overset{(L)}{g}{}_z^2]+\sin 2\alpha(\frac{1}{2}\overset{(L)}{d}_3+\overset{(L)}{d}_4)\frac{\cos\theta\sin\theta}{\cos^2(\alpha+\theta)}\overset{(L)}{g}{}_z^2+$$

$$+\sin^2\alpha(\overset{(L)}{d}_1-\overset{(L)}{d}_4\frac{\sin^2\theta}{\cos^2(\alpha+\theta)}\overset{(L)}{g}{}_z^2)\}\overline{\varphi}+\{\cos^2\alpha\frac{\cos\theta}{\cos(\alpha+\theta)}i\overset{(T)}{g}_z(\overset{(T)}{d}_1+\overset{(T)}{d}_2+\overset{(T)}{d}_3-\overset{(T)}{d}_4\frac{\cos^2\theta}{\cos^2(\alpha+\theta)}\overset{(T)}{g}{}_z^2)-$$

$$-\sin 2\alpha\frac{\sin\theta}{\cos(\alpha+\theta)}i\overset{(T)}{g}_z(\frac{1}{2}\overset{(T)}{d}_3-\overset{(T)}{d}_4\frac{\cos^2\theta}{\cos^2(\alpha+\theta)}\overset{(T)}{g}{}_z^2+\sin 2\alpha\frac{\cos\theta}{\cos(\alpha+\theta)}i\overset{(T)}{g}_z(\overset{(T)}{d}_1-\overset{(T)}{d}_4\frac{\cos^2\theta}{\cos^2(\alpha+\theta)}\overset{(T)}{g}{}_z^2)\}\overline{w}; \tag{18}$$

$$\overline{\sigma}_{zx}=\{\frac{1}{2}\sin 2\alpha[\overset{(L)}{d}_2-(\overset{(L)}{d}_3+\overset{(L)}{d}_4)\frac{\cos^2\theta}{\cos^2(\alpha+\theta)}\overset{(L)}{g}{}_z^2+\overset{(L)}{d}_4\frac{\sin^2\theta}{\cos^2(\alpha+\theta)}\overset{(L)}{g}{}_z^2]-$$

$$-(1-2\sin^2\alpha)(\frac{1}{2}\overset{(L)}{d}_3+\overset{(L)}{d}_4\frac{\cos\theta\sin\theta}{\cos^2(\alpha+\theta)}\overset{(L)}{g}{}_z^2\}\overline{\varphi}+\{\frac{1}{2}\sin 2\alpha\frac{\cos\theta}{\cos(\alpha+\theta)}i(\overset{(T)}{d}_2+\overset{(T)}{d}_3-\overset{(T)}{d}_4\frac{\cos^2\theta-\sin^2\theta}{\cos^2(\alpha+\theta)}\overset{(T)}{g}{}_z^2)+$$

$$+(1-2\sin^2\alpha)\frac{\sin\theta}{\cos(\alpha+\theta)}i(\frac{1}{2}\overset{(T)}{d}_3-\overset{(T)}{d}_4\frac{\cos^2\theta}{\cos^2(\alpha+\theta)}\overset{(T)}{g}{}_z^2)\}\overline{w}.$$

So, the components of the displacements and stresses, presented in the boundary conditions (the local coordinate system $\{\overline{z},\ \overline{x}\}$ is connected with the boundary), are determined by relationships (16) – (18), in which the projections of the wave vector $\overline{g}$ in the coordinate systems $\{\overline{z},\ \overline{x}\}$ and $\{x^J,\ x^K\}$ are connected by relationships (15).

Matrices C, used in [2], characterize the wave properties of layers and present themselves main elements of specific solutions of the boundary-value problems for layered constructions in the case, when axes $\bar{z}$, $\bar{x}$ create angle α with the frame components $\bar{e}^J$, $\bar{e}^K$; they are defined as

$$C = \begin{pmatrix} c_{(1)} & c_{(2)} \\ c_{(3)} & c_{(4)} \end{pmatrix}, \tag{19}$$

or

$$C = \begin{pmatrix} C_{(1)} = \begin{pmatrix} C_{11} & C_{12} \\ C_{21} & C_{22} \end{pmatrix} & C_{(2)} = \begin{pmatrix} -C_{11} & C_{12} \\ C_{21} & -C_{22} \end{pmatrix} \\ C_{(3)} = \begin{pmatrix} C_{31} & C_{32} \\ C_{41} & C_{42} \end{pmatrix} & C_{(4)} = \begin{pmatrix} C_{31} & -C_{32} \\ -C_{41} & C_{42} \end{pmatrix} \end{pmatrix} \begin{pmatrix} e^{i\overset{(L)}{g}_z z} & 0 & 0 & 0 \\ 0 & e^{i\overset{(T)}{g}_z z} & 0 & 0 \\ 0 & 0 & e^{-i\overset{(L)}{g}_z z} & 0 \\ 0 & 0 & 0 & e^{-i\overset{(T)}{g}_z z} \end{pmatrix} e^{i g_x x};$$

$$g_x = \overset{(T)}{g}_x = \overset{(L)}{g}_x; \tag{20}$$

$$C_{11} = i\overset{(L)}{g}_z(\overset{(L)}{D}_1 + \frac{\cos\theta\cos\alpha}{\cos(\alpha+\theta)}\overset{(L)}{D}_2); \; C_{12} = \overset{(T)}{g}_z(-\cos\theta\,\overset{(T)}{D}_1 + \frac{\cos\alpha}{\cos(\alpha+\theta)});$$

$$C_{21} = i\overset{(L)}{g}_z[\frac{\cos\theta\sin\alpha}{\cos(\alpha+\theta)}(\overset{(L)}{D}_1 + \overset{(L)}{D}_2) + \frac{\sin\theta\cos\alpha}{\cos(\alpha+\theta)}\overset{(L)}{D}_1];$$

$$C_{22} = \overset{(T)}{g}_z[-\frac{\sin(\alpha+\theta)\cos\theta}{\cos(\alpha+\theta)}\overset{(L)}{D}_1 + \frac{\sin\alpha}{\cos(\alpha+\theta)}];$$

$$C_{31} = \cos^2\alpha[\overset{(L)}{d}_1 + \overset{(L)}{d}_2 - (\overset{(L)}{d}_3 + \overset{(L)}{d}_4)\frac{\cos^2\theta}{\cos^2(\alpha+\theta)}\overset{(L)}{g}{}_z^2] + \sin 2\alpha(\frac{1}{2}\overset{(L)}{d}_3 + \overset{(L)}{d}_4)\frac{\cos\theta\sin\theta}{\cos^2(\alpha+\theta)}\overset{(L)}{g}{}_z^2 +$$

$$+\sin^2\alpha(\overset{(L)}{d}_1 - \overset{(L)}{d}_4\frac{\sin^2\theta}{\cos^2(\alpha+\theta)}\overset{(L)}{g}{}_z^2);$$

$$C_{32} = \cos^2\alpha\frac{\cos\theta}{\cos(\alpha+\theta)} i\overset{(T)}{g}_z(\overset{(T)}{d}_1 + \overset{(T)}{d}_2 + \overset{(T)}{d}_3 - \overset{(T)}{d}_4\frac{\cos^2\theta}{\cos^2(\alpha+\theta)}\overset{(T)}{g}{}_z^2) -$$

$$-\sin 2\alpha\frac{\sin\theta}{\cos(\alpha+\theta)} i\overset{(T)}{g}_z(\frac{1}{2}\overset{(T)}{d}_3 - \overset{(T)}{d}_4\frac{\cos^2\theta}{\cos^2(\alpha+\theta)}\overset{(T)}{g}{}_z^2) +$$

$$+\sin^2\alpha\frac{\cos\theta}{\cos(\alpha+\theta)}i\overset{(T)}{g}_z(\overset{(T)}{d}_1-\overset{(T)}{d}_4\frac{\sin^2\theta}{\cos^2(\alpha+\theta)}\overset{(T)}{g}{}_z^2);$$

$$C_{41}=\frac{1}{2}\sin 2\alpha[\overset{(L)}{d}_2-(\overset{(L)}{d}_3+\overset{(L)}{d}_4)\frac{\cos^2\theta}{\cos^2(\alpha+\theta)}\overset{(L)}{g}{}_z^2+\overset{(L)}{d}_4\frac{\sin^2\theta}{\cos^2(\alpha+\theta)}\overset{(L)}{g}{}_z^2]-$$

$$-(1-2\sin^2\alpha)(\frac{1}{2}\overset{(L)}{d}_3+\overset{(L)}{d}_4\frac{\cos\theta\sin\theta}{\cos^2(\alpha+\theta)}\overset{(L)}{g}{}_z^2;$$

$$C_{42}=\frac{1}{2}\sin 2\alpha\frac{\cos\theta}{\cos(\alpha+\theta)}i(\overset{(T)}{d}_2+\overset{(T)}{d}_3-\overset{(T)}{d}_4\frac{\cos^2\theta-\sin^2\theta}{\cos^2(\alpha+\theta)}\overset{(T)}{g}{}_z^2)+$$

$$+(1-2\sin^2\alpha)\frac{\sin\theta}{\cos(\alpha+\theta)}i(\frac{1}{2}\overset{(T)}{d}_3-\overset{(T)}{d}_4\frac{\cos^2\theta}{\cos^2(\alpha+\theta)}\overset{(T)}{g}{}_z^2). \qquad (21)$$

When $\alpha = 0$, the elements (21) of the matrix C coincides with corresponding elements of this matrix in the case, when normal to the boundary coincides with axis of the material symmetry; these elements are presented by formulas (3.82) in [2]. If $\alpha = \pi/2$, then we have the case, when the symmetry axis is tangential to the interface and formulas (21) coincide with expressions (3.89) – (3.91) in [2].

By knowing the expressions for matrices C, it is possible to construct solutions for various problems, by using the methods, developed in [2].

The described approach has been successively used in analysis of diagnostic results of state of the layered constructions, manufactured from transverse-isotropic materials.

REFERENCES

[1] Sizov, V. P. *Izvestia RAN. Mekhanika Tverdogo Tela.* 1988, *No. 5*, 55-58.

[2] Sizov, V. P.; Miroshnichenko, I. P. *Excitation of Elastic Waves in Layered Anisotropic Constructions.* LAP LAMBERT Academic Publishing. 2012. pp. 1 - 270.

In: Physics and Mechanics of New Materials …
Editors: Ivan A. Parinov and Shun-Hsyung Chang
ISBN: 978-1-62618-535-7

Chapter 18

THE ANALYTICAL SOLUTION OF AXISYMMETRIC CONTACT PROBLEM ABOUT TORSION FOR THE COATINGS OF COMPLICATED STRUCTURE

***Sergey Aizikovich*[*a], *Andrey Vasiliev*[†a], *Sergey Volkov*[a], *Boris Mitrin*[‡a] and *Yue-Sheng Wang*[b]**
[a]Scientific Educational Center "Materials", Don State Technical University, Rostov-on-Don, Russia
[b] Institute of Engineering Mechanics, Beijing Jiaotong University, Beijing, People's Republic of China

ABSTRACT

Contact problem about torsion by a circular indenter with a flat base of a half-space with inhomogeneous coating is investigated. Approximated analytical solution of the problem is developed. The obtained formulas are asymptotically exact both at large and small values of characteristic dimensionless geometrical parameter of the problem. Inhomogeneous coatings of complicated periodical structure are analyzed.

1. INTRODUCTION

The problem about torsion of a circular indenter attached to homogeneous isotropic elastic half space was first formulated and solved by Reissner and Sagoci (1944). Rostovtsev (1955) and Sneddon (1966) independently developed an alternative technique to reduce the problem to the solution of integral equation using Hankel integral transformation.

The problem for inhomogeneous isotropic half-space was first addressed in works of Kassir (1970) and Dhaliwal and Singh (1978) for the case of power variation of the shear

* e-mail: saizikovich@gmail.com
† e-mail: andre.vasiliev@gmail.com
‡ e-mail: boris.mitrin@gmail.com

modulus with depth. Aizikovich (1978) proposed an approximated analytical solution for the problem for isotropic half-space coated by a layer with arbitrary variation of the shear modulus.

Selvadurai (1982) solved the Reissner-Sagoci problem for a transversely isotropic elastic half-space when the rotation of the bonded circular indenter is induced by a concentrated couple located along the axis of symmetry.

Grilitsky (1961) addressed the static problem about torsion of a indenter attached to an isotropic two-layer media and orthotropic layer. This problem was independently considered by Tang (1979) for a layer of large thickness and Erguven (1986).

More recently, Liu and Wang (2009) considered functionally-graded half-space using piecewise-linear approximation of kernel's transform to reduce the problem to an integral equation that was solved by Krenk method (see Krenk, 1975). Xiong, et al. (2005) examined the torsion of transversely isotropic piezoelectric homogeneous half-space.

2. Torsion of the Indenter on a Half-Space with the Coating: Formulation and Solution of the Problem

Non-deformable indenter with flat circular base is rigidly linked to the boundary Γ of elastic inhomogeneous half-space Ω by the surface $z = 0, r \leq a$ (in cylindrical system of coordinates r,φ,z, Figure 1). The indenter is subjected to torque M with the axis coinciding with z.

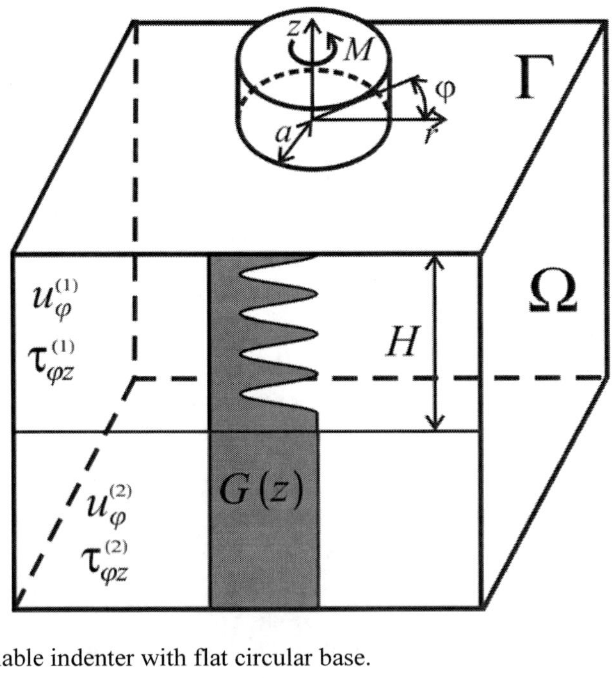

Figure 1. Non-deformable indenter with flat circular base.

Under the action of this torque, the indenter twists about z-axis on angle ε, which leads to the torsion strain in Ω. The shear modulus G of the half-space varies with depth as

$$
\begin{aligned}
&1)\ G=G_0(z),\ -H\le z\le 0;\\
&2)\ G=G_1=G_0(-H),\ -\infty<z<-H.
\end{aligned}
\tag{1}
$$

Outside the indenter, surface Γ is traction-free. Boundary conditions can be formulated as following

$$
z=0:\ \sigma_z=\tau_{rz}=0,\ \begin{cases}\tau_{z\varphi}=0, & r>a;\\ u_\varphi=r\varepsilon, & r\le a.\end{cases}
\tag{2}
$$

The stresses vanish for $r\to\infty$ and $z\to-\infty$.

The coating and the substrate are assumed to be glued without sliding, so that the continuity conditions

$$
z=-H:\ \tau_{z\varphi}^{(1)}=\tau_{z\varphi}^{(2)},\ u_\varphi^{(1)}=u_\varphi^{(2)}
\tag{3}
$$

are satisfied. Hereafter, superscripts (1) and (2) correspond to the coating and to the substrate respectively.

Contact tangential stresses under the indenter

$$
\tau_{z\varphi}\big|_{z=0}=\tau_a(r),\ \ r\le a
\tag{4}
$$

as well as the connection between the torque and twisting angle are to be determined.

Using integral transformation technique the problem is reduced to the solution of the following integral equation:

$$
\int_0^1\tau(\rho)\rho d\rho\int_0^\infty L(u)\mathrm{J}_1(ur\lambda^{-1})\mathrm{J}_1(u\rho\lambda^{-1})du=\lambda G_0(0)r\varepsilon,\ \ r\le 1,
\tag{5}
$$

where $\tau(\rho)=\tau_a(\rho a)$, $\lambda=H/a$ is the geometrical parameter of the problem, and $L(u)$ is the kernel transform of the integral equation.

It is proved (Aizikovich, Alexandrov, 1982) that the kernel transform can be represented in the form:

$$
L(u)=\Pi_N(u)+\Sigma_\infty(u),
\tag{6}
$$

The following notations are used:

$$
\Pi_N(u)=\prod_{i=1}^{N}\frac{u^2+A_i^2}{u^2+B_i^2},
\tag{7}
$$

$$\Sigma_M(u) = \sum_{k=1}^{M} \frac{c_k |u|}{u^2 + D_k^2}$$

and $A_i, B_i, D_k \in C$; $C_k \in R$ are certain constants. For the kernel transform $L_N(u)$ in the form (7) using bilateral asymptotic method as (see Aizikovich, 1982 for detail) contact tangential stresses have the form:

$$\tau(r) = \frac{4G_0(0)\varepsilon}{\pi} \left\{ L^{-1}(0) \frac{r}{\sqrt{1-r^2}} + \sum_{i=1}^{N} C_i Z(r, A_i \lambda^{-1}) \right\}, \tag{8}$$

where constants C_i are the solutions of the system of linear algebraic equations:

$$\sum_{i=1}^{N} C_i p\left(\frac{A_i}{\lambda}, \frac{B_k}{\lambda}\right) + \frac{1 + B_k \lambda^{-1}}{L_N(0) B_k{}^2 \lambda^{-2}} = 0 \tag{9}$$

The following notations are used:

$$Z(r, A) = \frac{\operatorname{sh} Ar}{r} + \frac{r \operatorname{sh} A}{S(1,r)} - Ar \int_r^1 \frac{\operatorname{ch} At dt}{S(t,r)};$$

$$p(A, B) = (A \operatorname{ch} A + B \operatorname{sh} A) \cdot \left(B^2 - A^2\right)^{-1};$$

$$S(t,r) = \sqrt{t^2 - r^2}\left(t + \sqrt{t^2 - r^2}\right). \tag{10}$$

Solution (8) is asymptotically exact at both large and small values of characteristic geometrical parameter λ of the problem (Aizikovich, 1982). Approximation of function $L(u)$ by expression (7) is described in (Aizikovich, 1982).

3. Inhomogeneous Coatings of Periodic Structure

Let us suppose, that the shear modulus (1) of inhomogeneous coating varies in depth according to the following rules:

$$f(z) = \frac{f_0 + 1}{2} - \frac{f_0 - 1}{2} \cos\left(2\pi k \frac{z}{H}\right), \tag{11}$$

or

$$f(z) = \frac{f_0 + 1}{2f_0} + \frac{f_0 - 1}{2f_0}\cos\left(2\pi k\frac{z}{H}\right). \tag{12}$$

Here $f_0 > 0$ is the parameter of inhomogeneity which describes the ratio between the shear modulus on the coating to that of the substrate. We will consider the case $f_0 = 3.5$

Variation of shear modulus in depth for rules (11), (12) for the case $k = 1$ are given in Figure 2.

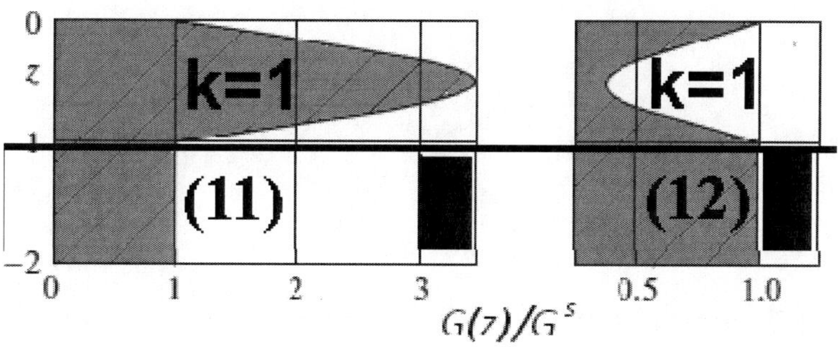

Figure 2. Variation of shear modulus in depth for rules (11), (12), k=1.

The rules (11), (12) describe a medium in which the shear modulus varies sinusoidally. The parameter k specifies the number of sinusoidal waves and corresponds to the number of changes of sign of the variation gradient of elastic properties of the medium.

In these coatings softer «layers» interleaves with harder «layers». Thereby elastic properties change continuous; this implies, that such coatings would avoid layering effect on the layer's joint border. Such coatings are under active studying.

Coatings' properties, described by the rules (11), (12), appear because of neighboring layers' diffusion in multilayer coatings. Diverse technologies of such coatings' creation are described in literature (vacuum deposition, PVD methods and others).

Figure 3 shows kernel transforms for the rules (11), (12) in the cases of $k = 1, 2, 4, 50$. Curves 1, 2, 3, 4 correspond to the rule (11), cases $k = 1, 2, 4, 50$, curves 5, 6, 7, 8 correspond to the rule (12), cases $k = 1, 2, 4, 50$. Chart shows the dynamics of kernel transform's changing with oscillation number increase. Let's point following properties:

(i) kernel transforms have single extreme point for all k values, we shall call it u_0. It is maximum point for rule (11) and minimum point for rule (12);
(ii) for $u < 0.5$ kernel transforms are similar for any value of k;
(iii) with increasing oscillation numbers (parameter k) extreme point u_0 moves rightwards in u-direction;
(iv) with increasing oscillation numbers (parameter k) the value of function's extreme $L(u_0)$ is increases for rule (11) and decreases rule (12). The values $L(u_0)$ are close for cases $k = 10$ and $k = 50$, i.e. it can be supposed, that they converge to some limit. For example, according to rule (11) $L(u_0) = 1.705$ for $k = 50$, according to rule (12) $L(u_0) = 0.488$ for $k = 50$;
(v) with the increase of oscillation number the interval of variable u values where kernel transform significantly differs from 1 increases.

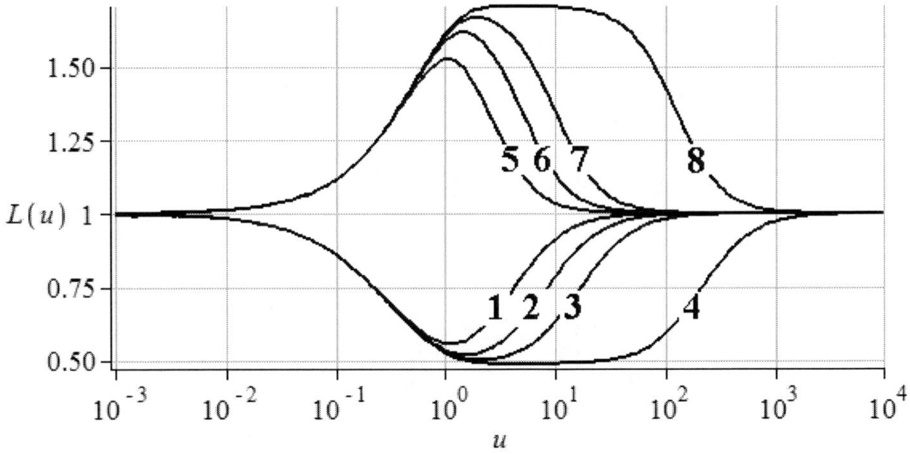

Figure 3. Kernel transforms for rules (11), (12) for k = 1, 2, 4, 50.

Table 1. Measure of accuracy of tranform's approximation for rules (11), (12).

	$k = 1$	$k = 2$	$k = 4$	$k = 10$	$k = 50$
rule (11)	0.32%	0.26%	0.5%	0.19%	1.1%
rule (12)	0.53%	0.58%	0.47%	0.19%	0.52%

Table 1 contains approximation's errors of kernel transform by expression (7) for rules (11) and (12). Approximation error is defined by the formula:

$$\Delta_L(u) = \left|\frac{L_N(u)}{L(u)} - 1\right| \times 100\% \cdot \qquad (18)$$

Let us introduce the function:

$$\tau_{rel}(\lambda, r) = \frac{\tau(\lambda, r)}{\tau_{\hom}(r)} \qquad (19)$$

Here $\tau_{\hom}(r)$ is the contact stresses of homogeneous half-space with shear modulus equal to shear modulus's value on the surface, i.e. $G = G(0)$. $\tau_{rel}(\lambda, r)$ characterizes distribution of contact tangential stresses under the indenter for concerned inhomogeneous half-space in comparison with homogeneous half-space.

Figure 4 represents $\tau_{rel}(\lambda, r)$ values for rule (11). Surfaces 1, 2, 3 are correspond to the cases k = 1, 2, 10, $\tau_{rel}(\lambda, r)$ for rule (12), k = 1, 2, 10 (notations are similar) are shown in Figure 5.

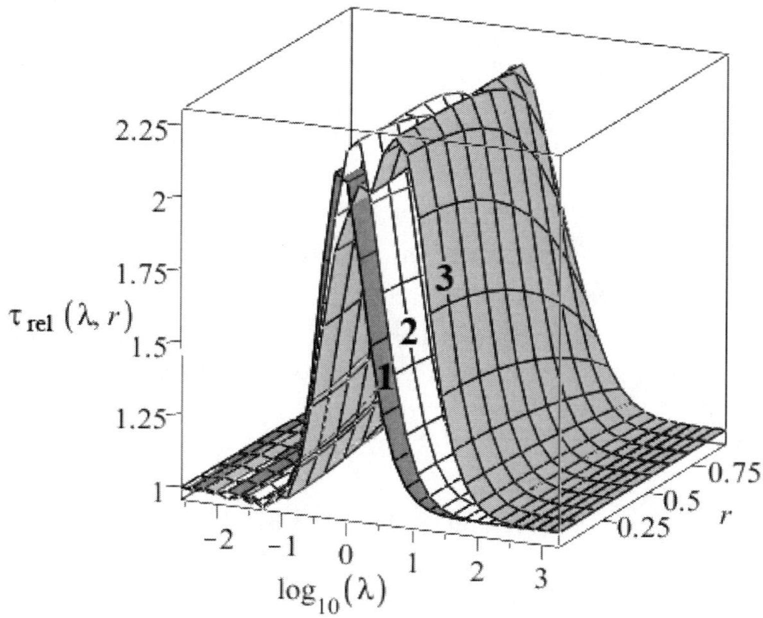

Figure 4. Contact stresses for rule (11), k = 1 (surface 1), k=2 (surface 2), k=10 (surface 3).

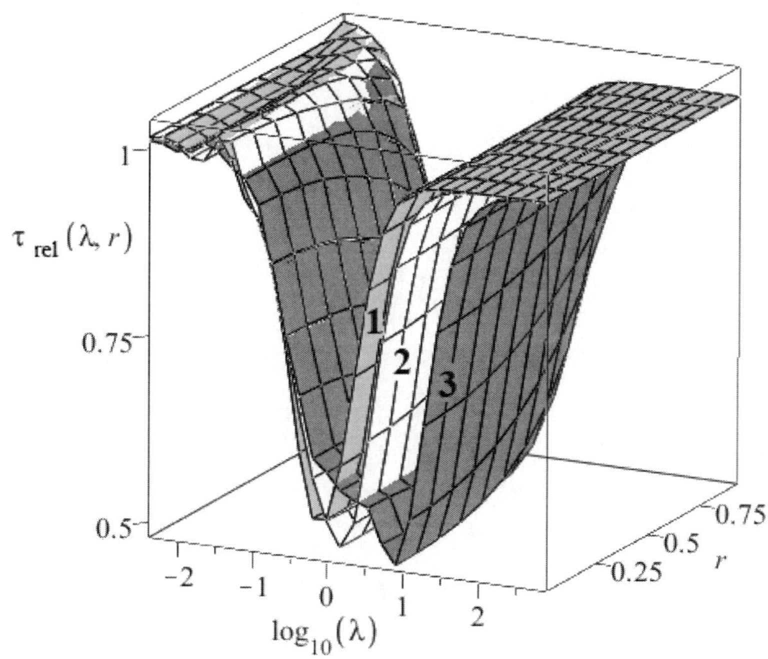

Figure 5. Contact stresses for rule (12), k = 1 (surface 1), k=2 (surface 2), k=10 (surface 3).

4. Conclusion

Following conclusions can be made due to the figures:

1. For every r function $\tau_{rel}(\lambda, r)$ has single extreme point. Let us call it λ_0;
2. for $\lambda < 1/3$ distribution of contact stresses for different values of parameter k differ slightly;
3. with increasing oscillation number (k parameter), extreme point λ_0 moves rightwards in the λ -direction;
4. with increasing oscillation number, the function's extreme value $\tau_{rel}(\lambda_0, r)$ distances from 1 (increases for rule (12), decreases for rule (11));
5. with increasing oscillation number range of λ where $\tau_{rel}(\lambda, r)$ significantly differs from 1 increases (λ range, where distribution of contact stresses for inhomogeneous half-space significantly differs from distribution of contact stresses for homogeneous half-space).

Acknowledgments

The work is fulfilled with support of Russian Foundation for Basic Research (11-08-91168-GFEN_a ,12-07-00639_a), and Russian GK Nos. 11.519.11.3015, 11.519.11.3028, P1107

References

Aizikovich, S. M. Torsion of a non-homogeneous half-space by a circular indenter, *In: Analysis of Shells and Plates*, Rostov Institute for Civil Engineering Press, Rostov-on-Don, 1978, 156-169 (in Russian).

Aizikovich, S. M. *J. Appl. Math. Mech. (PMM),* 1982, *vol. 46*, 116-124.

Aizikovich, S. M.; Alexandrov, V. M. *Soviet Phys. Dokl.* 1982, *vol. 27*, 765-767.

Babeshko, V. A.; Glushkov, E. V.; Glushkova, N. V. *Numerical Mathematics and Math. Physics Journal*. 1987, *vol. 27*, 93–101 (in Russian).

Dhaliwal, R. S.; Singh, B. M. *Int. J. Engng Sci.* 1978, *vol. 16*, 649-658.

Erguven, M. E. *Int. J. Engng .Sci.* 1986, *vol. 24*, 1501-1509.

Grilitskii, D. V. *Prikladnaja Mekhanika.* 1961, *vol. 7*, 89-94 (in Russian).

Reissner, E.; Sagoci, H. F. *J. Appl. Phys*. 1944, *vol. 15*, 652-654.

Kassir, M. K. *Int. J. Engng Sci*. 1970, *vol. 8*, 875-885.

Krenk, S. *Quart. Appl. Math.* 1975, *vol. 33*, 225–232.

Kudinov, V. V. *Plasma Sprayed Coatings*, Nauka, Moscow, 1977 (in Russian).

Rostovtsev, N. A. *Prikladnaya Mathematka and Mekh*anika, 1955, *vol. 19*, 55-60.

Selvadurai, A. P. S. *Lnt. Appl. Engng Sri*. 1982, *vol. 20*, 1365-1372.

Sneddon, I. N. *Proc. Glasgow Math. Assoc*. 1966, *vol. 7*, 136–144.

Tang, R. C. *Fibre Science and Technology,* 1979, *vol. 12*, 187-200.

Tie-Jun, Liu; Yue-Sheng, Wang, *Mechanics Research Communications*, 2009, *vol. 36*, 322–329.

Xiong, Su-ming; Ni, Guang-zheng; Hou, Peng-fei *Journal of Zhejiang University SCIENCE A,* 2005, *vol. 6*, 986-989.

In: Physics and Mechanics of New Materials …
Editors: Ivan A. Parinov and Shun-Hsyung Chang
ISBN: 978-1-62618-535-7

Chapter 19

ANALYTICAL SOLUTION OF THE CONTACT PROBLEM FOR THE GRADED COATINGS OF NON-DEFORMABLE BASE, TAKING INTO ACCOUNT INDEPENDENT CHANGES, ELASTIC MODULUS AND POISSON'S RATIO

I. S. Trubchik*

Vorovich Mechanics and Applied Mathematics Research Institute,
South Federal University, Rostov-on-Don, Russia,
Department of Computer Science and Computer Engineering Don
State Technical University, Rostov-on-Don, Russia

ABSTRACT

The contact problems for a layer or wedge covering the non-deformable base are investigated for the case, when elastic properties of the coating are arbitrary continuously-differentiable functions of its thickness. Effective numerical method was developed by using bilateral asymptotic method, described in papers of S. M. Aizikovich, V. M. Alexandrov, etc. [1, 2].

The specific idea of this method is to present the kernel transform for the dual integral equations by the function of special form convenient for the obtaining the solution in analytical form.

This method permits to obtain the problem solution for the wide class of the heterogeneity laws, in particular in the case, when the elastic modulus and Poisson's ratio vary arbitrarily with the normal direction to the contact surface. The analytical expression for the contact stresses is obtained for the integral equation with kernel of specific form constructed numerically with pre-given accuracy. The separate effect of arbitrary variation of the elastic modulus and Poisson's ratio is examined for functionally graded coatings with complicated structure.

* e-mail: trubchik@math.rsu.ru

1. INTRODUCTION

Characterization of mechanical and other physical properties of heterogeneous materials is very important in the material sciences. In contrast to homogeneous materials, it is necessary to determine not only average values of the investigated properties, but also spatial distributions of the properties, that requires much greater amount of experimental work and makes such investigations quite expensive. Therefore, it is necessary to use alternative methods for the material characterization, based on computer modeling of the material structure. Here, one analytical model of heterogeneous material is described. This model uses a concept of the functionally graded materials (FGMs), which represent multiphase composites with continuous variation in the concentrations of components in the volume. It leads to the smooth changing mechanical properties of the material from one point to other. Originally, the FGMs have been proposed as some alternative to uniform materials and interphase layers in details of aerospace vehicles, subjected to high temperatures. It has been shown that coatings with variable elastic properties had greater resistance to wear and contact stress effects. Moreover, machine parts, manufactured from FGMs show significantly longer lifetime than conventional ceramic ones.

These facts served as a motivation for a number of publications on the contact problems for the functionally graded materials. Failure of machine parts is mostly initiated near the free surfaces due to the stress concentration. To reduce the effect of wear and fatigue, caused by high loads, special coatings are required. FGM provides a reasonable solution to avoid drastic changes of material properties between coating and substrate, that usually leads to accumulation of cracks and other defects in the contact zone.

Therefore, the problem of non-destructive control and monitoring of elastic properties of machine parts, coated by FGMs becomes especially important now due to the development of modern technologies of protection of the materials and goods, connected with laser deposition, thermal spray or PVD technology. Review of scientific information sources for last decade on the investigated problem led to the next conclusion. In order to solve the contact and mixed problems for semi-infinite domains, such as layer rigidly fixed to a deformable base or lying without friction on the non-deformable foundation, and the wedge rigidly fixed on one of its surfaces, the following problems were studied:

i. the homogeneous strip or layer with various fixing conditions and different shapes of indenters, indented on the contact surface (flat, parabolic, spherical and oblique dies) [3–6];
ii. the layered base (sandwich structures) [7–9, 17, 18];
iii. the functionally graded media [10–14];
iv. the problems taking into account the thermal influence for both homogeneous and inhomogeneous media [15–19, 22];
v. the problems for porous media and media with cracks and inclusions [18–23].

Among different methods, the numerical methods (in particular, the finite-element method, etc.) are most intensively used methods in the last years because of the complexity of tasks [23]. Several papers were devoted to the known asymptotic methods, such as “large λ” and “small λ“. There are different analytical approaches to the solution of contact and mixed problems for homogeneous and functionally graded materials [24–26]. The present paper addresses to the semi-analytical method, developed by using bilateral asymptotical method of S. M. Aizikovich, V. M. Alexandrov, et al. [1, 2, 24, 25]. This method permits to obtain the analytical solution for the wide class of the heterogeneity laws, in particularly in the case, when the elastic modulus and Poisson's ratio vary arbitrarily with the normal direction to the contact surface.

2. General Statement of Problems

2.1. Formulation of Contact Problems for the Functionally Graded Layer and Strip

The next static contact problems are discussed, namely: (i) the shear by the flat punch $|x| < a$, $-\infty < z < \infty$ of the inhomogeneous layer on depth, rigidly fixed to the non-deformable foundation (S1), and (ii) the indentation of the inhomogeneous strip on depth by non-deformable punch with width of $2a$ and indenter shape $\gamma(x)$ (S2).

It is assumed, the shear force, P, is distributed on unit length of punch in the parallel direction to z-axis, so it is created punch displacement, δ, along the z-axis, causing the pure-shear deformation in the layer (S1) or the indentation force, P, causes the punch displacement along the x-axis on the δ value (S2) (Figure 1).

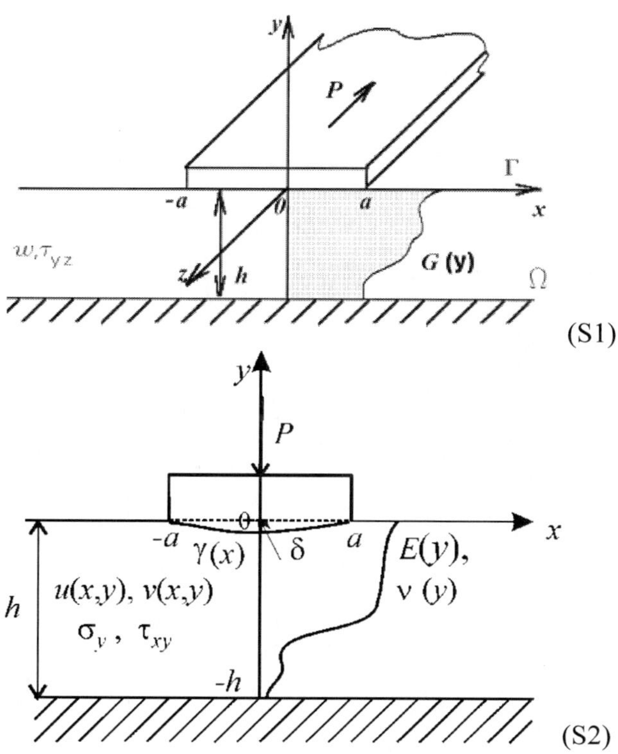

Figure 1. Schemes of contact problems for the inhomogeneous layer (S1) and strip (S2).

The Poisson's ratio, ν, shear modulus of the layer, G, Young's modulus of the strip, E, are arbitrary functions of coordinate y:

$$E = E_0 f(y);\; G = G_0 f(y);\; \nu = \nu_0 f(y);\; -h \le y \le 0, \tag{1}$$

where $f(y)$ is the continuously differentiable function, no equal to zero for the contact zone, and gradient of elastic properties into layer may be positive, negative or changes its sign with depth of the layer.

We assume a frictionless interaction of the punch with elastic layer, and surfaces outside the contact zone are free of stresses. It is required to determine the distribution of the contact shear stresses (S1) or normal stresses (S2) under the punch.

$$\tau_{yz}(x,0,z) = -\tau(x);\; \sigma_y(x,0,z) = -q(x);\; |x| \le a,\, z = 0, \tag{2}$$

where τ_{yz}, σ_y are the continuous functions of the coordinates corresponding to shear and normal stresses on the upper surface $z = 0$ of the layer or strip, respectively.

2.2 Reducing the Problem for the Heterogeneous Layer to Solution of Integral Equation

By substituting the Fourier integral transforms

$$\begin{aligned} u(x,y) &= \frac{1}{2\pi}\int_{-\infty}^{\infty} U(\alpha,y)e^{-i\alpha x}d\alpha; \\ v(x,y) &= \frac{1}{2\pi}\int_{-\infty}^{\infty} V(\alpha,y)e^{-i\alpha x}d\alpha; \\ w(x,y) &= \frac{1}{2\pi}\int_{-\infty}^{\infty} W(\alpha,y)e^{-i\alpha x}d\alpha \end{aligned} \tag{3}$$

into Lamé equilibrium equations in terms of displacements, and taking into account the relations of Hooke's law, the contact problems can be reduced, in terms of the unknown normal contact stresses under the punch $q(x)$ to the Fredholm integral equation:

$$\int_{-1}^{1}\varphi(\xi)k\left(\frac{\xi - x}{\lambda}\right)d\xi = 2\pi C_0 g(x'),\; |x'| \le 1,\; k(t) = \int_0^{\infty} L(\alpha)\alpha^{-1}\cos(\alpha t)d\alpha, \tag{4}$$

where $x' = x/a$; $\lambda = h/a$; $\varphi(\xi) = q(x')/C_0$ is the unknown function of contact stresses. Here, $g(x') = \varepsilon/a$; $C_0 = G(0)$ (for the case S1); $g(x') = (-\delta + \gamma(x))/a$, $C_0 = \theta_0$ (for the case S2),

$$\theta_0 = \theta(0) = \frac{8E(0)}{(7-\kappa)(1+\kappa)},\; \kappa = 3 - 4\nu(0).$$

In particular, the equations of spatial elasticity theory in the Cartesian coordinates at substitution of the Hook's law relations are reduced to the set of ordinary differential equations with variable coefficients for the transform functions of the Fourier integral transforms. In this case, there are the ordinary differential equations of second order for anti-plane problem S1 and the ordinary differential equations of fourth order for plane problem S2. It is convenient to rewrite these systems in vector form as

$$d\vec{\mathbf{w}}(\alpha, y)/dy = \mathbf{A}\vec{\mathbf{w}}(\alpha, y); \; -h \le y \le 0. \tag{5}$$

Here A is the square matrix with variable elements. Vector $\vec{\mathbf{w}}(\alpha, y)$ for the S1 case has the form:

$$\mathbf{A} = \begin{Vmatrix} 0 & 1 \\ \alpha^2 & -G'(y)/G(y) \end{Vmatrix}; \; \vec{\mathbf{w}} = \begin{Vmatrix} W \\ W' \end{Vmatrix}$$

and for the S2 case, it has the other form:

$$\mathbf{A} = \begin{Vmatrix} 0 & 1 & 0 & 0 \\ \alpha^2 \dfrac{2(1-\nu)}{1-2\nu} & F_1(y) & \alpha \cdot F_1(y) & -\alpha \dfrac{1}{1-2\nu} \\ 0 & 0 & 0 & 1 \\ \alpha \cdot F_2(y) & \alpha \dfrac{1}{2(1-\nu)} & \alpha^2 \dfrac{1-2\nu}{2(1-\nu)} & -F_3(y) \end{Vmatrix}; \; \vec{\mathbf{w}} = \begin{Vmatrix} U \\ U' \\ V \\ V' \end{Vmatrix};$$

$$F_1(y) = -\frac{E'}{E} + \frac{\nu'}{1+\nu}; \; F_2(y) = \frac{E'}{E} \cdot \frac{\nu}{1-\nu} + \frac{\nu'(1+2\nu^2)}{(1-\nu^2)(1-2\nu)};$$
$$F_3(y) = \frac{E'}{E} + \frac{2\nu'\nu(2-\nu)}{(1-\nu^2)(1-2\nu)}; \; E' = E'(y), E = E(y); \; \nu' = \nu'(y); \; \nu = \nu(y).$$

The boundary conditions for Equation (5) present the conditions on upper and lower layer surfaces.

We shall search the solution of the vector differential equation (5) in the form of linear combination of vectors of the fundamental system of solutions as

$$\vec{\mathbf{w}}(\alpha, y) = \sum_{i=1}^{r} C_i(\alpha)\vec{\mathbf{a}}_i(\alpha, y)e^{\gamma_i y}; \; -h \le y \le 0. \tag{6}$$

Here $\gamma_{1,2} = -\alpha$, $\gamma_{3,4} = \alpha$, r is the rank of matrix **A**. One is constructed by using the method developed in [27]. The method is based on the isolation of the exponential components in Green's matrix construction. It allows one to develop the stable numerical algorithm.

The vectors $\vec{\mathbf{a}}_i(\alpha, y)$ are changed along y-axis slower than exponential function and obtained from the set of Cauchy problems for distinct values of α

$$d\vec{\mathbf{a}}_i(\alpha,y)/dy = \mathbf{A}\vec{\mathbf{a}}_i(\alpha,y) - \gamma_i\vec{\mathbf{a}}_i(\alpha,y);\ -h \le y \le 0;\ i = 1,\ldots,r\,. \tag{7}$$

The initial conditions for Equation (7) are obtained from the appropriate solution for homogeneous medium on lower surface of the layer, $y = -h$.

So, the next expressions were obtained for numerical definition of transform functions $L(\alpha)$ of the dual integral equation (4):

$$\text{(S1): } L(\alpha,0) = |\alpha| \frac{a_1^1 - a_2^1\, e^{-2|\alpha|h}}{a_1^2 - a_2^2\, e^{-2|\alpha|h}};$$

$$\text{(S2): } L(\alpha) = \sum_{i=1}^{4} C_i(\alpha) a_i^3,\ a_i^j = a_i^j(\alpha,0)\,. \tag{8}$$

2.3. Formulation of Contact Problems for the Functionally Graded Wedge

The static contact problem for an elastic wedge-shaped region with opening angle, α ($0 \le \varphi \le \alpha$) is considered. The cylindrical coordinate system (r, φ, z) is associated with this area. We assume that the surface of the wedge, $\varphi = 0$, is rigidly fixed, and solid flat stamp acts on the surface, $\varphi = \alpha$. The shear force, P_s, acts on unit length of the stamp parallel to the z-axis, under action of which the stamp moves in the direction of the z-axis on distance, ε, causing the wedge deformation in the form of pure shear (the problem K1); or stamp is impressed in the wedge by the force, P (here the force acts on the unit length of the die), applied at a distance H from the vertex of the wedge (the problem K2). Let's assume that the friction force between the stamp and the surface of the wedge is absent and the surface of the wedge $\varphi = \alpha$ is not loaded outside the stamp. The width of the contact area of the stamp with the wedge is defined by the inequality, $a \le r \le b$, and shape of the stamp base in the contact zone is determined by the function $\gamma(r)$ (see Figure 2).

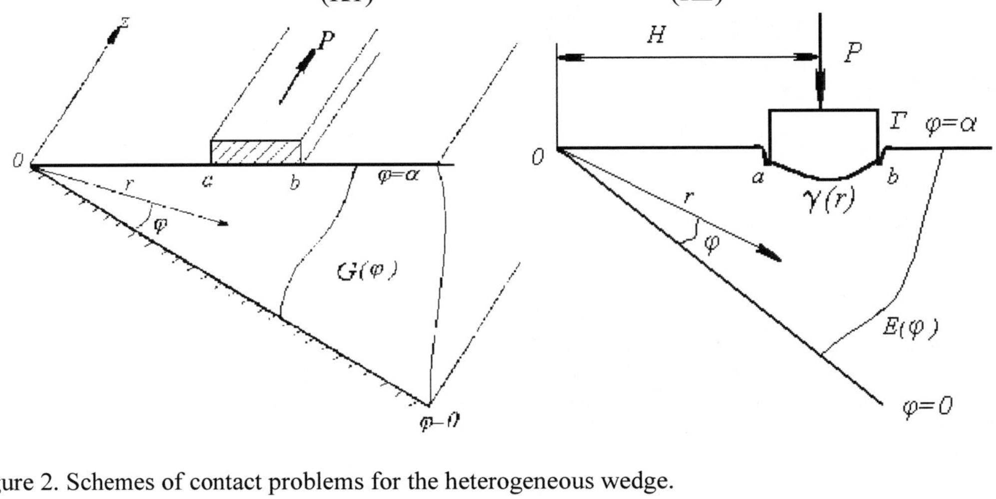

Figure 2. Schemes of contact problems for the heterogeneous wedge.

For the both problems, the Poisson's ratio, $\nu =$ const, the shear modulus and Young's modulus of the wedge vary according to the law:

$$G = G_\alpha f(\varphi);\ E = E_\alpha f(\varphi);\ \nu = \nu_\alpha f(\varphi);\ 0 \le \varphi \le \alpha\,, \tag{9}$$

where $f(\varphi)$ is the arbitrary positive-definite continuously-differentiable function everywhere in the domain. The gradient changes of elastic properties of the wedge can be positive, negative or change sign at $0 \le \varphi \le \alpha$.

The boundary conditions of the problems (K1) and (K2) under these assumptions are defined as

$$(K1): w(r,0,z) = 0;\ (\text{K2}): u(r,0) = v(r,0) = 0;\ a \le r \le b,\ \varphi = 0; \tag{10}$$

$$\begin{aligned}&(K1):\ \tau_{\varphi z}(r,\alpha,z) = 0;\ 0 < r < a,\ r > b;\quad w(r,\alpha,z) = \varepsilon;\ a \le r \le b;\\ &(K2):\ \tau_{r\varphi}(r,\alpha) = 0; \sigma_\varphi(r,\alpha) = 0; 0 < r < a,\ r > b; v(r,\alpha) = -\delta(r); a \le r \le b.\end{aligned} \tag{11}$$

It is required to determine the distribution of tangential contact stresses (in the problem K1) and normal contact stresses (in the problem K2) under the stamp:

$$\begin{aligned}&(K1):\ \tau_{\varphi z}(r,\alpha,z) = -\tau(r);\\ &(K2):\ \sigma_\varphi(r,\alpha) = -q(r);\quad a \le r \le b.\end{aligned} \tag{12}$$

By substituting the Mellin transform

$$\vec{\mathbf{U}} = \frac{1}{2\pi i}\int_{c-i\infty}^{c+i\infty} \vec{\mathbf{V}}(u,\varphi) r^{-u} du;\ \vec{\mathbf{U}} = \begin{pmatrix} u(r,\varphi) \\ v(r,\varphi) \\ w(r,\varphi) \end{pmatrix};\ \vec{\mathbf{V}} = \begin{pmatrix} U(u,\varphi) \\ V(u,\varphi) \\ W(u,\varphi) \end{pmatrix} \tag{13}$$

into the Lamé equilibrium equation in terms of displacements, taking into account the relations of the Hooke's law, the contact problems can be reduced, in terms of the unknown normal contact stresses under the punch to the Fredholm integral equation, which is similar to Equation (4), where $k(t) = \int_0^\infty L_1(u,\alpha) u^{-1} \cos(ut) du$; $t = \ln(\rho / r)$; $\xi = \lambda \ln(\rho / a) - 1$; $x = \lambda \ln(r / a) - 1$; $\lambda = 2(\ln(b / a))^{-1}$; $\varphi(\xi)$ are the unknown function of contact stresses, $\theta_\alpha = G_\alpha /(1 - \nu)$; $G_\alpha = G(\varepsilon)$ is the shear modulus on the upper surface of the wedge.

Here $\varphi(\xi) = \rho\tau(\rho)$. $g(x) = G_\alpha \lambda\varepsilon$ (for problem K1); $\varphi(\xi) = \rho q(\rho)$, $g(x) = G_\alpha \lambda\varepsilon$ (for problem K2). The dimensionless parameter λ ($0 < \lambda < \infty$) characterizes position of the stamp on the wedge surface. If parameter λ is large (small), then the stamp is located relatively far (close) from the tip of the elastic wedge. Note that the integral equation

(4) at $\alpha \to 0$ and fixed value of $H = ag\alpha$ reduces to integral equation of the contact problem for an elastic layer (or strip) [3].

To complete the statement of the contact problems for inhomogeneous elastic wedge and reducing them to the integral equation (4), it should be added the equilibrium condition of the stamp:

$$(K1): P_s = \int_a^b \tau(r)dr;$$
$$(K2): P = \int_a^b q(r)dr. \tag{14}$$

For this condition, it should be determined (at $\alpha = \pi$ and $\alpha = 2\pi$) a relation between tangential force, P_s, and stamp displacement along the wedge surface ε (for the problem K1) or pressing force, *P*, and stamp displacement in the wedge, δ (for the problem K2). For the correct formulation, the condition $\tau(r) \ge 0$ or $\sigma(r) \ge 0$ must be fulfilled into range of $a \le r \le b$.

2.4. Reducing the Problems for the Inhomogeneous Wedge to Solution of Integral Equation

The method of integral transforms and its generalization to modeling functions are used for definition of transformants of the integral equation kernels in the coordinate systems, differing of Cartesian one [2]. The problems are reduced to the solution of ordinary differential equations of second order (in anti-plane problem K1) or a set of second order differential equations with variable coefficients (in plane problem K2) relatively the integral transforms of the Mellin type for the displacements:

$$\frac{d\vec{\mathbf{w}}(u,\varphi)}{d\varphi} = \mathbf{A}\vec{\mathbf{w}}(u,\varphi),\ \varphi \in [0,\alpha]. \tag{15}$$

In the problem K1, rank of matrix **A** is equal to two and in the problem K2, the rank equals to four. The matrix elements are the variables of *z*. The boundary conditions for Equation (15) are found from conditions on upper and lower surfaces of the wedge. In particular, for the problem K1, matrix **A** has the form:

$$\mathbf{A} = \begin{Vmatrix} 0 & 1 \\ s^2 & -\kappa(\varphi) \end{Vmatrix};\ \vec{\mathbf{w}} = \begin{Vmatrix} \mathrm{W}(u,\varphi) \\ \mathrm{W}'(u,\varphi) \end{Vmatrix};\ \kappa(\varphi) = \frac{G'(\varphi)}{G(\varphi)},$$

where prime denotes the derivative with respect of φ. The boundary conditions are found as $z_1(u,\alpha) = 0,\ z_2(u,\alpha) = 1$.

For the problem K2 we have

$$\frac{d\vec{\mathbf{w}}}{d\varphi} = \mathbf{A}\vec{\mathbf{w}}; \tag{16}$$

where

$$\mathbf{A} = \begin{Vmatrix} 0 & 1 & 0 & 0 \\ -\frac{2(1-\nu)}{1-2\nu}\left(s^2-1\right) & -F_1(\varphi) & (s+1)F_1(\varphi) & \frac{s+1}{1-2\nu}+2 \\ 0 & 0 & 0 & 1 \\ F_2(\varphi) & \frac{s-3+4\nu}{2(1-\nu)} & -\frac{1-2\nu}{2(1-\nu)}\left(s^2-1\right) & -F_3(\varphi) \end{Vmatrix}; \quad \vec{\mathbf{w}} = \begin{Vmatrix} U(u,\varphi) \\ U'(u,\varphi) \\ V(u,\varphi) \\ V'(u,\varphi) \end{Vmatrix};$$

$$F_1(\varphi) = \frac{E'}{E} - \frac{\nu'}{\nu+1}; \; F_2(\varphi) = \frac{E'}{E}\left(s\frac{\nu}{1-\nu} - 1\right) + \frac{\nu'}{1-\nu^2}\frac{s\left(1+2\nu^2\right)+2\nu(\nu-2)}{1-2\nu};$$

$$F_3(\varphi) = \frac{E'}{E} - \frac{2\nu\nu'(\nu-2)}{\left(1-\nu^2\right)\left(1-2\nu\right)}; \; \nu = \nu(\varphi), \; E = E(\varphi).$$

The boundary conditions for this case are found as

$$w_1(u,0) = w_3(u,0) = 0;$$

$$w_2(u,\alpha) - (u+1)w_3(u,\alpha) = 0;$$

$$(5-3u+\kappa(u-1))w_1(u,\alpha) + (5-\kappa)w_4(u,\alpha) = \frac{4u}{\kappa+1}.$$

Further, the solution of the differential equation can be presented by a linear combination of fundamental solutions:

$$\vec{\mathbf{w}} = \sum_{i=1}^{r} C_i(u)\vec{\Psi}_i^F(u,\varphi), \tag{17}$$

where $\vec{\Psi}_i^F(u,\varphi) = \mathbf{T}_i(u,\varphi)\vec{\Psi}_i(u,\varphi)$, $\mathbf{T}_i(u,\varphi)$ are the diagonal matrices, along the diagonal of which the components of vectors $\vec{\mathbf{t}}_i(u,\varphi) = \left(t_i^1(u,\varphi), t_i^2(u,\varphi), \ldots, t_i^r(u,\varphi)\right)$ are located.

Vectors $\vec{\Psi}_i(u,\varphi)$ are the eigenvectors of the matrices $\mathbf{A}$ at the $k(\varphi) = 0$ (for the case of homogeneous medium), $\vec{\mathbf{t}}_i(u,\varphi)$ are the modeling vector-functions, defined by the medium heterogeneity, $C_i(u)$ are the some coefficients independent of φ and determined by solving a set of linear algebraic equations, obtained from the boundary conditions. This method allows us to introduce the components $\vec{\Psi}_i(u,\varphi)$ into series (17) and to find numerical solution. The eigenvectors $\vec{\Psi}_i(u,\varphi)$ of the matrix **A** for the problem K1 have the forms:

$$\vec{\Psi}_1(s,\varphi) = (\sin s\varphi, s\cos s\varphi),\ \vec{\Psi}_2(s,\varphi) = (\cos s\varphi, -s\sin s\varphi)$$

and for the problem K2 we have

$$\vec{\Psi}_1 = [\sin(s+1)\varphi;\ (s+1)\cos(s+1)\varphi;\ \cos(s+1)\varphi;\ -(s+1)\sin(s+1)\varphi];$$
$$\vec{\Psi}_2 = [\cos(s+1)\varphi; -(s+1)\sin(s+1)\varphi;\ \sin(s+1)\varphi;\ (s+1)\cos(s+1)\varphi];$$

$$\vec{\Psi}_3 = [\sin(s-1)\varphi;\ (s-1)\cos(s-1)\varphi;\ \cos(s-1)\varphi;\ -(s-1)\sin(s-1)\varphi];$$
$$\vec{\Psi}_4 = [\cos(s-1)\varphi; -(s-1)\sin(s-1)\varphi;\ \sin(s-1)\varphi;\ (s-1)\cos(s-1)\varphi].$$

The modeling vector-functions $\vec{\mathbf{t}}_i(u,\varphi)$ are determined from the Cauchy problem for the fixed values of u:

$$\frac{d\vec{\mathbf{t}}_i(u,\varphi)}{d\varphi} = \mathbf{B}_i\vec{\mathbf{t}}_i(u,\varphi) \quad (i = 1,\ldots,r),\ 0 \le \varphi \le \alpha. \tag{18}$$

Initial conditions for the Cauchy problem are determined from the homogeneous solution of Equation (16), in particular for the problem K1 we obtain

$$\vec{\mathbf{t}}_1(u,0) = (1;1);\ \vec{\mathbf{t}}_2(u,0) = (1;1),$$

and for the problem K2 we have

$$\vec{\mathbf{t}}_1(u,0) = [u+\kappa;\ u+\kappa;\ -(u-\kappa); -(u-\kappa)];$$
$$\vec{\mathbf{t}}_2(u,0) = [u+\kappa;\ u+\kappa;\ u-\kappa;\ u-\kappa];$$
$$\vec{\mathbf{t}}_3(u,0) = [u-1;\ u-1;\ -(u-1); -(u-1)];$$
$$\vec{\mathbf{t}}_4(u,0) = [u-1;\ u-1;\ u-1; u-1],$$

where $\kappa = 3-4\nu(0)$, the matrix $\mathbf{B}_i$ has the form: $\mathbf{B}_i = \vec{\Psi}_i^{-1}\left(\mathbf{A}\vec{\Psi}_i - \mathbf{E}\frac{d\vec{\Psi}_i}{d\varphi}\right)$.

So, the next expressions are obtained for numerical definition of transform functions $L_1(u,\alpha)$ of the dual integral equation (4):

(K1): $L_1(u,\alpha) = th(uA\alpha)t_1^1 / t_1^2$;

$$\text{(K2):}\quad \begin{aligned} &L_1(u,\alpha) = V(iu,\alpha) = a_1 t_1^3 l + a_2 t_2^3 k + a_3 t_3^3 n + a_4 t_4^3 m; \\ &l = \cos(iu+1)\alpha;\ k = \sin(iu+1)\alpha; n = \cos(iu-1)\alpha;\ m = \sin(iu-1)\alpha, \end{aligned} \tag{19}$$

where $t_i^3 = t_i^3(u,\alpha)$ are the solutions of Cauchy problems (18), $a_i = a_i(u)$ is the solution of the system of linear algebraic equations, obtained from the boundary conditions of the problem K2.

2.5. Approximation of the Transform Functions

2.5.1. General Properties of the Transform Functions

When the following conditions:

$$\min_{z\in(-\infty;0]} \Theta(z) \geq c_1 > 0;\ \max_{z\in(-\infty;0]} \Theta(z) \leq c_2 < \infty;\ \Theta(z) = \frac{E(z)}{2(1-\nu^2(z))} \tag{20}$$

are satisfied, where $E(z)$ is Young modulus and $\nu(z)$ is Poisson's ratio, then it could be shown [29], that the transform functions $L(\alpha)$ of above-mentioned problems have the following general properties:

$$\begin{aligned} &L(\alpha) = A|\alpha| + B\alpha^2 + O(\alpha^3),\ \alpha \to 0;\ A = \lim_{\alpha\to 0} L(\alpha)\alpha^{-1}; \\ &L(\alpha) = 1 + C|\alpha|^{-1} + D\alpha^{-2} + O(\alpha^{-3}),\ \alpha \to \infty, \end{aligned} \tag{21}$$

where A, B, C, D are some constants.

For multilayer media the property of a compliance function similar to (21) was pointed by Privarnikov [10, 11]. The properties of $L(\alpha)$ for the graded layer at the small value of α ($\alpha \to 0$) are differed essentially from the layered media. The next proposition maybe proved: if the function $L(\alpha)$ has the properties (21), then one could be approximated by the specific form:

$$L(u) = L_N(u) = \text{th}\, Au \prod_{n=1}^{N} \frac{u^2 + \delta_n'^2}{u^2 + \gamma_n'^2}. \tag{22}$$

2.5.2. Auxiliary Theorems, Concerning Analytical Approximation of the Functions

Now we prove that the function $L(u)$ with the properties (21) maybe expressed in the form:

$$L(u) = thAu[L_{\Pi}^{N}(u) + L_{\Sigma}^{\infty}(u)], \tag{23}$$

where

$$L_{\Pi}^{N}(u) = \prod_{i=1}^{N} \frac{u^2 + A_i^2}{u^2 + B_i^2}, \ (B_i - B_k)(A_i - A_k) \neq 0, \ i \neq k;$$

$$L_{\Sigma}^{M}(u) = \sum_{k=1}^{M} \frac{c_k |u|}{u^2 + D_k^2}.$$

We have for $u \to 0$

$$L_{\Pi}^{N}(u) = \prod_{i=1}^{N} \frac{A_i^2}{B_i^2} + O(u^2); \ L_{\Sigma}^{M}(u) = |u| \sum_{k=1}^{M} \frac{c_k}{D_k^2} + O(|u|^3), \tag{24}$$

and for $u \to \infty$

$$L_{\Pi}^{N}(u) = 1 + O(u^{-2}); \ L_{\Sigma}^{M}(u) = |u|^{-1} \sum_{k=1}^{M} c_k + O(|u|^{-3}). \tag{25}$$

We use the next lemma from Reference [30].

Lemma 1. If an even real function $\varphi(u)$ is continuous on a whole real axis and vanishes at infinity, then one can be approximated in $C_{(-\infty,\infty)}$ by a series of functions of the form:

$$\varphi_k(u) = (u^2 + \eta_k^2)^{-1}. \tag{26}$$

We choose constants A_i, B_i, $(i = 1, \ldots, N)$ in (23) such that $A \prod_{i=1}^{N} A_i^2 B_i^{-2} = C$ and consider the function:

$$L_s(u) = \frac{L(u) - L_{\Pi}^{N}(u)}{|u| th(Au)}. \tag{27}$$

Based on the properties (20), (21) and expression (26), it is followed that function $F(u) = L(u)/(|u| thAu)$ has the next property: $F(u) = |u|^{-1} + O(u^{-2}),\ u \to \infty$. Hence $L_s(u)$ satisfies the condition of Lemma 1. This means that the following representation is valid:

$$L_s(u) = \sum_{k=1}^{\infty} c_k \left(u^2 + \eta_k^2\right)^{-1}, \tag{28}$$

or from expressions (27) and (28), we obtain

$$L(u) = th(Au)\left[L_{\Pi}^{N}(u) + u\sum_{k=1}^{\infty} c_k \left(u^2 + \eta_k^2\right)^{-1}\right] \tag{29}$$

This fact proves the following theorem:

Theorem 1. In condition, when, the transform function $L(u)$ possesses the properties (21), there is its approximation, expressed in the form (23).

The constant A in (29) can be determined with sufficient accuracy by calculating the value of transformant $K(u) = L(u)/|u|$ for small values of argument u (about 10^{-3}). To improve the approximation, it is introduced an additional factor K_A, which determined for each heterogeneity law accordingly to the criterion of minimal relative error of the approximation in the range $0.01 < u < 30$.

A numerical realization of the improved approximation of $L(u)$ by functions of the type (22) can be achieved successfully by using the algorithm described in [1].

3. Development of Analytical Methods for Solving Integral Equation, Obtained by Fourier and Mellin Transforms

Below, by using the expressions:

$$\Phi(u) = \int_{-1}^{1} \varphi(\xi) e^{iu\xi} d\xi; \quad \int_{-\infty}^{\infty} \Phi(u) e^{-iux} du = \begin{cases} 2\pi g(x), & |x| \le 1; \\ 0, |x| > 1, \end{cases} \tag{30}$$

the integral equation (4) can be re-written in the form of dual integral equation:

$$\begin{cases} \int_{-\infty}^{\infty} \Phi(u) K(u) e^{-iux} du = 2\pi g(x), \ |x| \le 1; \\ \int_{-\infty}^{\infty} \Phi(u) e^{-iux} du = 0, \quad |x| > 1, \end{cases} \tag{31}$$

with $K(u)$, approximated by an expression of the form:

$$K(u) \approx \frac{\text{th}\, Au}{u} \prod_{n=1}^{N} \frac{u^2 + \delta_n'^2}{u^2 + \gamma_n'^2} = \frac{\text{th}\, Au}{u} \frac{P_1(u^2)}{P_2(u^2)}. \tag{32}$$

Here $P_1(u^2)$, $P_2(u^2)$ are the entire functions (polynomials of degree $2N$, where $\pm i\gamma_n'$, $\pm i\delta_n'$ are the zeros of these polynomials). We substitute approximation (32) in the integral equation (31), make the change of variables, $u = \lambda\beta$, and introduce the notations $\delta_n' = \delta_n\lambda$; $\gamma_n' = \gamma_n\lambda$; $\lambda = h/a$. Then the kernel of the integral equation (31) will have the form:

$$k(t) = \int_{-\infty}^{\infty} \frac{\operatorname{th} A\lambda\beta}{\lambda\beta} \prod_{n=1}^{N} \frac{\beta^2 + \delta_n'^2}{\beta^2 + \gamma_n'^2} e^{-i\beta t}\, d\beta. \tag{33}$$

We assume that the function $g(x)$ can be represented by Fourier-series as

$$g(x) = \frac{a_0}{2} + \sum_{k=1}^{\infty} \left(a_k \cos k\pi x + b_k \sin k\pi x\right). \tag{34}$$

By dividing the function $g(x)$ on even and odd parts:

$$g(x) = g_+(x) + g_-(x);\ g_+(x) = \frac{a_0}{2} + \sum_{k=1}^{\infty} a_k \cos k\pi x;\ g_-(x) = \sum_{k=1}^{\infty} b_k \sin k\pi x,$$

we apply the method of superposition and obtain even and odd parts of solution of the dual integral equation, respectively.

Then, by using the method of [7], we obtain an expression for the stresses in the form:

$$\varphi_+(x) = \frac{a_0}{2} \frac{\theta}{Q_{-\frac{1}{2}} \sqrt{2(\operatorname{ch}\theta - \operatorname{ch} x\theta)}} + \sum_{k=1}^{\infty} a_k \frac{\operatorname{th}(A\lambda k\pi)}{L(\lambda k\pi)} \Psi\left(-\frac{1}{2}, -\frac{1}{2} + i\frac{k\pi}{\theta}, x\right) + \\ + \sum_{n=1}^{N} C_n \Psi\left(-\frac{1}{2}, -\frac{1}{2} + \frac{\delta_n}{\lambda\theta}, x\right), \tag{35}$$

where

$$\Psi(u,v,x) = \theta \operatorname{sh}\theta \frac{Q_u P_v^1 - P_v Q_u^1}{Q_u \sqrt{2(\operatorname{ch}\theta - \operatorname{ch}\theta x)}} - \theta^2 \left(v + \frac{1}{2}\right)^2 \int_x^1 \frac{P_v(\operatorname{ch}\theta\tau)\operatorname{sh}\theta\tau}{\sqrt{2(\operatorname{ch}\theta\tau - \operatorname{ch}\theta x)}} d\tau, \qquad \theta = \frac{\pi}{A\lambda},$$

$P_u^\mu \equiv P_u^\mu(\operatorname{ch}\theta)$, $Q_u^\mu \equiv Q_u^\mu(\operatorname{ch}\theta)$ are the associated Legendre functions.

The constants C_n are founded from the following linear algebraic set of equations (summing over n from 1 up to N) :

$$\sum_{n=1}^{N} x_n = f_m + \sum_{n=1}^{N} a_{mn} x_n;\quad m = 1, \ldots, N, \tag{36}$$

where

$$x_n = Q^1_{-\frac{1}{2}} P_{-\frac{1}{2}+\frac{\delta_n}{\lambda\theta}} C_n; \; a_{mn} = \frac{Q_{-\frac{1}{2}}}{Q^1_{-\frac{1}{2}}} R\left(-\frac{1}{2}+\frac{\delta_n}{\lambda\theta}, -\frac{1}{2}+\frac{\gamma_m}{\lambda\theta}\right);$$

$$f_m = \frac{P_2(\varepsilon^2)}{P_1(\varepsilon^2)} \frac{\varepsilon^2}{Q_{-\frac{1}{2}+\frac{\gamma_m}{\lambda\theta}}} E\left(-\frac{1}{2}+i\frac{\varepsilon}{\theta}, -\frac{1}{2}+\frac{\gamma_m}{\lambda\theta}, -\frac{1}{2}\right);$$

$$R(u,v) = \frac{\left(u+\frac{1}{2}\right)^2 P_u Q_v^1 - \left(v+\frac{1}{2}\right)^2 P_u^1 Q_v}{(u-v)(u+v+1) P_u Q_v};$$

$$E(u,v,w) = Q_w T(u,v) - Q_v T(u,w);$$

$$T(u,v) = \frac{P_u Q_v^1 - Q_v P_u^1}{\left[\left(u+\frac{1}{2}\right)^2 - \left(v+\frac{1}{2}\right)^2\right]\theta^2};$$

$$h(u,v) = P_u Q_v^1 - Q_v P_u^1; \; H(u,v) = \frac{P_u P_v^1 - P_u^1 P_v}{\left[-\left(u+\frac{1}{2}\right)^2 + \left(v+\frac{1}{2}\right)^2\right]\theta^2};$$

The odd part of the solution has the form:

$$\varphi_-(x) = -\operatorname{sh} x\theta \left[\sum_{k=1}^{\infty} b_k \frac{\operatorname{th}(A\lambda k\pi)}{L(\lambda k\pi)} \mathrm{S}\left(-\frac{1}{2}+i\frac{k\pi}{\theta}, x\right) + \right.$$
$$\left. + \sum_{n=1}^{N} D_n \delta_n \lambda^{-1} \mathrm{S}\left(-\frac{1}{2}+\frac{\delta_n}{\lambda\theta}, x\right)\right], \tag{37}$$

where

$$\mathrm{S}(u,x) = \frac{\mathrm{P}_u(\operatorname{ch}\theta)}{\sqrt{2(\operatorname{ch}\theta - \operatorname{ch} x\theta)}} - \vartheta \int_x^1 \frac{\mathrm{P}_u^1(\operatorname{ch}\theta\tau)}{\sqrt{2(\operatorname{ch}\theta\tau - \operatorname{ch}\theta x)}} d\tau.$$

We find the coefficients D_n from the set of equations:

$$\sum_{n=1}^{N} a_{mn} x_n = f_m; \; m = 1, \ldots, N. \tag{38}$$

Here

$$x_n = D_n \delta_n \lambda^{-1}, \ a_{mn} = \mathrm{T}\left(-\frac{1}{2}+\frac{\delta_n}{\lambda\theta}, -\frac{1}{2}+\frac{\gamma_m}{\lambda\theta}\right);$$

$$f_m = \sum_{k=1}^{\infty} b_k \frac{\mathrm{th}(A\lambda k\pi)}{L(\lambda k\pi)} \mathrm{T}\left(-\frac{1}{2}+i\frac{k\pi}{\theta}, -\frac{1}{2}+\frac{\gamma_m}{\lambda\theta}\right).$$

The function $T(u, v)$ is the same as in Equation (36).

The general solution of the original integral equation (31) has the form in accordance with the superposition principle:

$$\varphi(x) = \varphi_+(x) + \varphi_-(x). \tag{39}$$

The relation between the impressing force and the upsetting of the stamp is obtained by substituting Equation (35) for the even part of solution into the equilibrium condition for the stamp:

$$P = \int_{-1}^{1} \varphi_+(x) dx. \tag{40}$$

Hence

$$P = \frac{\pi\delta}{\mathrm{Q}_{-\frac{1}{2}}} \left[\frac{a_0}{2} \mathrm{P}_{-\frac{1}{2}} L^{-1}(0) + \sum_{k=1}^{\infty} a_k \frac{\mathrm{th}(A\lambda k\pi)}{L(\lambda k\pi)} \mathrm{P}_{-\frac{1}{2}+i\frac{k\pi}{\theta}} + \sum_{n=1}^{N} C_n \mathrm{P}_{-\frac{1}{2}+\frac{\delta_n}{\lambda\theta}} \right], \tag{41}$$

where a_k $(k = 0,1,2,\ldots)$ are the Fourier coefficients of the function $g(x)$, i.e. $a_k = \int_{-1}^{1} g(x)\cos k\pi x dx; \ k = 0,1,2,\ldots$

In particular, for the case of flat stamp and simple approximation for $N = 1$ in (22), the following expression for contact stresses and relation between the impressing force and the upsetting of the stamp will be useful in the set of practical calculations in the form:

$$\begin{aligned} \varphi(x) = C\{ & \vartheta F_1 (2(\mathrm{ch}\vartheta - \mathrm{ch}\vartheta x))^{-0,5} (Q_{-0,5} L(0))^{-1} + \\ & + (F_2 \delta^2 \lambda^{-2} \mathrm{sh}^{-1}\vartheta (\delta^2 F_3 + \gamma^2 F_4)^{-1}) I(x)\} \end{aligned} \tag{42}$$

where

$$\omega = -0.5 + \delta/(\lambda\vartheta); \ \upsilon = -0.5 + \gamma/(\lambda\vartheta); \ F_1 = 1 - h(\omega; 0.5);$$

$$F_2 = (\delta^2 - \gamma^2) Q_\upsilon; \ F_3 = P_\omega G(-0.5; \upsilon), F_4 = Q_\upsilon h(\omega; -0.5);$$

$$h(a,b) = P_a(\operatorname{ch}\vartheta)Q_b^1(\operatorname{ch}\vartheta) - P_a^1(\operatorname{ch}\vartheta)Q_b(\operatorname{ch}\vartheta);$$

$$G(a,b) = Q_a(\operatorname{ch}\vartheta)Q_b^1(\operatorname{ch}\vartheta) - Q_a^1(\operatorname{ch}\vartheta)Q_b(\operatorname{ch}\vartheta);$$

$$I(x) = \int_x^1 P_\omega(\operatorname{ch}\vartheta\tau)\left(2(\operatorname{ch}\vartheta - \operatorname{ch}\vartheta x)\right)^{-0.5} \operatorname{sh}\vartheta\tau d\tau;$$

$$P = \frac{\pi\delta}{Q_{-0,5}}\left[P_{-0.5}L^{-1}(0) - (\delta^2 - \gamma^2)Q_\omega(\operatorname{ch}\vartheta)P_\omega(\operatorname{ch}\vartheta)\operatorname{sh}^{-1}\vartheta \cdot Z^{-1}(-0.5;\omega)\right], \tag{43}$$

where $Z(-0.5;\omega) = \delta^2 G(-0.5;\omega)P_\omega(\operatorname{ch}\vartheta) - \gamma^2 Q_\omega(\operatorname{ch}\vartheta)h(\omega;-0.5)$.

In general case with $L(u)$ in the form (23), the solution of (31) is constructed by successive approximations.

In [2], it has been proved the next Theorem 2 by using the method, known in the perturbation theory and based on the method of subsequent approximations.

Theorem 2. Equation (31) is uniquely solvable in the space $C^{(k)}_{k+\frac{1}{2}}(-1,1)$, for $L(u)$ in the form (22), if $g(x) \in B^{\frac{1}{2}+\varepsilon}_{k+2}(-1,1)$, $\varepsilon > 0$, for $0 < \lambda < \lambda^*$ and $\lambda > \lambda_0$, where λ^*, λ_0 are the certain fixed values of λ, and the bound

$$\|\varphi(x)\|_{C^{(k)-}_{k+\frac{1}{2}}(-1,1)} \le m(\Pi_N, \Sigma_\infty, k) \cdot \|g\|_{B^{\frac{1}{2}+\varepsilon}_{k+2}(-1,1)} \tag{44}$$

is obtained.

So, it was found from [4], that the solutions obtained are two-sided asymptotically accurate solutions of initial integral equation (4) both as for small as for large values of dimensionless geometrical parameter λ. A comparison of the analytical approximate solutions with known solutions for a homogeneous wedge of the monograph [25] showed an error is lesser than 1% for $\lambda < 1$ and $\lambda > 2.8$. The greatest divergences (more than 10%) are observed at $\lambda = 2$. The explicit analytic form of the solutions provides the possibility of their widespread use for estimating both the mechanical (stiffness function), and protection properties (heat resistance, wear resistance) of functionally graded coatings of complicated structure.

4. Numerical Results

The effectiveness of above method is illustrated by the examples of contact problems for a semi-infinite homogeneous and heterogeneous media and regions with angular points.

4.1. Numerical Analysis of the Contact Problem Solution for the Strip with Independently Varying Young's Modulus and Poisson's Ratio

In the numerical experiment, it is analyzed mutual influence of continuously changing on depth Young's modulus and Poisson's ratio on stress-strain state of the layer at the impressing of the plain stamp in the layer. It is considered the simplest case of continuous change of the properties on depth, namely their uniform decreasing or increasing. It is assumed, that the values of Young's modulus and Poisson ratio change in two times compared to substrate, and change of Young's modulus is reverse proportional to change of Poisson ratio.

$$\begin{aligned}
&(f_0): E(y)=1, v(y)=0.2;\\
&(f_1): E(y)=1-y,\ v(y)=0.2;\\
&(f_2): E(y)=1+0.5y,\ v(y)=0.2;\\
&(f_3): E(y)=1,\ v(y)=0.2+0.1y;\\
&(f_4): E(y)=1,\ v(y)=0.2-0.2y;\\
&(f_5): E(y)=1+0.5y,\ v(y)=0.2-0.2y;\\
&(f_6):: E(y)=1-y, v(y)=0.2+0.1y;\ -1\le y\le 0.
\end{aligned} \tag{45}$$

In Figure 3 the six different types of functions $f_i(y)$, determining the character of the variation of elastic parameters are presented.

In Figure 4, are presented relative values of transformants of kernels $F_i(\alpha) = L_i(\alpha)/L_0(\alpha)$ of the integral equation, where $L_0(\alpha)$ is transformant for homogeneous medium for all of six cases of the changing mechanical properties with depth. The figure shows clearly that the decreasing of Poisson's ratio with depth at constant Young's modulus does not affect on the elastic properties of strip. In particular, the curve of the transformant in the case of homogeneous strip is practically the same as the transformant for the heterogeneity law $f_3(y)$.

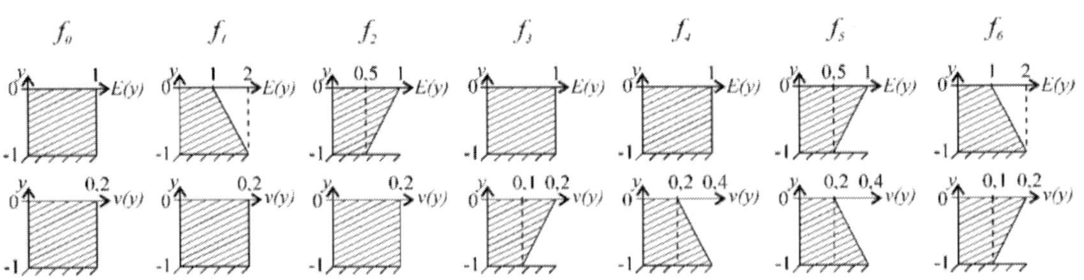

Figure 3. Heterogeneity laws for realization of numerical experiments in the contact problem (S2) on impressing of the flat stamp in inhomogeneous strip, rigidly fixed with non-deformable foundation.

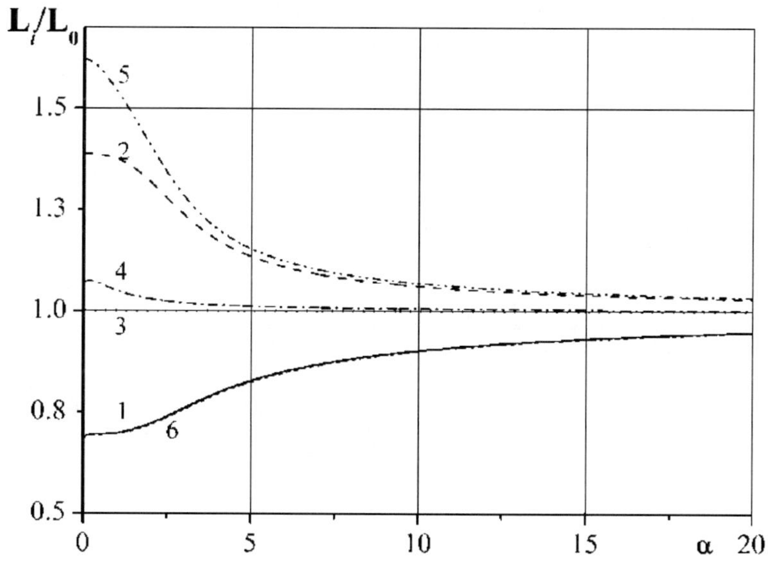

Figure 4. Relative values of transformants of kernels $F_i(\alpha)$ for the heterogeneity laws (45).

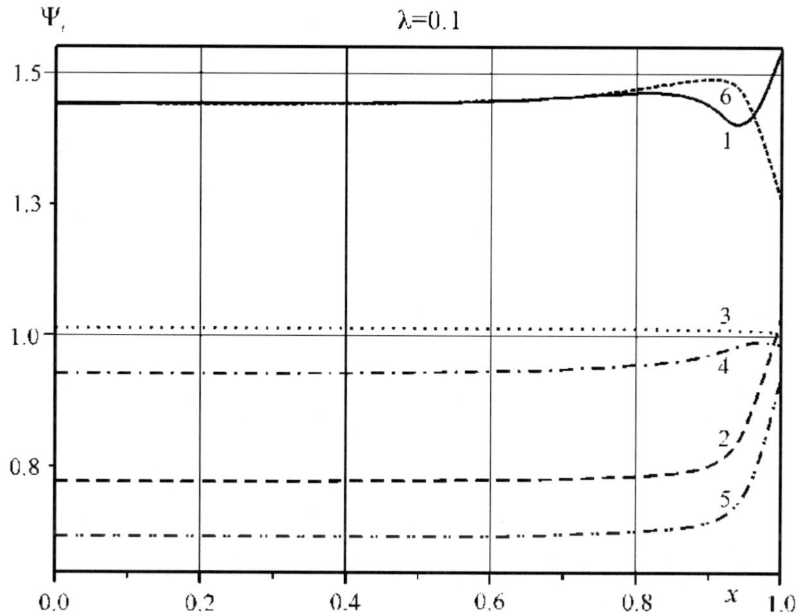

Figure 5. Distributions of relative contact stresses $\Psi_i(x) = \varphi_i(x)/\varphi_0(x)$ for the heterogeneity laws $f_i(y)$ at $\lambda = 0.1$.

An increasing of Poisson's ratio with depth in similar case influences insignificantly the form of transformant of the integral equation, and hence solution form of the considered contact problem. In other cases, the combinations of independent changes of Young's modulus and Poisson's ratio render significant influence on the solution form of this contact problem. The digitals near the curves point number of corresponding heterogeneity law (45).

Table 1. The results of numerical approximation of function $L(u)$ by expression (22) for $N = 1$

The number of heterogeneity law in Figure 3	The coefficients of the approximation			Maximal relative error of approximation, %
	AK_A	δ_i	γ_i	
0	0.55108	0.96036	1.04128	3.0
1	0.77652	1.97007	3.04558	5.3
2	0.40647	5.02871	3.97717	5.3
3	0.55108	0.95726	1.04465	3.1
4	0.45088	4.59958	4.34822	3.1
5	0.51485	4.92910	4.05754	3.8
6	0.47593	4.06454	4.92061	5.0

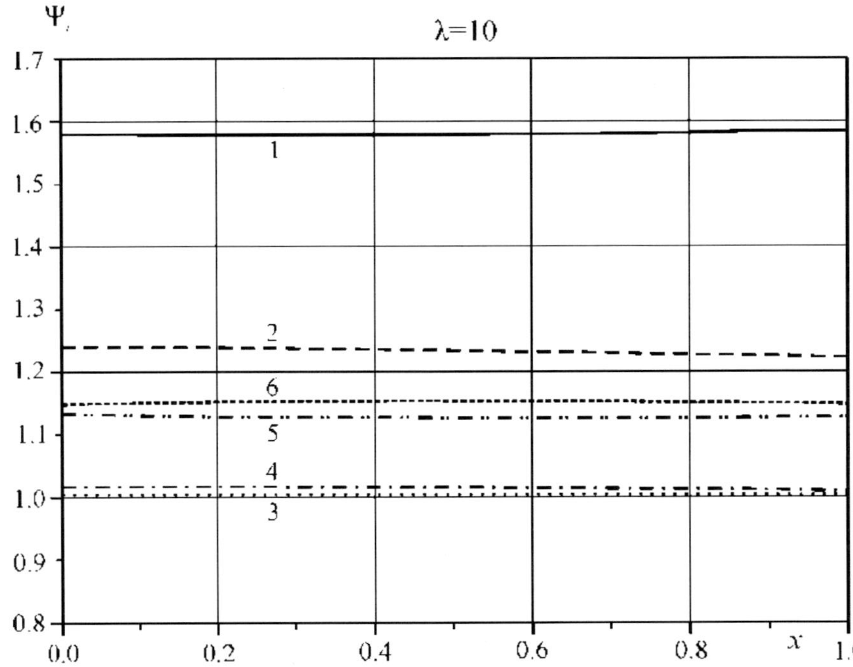

Figure 6. Distributions of relative contact stresses $\Psi_i(x) = \varphi_i(x)/\varphi_0(x)$ for the heterogeneity laws $f_i(y)$ at $\lambda = 10.0$.

Figures 5 and 6 present distributions of relative contact stresses $\Psi_k(x) = \varphi_k(x)/\varphi_0(x)$; $k = 1,..6$ for heterogeneous strip for different combinations of the heterogeneity laws $f_i(y)$ of the elastic moduli at small and great values of the dimensionless parameter $\lambda = h/a$, but Figure 7 shows similar distribution for the ratio of impressing force to corresponding value in homogeneous case.

The digits near the curves point the number of heterogeneity law (45).

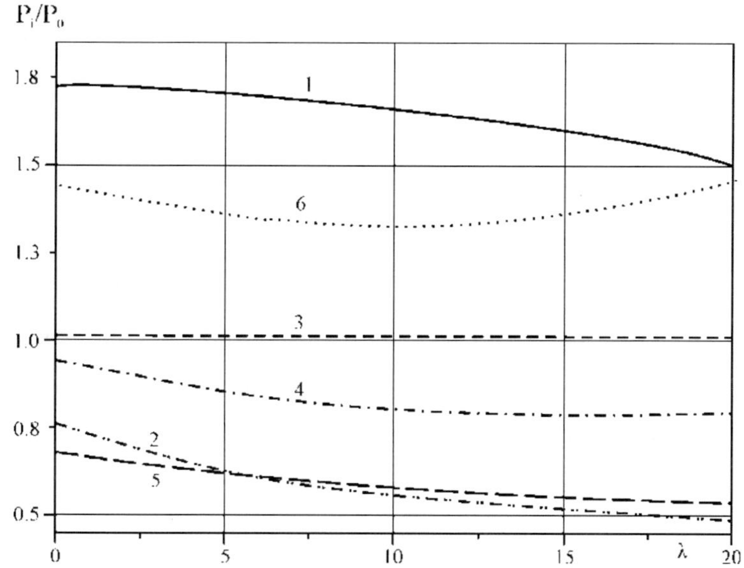

Figure 7. Distributions of relative values of $P_i(\lambda)/P_0(\lambda)$ for the heterogeneity laws $f_i(y)$, $i = 1,..,6$.

The contact stresses are calculated by using formula (42), but the dependence of the stamp displacement on the impressing force computed by formula (43) at $N = 1$. Table 1 depicts approximation factors for the transform function $L(u)$, calculated in the case of contact problem for strip, rigidly fixed with non-deformable foundation, heterogeneity of elastic properties of which is defined by the laws (45) at $N = 1$. Moreover, the table presents the maximal relative errors of these approximations.

By analyzing the curves, shown in above figures, we note that the contact pressure at the constant Young's modulus, E, and Poisson's ratio, ν, linearly decreasing from 0.2 at surface down to 0.1 on depth, does not differ from homogeneous case. By decreasing Poisson's ratio from 0.2 at surface down to 0.1 on depth, there is only weak diminishing of contact pressure (on value of 0.95). However, by changing Young's modulus, the significant influence on character of distribution of the contact stresses is caused namely by decreasing the Poisson's ratio on depth.

The analysis of curves, presented in Figure 7, demonstrates that change with depth of the Poisson's ratio influences the dependence of the stamp displacement on the impressing force only in the case of growth of the Poisson's ratio with depth at the constant Young's modulus. The decreasing of the Poisson's ratio with depth renders significant influence on the solution form of the contact problem only in the case of simultaneous changing of the Young's modulus on the strip depth.

4.2. Numerical Analysis of Solution for Changing Young's Modulus and Constant Poisson's Ratio in Contact Problem for Wedge

By conducting numerical experiments, we shall suggest that for all possible types of the heterogeneity laws, the ratio of Young's modulus at upper face of the wedge to its value in any internal point (designed through n_c) be into range $1/3.5 < n_c < 3.5$, but the values of Poisson's ratio should be into range of $-1 < \nu < 0.5$ for any angle $\varphi \in [0; \pi/4]$.

Then we consider the next types of heterogeneity:

$$G/G_0 = f_a(\varphi) = 2.25 - 1.25\cos(8a\varphi),\ 0 < \varphi < \pi/4; \quad (46)$$

$$G/G_0 = g_a(\varphi) = 1/f_a(\varphi),\ 0 < \varphi < \pi/4. \quad (47)$$

Figure 8 presents the curves $L(s)/L_0(s)$, obtained in one coordinate plane for the heterogeneity laws $f_a(\varphi)$ and $g_a(\varphi)$, which correspond to the heterogeneity laws (45).

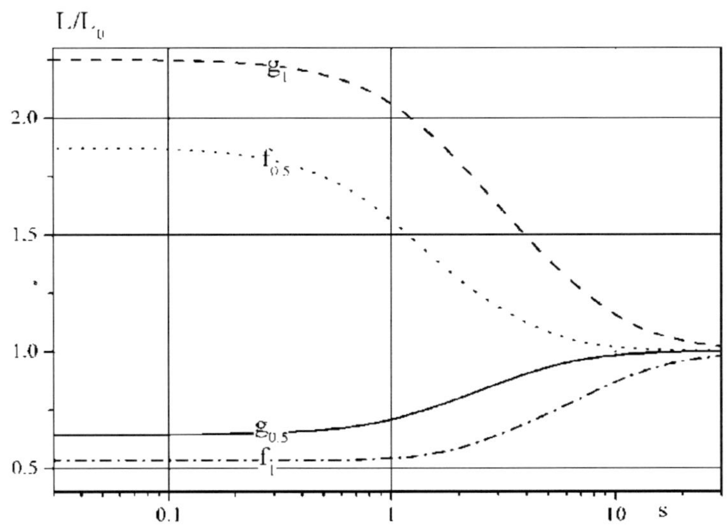

Figure 8. Relative transformants $F_i(s)$ for heterogeneity laws (46), (47) for $a = 0.5$ and $a = 1$.

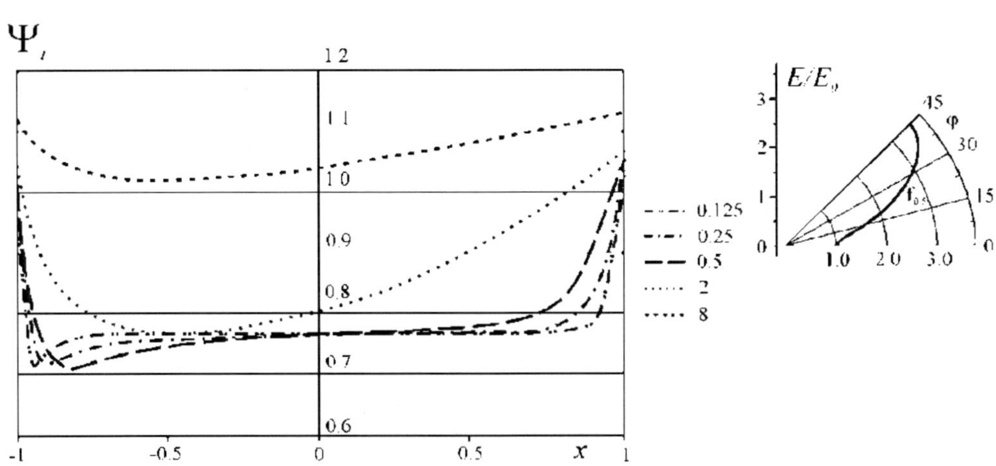

Figure 9. Relative contact stresses $\Psi_i(x) = \varphi_i(x)/\varphi_0(x)$ for heterogeneity law $f_{0.5}(\varphi)$ at various values of λ.

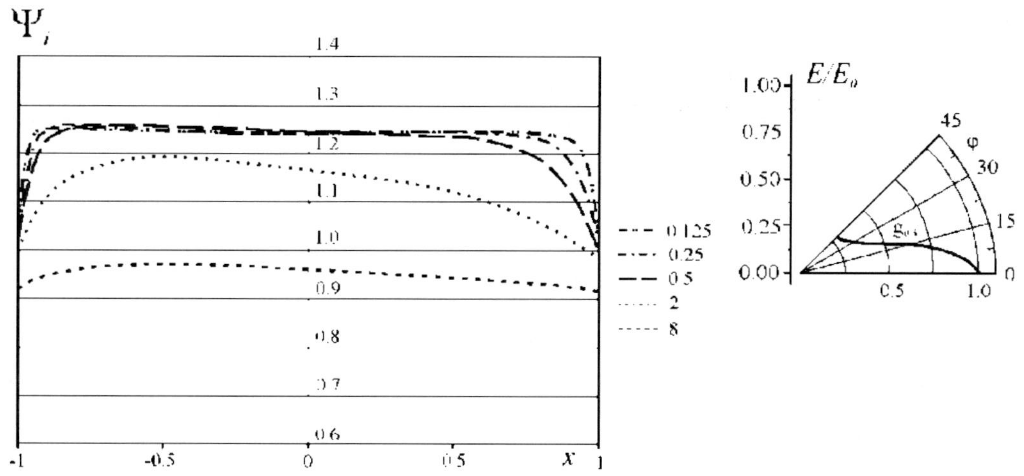

Figure 10. Relative contact stresses $\Psi_i(x) = \varphi_i(x)/\varphi_0(x)$ for heterogeneity law $g_{0.5}(\varphi)$ at various values of λ.

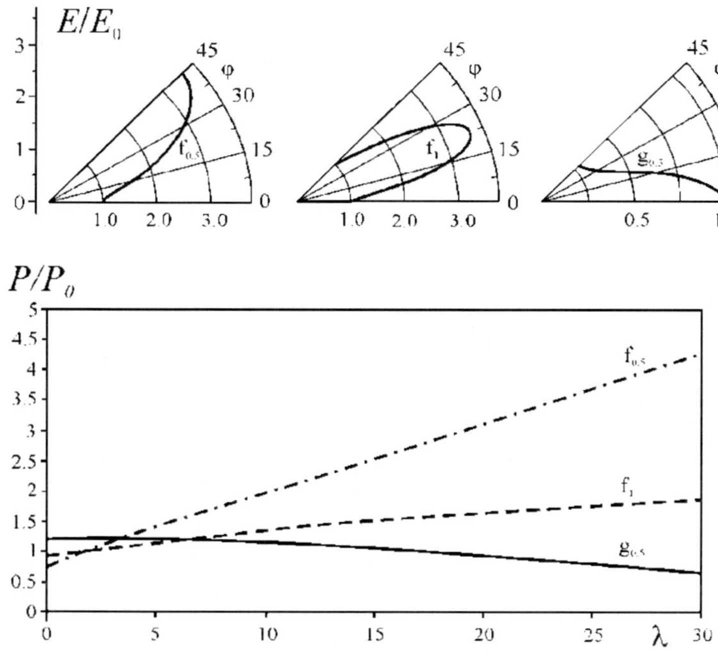

Figure 11. Dependences of relative impressing force P/P_0 on character parameter λ of the problems for different laws of wedge heterogeneity.

Figures 9, 10 show the plots of relative values of $\Psi_i(x) = \varphi(x)/\varphi_0(x)$, characterizing the distribution of contact normal stresses $\varphi(x)$ under stamp, impressed by unit force in heterogeneous semi-space with Young's modulus, expressed by (27), (28) at a = 0.5 and various values of $\lambda = 2(\ln(b/a))^{-1}$. Poisson's ratio, ν, in this example equals to 0.333. The value of $\varphi_0(x) = rq_0(r)$ presents a dimensionless distribution of contact stresses under stamp for homogeneous wedge at $E(\varphi) = E_0$. Numbers at the curves correspond to the value of λ, but the values of $\varphi(x)$ are defined by formula (42) at $N = 1$.

Finally, Figure 11 presents dependences of relative impressing force P on character parameter λ of the problems for different laws of wedge heterogeneity.

ACKNOWLEDGMENTS

The work was carried out with partial financial support from Russian Foundation for Basic Research (Project codes 10-08-00839-*a*, 11-08-91168_GFEN-*a*) and in most part by the Federal Goal Program No. P1107.

REFERENCES

[1] Aizikovich, S. M.; Alexandrov, V. M.; Kalker, J. J.; Krenev, L. I.; Trubchik, I. S. *Int. J. of Solids and Structures*. 2002, *vol. 39(10)*, 2745–2772.
[2] Aizikovich, S. M.; Alexandrov, V. M.; Trubchik, I. S. In: *Operator Theory: Advances and Applications*. 2009, *vol. 191*, 3–17.
[3] Kul'chyts'kyi-Zhyhailo, R.; Rogowski, G. *Material Science*. 2009, *vol. 45*, 845–858.
[4] Huand, C. S.; Leissa, A. W. *Composite Struct*. 2008, *No. 82*, 488–498.
[5] Korsunsky, A. M.; Constantinescu, A. *Thin Solid Films*. 2009, *vol. 517*, 4835–4844.
[6] Chebakov, M. I. *J. of Applied Mathematics and Mechanics*. 2005, *vol. 69*, 296–304.
[7] Zelentsov, V. B. *J. of Applied Mathematics and Mechanics*, 2002, *vol. 66*(5), 841−855.
[8] Fuqian, Yang. *Theoretical and Applied Fracture Mechanics*. 1997, *No. 26*, 211–217.
[9] Jaffar, M. J. *J. Mech. Phys. Solids*. 1988, *vol. 36(4)*, 401−416.
[10] Lamzuk, V. D.; Privarnikov, A. K. In: *Stability and Durability of Construction Elements*. (in Russian), Dnepropetrovsk University, Dnepropetrovsk, 1978, *No. 1*, 1−64., *No. 2*, 1−68.
[11] Privarnikov, A. K. In: *Stability and Strength of Structural Elements* (in Russian), Dnepropetrovsk University, Dnepropetrovsk, 1973, 27–45.
[12] Kovalenko, Ye. V. *J. of Applied Mathematics and Mechanics*, 1999, *vol. 63(1)*, 109–116.
[13] Liangchi, Zhang; Zhongqin, Lin. *Int. J. Solids Structures*, 1997, *vol. 34(3)*, 327–339.
[14] Hyung, Jip Choi. *J. of Mech. Sci. and Techn*. 2009, *vol*. 23, 2703–2713.
[15] Pauk, V. J. *Int. J. of Heat and Mass Transfer*. 1999, *No. 42*, 2583–2589.
[16] Yevtushenko, A. A.; Pyr'ev, Yu. A. *J. Appl. Mech*. 1999, *vol. 63(5)*, 795–801.
[17] Manabu, Ohmichi; Naotake,Noda; Masayuki, Ishihara. *Acta Mech*. 2010, *vol. 212*, 219–232.
[18] Frankle, M.; Munz, D.; Yang, Y. Y. *Int. J. Solids and Struct*. 1996, *vol. 33(14)*, 2039–2054.
[19] Krasnyuk, P. P. *Int. Appl. Mech*. 2000, *vol. 36(1)*, 118–127.
[20] Hyung, Jip Choi. *Int. J. Solids Structures*. 1996, *vol. 33(28)*, 4100–4117.
[21] Yong-Dong, Li; Kang, Yong Lee. *Arch. Appl. Mech*. 2008. *vol. 78*, 855–866.

[22] Naotake, Noda; Zhi-He, Jin. *Int. J. Solids Structures*, 1993, *vol. 30(8)*, 1039–1056.
[23] Paul, E. W.; Labossiere, Martin L. Dunn. *J. of Mech. and Phys. of Solids*. 2001, *No. 49*, 609–634.
[24] Alexandrov, V. M.; Vorovich, I. I. *PMM*, 1960. *vol. 24(2)*. 323–333 (in Russian).
[25] Vorovich, I. I.; Alexandrov, V. M.; Babeshko, V. A. *Nonclassical Mixed Problems of Elasticity Theory*. Nauka, Moscow, 1974, 1 – 456 (in Russian).
[26] Noor, M. A. *Comput. Modeling*, 1989, *vol. 12(9)*, 1079–1084.
[27] Babeshko, V A,; Glushkov, E. V.; Glushkova, N. V. *Numerical Mathematics and Math. Physics Journal*, 1987. *vol. 27(1)*, 93–101.
[28] Alexandrov, V. M.; Romalis, B. L. *Contact Problems in Mechanical Engineering*, Mashinostroenie, Moscow, 1986, 1 – 176 (in Russian).
[29] Aizikovich, S. M.; Aleksandrov, V. M. *Dokl. Akad. Nauk SSSR*, 1982, *vol. 266(1)*, 40–43.
[30] Akhiezer, N. I. *Lectures on the Theory of Approximations*. Nauka, Moscow, 1965 (in Russian).

In: Physics and Mechanics of New Materials …
Editors: Ivan A. Parinov and Shun-Hsyung Chang
ISBN: 978-1-62618-535-7

Chapter 20

IDENTIFICATION OF STIFFNESS PARAMETERS OF THE NONLINEAR ELASTIC REINFORCED BEAM

***V. O. Kaledin*[1]*, O. K. Kaznacheeva*[2*] *and O. A. Burtseva*[2]**
[1]Novokuznetsk Institute (the branch) of Kemerovo State University, Novokuznetsk, Russia;
[2]The South-Russian State Technical University (Novocherkassk Polytechnicnical Institute), Novocherkassk, Russia

ABSTRACT

We proposed a new identification method of stiffness parameters for reinforced concrete beam. We also stated and solved the next tasks: (i) developed a model of static deformation of the construction with variable parameters of stiffness to calculate the state variables under different influences; (ii) built an approximation of the response function with variable physical and mathematical parameters of material, which defined the construction stiffness, with aim to improve an economic design of the parameters; (iii) developed calculation algorithms for point and interval estimates of the required stiffness parameters by using the data of natural physical experiments; (iv) created calculation programs of the identification.

1. IDENTIFICATION METHOD OF STIFFNESS PARAMETERS

The reinforced constructions have obtained wide application in machine-building engineering and civil engineering. Non-destructive testing of strength of these constructions is difficult because of physical and mathematical characteristics instability, especially, at variable on volume the reinforcing schemes and impossibility to directly measure the geometric parameters which define stiffness. We can improve diagnostics by defining the actual physical and mathematical characteristics of component materials in construction.

[*] Email: kazn_olga@mail.ru

We examine two-level model, which consists of finite-element model of deformation (basic model) and based on the finite-element modeling the reduced model with given response functions.

In the basic model, the static deformation is described by nonlinear system of algebraic equations of high order:

$$\boldsymbol{H}\, q = r\,, \tag{1}$$

where H is the stiffness matrix, q is the generalized displacement vector, r is the external loading vector.

The stiffness matrix for nonlinear elastic material has the next structure:

$$\boldsymbol{H}_E = \int_V \Phi^{\mathrm{T}}(x)\boldsymbol{D}^{\mathrm{T}}\boldsymbol{E}_{cr}\boldsymbol{D}\,\Phi(x)\,dV\,, \tag{2}$$

where the integrating is fulfilled on the volume of construction.

Here $\boldsymbol{E}_{cr}$ is the matrix of secant moduli of elasticity, depending on deformation of material, $\Phi(x)$ is the vector-column of the basic interpolation functions, $\boldsymbol{D}$ is the differentiation operator which presences in the kinematic relations (the dependencies between displacements and deformations) as

$$\delta\,\varepsilon(x) = \boldsymbol{D}\,\delta u(x);\;\; \delta u(x) = \Phi(x)\delta q\,, \tag{3}$$

where $\delta u(x)$ is the variation of displacement, depending on coordinates of point.

The analytical model of measurement system allows one to express the deterministic component of measurement vector $\boldsymbol{Z}$ through generalized displacements q in the form:

$$\boldsymbol{Z}(p) = \boldsymbol{C}q(p)\,. \tag{4}$$

Further we propose that this link will be linear.

Quality criterion of identification should estimate the deviation of sensor reading from the actual measured data (Z^*), which are calculated by means of the model (4). The simplest criterion maybe built as a weighted sum of deviations of the calculated and measured data:

$$\varphi = (\boldsymbol{Z}^* - \boldsymbol{C}q)^T \boldsymbol{V}(\boldsymbol{Z}^* - \boldsymbol{C}q) \rightarrow \min\,, \tag{5}$$

where V is the matrix of the weight coefficients.

Minimum of the criterion (5) is found by varying the parameter p. In this case, the response function q is found by solving nonlinear algebraic set of equations (1) with constant matrix of coefficients (2).

The dependence of the generalized displacements on variable parameters of stiffness is determined by fractional rational function as

$$q(p) = A_0 + A_1 \frac{1}{p_1 - \lambda_1} + A_2 \frac{1}{p_2 - \lambda_2} + \ldots + A_s \frac{1}{(p_1 - \lambda_1)^{s_1} (p_2 - \lambda_2)^{s_2} \cdot \ldots} \quad (6)$$

where λ_i are the roots of the characteristic equation:

$$\det(\boldsymbol{H}_0 + \lambda_i \boldsymbol{H}_i) = 0 . \quad (7)$$

In order to calculate the expansion coefficients we should introduce new dimensionless factors ξ_i as

$$\xi_i = \frac{1}{p_i - \lambda_i} . \quad (8)$$

We examined the beam (Figure 1), which was made of the ceramsite concrete B 20 and included one set of the armature A-III. We supposed, that the beam has been subjected to a fire influence during its work. It led to thermal destruction of upper and/or lower surface layers.

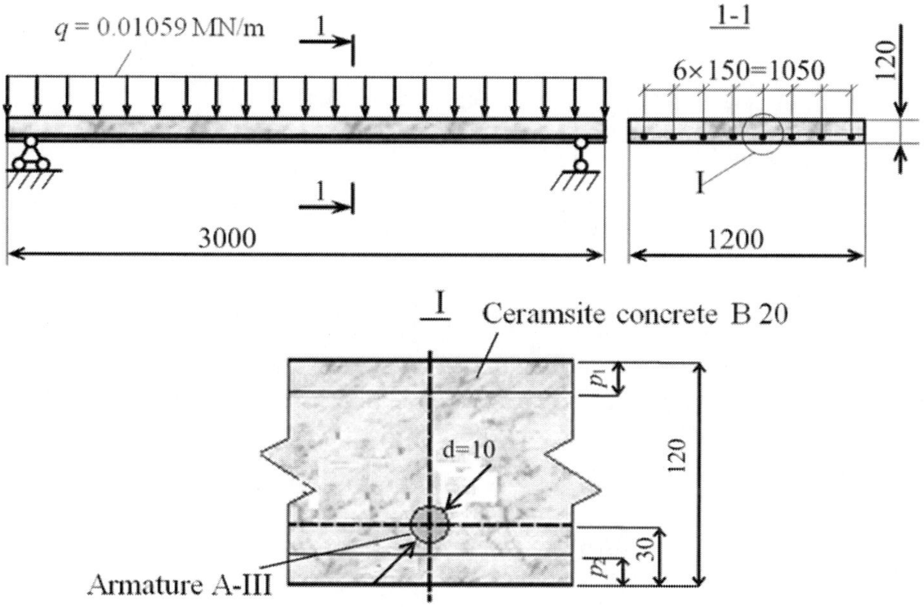

Figure 1. Scheme of beam loading and fastening in different sections (all linear sizes are measured in millimeters).

In this case, the thickness of damaged layers, which adjacent with the upper and lower sides of the section, was variable parameter. This example was a model, for which the computation experiment was carried out. Displacements and stresses of beam with given thickness of damaged layers ($p_1 = 0.01$ m, $p_2 = 0.015$ m) were calculated by using the finite-element method. The calculated variables were taken out as the mathematical expectations of the sensor data.

Finite-element model of deformation of the reinforced beam was based on the next physical constitutive law:

$$\sigma = \begin{cases} \dfrac{\sigma_{\mathrm{pl}}}{\varepsilon_{\mathrm{pl}}}(\varepsilon - \varepsilon_0), & \varepsilon - \varepsilon_0 \le \varepsilon_{\mathrm{pl}}; \\ \sigma_{\mathrm{pl}} + \dfrac{\sigma_{\mathrm{e}} - \sigma_{\mathrm{pl}}}{\varepsilon_{\mathrm{e}} - \varepsilon_{\mathrm{pl}}}(\varepsilon - \varepsilon_0 - \varepsilon_{\mathrm{pl}}), & \varepsilon_{\mathrm{pl}} < \varepsilon - \varepsilon_0 \le \varepsilon_{\mathrm{e}}, \end{cases} \tag{9}$$

where σ_{pl} and $\varepsilon_{\mathrm{pl}}$ is the proportionality limit of stress and strain, respectively, σ_{e} and ε_{e} is the elasticity limit of stress and strain, respectively, ε_0 is the initial strain. Obviously, these parameters are different in the cases of tension and compression (see Figure 2).

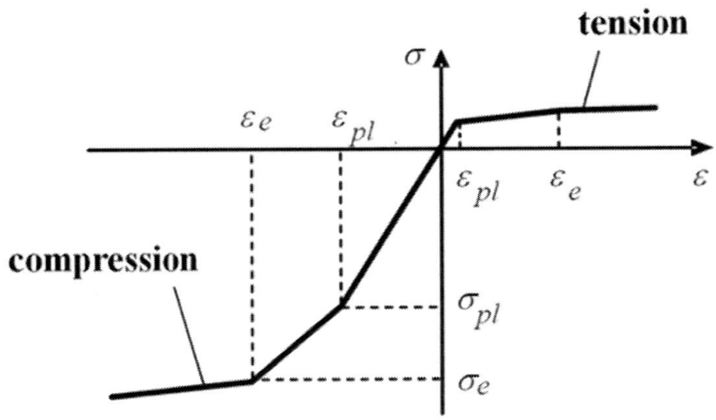

Figure 2. Diagram approximation of concrete nonlinear elastic deformation.

These characteristics are known for famous kinds of concrete and armature steel. They are enough stable [1]. However, in emergency conditions, for example in condition of fire influence, they can deteriorate.

2. Computation Experiments

Further, we carried out a set of calculations for definition of p_1 and p_2. In these calculations, the sensor characteristics were taken as generated arbitrary numbers distributed on normal law with the mean-square deviations of 0.1 mm for displacements and 0.03 MPa for stresses.

We present matrix equation of equilibrium as

$$(\boldsymbol{H}_0 + p_1\boldsymbol{H}_1 + p_2\boldsymbol{H}_2)q = r\,. \tag{10}$$

In this case, the matrices H_0, H_1, H_2 depend on the deformations. The set of equations (10) is nonlinear.

We define the error of the response function at calculation using the basic model. First, we must estimate a convergence speed of solution of the boundary-value problem at the finite-element mesh refinement and a convergence speed of algorithm for the set of nonlinear equations (10).

The asymptotic estimation of error of the boundary problem approximation is defined in the form [2]:

$$\Delta y(N_e) \le C N_e^{-m}, \tag{11}$$

where C is a constant, N_e is the number of finite elements along the beam length, m is the convergence order.

We use the Runge's rule in conditions of known convergence order [3]:

$$\Delta y^{(N)} \le \frac{\left| y^{(N)} - y^{(2N)} \right|}{1 - 2^{-m}}, \tag{12}$$

where in numerator, we have difference of solutions on the FEM meshes with N and $2N$ elements.

Effective convergence order is defined by set of calculations with successive doubling of the finite elements. The results of calculations are demonstrated in Figure 3. The number of finite elements along the beam axis is presented by the horizontal axis. The calculated maximum deviation is present by the vertical axis. Figure 3b presents in logarithmic coordinates difference of numerical solutions at successive doubling FEM elements in dependence on the number of the elements.

The points in this figure correspond to actual calculations. The solid line is the straight line, obtained by the method of the least squares. Angular coefficient of this straight line (with the sign of minus) is equal to effective convergence order (one is equal to 1.933). The dotted line presents the values, which correspond to the theoretical convergence order of the linear problem (one is equal to 2).

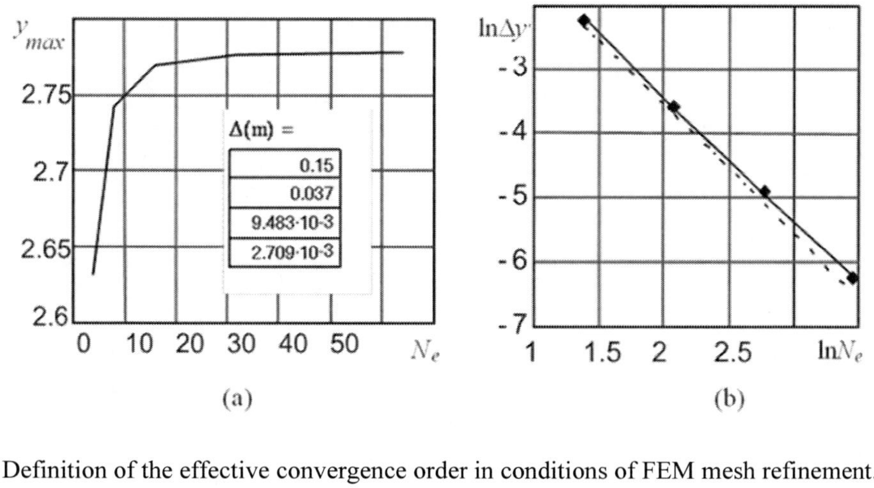

Figure 3. Definition of the effective convergence order in conditions of FEM mesh refinement.

Further, we examine the second component of the error for the numerical solution, namely an error, connected with solution of the nonlinear equations set. The asymptotic estimation of the error arising at solution of the nonlinear equations set by using iteration method is expressed by the inequality:

$$\Delta y(K) \leq Cq^{K}, \tag{13}$$

where K is the number of iterations, q is the effective convergence speed.

For definition of actual convergence speed, we depict in Figure 4 the dependence (in logarithmic coordinates) of the error for the closed equations system on number of iterations.

By using the method of the least squares, we found value of q = 0.875 after approximation of this curve by linear function.

Thus, by using inequality (13), we can estimate the number of iterations, which is necessary to obtain the given admissible error, and the required number of finite elements is calculated by using inequality (11).

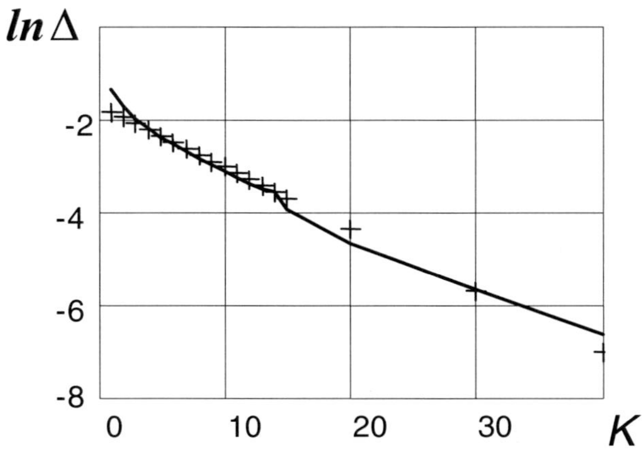

Figure 4. Definition of the effective speed of convergence for solution of set of the nonlinear equations.

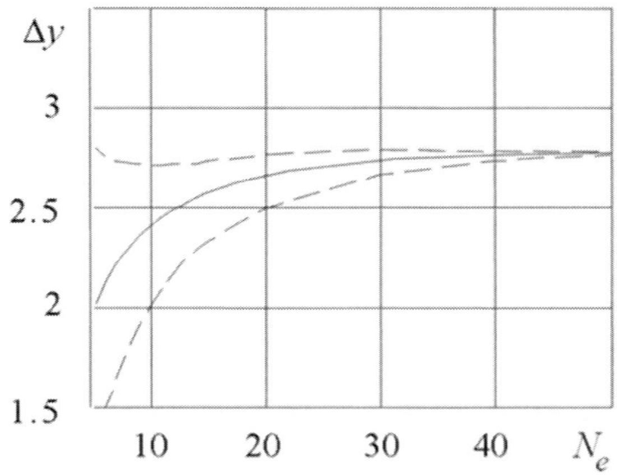

Figure 5. Total computation error of response function in dependence on number of finite elements for 500 iterations.

Total error in the case of 500 iterations is represented in Figure 5. The dependence of the numerical solution on the number of finite elements is shown by solid curve, upper and low bounds of exact solution are shown by dotted lines. We can see that numerical solution gives the lower estimate for deflection, that corresponds to the analytical solution.

The approximation of the response function is calculated by two ways by using (i) polynomials, and (ii) fractional rational functions.

The polynomial approximation was constructed by means of the complete two-factor experiment, in which two levels of significance are taken into account. Minimal level of each factor was assumed to be equal to zero, that corresponded to undamaged section of the beam; maximum level was assumed to be equal to 0.02 m.

When we built the fractional rational approximation, the roots of characteristic equation were found by next way: stiffness matrix degenerated, if the sum of thicknesses of the damaged layers was equal to the height of section. Thus, basic fractional rational functions may be selected as

$$f_1 = \frac{p_1}{p_1 - h}, \quad f_2 = \frac{p_2}{p_2 - h}, \quad f_3 = \frac{p_1 p_2}{p_1 + p_2 - h}. \tag{14}$$

The numerical experiment was carried on the base of finite-element modeling. As the responses we chose the deflections in 9 points, which were uniformly distributed on the whole beam san and also stresses on the upper surface of the central cross-section. Further, we demonstrate the results of approximation of the maximal deflection; the polynomial and fractional rational approximations are shown in Figure 6a. The values, which were calculated by using the basic model, are shown by points. Note that when we choose only deflection as the response, model was sensitive to one parameter p_2 and insensitive to parameter p_1.

The polynomial approximation error was lesser, when we considered greater number of points in the numerical experiment and respectively the higher polynomial order.

The error of 2 % (relatively to the results, based on the basic model) is reached by using quadratic polynomial, that requires no lesser than 9-point plan. At the same time, the error of fractional rational approximation is no greater than 1% even at the 4-point plan. The errors of fractional rational approximation and fractional bilinear approximation had a small difference. By estimating, the minimization of the quadratic criterion of quality gives the point estimates $p_1 = 0.0104$ m, $p_2 = 0.014$ m. They differ from really used estimates of the basic model on 4 % and 6.5 %, respectively.

For interval estimate of stiffness parameters we used the numerical modeling of random error of measurements. We generated the sampling from 10,000 random vectors, each component of which was uniformly distributed into symmetrical range relatively of zero. The half-width of the range was equal to the measurement error. For transducers of displacements, the measurement error was assumed to be equal to 0.1 mm, and for sensors of stresses – 2% of the nominal stress value. We calculated the point estimate of stiffness parameters at every generated combination of the "measured data".

The results of statistic treatment of the obtained samplings are shown in Figure 7.

The obtained empirical distribution coincided with the normal one at the significance level of 5 %. The confidence intervals were defined by using the quantiles of the normal distribution with the confidence probability of 99.7%.

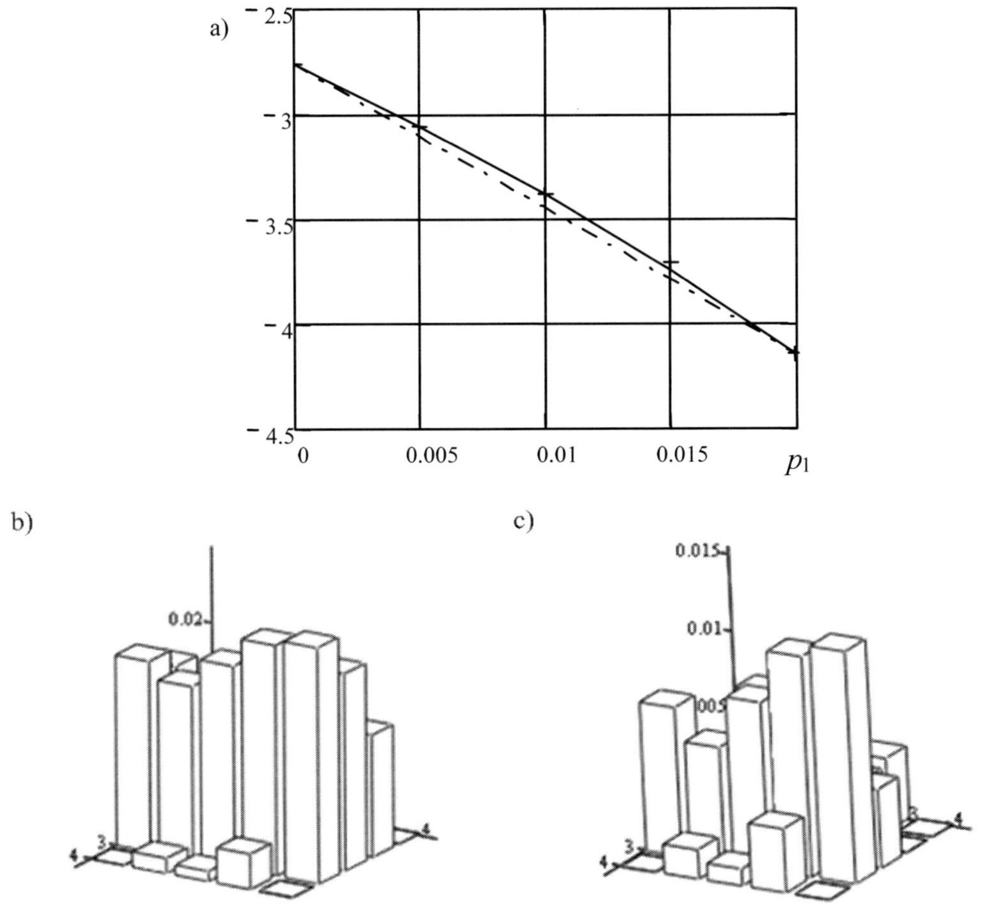

Figure 6. Approximation of response function: (a) plots of polynomial (dotted line) and fractional rational (solid line) functions, (b) error of polynomial approximation (9-point plan), (c) error of fractional rational approximation (4-point plan).

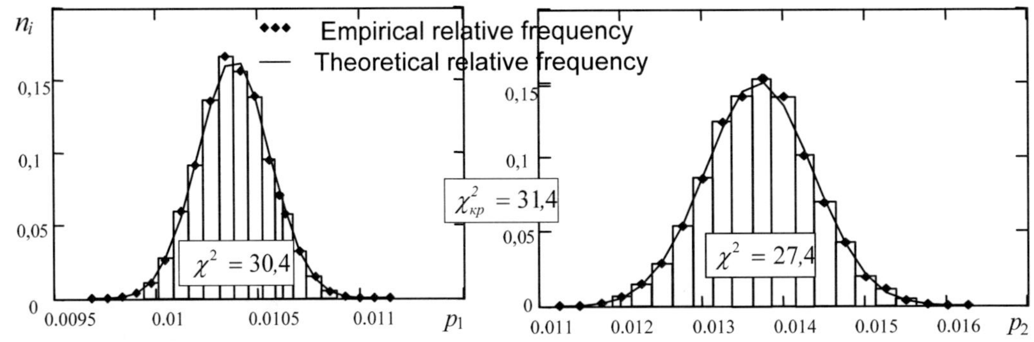

Figure 7. The results of statistic treatment of obtained samplings.

The mathematical expectation of the parameter p_1 was equal to 0.0104, and for p_2 corresponded to 0.0137, but the confidence intervals were defined as $0.0099 < p_1 < 0.011$, and $0.011 < p_2 < 0.016$.

Thus, the proposed identification method allows one to estimate the stiffness parameters of reinforced beam. The values of these parameters are confirmed by the numerical experiments.

References

[1] Davidkin, N. F.; Strahov, V. L.; *Fire Resistance of Underground Resistance*; Moscow, 1998; pp. 1 – 296.

[2] Streng, G.; Fix, G.; *Theory of Finite Elements Method*; Mir, Moscow, 1977; pp. 1 – 350.

[3] Godunov, S. K. *Advances of Mathematical Science*. 1961, *vol. 16(13)*, 171–174.

IV. Applications of New Materials

In: Physics and Mechanics of New Materials … ISBN: 978-1-62618-535-7
Editors: Ivan A. Parinov and Shun-Hsyung Chang

Chapter 21

Modeling and Optimal Design of Power High Stroke Piezoelectric Actuators for Rotorcraft Applications

S. Shevtsov*, M. Flek and I. Zhilyaev
Aircraft Engineering Laboratory at South Center
of Russian Academy of Science, Rostov-on-Don, Russia
Rostvertol Helicopters, Rostov-on-Don, Russia†

Abstract

Because of helicopter main rotor is a source of external noise, cabin noise and intensive vibrations, the problem of active vibration preventing and suppression attracts attention of many engineers and scientific investigators last three decades. The mentioned vibrations are caused by the non-symmetrical and unsteady rotor aerodynamics. Last years to realize the advanced helicopter rotor control the Active Trailing Edge (ATE) or Active Flap (AFC) concepts have been proposedЮ According to AFC concept the small flap which is placed on the distance close to 0.7 of the blade radius, can deflect upwards or downwards, and due to this the blade changes the aerodynamic properties, and vibration level can be decreased.

Very effective and structurally simple solution for driving whole blade or active flap, is the use of power piezoelectric devices. Such piezoelectric actuator drives an active flap hosted on the helicopter rotor blade to change its aerodynamic properties at rotor rotation. As the flap deflection properly synchronized with the azimuthal angle of rotor blade, such active flap control devices allow us to significantly reduce the noise and vibration level of helicopter.

Despite the inherent ability to create of very high forces, all piezoelectric actuators have some drawbacks, among which an incapability to produce enough large displacement, and bad withstanding an external tensile stress. These limitation is forced to develop the different structures for the stroke amplification. This article concern a structural optimization problem for a flextensional piezoelectric actuator which consists

* E-mail: aeroengdstu@list.ru. 41, Tchekhov Str., Rostov-on-Don, 344006, Russia.
† 5, Novatorov Str., Rostov-on-Don, 344038, Russia.

of the high power piezoelectric stack and polymeric composite shell intended for amplification of the stroke. To formulate an optimization problem we parametrize the shape of the amplificator's shell by the rational Bezier curves, which parameters (coordinates and weights of the control points) are changed iteratively by genetic algorithm according to the fitness function value calculated by the finite element model of the transducer with varied geometry. We used the fitness function as the shell's deflection under action of external force. Finally, we demonstrate some optimization results, which confirmed the proposed approach.

1. INTRODUCTION

Reducing high frequency dynamic structural loads that increase the wear of components by minimizing the corresponding disturbances, is an important aim for aerospace designers [1-3]. Besides, the modern aircraft must meet the strong requirements to limits for emitted noise. The helicopter vibrations are mainly excited by periodic forces generated by the main rotor and transferred through the rotor hub and the gearbox into the fuselage. Most intensive vibrations are generated at forward flight and because of the blade-vortex interaction (BVI). At the forward flight with the airspeed V the chordwise velocity U of a blade element is given by

$$U = \Omega r + V \sin\psi , \tag{1}$$

where ψ is the blade azimuthal position, Ω is the rotor angular velocity, and r is the blade element distance from the rotor axis. On the retreating side the component of rotational velocity and horizontal velocity are subtractive and a region of reversed flow will exist. Since the flow in the reversed flow region is from trailing edge to leading edge, it has the effect of pushing the rotor around the mast thereby extracting power from the airflow. BVI is an example of an intense vertical velocity field with high velocity gradients. While passing an airfoil at a predetermined distance, a convecting vortex of positive circulation produces a downwash velocity while upstream of the blade (airfoil), and this changes to an upwash as it moves downstream. This situation leads to a rapidly and continuously changing angle of attack, resulting in highly unsteady aerodynamic loads [4].

Different approaches have been explored to achieve a reduction of the helicopter vibration. Passive means, either located in the rotating or fixed system, aim at exclude the transfer of the exciting forces into the fuselage by especially tuned absorbers. Research activities of last 40 years showed the high potential of active rotor control with a higher harmonic blade pitch variation for vibration reduction [4]. In this case a blade pitch oscillation with harmonics of the rotor rotational frequency is superimposed to the collective and cyclic blade pitch (see Figure 1, a). The Higher Harmonics Control (HHC) provides that the control signal was fed into actuators located beneath the swashplate. The promising theoretical and experimental results led to the development of the Individual Blade Control (IBC) system (see Figure 1, b), where the original pitch links located above the swashplate are replaced by power actuators which change the incidence (or pitch) angle of the blades. Both technologies work with a higher harmonic excitation of the blade pitch at the blade root, which is a very energy-consuming, and also can lead to the blades excitation on first twisting vibration mode [3]. With an active trailing edge (ATE) flap the excitation is not introduced at the blade root, but localized at a distance 75 – 90% of the blade radius (see Figure 2). ATE concept can be implemented in the form of flexible or turned trailing edge flap, and as the locally morphing airfoil [8].

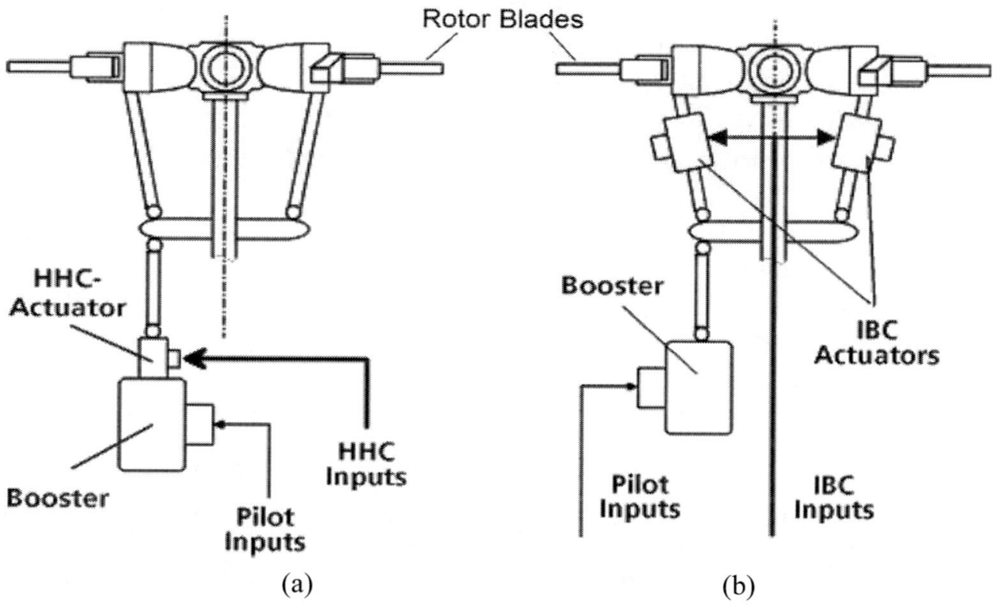

Figure 1. HHC *vs* IBC concepts [3].

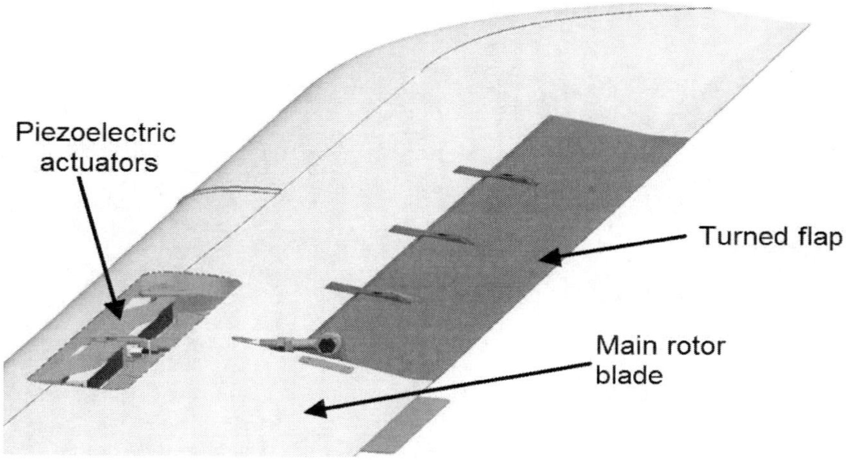

Figure 2. SMART rotor blade with piezoelectric actuator driving the trailing edge flap [6].

The Actively Controlled Flaps are generally 15% of a chord length and most known structures are driven by the power piezoelectric transducers. Due to the high operating forces and frequency, smart materials such as piezoelectric actuators are well adapted for this task, but relatively large displacements require some sort of mechanical amplification of these devices movement. Because of the very small displacements created by the piezoelectric devices some design solutions to amplify the stroke of PZT actuators are proposed [7, 9].

The most important requirements to the piezoelectric actuator for the active flap design are the following [2]:

i. big force and large displacement in compact sizes,
ii. high resolution within the micrometer range,
iii. very short response time less than 1 ms,
iv. life time greater than 1010 cycles,
v. low voltage supply less than 150V DC, no backlash and no play,
vi. low power consumption when statics,
vii. severe environment compatibility (broad temperature and moisture ranges).

Among the known structural concepts of actuators most adoptable properties had the flextensional actuator which consist of the high power piezoelectric stack and shell which supply the stroke amplification (see Figure 3).

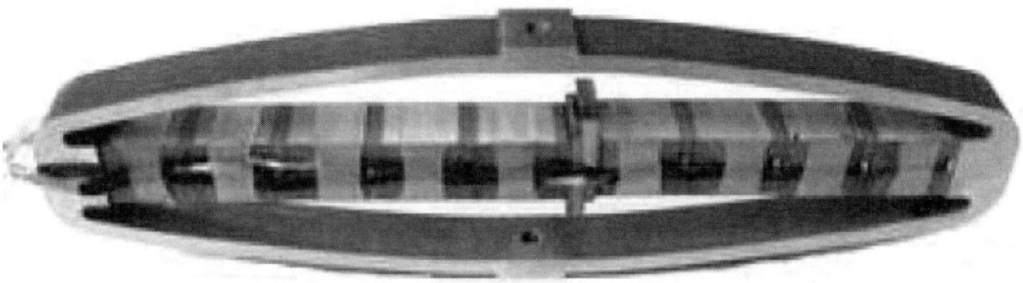

(a)

(b)

Figure 3. Piezoelectric amplified stack with steel (a) and polymeric composite (b) shell [2].

These actuators provide a high stiffness and relatively large displacements. On the other hand, their mass is not optimized and the metallic shell implies a large penalty on the mass. In order to improve the mass of these flextensional actuators a carbon composite is used instead of metal for the frame. Figure 4 demonstrates the mounting of actuators on the rotor blade and shows the transmission of aerodynamic forces to the actuator through the levers. These forces deflect the flap in opposite direction to its active deflection. Because of acting aerodynamic forces a higher flap deflections require larger and therefore heavier actuators, but additional mass is needed to equilibrate the center of mass of the blade close to active flap mounting.

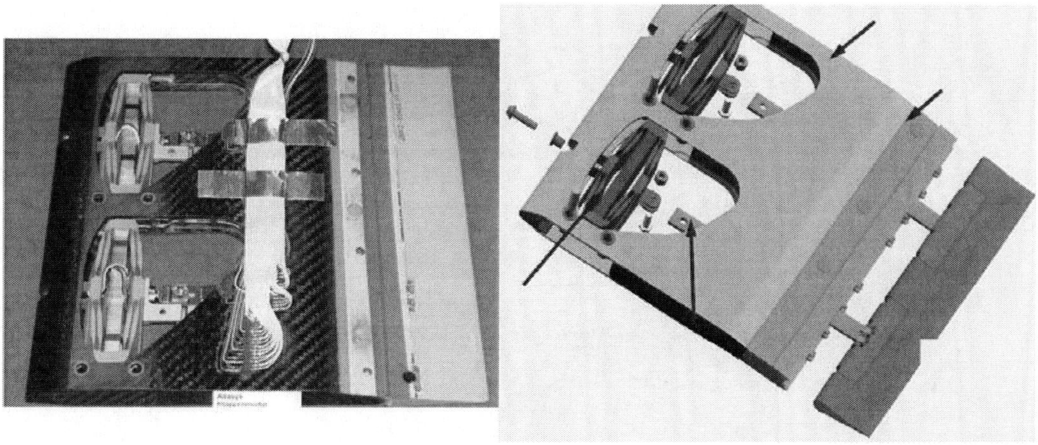

Figure 4. Mounting of active flap and driving piezoelectric actuators on the rotor blade [6].

In order to realize the active flap on the real-world five-bladed helicopter we need to optimize the mechanical properties of the considered flextensional actuator. Primarily we investigate by the finite element method (FEM) the properties of actuator with most widely used elliptic composite shell. Next we describe the shape of the shell by the rational Bezier curves, which allow the good flexibility and to simultaneously optimize the stiffness and operating stroke. An optimization problem will be stated with the given performance of piezoelectric material and the constraints will be imposed on the acting external force as well as the coordinates of the points controlling the Bezier curve behavior. As the fitness function we accept the actuator's stroke at permissible operating voltage on the PZT stack and given external force. Because of total number of degrees of freedom for this problem is very big, we use genetic algorithm implemented in the soft package MATLAB (Genetic Algorithm Toolbox) and simplified one-quarter 2D FEM model of actuator worked in Comsol Multiphysics. Finally we demonstrate the effectiveness of the developed optimization method and significantly improved properties of the optimized actuator.

2. Finite Element Modeling of Flextensional Actuator with Elliptic Shell

We have investigated only actuators with shell made of glass fiber epoxy polymeric composite which had the longitudinal Young modulus $3 \cdot 10^{10}$ Pa and density 1850 kg/m^3. Technology of the shell manufacturing includes the winding of unidirectional glass-fiber tape on a mandrel with appropriate shape, then curing in autoclave until the complete solidification. The size of the finished shell should provide a prestress for the assembled piezoelectric stack for a double-side acting actuator. Stack was constructed of a multilayer piezoelectric ceramic PZT-5H, the thickness of each layer, polarized in thickness, was 0.5 mm. Driving electric potential was taken as 400 V. Geometry of a typical FEM model is presented in Figure 5.

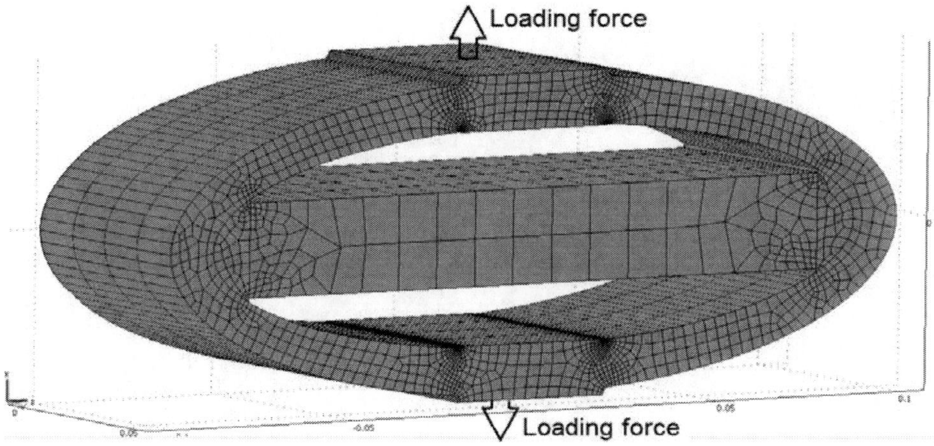

Figure 5. Geometry of actuator's FEM model with elliptic shell and varying ratio of the ellipse semiaxes.

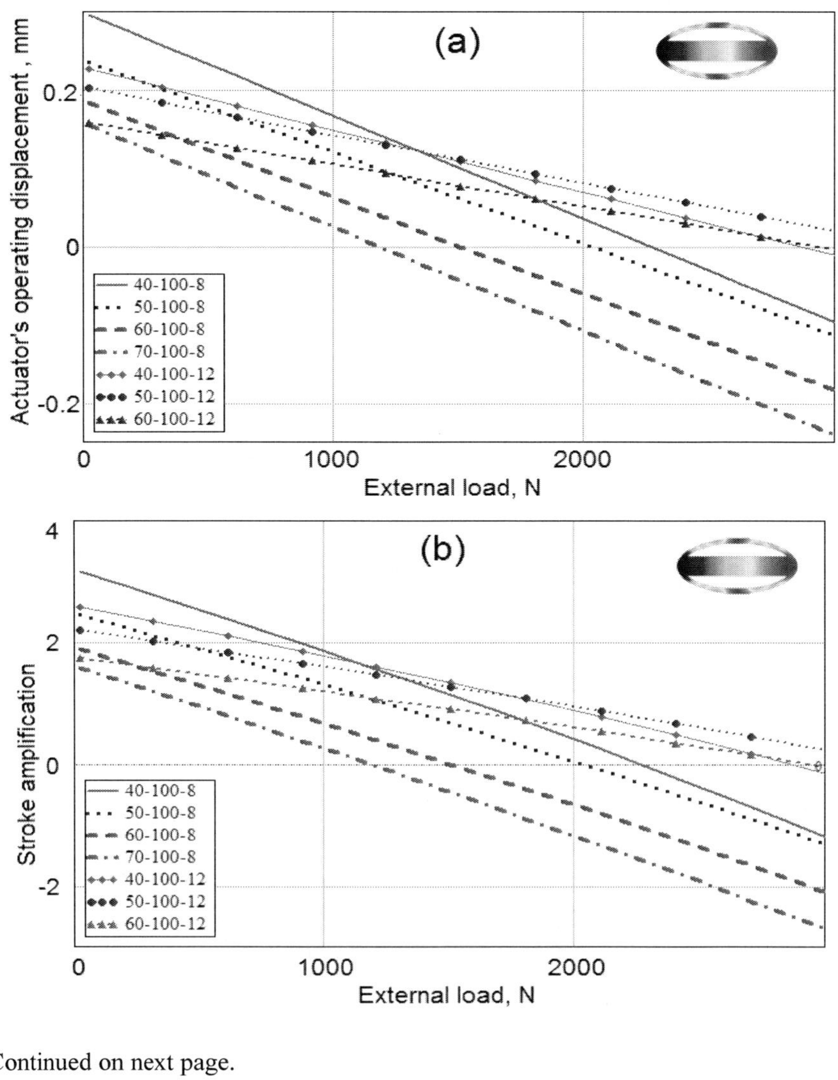

Figure 6. Continued on next page.

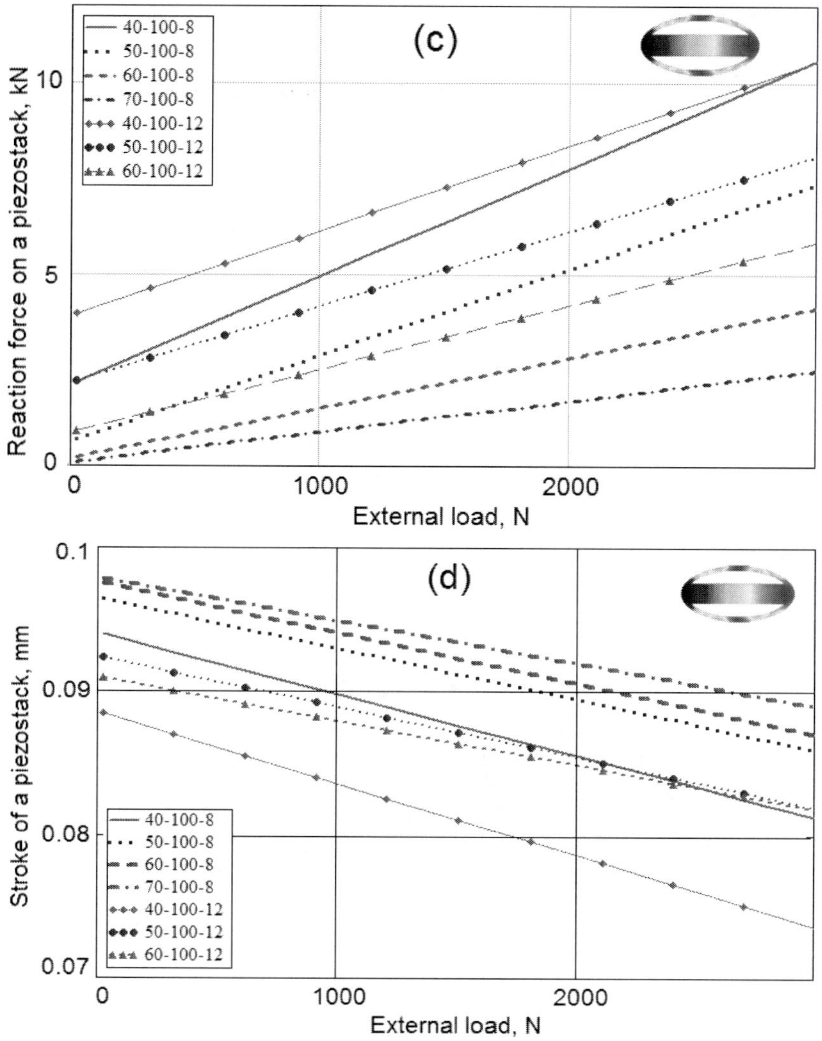

Figure 6. Dependencies of the actuator's stroke (a), amplification ratio (b), and reaction force applied to the PZT stack (c), and compressive deformation of the stack under this reaction force (d) on the external force for the different dimensional parameters of composite shell.

In our numerical experiments we used shell geometry with the thickness 8 mm and 12 mm; major semi-axis was always 100 mm, whereas the small semi-axis took values 40, 50, 60, 70 mm.

The FEM static analysis was performed as follows. After applying of the driving potential the stack lengthened, deforming the shell and causing it to contraction in the vertical direction. Then to the executive surfaces of the shell (small quadrilateral planes) the gradually increasing tensile force was applied. When the structure was loading the values of operating stroke, applied reaction force and deformation of piezoelectric stack were monitored. As soon as the executive surfaces displacement returns to zero, the blocking force is recorded.

Then to the executive surfaces of the shell (small quadrilateral planes) the gradually increasing tensile force was applied. When loading, the values of operating stroke, applied reaction force and deformation of piezoelectric stack were measured. Some simulation results are presented in Figure 6, where, for example, the acronym 40-100-8 means as follows: semi-major axis of 100 mm, small axis of 40 mm, thickness 8 mm.

One can see from the plots (Figure 6, a, d) that the total compliance of actuator is determined by the shell, not PZT stack that has a greater stiffness. Indeed, total displacement of actuator (40-100-8) at the loading force 2 kN consists of 0.23 mm, whereas deformation of PZT stack is equal to 0.01 mm, only. This can be explained by the results of FEM modeling. Figure 7 shows von Mises stress distribution in shell before and after applying the external tensile force. As one can see the stress-strain state along the shell is non-homogeneous, thus we can enhance the actuator performance by varying the shell thickness distribution.

As might be expected, thicker shell has more stiffness, but lesser free stroke. Dependence of the actuator's parameters on the ratio of the elliptic shell axes is also essential. The more shallow shells provide greater amplification, but lesser ability to counteract external loads. This fact is illustrated in Figure 8, where dependencies of the amplification ratio and blocking force for the different shell dimensions are presented.

These preliminary results justify the need and opportunity to optimize the shell's geometry. This optimization can be performed by varying the shape of the generatrix and thickness distribution of the shell.

Figure 7. Deformed shape and von Mises stress distribution in polymeric composite shell before (above) and after (below) applying the tensile force to the upper and lower plate of shell (Results of static FEM simulation. Von Mises stress is illustrated by the colour: most intensive red colour corresponds to the value 35 MPa. All deflections are increased in 50 time).

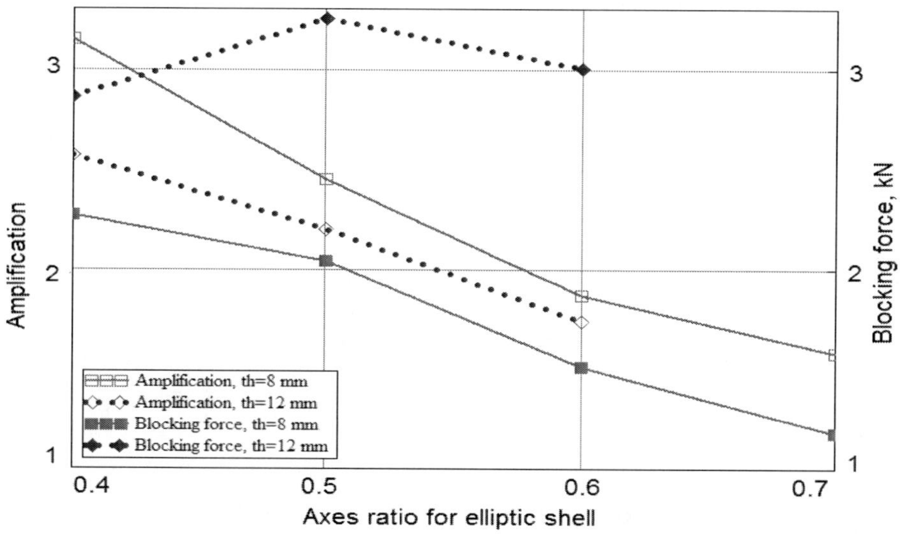

Figure 8. Amplification ratio and blocking force of the flextensional actuator as the functions of the different axes ratio for the elliptic composite shell with thickness 8 mm (solid lines) and 12 mm (dotted lines).

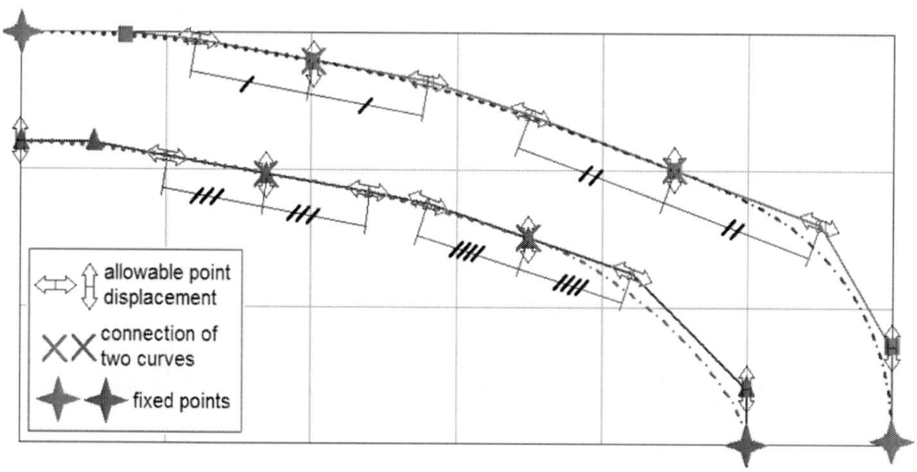

Figure 9. Representation of the shell's profile by the 3rd order rational Bezier curves.

3. GA Based Optimization of the Shell Geometry

As the objective function is advisable to choose the operating stroke h_{op} at the given external load F_{act} (in our case this value is 600 N). To describe the shape of the generatrix with the necessary flexibility we use the rational Bezier curves because of these lines are inherent for the used FEM software Comsol Multiphysics. According to the adopted assumptions the one-quarter of the symmetrical shell can be presented as two branches, each of which consists of three 3rd order Bezier curves (see Figure 9). The rational Bezier curve of n-th order, defined by the $(n+1)$ control points P^j, is described by the equation:

$$R(u)=\sum_{i=0}^{n} w_i B_n^i(u) P^i \Big/ \sum_{i=0}^{n} w_i B_n^i(u), \tag{2}$$

where $B_n^i(u)$ are the Bernstein polynomials, defined as

$$B_n^i(u)=\frac{n!}{i!\cdot(n-i)!}u^i(1-u)^{n-i}, \tag{3}$$

w_i, $i=0,1...n$ are the weights of the control points, and u is the parameter runs through the values from 0 to 1.

At optimization process, external dimensions of the shell remain to be unchanged, so, only shape of the generatrices is optimized. In order to supply C^1 continuity at the connection points (denoted by "×") it is necessary to impose additional constraints on the location of the control points, which are adjacent to the point connection of two curves. If P^c is the vector coordinates of connection point, P^{lc} is the vector coordinates of a point adjacent to the left-to-point connection, then vector coordinates of a point adjacent to the right-to-point connection is

$$P^{cr}=2P^c-P^{lc} \tag{4}$$

Figure 9 shows the equality of the corresponding edges of the control polygon. The placement on the vertical (horizontal) lines of points adjacent to the terminal points (denoted as "+") of the generatrices allows us to automatically satisfy condition of C^1 continuity due to symmetry of the shell (see Figure 9). So, we have 21 degrees of freedom for the points coordinates, and 18 degrees of freedom for the weights, at last, total number of DoFs is 39.

All control points positions are constrained by the system of inequalities:

$$\begin{cases} X_i > X_{(i-1)}, & i\in[1;(n-1)]; \\ Y_i < Y_{(i-1)}, & i\in[2;n], \end{cases} \tag{5}$$

by the equalities for connection points from the relationship (4)

$$\begin{cases} X_0=0; \quad Y_0=b^{in,out}; \\ Y_1=b^{in,out}; \\ X_n=a^{in,out}; \quad Y_0=0; \\ X_{n-1}=b^{in,out}, \end{cases} \tag{6}$$

where the dimensions of the external generatrix a^{out}, b^{out} are fixed, but the dimensions of the internal generatrix are satisfied by the relationships:

$$\begin{cases} a^{out} > a^{in} = \text{fixed} \\ b^{out} > b^{in} = \text{varied} \end{cases} \tag{7}$$

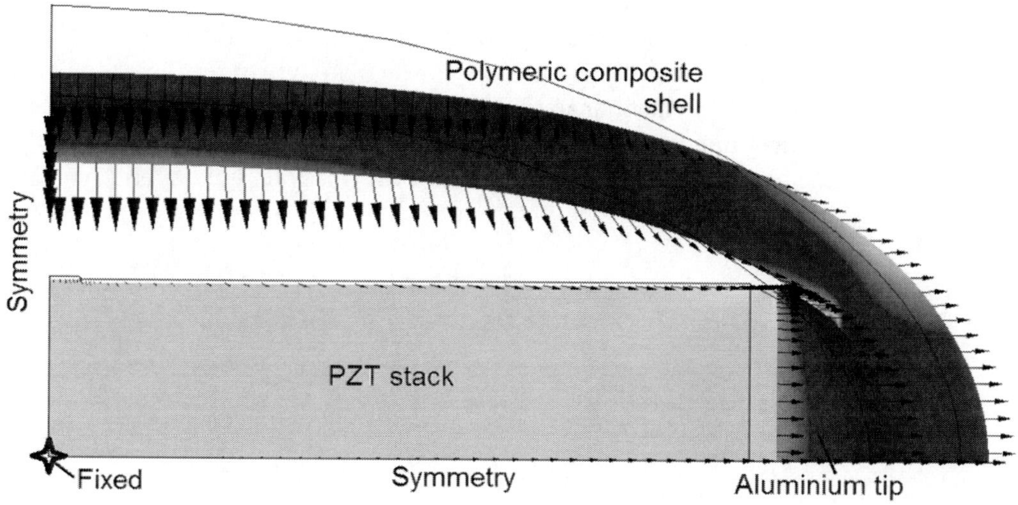

Figure 10. Geometry and displacements field for optimized FEM model of the flextensional actuator.

All weights also are constrained by the system of inequalities. Each of the three Bezier curves, which form the generatrix, is described by its own equation (2). But to ensure the smoothness of generatrix, end points of one curve and the initial point of the connecting curve have the same weight. So, we have the following system of constraints for weights:

$$\begin{cases} w_0 = 1; \\ 0 < w_i < 5, \quad i \in [1; n]. \end{cases} \tag{8}$$

After a number of numerical simulations some additional restrictions are introduced to narrow the range of the coordinates of control points

$$\sum_{k=1}^{i} \xi\left(k, a^{in,out}\right) - 0.25\xi\left(i, a^{in,out}\right) < X_i < \sum_{k=1}^{i} \xi\left(k, a^{in,out}\right) + 0.25\xi\left(i, a^{in,out}\right),$$

$$\text{where} \quad \xi\left(i, a^{in,out}\right) = \frac{a}{372}\left[64 - (i-1)^2\right], \quad i = 1, \ldots n-1. \tag{9}$$

These constraints allow us to agree the density of the control points with the curvature of the generatrix, and thereby significantly reduce the search area. This is very important for a problem with large number of DoFs. Similar optimization problems can be successfully solved only by the genetic algorithm. Therefore we have used the Genetic Algorithm Toolbox

MATLAB which has the advanced optimization means and direct access to the FEM computation.

To reduce the computational cost, our FEM model was very simplified. Due to symmetry of the shell we consider one-quarter of actuator's structure (see Figure 10). All material properties and displacements are considered as linear. Quadrilateral FEM mesh consists of near 500 elements. FEM model works in the Structural Mechanics – Piezoelectric mode. At each iteration step, GA Toolbox rebuilds the shell geometry, remeshes it, and performs the parametric static analysis. The calculated value of the actuator stroke is passed to the GA Toolbox, which changes the value of the design variables according to the constraints. Whole computation flowchart is controlled by MATLAB which in turn refers to modules that perform the parametric finite element analysis. These modules are implemented as standard MATLAB's m-files. Most important GA settings are: population size is 50, elite count is 5, crossover is configured as scattered, mutation – as adaptive feasible, used hybrid function is "fmincon".

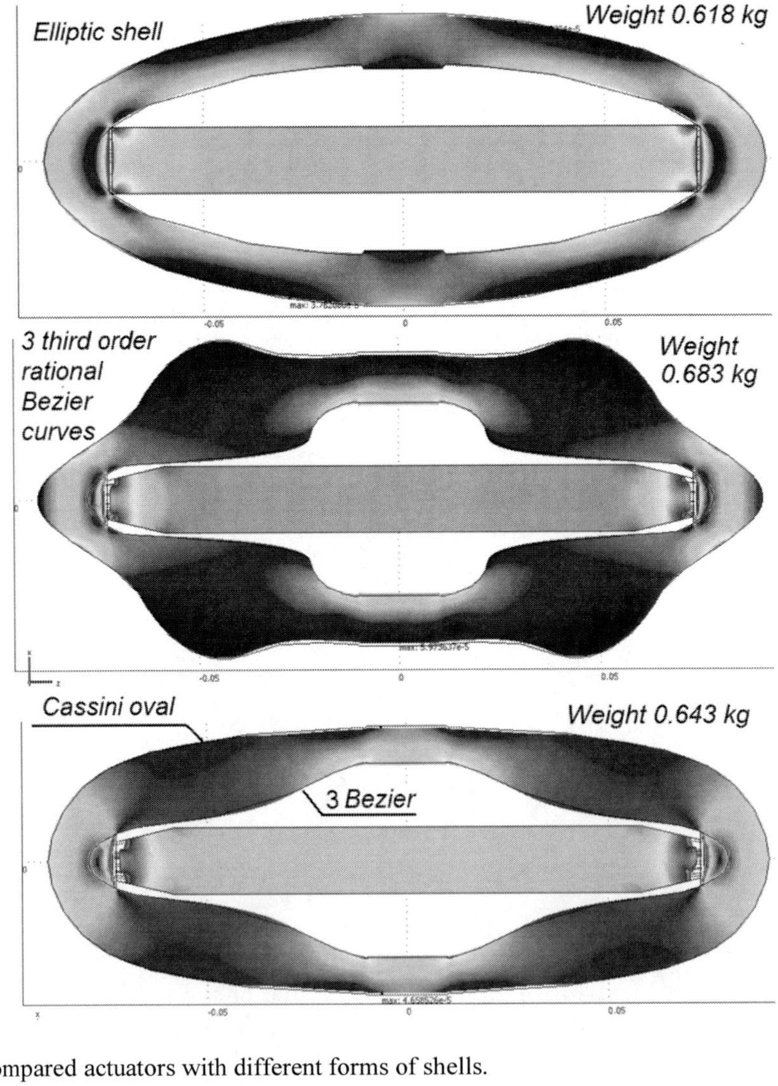

Figure 11. Compared actuators with different forms of shells.

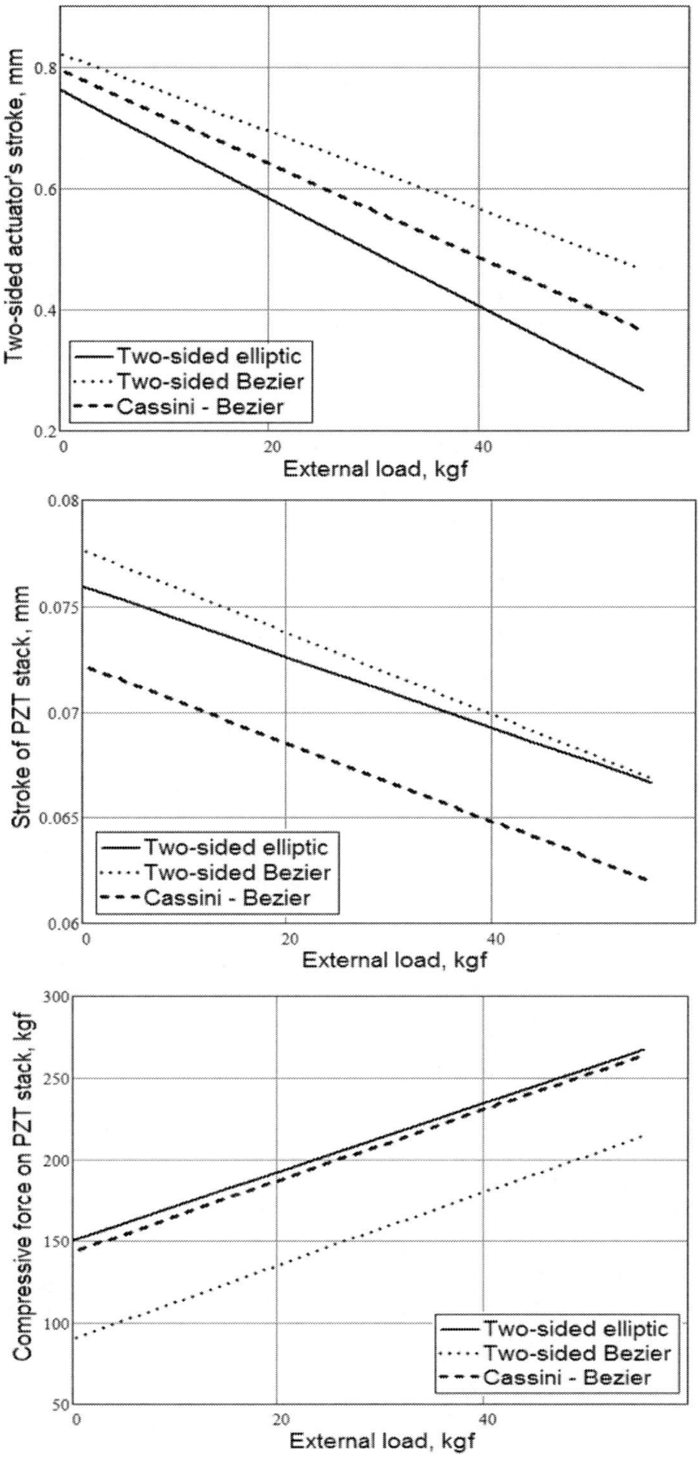

Figure 12. Performance parameters of three compared flextensional actuarors with different shells.

4. Results and Discussion

The developed optimization technique has been successfully applied to the different flextensional actuators with polymeric composite shell of the same outer dimensions $180\times70\times24$ mm^3. Multilayered PZT stacks with dimensions $144\times24\times16$ mm^3 were controlled by the same voltage (electric field 400 V/mm). All optimized actuators with improved parameters have a more complicated shape of the generatrix and a varied thickness along it. For the case when the shape of external generatrix is restricted by the smooth curves with no kinks (structural restrictions due to complexity of mould for the shell cure) all outer control point were fixed in positions appropriate to the Cassini oval [12] which is described by the equation:

$$r^4 - 2a^2r^2\cos 2\theta = b^4 - a^4, \tag{10}$$

where the parameters a, b, which determine the shape of curve, are linked by the relationship $b/a > 1$, and spatial coordinates vary with the polar angle θ as $x = r\cos\theta$; $y = r\sin\theta$.

Three actuators with different shells have been compared: with elliptic, two-sided Bezier, and Cassini – Bezier generatrices (see Figure 11).

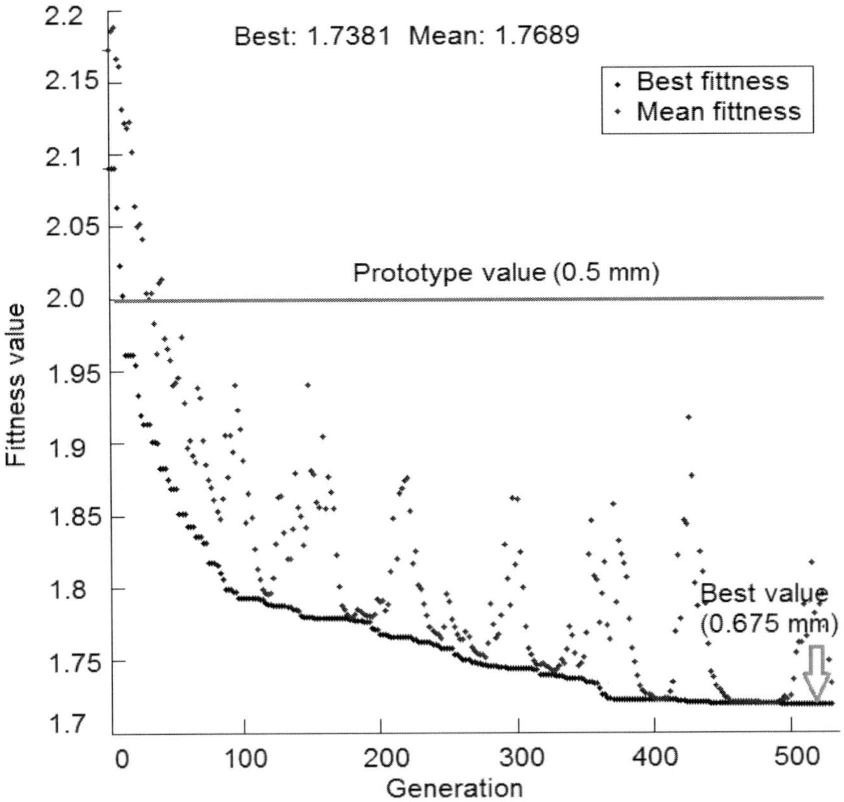

Figure 13. Diagram of the best fitness convergence.

The following plots (see Figure 12) shows comparative data of three investigated similar size actuators: with the elliptic shell and with shells optimized by the proposed method. These data demonstrates several best features of the optimized actuators. Actuator with two generatrices in the form of optimal Bezier curves is the best. At the load 40 kgf it provides a stroke half times greater than the actuator with a simple elliptical shell. In addition, at the same external load, it transmits a much smaller force on the piezoelectric stack that allows us to reduce its cross-section and hence the total mass of actuator. These advantages are crucial, since the power actuators are designed to move against the acting forces and should not change the mass balance of rotor blade.

Actuator with Cassini – Bezier shell has the intermediate parameters, but more preferable compared to simple elliptic shell.

The total computer time engaged by the process of optimization (about 500 epochs) was 5 – 6 hours on i5 computer. To illustrate character of optimization process the typical diagram of the fitness function convergence is presented in Figure 13. Here the solid red line shows the value of stroke for the actuator with elliptic shape. For all investigated structures the optimization using genetic algorithm and chosen fitness function ensures good repeatability and finite element mesh independence.

The developed method has been used to design the actuators for helicopter rotor control.

Conclusion

To provide an actively controlled flap for a helicopter rotor blade an optimization method for flextensional piezoelectric actuator has been developed and tested. This method is based on the representation of the shape of polymeric composite shell by the rational Bezier curves, whose parameters (coordinates and weights of the control points) are the design variables for the optimization algorithm. The optimization process is controlled by the genetic algorithm Toolbox MATLAB, which access a finite element model of device, calculates the objective functional (the stroke of actuator against given external load), and modifies design variables to achieve the optimum. The effectiveness of the proposed method is examined by comparative analysis of the prototype structure with elliptic shape of the composite shell. Our results prove the usefulness of the developed optimization methodology and its attractiveness not only for the aircraft applications but also in many areas of electromechanics, automatics and nanopositioning.

Acknowledgements

The authors wish to acknowledge the support of the presented work by the Russian National Helicopter Center, and Russian Foundation for the Basic Research (Grant 10-08-13300).

REFERENCES

[1] *Actively Controlled Flap*. Available at: http://www.smartrotorsystems. ca.

[2] *Flaps Piezo Mechanisms* - v1-3-2, 2010. Cedrat Technologies. Available at: www.cedrat.com.

[3] Höfinger, M.; Leconte, P. Development of an Active Flap Rotor Model. *Proc. 6th ONERA-DLR Aerospace Symposium.* Berlin, Germany, 2004, 14.

[4] Kessler, C. *CEAS Aeronaut J.* 2011, vol. 1, 3.

[5] Kessler, C. *CEAS Aeronaut J.* 2011, vol. 1, 23.

[6] Kloepel, V. Advanced Reduction of Helicopter Noise & Vibrations (Eurocopter on EADS). *Proc. Int. Conf. Cooperation with the New EU Member States in Aeronautics Research.* Berlin, Germany, 2006, 14.

[7] Mangeot, C.; Andersen, B.; Hilditch, R. *New Actuators for Werospace*. Available at: www.noliac.com.

[8] Maucher, C. K.; et al. Actuator Design for the Active Rrailing Edge of a Helicopter Rotor Blade. *Proc. 33rd ERF.* Kazan, Russia, 2007, 13.

[9] *Noliac Standard Products*. Available at: www.noliac.com.

[10] Viswamurthy, S. R.; Ganguli, R. J. *Vibration and Control*, 2008, vol. 14, 1175.

[11] Singh, K. *The Servo Flap Controlled Rotor*. Available at: www.helis.com.

[12] Lawrence, J. D. *A Catalog of Special Plane Curves*. Dover Publ., 1972, 155.

In: Physics and Mechanics of New Materials …
Editors: Ivan A. Parinov and Shun-Hsyung Chang
ISBN: 978-1-62618-535-7

Chapter 22

DEVELOPMENT OF NEW PIEZOELECTRIC MATERIALS AND TRANSDUCER DESIGNS FOR ENERGY HARVESTING DEVICES

***A. N. Rybyanets**, Y. N. Zakharov, I. P. Raevskii*[1], *V. A. Akopjan, E. V. Rozhkov and I. A. Parinov*[2]**

[1]Institute of Physics, Southern Federal University Rostov-on-Don, Russia
[2]Institute of Mechanics and Applied Mathematics, Southern Federal University Rostov-on-Don, Russia

ABSTRACT

In this paper a comprehensive review and critical comparison of different piezoelectric materials and devices designs for energy harvesting applications were presented. Various types of energy harvesting piezoelement and transducer designs including multilayer stacks, cantilevers, flex-tensional elements, and shear-mode elements and stacks were examined. Empirical criteria of piezoelectric materials efficiency for energy harvesting devices were discussed. Several types of hot-pressed and conventionally sintered PZT piezoceramics, porous PZT piezoceramics, and multilayer PZT/PZT ceramic piezocomposites with diverse ferroelectric "hardness" were fabricated, optimized and tested. Systematical experimental results comprising electric impedance and capacitance measurements, cyclic loading at different load resistances, frequencies, and mechanical stresses for various types of piezoelements were presented. Original design concept for energy harvesting devices with higher overall efficiency, stability, life-time and low self-cost, based on both new materials (piezoelectric, pyroelectric and electret ceramics and composites) and new physical principles (giant field-induced piezoelectric and pyroelectric responses, multimode vibrations and complex deformations) were discussed.

* E-mail: arybyants@gmail.com

1. INTRODUCTION

The history of energy harvesting dates back to the windmill and the waterwheel. People have searched for ways to store the energy from heat and vibrations for many decades. One driving force behind the search for new energy harvesting devices is the desire to power sensor networks and mobile devices without batteries. Energy harvesting is also motivated by a desire to address the issue of climate change and global warming [1].

Energy harvesting is the process by which energy is derived from external sources or human activity (e.g., solar power, thermal energy, wind energy, salinity gradients, and kinetic energy), captured, and stored for small, wireless autonomous devices, like those used in wearable electronics and wireless sensor networks. The energy source for energy harvesters is present as ambient background and is free. Main energy sources for harvesting devices are listed below [1, 2, 5]:

1. Ambient-radiation sources
2. Biomechanical harvesting
3. Photovoltaic harvesting
4. Electrostatic (capacitive) harvesting
5. Blood sugar energy harvesting
6. Tree metabolic energy harvesting
7. Noise harvesting
8. Engine vibrations harvesting
9. Piezoelectric and pyroelectric energy harvesting etc.

In last decade piezoelectric and pyroelectric energy harvesting principles attract growing attention because of high energy density at "mechanical to electric" energy conversion and ease of application [1, 2, 6].

The piezoelectric effect converts mechanical strain into electric current or voltage. This strain can come from many different sources: human and auto motion, low-frequency seismic vibrations, and acoustic noise. Most piezoelectric electricity sources produce power on the order of mill watts, too small for system application, but enough for hand-held devices such as some commercially available self-winding wristwatches.

The pyroelectric effect converts a temperature change into electric current or voltage. Like piezoelectricity, pyroelectricity requires time-varying inputs and suffers from small power outputs in energy harvesting applications.

Piezoelectric and pyroelectric transducers materials can be used as a means of transforming ambient influences into electrical energy that can then be stored and used to power other devices. However, the energy produced by these transducers materials is in many cases far too small to directly power an electrical device. Therefore, much of the research into power harvesting has focused on development of a more effective materials, transducers as well as methods of accumulating the energy. Piezoelectric converters are prominent choice for mechanical to electric energy conversion because in general the energy density is three times higher as compared to electrostatic and electromagnetic. The efficiency of the piezoelectric conversion process in resonance condition is dependent upon the coupling coefficient and mechanical quality factor of the piezoelectric. In the off-resonance condition, the product ($d \cdot g$) is directly related to the energy density. The analysis shows that the piezoceramic and pyroelectric materials properties have been a limiting factor in transducer efficiency. The modifications of these properties along with the development of new transducers designs will yield dramatic improvements in performance and efficiency of energy harvesting devices.

In this paper, a comprehensive review and critical comparison of different piezoelectric materials and devices designs for energy harvesting applications, as well as systematical experimental results for various types of piezoelectric materials and piezoelements were presented.

2. Piezoelectric Energy Harvesting

2.1. Piezoelectric Energy Harvesting Devices

Vibration-based energy harvesting has received growing attention over the last decade [1-7]. The research motivation in this field is due to the reduced power requirement of small electronic components, such as the wireless sensor networks used in passive and active monitoring applications. The ultimate goal in this research field is to power such small electronic devices by using the vibration energy available in their environment. If this can be achieved, the requirement of an external power source as well as the maintenance costs for periodic battery replacement and the chemical waste of conventional batteries can be reduced.

The main advantages of piezoelectric materials in energy harvesting (compared to using the other two basic transduction mechanisms) are their large power densities and ease of application. The power density versus voltage comparison (due to Cook-Chennault et al. [3]) shows that piezoelectric energy harvesting covers the largest area in the graph with power density values comparable to those of thin-film and thick-film lithium-ion batteries and thermoelectric generators. Voltage outputs in electromagnetic energy harvesting are typically very low and often multistage post-processing is required in order to reach a voltage level that can charge a storage component. In piezoelectric energy harvesting, however, usable voltage outputs can be obtained directly from the piezoelectric material itself. When it comes to electrostatic energy harvesting, an input voltage or charge needs to be applied so that the relative vibratory motion of the capacitor elements creates an alternating electrical output [4, 5]. On the other hand, the voltage output in piezoelectric energy harvesting emerges from the constitutive behavior of the material, which eliminates the requirement of an external voltage input. As another advantage, unlike electromagnetic devices, piezoelectric devices can be fabricated both in macro-scale and micro-scale due to the well-established thick-film and thin-film fabrication techniques. Because of the small electrical output, piezoelectric generators have been largely neglected for power generation. But with advances in integrated circuit technology this is no longer an issue. Other problems with piezoelectric generators are application specific so that a common solution does not exist. Hence, power requirements are different for a variety of modules, thus suitable circuitry is required to adapt to various applications.

Most piezoelectric energy harvesters are in the form of cantilevered beams with one or two piezoceramic layers (i.e., a unimorph or a bimorph). In the case of direct force excitation or patches excited by surface strain fluctuations the stacks and the Moonie or Cymbal transducers are used [1].

From an electrical engineering point of view, the alternating voltage output should be converted to a stable rectified voltage through a rectifier bridge and a smoothing capacitor (which constitute an AC–DC converter) for charging a small battery or a capacitor by using the harvested energy. Often a second stage (DC–DC converter) is employed to regulate the voltage output of the rectifier so that the power transfer to the storage device can be maximized.

Two main types of piezoelectric energy harvesting devices are cantilever beam with tip mass, and single layer or multilayer piezoelectric stack subjected to transverse vibration excited at the base. There are two common modes utilized for piezoelectric energy harvesting: 33-mode (stack actuators) and 31-mode (bimorphs). In 33-mode, the direction of applied stress (force) and generated voltage is the same, while in 31-mode the stress is applied in axial direction but the voltage is obtained from perpendicular direction. Conventionally, the 31-mode has been the most commonly used coupling mode, however the 31-mode yielded a lower coupling coefficient (k) than the 33-mode. The power output of piezoelectric system is the highest if the system is operating at natural frequency which dictates the selection of material and dimensions [13].

Piezoelectric transducer selection plays a significant role in the efficient energy harvesting from a given vibration spectrum. The transducer design is tailored for a specific frequency range (ideally the working frequency coincides with the resonance frequency of transducer) and mounting mechanism should be designed such that an effective transfer of stress between the vibration source and transducer structure can be obtained. Transducers such as bimorph, rainbows, unimorph, and disk benders can be strained to maximum at very low magnitudes of stress and thus they are not suitable to harvest energy from high amplitude vibrations. These transducers have lower operating frequency where the resonance can be easily changed by varying the mounting mechanism and adding extra mass to the transducer. On the other hand, transducers such as stack, mechanically amplified composite structures, and raw ceramic require large stresses to provide significant deflection or to generate electric charge and thus they are not suitable to harvest energy from low amplitude vibrations. Cymbal and Moonie structures fall in the middle of the stress – strain curve and as such present an interesting possibility to harvest energy over wide band. Mechanical impedance of the transducer plays an important role in effectively transferring the stress from the source to the generator.

Keawboonchuay [8] studied the relationship between the power output of a bulk piezoelectric material and the thickness to area ratio. The PZT materials used in this investigation were PZT5H and PZT5A. Both theoretical and experimental results have shown that power output linearly increase with the thickness to area ratio of the piezoelectric element. The peak power output of the generator ranges from 7-28 mW or 173 mW/cm^3. In [9] the power generation characteristics of piezoelectric generator under quasi-static (stress duration > 100 ms) and dynamic (stress duration < 10 ms) stress condition were investigated. It was found from simulation that a quasi-static stress produced a bidirectional voltage output, and a dynamic stress produced a unidirectional voltage output. It was also found that the dynamic stress produced a 10 times higher voltage output than the quasi-static stress because the piezoelectric capacitance has enough time to charge in the quasi-static case.

Poulin et al. [10] investigated the power generation of a PZT ceramic bar polarized in the longitude direction embedded at one end and vibrating at 33-mode. The maximum power output is 7.7 μW or 86 W/m^3, and the corresponding conversion coefficient is about 0.52 at resonance. The maximum conversion coefficient is about 0.76, and the corresponding power output is 5.6 μW or 63 W/m^3 at resonance.

To improve the efficiency of the piezoelectric generation using the d_{33}-mode the multilayer ceramic stacks are used [11]. Each stack comprises several piezoceramic discs, each having the same thickness and surface area and poled in alternate direction. Common electrodes are usually used between faces of adjacent elements. These stacks may also be connected in parallel into large arrays [11, 12].

Platt et al. [14] investigated power harvesting using approximately 145 PZT wafers stacked mechanically in series, but electrically in parallel, to form a 1.0 cm square stack with a height of 1.8 cm. A solid monolithic cylinder of PZT with a diameter of 1.0 cm and a height of 2.0 cm was tested for comparison. The monolithic cylinder had a low capacitance of about 47 pF and a very high open circuit voltage of around 10 kV. The PZT stack, however, had an increased capacitance in the range of 1–10 μF and a decreased open circuit voltage of around 30 V. Through experimentation it was found that stacked and monolithic PZT of the same geometry produce the same power if the load resistance is matched to the physical system, but that the matching load is in the kΩ range for stacked configurations and in the GΩ range for monolithic elements. It was concluded that both the voltage output and the matching resistive load are much more manageable in a PZT stack than in a monolithic configuration, thus making the stack a more useful option. Baker et al. [15] have shown that, for three different types of piezoelectric materials, the 31-mode has a lower coupling coefficient (k) than the 33-mode. Upon comparing a piezoelectric stack operating in the 33-mode to a cantilever beam operating in the 31-mode of equal volumes, however, it was observed that, although the stack was more robust and had a higher coupling coefficient, the cantilever produced two orders of magnitude more power when subjected to the same force. This result is due to the high mechanical stiffness in the stack configuration which makes straining of the material difficult. It was concluded that in a small force, low vibration level environment, the 31 configuration cantilever proved most efficient, but in a high force environment, such as a heavy manufacturing facility or in large operating machinery, a stack configuration would be more durable and generate useful energy. Similar conclusions were made by Roundy et al. [16] who showed that the resonant frequency of a system operating in the 31-mode is much lower, making the system more likely to be driven at resonance in a natural environment, thus providing more power.

Cantilevers may be of monomorph and bimorph type. Ng and Liao [17] studied two types of bimorphs along with a unimorph piezoelectric harvester. The unimorph consisted of a single piezoelectric patch mounted to a metallic cantilever beam. The first bimorph, referred to as the series triple layer, consisted of two piezoelectrics with a metallic layer sandwiched between them. The piezoelectric patches were connected electrically in series. The second bimorph, called the parallel triple layer, was the same as the series triple layer except that the piezoelectrics were connected electrically in parallel. Findings showed that under low load resistances and excitation frequencies the unimorph generated the highest power, under medium load resistances and frequencies the parallel triple layer had the highest power output, and under high load resistances and frequencies the series triple layer produced the greatest power. This result is due to the concept that maximum power transfer from the piezoelectric device occurs when the load resistance is matched to the impedance of the piezoelectric device. A series connection increases the device impedance, leading to more efficient operation at higher loads, as was found.

Much work has been done on the optimal structure of cantilever-type transducers e.g. by varying the proof mass, length, and width. Some of these works have been reviewed in [2]. Smoothening of the strain distribution of the beam is a practical method to improve its efficiency. Roundy et al. [18] have shown that, with an increasingly trapezoidal shaped cantilever, the strain can be more evenly distributed throughout the structure as opposed to a rectangular beam that contains a non-uniform strain distribution. Also stated was that, for the same volume of PZT, a trapezoidal cantilever can generate more than double energy than a rectangular beam.

Thus a by using a trapezoidal configuration, a smaller and less expensive harvester could be used to satisfy a given power requirement. Further improvement can be achieved e.g. by using triangle-shaped beams [19]. Such 12 mg PZT bimorph actuator was reported to have energy density greater than 2 $J \cdot kg^{-1}$. This gives a performance increase of an order of magnitude or greater compared to existing commercially available piezoelectric bending actuators. A triangular cantilever with base and height dimensions equal to the base and length dimensions of a rectangular beam have a higher strain and maximum deflection for a given load. Higher strains and deflections in piezoelectric materials translate to higher power outputs; therefore, a triangular cantilever beam will produce more power per unit area than a rectangular one.

In [20] a right-angle cantilever is reported that can dramatically smooth the strain distribution. The device has an auxiliary beam in the orthogonal direction of the main beam. The main beam, which can be regarded as a traditional piezoelectric cantilever, is fixed to a vibration base. One end of the auxiliary beam is fixed to the top of the main beam and the other end is installed with a proof mass. Usually, the auxiliary beam is fabricated with a very high moment of inertia. When the base vibrates in the z-direction, the proof mass will cause a torque to any position of the main beam. Thus a longitudinal uniform strain distribution in the piezoelectric element can be obtained, which means the piezoelectric element should be fully extended. The maximum energy output value for the traditional cantilever obtained at the initial resonance frequency (~ 40 Hz) and at sinusoidal vibration amplitude of 2.5 $m \cdot s^{-2}$ is approximately 8.22 mW while for the right-angle cantilever it is 13.54 mW. It must be noted that, under this condition, the maximum strain in PZT sheet in right-angle cantilever is lower than that in the traditional cantilever. This is why the maximum output power from a right-angle cantilever is not much higher than that from the traditional one.

Very attractive for piezoelectric harvesting applications are so-called low profile transducers characterized by light weight, flexibility, easy mounting, large response, and low frequency operation (the terminology "low-profile" has been used to describe the transducer structure which posses all these qualities [6]). Pre-stressed transducers such as THUNDER (THin layer UNimorph DrivER) consist of a sandwich structure where the bottom layer is stainless steel, the middle layer PZT ceramic, and the top layer is aluminum. LaRC™-SI is used as the adhesive between the layers. During the cooling of the whole structure from high temperature (~300 °C), the mismatch in coefficients of thermal expansion causes the metal and ceramic layers to contract at different rates, imparting a characteristic bend and leaving the ceramic in pre-stress condition. Thunder elements have been used to harvest the energy from the walking motion with a net transfer of 2 mJ per step. Another transducer has been developed on the similar principle by Konkuk University called as LIPCA, where the metallic layers have been substituted with fiberglass and unidirectional carbon. Experimental results have demonstrated that LIPCA provides approximately three times higher free displacement compared to the THUNDER actuator, probably having the highest displacement capabilities among the pre-stressed benders [6, 21, 22]. Further, LIPCA is lighter than THUNDER making it suitable for tapping vibrations.

Kim et al. [23] used for power harvesting a so-called piezoelectric 'cymbal' in which two dome-shaped metal end-caps are bonded on either side of a piezoelectric circular plate. By using cymbal end-caps, the stress applied to the piezoelectric material when compressed is more evenly distributed than in a conventional stack configuration. By distributing the stress throughout the piezoceramic material, the efficiency of the power harvester is increased as a larger amount of the material is actively generating energy. End-caps also allow greater forces and higher frequency loads to be applied to the structure, both of which help increase the power output. Moreover, Cymbal transducer provides higher $d_{eff} \cdot g_{eff}$ values than that of other transducer structures such as multilayer stacks and bimorphs. The actual piezoelectric coefficient of the cymbal is amplified several times due to the presence of cavity which allows the steel end-caps to serve as a mechanical transformer for transforming and amplifying a portion of the incident axial stress in the radial stresses of opposite sign. Thus, the d_{33} and d_{31} contributions of the PZT now combine to provide the effective d_{eff} value of the cymbal as:

$$d_{eff} = d_{33} + A|d_{31}|\,,$$

where A is amplification factor. The amplification factor A can be very large in the range of 10–100 depending on the design of the caps. For a cymbal of 12.7 mm diameter and 1.7 mm total thickness, for example, an effective piezoelectric constant more than 15000 pC/N has been reported [24]. Thus for a given volume of ceramic, a cymbal has a high electromechanical transduction rate and stress in comparison with a bimorph, and produces a displacement much higher than that of a multilayer stack. Under a force of 7.8 N at a frequency of 100 Hz, a 29 mm diameter, 1 mm thick piezoelectric plate with steel end-caps 0.3 mm thick, 17 mm in diameter, and 1 mm tall produced 39 mW of power into a load of 400 kΩ [23].

In a follow-up study, Kim et al. [25] subjected a cymbal transducer of 29 mm diameter, 1.8 mm thickness with 0.4 mm thick steel end-caps to a 70 N force at 100 Hz. The harvester generated 52 mW of power into a load of 400 kΩ. Results show that cymbal transducers are capable of withstanding high force applications while producing useable power. An efficient scheme to increase overall efficiency of an energy harvester is to ensure that it harvests power across varying vibration inputs. The primary problem of the current energy harvester's is its narrow bandwidth.

For example, the conventional piezoelectric harvester comprises a single piezoceramic cantilever beam, where the operating frequency is given by: $\omega = \sqrt{k/m}$ (k is the equivalent beam stiffness and m is the weighted sum of cantilever and tip mass). However, for minute departures from the beams resonance frequency, there is a significant drop in power levels produced. It should be pointed here that several simple schemes can be designed to achieve the wide-bandwidth operation, but most of these have fundamental drawbacks that power density decreases. Recent research is focused on developing transducer structures that exhibit multi-resonances which are closely spaced [1].

The concept of multimode transducers is very much applicable to the design of wide-band energy harvesters. The piezoelectric "Tunnel" transducer consists of two curved piezoelectric plates and a flat piezoelectric plate, all of them bonded together [26]. The asymmetry in the dimensions, shape, and poling of the three plates allows one to overlap the impedance curve closer to each other. This extends the range of the response of the piezoelectric. This concept can be easily applied to design the structures operating in the low frequency ranges to obtain wide-band spectrums.

In order to effectively harvest energy from the available environmental resources, the concept of multi-modal energy harvesting is being pursued. This concept improves upon the existing systems by integrating the following ideas: (i) in a given scenario, two or more energy sources may be available for harvesting energy such as solar, wind, and vibration and (ii) two different energy harvesting schemes can be combined in one system such that one assists the other.

A suitable example of multi-modal energy harvesting device is a system that combines electromagnetic and piezoelectric energy harvesting mechanisms. Such a device consists of piezoelectric transducer and electromagnetic system mutually contributing towards power generation [27]. The design consists of cantilever beam with piezoelectric plates bonded at maximum stress locations. At the tip of cantilever beam, a permanent magnet is attached which oscillates within a stationary coil fixed to the top of package. The permanent magnet serves two functions: acts as a proof mass for the cantilever beam and acts as a core which oscillates between the inductive coils resulting in electric current generation through Faraday's effect. The piezoelectric single crystal plates were bonded on both top and bottom of the beam. The plates were isolated from beam using an insulating epoxy. The output voltages of top and bottom row of piezoelectric were connected in parallel. The beam has tapered geometry with linear variation in dimension along the length such that its moment of inertia along the axis perpendicular to the direction of vibration also varies linearly. This allows the beam to exhibit higher sensitivity. When a cantilevered beam is subjected to external excitation, various modes occur at respective resonance frequencies depending on the geometry, modulus of elasticity, density, and boundary conditions. The first two vibration modes were considered for experimentation as they cover the desired low-operating range of 20–400 Hz for the dimension of cantilever ($25 \times 30 \times 125$ mm^3). The advantage of this design is that at constant acceleration, the output power from electromagnet is much higher at lower frequencies (first transversal resonance mode), whereas the output from piezoelectric is higher at higher frequencies (second transversal resonance mode). Thus, this combination allows obtaining significant magnitude of output power from the same device over a wide operating frequency range. Since the frequency content of a random vibration is wide, such a device can utilize multiple spectrum. The bonded piezoelectric plates were located at maximum stress position for the cantilever beam operating in the second resonance mode. The first resonance frequency of the beam was found to be 20 Hz. The output power of electromagnetic coil at the first resonance frequency across a resistive load of 120 Ω was found to be 8.3 mW. The output power from the top (S2 and S1) and bottom piezoelectric plates (S4 and S5) across a resistive load of 3.3 MΩ was found to be 19 and 15 μW, respectively. Piezoelectric plates attached at the end (S3 and S6) generated the higher power of the order of 120 μW in the first resonance mode. The second resonance frequency was found to be 300 Hz. At this frequency, the output power from electromagnetic coil reduced to 3.1 μW. The output power from top piezoelectric plates (S2 and S1) increased to 60 μW, whereas from the bottom plates (S4 and S5) reduced to 14 μW. It was found that the maximum power was obtained at 35*g* acceleration and 20 Hz frequency with a magnitude of 0.25 mW from all piezoelectric plates.

2.2. Piezoceramic Materials for Energy Harvesting

2.1.1. Piezoelectric Polycrystalline Ceramics

The primary factor for the selection of a piezoelectric material for an energy-harvesting device or a sensor application is the transduction rate. The magnitude of the transduction is governed by the effective piezoelectric strain constant (d), and the effective piezoelectric voltage constant (g). Enhancement in the magnitude of the piezoelectric voltage constant requires an optimization in the magnitude of the piezoelectric stress constant and dielectric constant. In general, the variation in the magnitude of piezoelectric stress constant and dielectric constant is similar and they increase or decrease simultaneously upon modification with dopants or processing.

Using the linear constitutive piezoelectric equations, the condition for obtaining large magnitude of $d \cdot g$ was derived to be as $|d| = \varepsilon^n$, where ε is the permittivity of the material and n is a material constant having a lower limit of 0.5 [28–30]. Table 1 shows the relationship between magnitude of n and g_{33} for various commercial compositions. It can be clearly seen from this data that as the magnitude of n decreases the magnitude of g_{33} increases. Islam and Priya [28, 29] have shown that high-energy density piezoelectric polycrystalline ceramic composition can be realized in the system $Pb(Zr_{1-x}Ti_x)O_3 - Pb[(Zn_{1-y}Ni_y)_{1/3}Nb_{2/3}]O_3$ (PZT – PZNN). The compositions investigated in their study can be represented as: $0.9Pb(Zr_{0.52}Ti_{0.48})O_3 - 0.1Pb(Zn_{1/3}Nb_{2/3})O_3$ [0.9PZT(52:48)–0.1PZN] + y wt%$MnCO_3$, where y varies from 0 wt% to 0.9 wt% and $0.9Pb(Zr_{0.56}Ti_{0.44})O_3 - 0.1Pb[(Zn_{0.8}Ni_{0.2})_{1/3}Nb_{2/3}]O_3$ [0.9PZT(56:44) – 0.1PZNN] + y mol% MnO_2, where y varies from 1 mol% to 3 mol%. The $d_{33} \cdot g_{33}$ values of the samples having composition 0.9PZT(56:44) – 0.1PZNN + 2 mol% MnO_2 (sintered in two steps at 1000–1100 °C) was found to be as 18456.2×10^{-15} m^2/N. This composition was also found to exhibit a high magnitude of g_{33} as 83.1 mV·m/N, corresponding to the magnitude of n as 1.126 [28-30].

The efficiency of ceramic element in energy harvesting applications is dependent on the electromechanical coupling coefficient (k) and mechanical quality factor (Q_m). The efficiency increases with the increase in mechanical quality factor (Q_m), coupling factor (k), and decrease in dielectric loss factor (tan δ). Recently, a quantitative model was proposed relating the efficiency (η) of the mechanical to electric conversion process to k and Q_m [31]:

$$\eta = \left(\frac{1}{2}\right)\frac{k^2}{1-k^2}\left(\frac{1}{Q_m}+\frac{1}{2}\frac{k^2}{1-k^2}\right)^{-1}$$

This relation is often reduced to a more simple form [32]:

$$\eta = k^2 Q_m$$

Table 1. Piezoelectric properties and energy-harvesting parameter of various commercially available piezoelectric ceramic materials [28–30]

Composition	$\varepsilon_{33}^{T}/\varepsilon_0$	d_{33}, pC/N	g_{33}, mV·m/N	$d_{33}\cdot g_{33}$, 10^{-15} m^2/N	n
Morgan Electroceramics					
PZT 701	425	153	41	6273	1.165
PZT 703	1100	340	30	10200	1.181
PZT 502	1950	450	25	11250	1.204
PZT 507	3900	700	20	14000	1.227
APC International					
APC 880	1000	215	25	5375	1.2
APC 840	1250	290	26.5	7685	1.198
APC 841	1350	300	25.5	7650	1.202
APC 850	1750	400	26	10400	1.203
APC 855	3400	620	21	12600	1.224
Ferroperm Piezoceramics					
Pz 24	400	190	54	10260	1.150
Pz 26	1300	300	28	8400	1.199
Pz 39	1780	480	30	14400	1.194
Pz 52	1900	420	25	10500	1.206
Pz 29	2900	575	23	13225	1.217
EDO Corporation					
EC 63	1300	295	24.1	7109.5	1.2
EC 65	1725	380	25	9500	1.205
EC 70	2750	490	20.9	10241	1.222
EC 76	3450	583	19.1	11135.3	1.228

The material figure of merit $\eta = k^2 Q_m$ is of course different from the figure of merit of the electromechanical system, the latter being defined by the product of the overall squared coupling coefficient and the overall quality factor of the device. The maximal stress and strain the materials can accept are also very important parameters to be considered, in order to avoid irreversible alterations.

The piezoelectric constant g, which defines the voltage response in open circuit condition as a function of the mechanical strain, helps determining the inter-electrode distance. A sample of different piezoelectric materials, together with their coupling coefficients and mechanical quality factors are presented in Table 2.

PZT ceramics, which are actually the most common piezoelectric materials, exhibit good properties for energy harvesting. Soft ceramics have high coupling coefficients and moderate mechanical quality factors, whereas hard ceramics have lower coupling coefficients and higher mechanical quality factors. Soft and hard ceramics have roughly the same figure of merit k^2Q_m, but higher stress levels tolerated by hard materials lead to higher power densities. Electromechanical properties of the PZT thin-films developed for MEMS (microelectromechanical systems) applications are very near to bulk ceramics properties (Table 3) [33].

Table 2. Coupling coefficient and mechanical quality factor for different materials

Piezoelectric material	k_{33}	k_{33}	Q_m
PZT soft ceramics	0.75	0.64	100
PZT hard ceramics	0.68	0.57	≥1000
PMN-0.25PT ceramic	0.67	0.31	280
PMN-0.25PT single crystal	0.93 (1)	0.75 (2)	360
$BaTiO_3$	0.7	0.59	400
PVDF	0.2	0.12	10
Quartz	0.15	0.1	10^4–10^6

(1) <001> oriented, (2) <110> oriented and <010> vibration axis.

Table 3. Coupling piezoelectric and dielectric properties of thin films (taken from [6])

Coefficients/ Figure of Merit	ZnO	AlN	PZT
e_{31} (C/m^2)	−1.0	−1.05	−8…−12
d_{31} (pm/V)	5.9	3.9	60…130
ε_{33}	10.9	10.5	300…1300
tan δ (at 1–10 kHz, 10^5 V/m)	0.01…0.1	0.003	0.01…0.03

Table 4. Piezoelectric properties of lead-free piezoelectric ceramic materials ([36-45]

System	d_{33}, pC/N	$\varepsilon_{33}^T/\varepsilon_0$	tan δ	k_p	T_d/T_c (°C)
NBT-KBT-BT	183	770	0.03	0.37	100/290
NBT-KBT-LBT	216	1550	0.03	0.4	160/350
KNN-$LiNbO_3$	235	500	0.04	0.42	-/460
KNN-$LiTaO_3$	268	570	0.01	0.46	-/430
KNN-$LiSbO_3$	283	1288	0.02	0.05	-/392
KNN-Li(NbO_3	308	1009	0.02	0.51	-/339
KNN-$LiNbO_3$	225	1058	0.03	0.36	-/304
KNN-$LiNbO_3$	220	1447	0.02	0.4	-
KNN-$LiNbO_3$	241	1316	0.09	0.4	-/306

T_d is the depolarization temperature; T_c is Curie temperature.

Outstanding coupling coefficients of PMN-PT single crystals and their high quality factors enable to get very high power densities [34]. However these materials are significantly more expensive than ceramics.

PVDF polymers have relatively low figure of merit, but their mechanical flexibility enables particular applications such as submarine "energy harvesting cells" [35]. Similar mechanical flexibilities and better electromechanical couplings can be reached with PZT fiber composites.

The selection of piezoelectric ceramic composition for a particular application depends on parameters such as operating temperature range ($-20 \leq T \leq 80$ °C), operating frequency range, external force amplitude, and lifetime (>10^6 cycles).

The operating temperature range is determined by the Curie temperature of material which for most of the Pb(Zr, Ti)O_3 ceramics is greater than 200°C.

Recently, there has been emphasis on utilizing lead-free materials in domestic and medical applications. Out of all the possible choices for lead-free ceramics, (Na, K)NbO_3 (KNN)-based ceramics such as KNN-$LiNbO_3$, KNN-$LiTaO_3$, KNN-$LiSbO_3$, KNN-Li(Nb, Ta, Sb)O_3, KNN-$BaTiO_3$ (BT), KNN-$SrTiO_3$, and KNN-$CaTiO_3$ have gained prominence mainly for two reasons: (i) piezoelectric properties exist over a wide range of temperature, and (ii) there are several possibilities for substitution and additions. Table 4 lists some of the prominent lead-free piezoelectric ceramic materials. Compositions based on KNN and $(Na_{1/2}Bi_{1/2})TiO_3$ (NBT) – $(K_{1/2}Bi_{1/2})TiO_3$ (KBT). NBT – KBT - based ceramics suffer from drawback that there is anti-ferroelectric phase transition at low temperatures that limits the operating range of transducers.

Nonstoichiometric $(K_{0.5}Na_{0.5})0.97(Nb_{0.96}Sb_{0.04})O_3$ ceramic composition with CeO_2 addition was reported to have very promising parameters and was already tested in a cantilever-structured piezoelectric energy harvesting device [46]. At the sintering temperature of 1100°C, for the 0.2wt% CeO_2 added specimen, the optimum values of density = 4.359 g/cm^3, k_p = 0.443, Q_m = 588, ε_r = 444, d_{33} = 159 pC/N, and g_{33} = 35 × 10^{-3} V·m/N, were obtained. A cantilever-structured piezoelectric energy harvesting device using 0.2wt% CeO_2-added lead-free $(K_{0.5}Na_{0.5})0.97(Nb_{0.96}Sb_{0.04})O_3$ ceramics and a rectifying circuit for energy harvesting were fabricated. Under an external vibration acceleration of 0.7 *g*, when the mass, the frequency of vibration generator, and matching load resistance were 2.4 *g*, 70 Hz, and 721 Ω, respectively, output voltage and power of piezoelectric harvesting device indicated the optimum values of 24.6 V_{rms} and 0.839 μW, respectively, that is suitable for application as the electric power source of a ubiquitous sensor network sensor node.

In the recent years barium titanate-based compositions attract more and more attention due to high piezoelectric performance. Using the novel chemical routs enables one to obtain a ceramics with higher parameters than by using the usual solid-state reaction route. $BaTiO_3$ (BT) ceramics obtained by oxalate co-precipitation route the room temperature piezoelectric d_{33} coefficient is 260 pC/N (that higher than the values usually published in the literature) and it shows a good stability under electrical solicitations [47]. This ceramics has room temperature permittivity value 2000 and g_{33} = 14.7 × 10^{-3} V·m/N.

Recently a lead-free piezoelectric ceramic system $Ba(Ti_{0.8}Zr_{0.2})O_3$–$(Ba_{0.7}Ca_{0.3})TiO_3$ was reported showing a surprisingly high piezoelectric coefficient of d_{33} ~ 620 pC/N at optimal composition [48]. Its phase diagram shows a morphortropic phase boundary (MPB) starting from a tricritical triple point of a cubic paraelectric phase (C), ferroelectric rhombohedral (R), and tetragonal (T) phases.

The high piezoelectricity of the MPB compositions is supposed to stem from the composition proximity of the MPB to the tricritical triple point, which leads to a nearly vanishing polarization anisotropy and thus facilitates polarization rotation between <001>T and <111>R states. 50BCT also exhibits the maximum converse piezoelectric coefficient dS/dE with a very high value of dS/dE ~ 1140 pm/V, exceeding that of all PZT ceramics (with dS/dE ~ 360–900 pm/V). The property anomaly at MPB of the BZT-BCT system is analogous to a similar phenomenon found in PZT and PMN-PT systems.

Later high piezoelectric properties were reported also for a similar ceramic system $Ba(Sn_{0.12}Ti_{0.88})O_3$-$x(Ba_{0.7}Ca_{0.3})O_3$ (BTS-xBCT), characterized by a phase boundary starting from a tricritical triple point of a paraelectric cubic phase, ferroelectric rhombohedral, and tetragonal phases [49]. The optimal composition BTS-30BCT exhibits a high piezoelectric coefficient d_{33} = 530 pC/N at room temperature.

In view of the abovementioned high piezoelectricity in another lead-free system BZT-BCT, which possesses a similar tricritical triple point in the phase diagram, it seems that forming a suitable phase boundary starting from a tricritical triple point could be an effective way to develop high-performance lead-free piezoelectrics.

2.1.2. Piezoelectric and Electrostrictive Polymers

Polyvinylidene fluoride (PVDF) is a semi-crystalline high-molecular weight polymer with repeat unit (CH_2-CF_2), whose structure is essentially head-to-tail, i.e., CH_2-CF_2-(CH_2-CF_2)n-CH_2-CF_2. PVDF is approximately half crystalline and half amorphous. In the semi-crystalline polymers such as PVDF, there are regions where the chains exhibit a short- and long-term ordering (crystalline regions).

A net dipole moment (polar phase) is obtained by applying a strong electric field at or above T_g and then is frozen in by cooling the material resulting in a piezoelectric-like effect. PVDF sensors have been successfully used as NDE transducers in pulse-echo, through-transmission, and acousto-ultrasonic techniques to monitor curing, and to detect porosity and crack propagation in different model joint geometries [30].

The Curie temperature of PVDF is nearly 1100°C, which also makes it useful for some elevated temperature applications. Some of the other piezoelectric polymers that are known today include: polyparaxylene, poly-bischloromethyloxetane (Penton), aromatic polyamides, polysulfone, synthetic polypeptide, and cyanoethyl cellulose.

PVDF has the advantage that it is mechanically strong, resistant to a wide variety of chemicals including acids and can be manufactured on a continuous reel basis. Another important property of PVDF is that it shows a strong piezoelectric response even at microwave frequencies. Furthermore, the very high constant g of PVDF at the level of -339×10^{-3} V·m/N, when compared with the other commonly used piezoelectrics, makes this polymer an ideal transduction material. Table 5 compares the piezoelectric properties of the PVDF with other piezoelectric materials.

Recently, research has focused on electrostrictive polymers where it is possible to induce large piezoelectric effect by applying high-DC bias field due to Maxwell stress.

Electrostrictive polymers have been discovered that generate large strain (above 5%) under moderate electric field intensity (400–800V on a 20 μm film). Poly(vinylidene fluoride-trifluoroethylene) or P[VDF:TrFe] copolymers have been shown to exhibit d_{31} of 8.0×10^{-11} pC/N, and polyurethane has d_{31} of 1.7×10^{-10} pC/N. A theoretical investigation conducted by Liu et al. [50] has shown that energy densities of the order of 0.221 J/cm^3 are possible in the constant field condition using polyurethane material.

Table 5. Comparison between commonly used piezoelectric materials and PVDF (taken from [3])

Material	Relative dielectric constant, $\varepsilon/\varepsilon_0$	Piezoelectric constant, d_{33}, pC/N	Piezoelectric voltage constant, g_{33} (10^{-3} V·m/N)
$BaTiO_3$	1700	191	12.6
Quartz	4.5	2.3 (d_{11})	50.0 (g_{31})
PVDF	13	−33	−339.0
PZT-4	1300	289	25.1

2.1.3. Composites

Ceramic-polymer composites due to their mechanical flexibility and better electromechanical couplings are widely studied as promising materials for energy harvesting devices.

The recent progress in the field of piezoelectric fiber composites and thin profile flexible transducers allows the implementation of energy harvesters in variety of scenarios including complex surfaces and noisy environments [51].

The transducers shown in this figure are attractive for energy harvesting applications for several reasons including light weight, flexibility, easy mounting, large response, and low frequency operation. Pre-stressed transducers, such as Thunder, consist of a sandwich structure where the bottom layer is stainless steel, the middle layer is PZT ceramic, and the top layer is aluminum. The pre-stress in the structure results in higher charge generation as compared to the ceramics, however, they are equally brittle. Fiber composites [Active Fiber Composite (AFC) and Macro Fiber Composite (MFC)] overcome the drawback of rigidity and brittleness by using high aspect ratio structure and interdigital electrode pattern. The fibers in MFCs are machined from low-cost piezoelectric ceramic wafers using a computer-controlled dicing saw.

MFCs can be operated in either d_{33}-mode or d_{31}-mode by designing the two different electrode patterns. It has been shown that MFC operating in d_{33}-mode has higher energy conversion rate but lower electrical current as compared to that d_{31}-mode [52]. Further, MFC transducers have been reported to have the reliability of over 10^9 cycles operating at maximum strain.

NASA LaRC's Radial Field Diaphragm (RFD) comprises of a thin circular piezoelectric ceramic disk sandwiched between two polyimide dielectric films with copper-etched spiral electrodes [53]. This novel electrode pattern induces an electric field into the piezoelectric ceramic that extends out radially from the center of the wafer. At low frequencies of DC ~ 10 Hz, RFDs have been shown to provide 2–3 times higher displacement as compared to conventional unimorphs [53].

Experimental demonstration of the RFDs in energy harvesting application has not yet been reported.

Table 6. Piezoelectric and physical properties of polycrystalline composite materials (taken from [29])

Parameter/ Material	$PbNb_2O_6$ (porosity 15%)	PZT/α-Al_2O_3	(Na, Li)NbO_3/ $LiNbO_3$	PZT porous piezoceramics
k_p	0	0	0.2	0 – 0.2
k_t	0.4	0.54	0.58	0.55 – 65
d_{33} (10^{-12}), C/N	65	240	100	100 – 700
$\varepsilon_{33}{}^T/\varepsilon_0$	160	400	120	100 – 6000
tan δ, %	0.6	2	2	0.2 – 3
$Q_M{}^t$	< 10	20	30	5 – 20
ρ (10^3), g/cm^3	5.7	6	4.5	3 – 7
T_C, °C	560	340	350	150 – 350
N_t, kHz·mm	1300	1500	2500	500 – 1600

QuickPack (rigid) and PowerAct (flexible) transducer developed by Mide Technology Corporation are packaged bimorphs in a protective skin with preattached electrical leads, producing a highly reliable component with no soldered wires (Website: http://www.mide.com). QuickPack and PowerAct transducer has been widely utilized in the energy harvesting applications in both on-resonance and off-resonance vibration conditions.

Recently 1-3 composites using as a filler PMN-PT single crystals [54] and lead-free $BaTiO_3$ [55] have been studied and exhibited high energy-harvesting characteristics. Other types of composites e.g. porous ceramics and ceramic-ceramic composites are also extensively studied (Table 6) [56].

3. Experimental Scenario

3.1. Object Selection and Search Strategy

Taking into account growing demands and specific technical requirements for new piezoceramic materials for energy harvesting devices the following search strategy was chosen:

- selection of prospective chemical compositions and piezomaterial types;
- parameter adjustment using composite technology;
- parameters adjustment using technological regimes;
- parameters adjustment using corrections of components content.

Several types of hot-pressed and conventionally sintered PZT piezoceramics based on multicomponent solid solutions, porous PZT piezoceramics, and multilayer PZT/PZT ceramic piezocomposites with diverse ferroelectric "hardness" were fabricated, optimized and tested.

Using of composite technology has permit to change the main parameters of composites in a wide range. Low porosity piezoceramics demonstrate high piezoelectric properties along with high mechanical stability and durability. Lower elastic stiffness of porous ceramics compared to dense analogs provides increased deformations and, as following, higher power efficiency at low stress conditions [57]. Multimaterial ceramic composites can be realized in different forms (multilayer structures, 3-3 or 3-0 connectivity types) and also provides a possibility to change or adjust piezo- and dielectric properties of resulting piezoelements [58, 59].

3.2. Methods of Measurements

Preliminary testing of prospective piezoelectric materials and elements comprising impedance spectra measurements and analysis, static and low frequency capacitance measurements, and quasi-static piezomoduli measurements at different frequencies was performed. All measurements were done at room temperature. Dielectric, piezoelectric and elastic parameters were measured using the Agilent 4294A Impedance Analyzer according IEEE Standard [60] and PRAP software [61, 62] methods. Measurements of piezoelectric moduli were made using APC d_{33}-tester and PiezoMeter System PIEZOTEST. Low frequency capacitance measurements were done on HIOKI 3522 hi-Tester.

Testing of the piezoelectric elements under static and dynamic mechanical stresses, as well as cyclic loading tests at different load resistances, frequencies, and mechanical stresses for various types of materials were performed using testing machine KU40kN. Loading regimes comprise step or sinus loadings at loading speed from 0 to 5 Hz (single or multicycles). Maximal mechanical load was set as 40000 N.

Electrical signals registration was performed on variable load resistance connected in parallel to the piezoelectric element (0.1 – 80 MOhm) using LeCroy digital Oscilloscope and PowerGraf software. The output power was estimated using the standard relation:

$$P_{out} = U^2_{rms}/R_{load}.$$

3.2. Ceramic Piezocomposites

3.2.1. Multilayer Structures

Different types of multilayer structures consisting of 2 and 3 layers of co-sintered piezoceramics with different dielectric and piezoelectric coefficients were fabricated using either conventional sintering or hot-pressing method.

Ceramic composite consisting of 2 layers of piezoceramics PZT1 and PZT2 can be represented by the equivalent circuit shown on Figure 1.

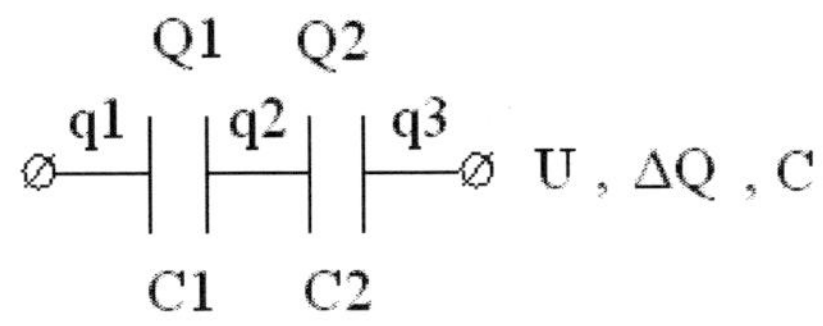

Figure 1. Equivalent circuit of 2-layers ceramic composite.

The electric charges generated by the first and second layers are:

$$Q_1 = d_{33}(1)\cdot F,\ Q_2 = d_{33}(2)\cdot F,$$

where F is the mechanical force (N).

The electric charges on electrodes are:

$$q_1 = Q_1;\ q_2 = Q_2 - Q_1;\ q_3 = Q_2.$$

The total capacitance of the composite C is defined as:

$$\frac{1}{C} = \frac{1}{C_1} + \frac{1}{C_2}.$$

The electric potential and charge differences are:

$$U = \frac{q_1 - q_2 - q_3}{2C_1} + \frac{q_1 + q_2 - q_3}{2C_2}$$

and

$$\Delta Q = \frac{UC}{2} = \frac{q_1(C_1 + C_2) + q_2(C_1 - C_2) - q_3(C_1 + C_2)}{2(C_1 + C_2)}.$$

Thereby the effective piezoelectric constant $d_{33}*$ and dielectric constant $\varepsilon^T{}_{33}{}^*$ of the composite can be found as:

$$d_{33}{}^* = \Delta Q/F$$

and

$$\varepsilon_{33}^{T*} = \frac{tC}{\varepsilon_0 S} = \frac{\varepsilon_1 \varepsilon_2 (t_1 + t_2)}{(\varepsilon_1 t_2 + \varepsilon_2 t_1)},$$

where ε_1 and t_1 are $\varepsilon^T{}_{33}$ and thickness for the first PZT1 layer, and ε_2 and t_2 are $\varepsilon^T{}_{33}$ and thickness for the second PZT2 layer.

For the ceramic composite which consist of 3 layers of piezoceramics PZT1, PZT2 and PZT3 the total capacitance C, electric potential U and charge differences ΔQ of the composite are defined as:

$$\frac{1}{C} = \frac{1}{C_1} + \frac{1}{C_2} + \frac{1}{C_3},$$

$$U = \frac{q_1 - q_2 - q_3 - q_4}{C_1} + \frac{q_1 + q_2 - q_3 - q_4}{C_2} + \frac{q_1 + q_2 + q_3 - q_4}{C_3}$$

and

$$\Delta Q = \frac{UC}{2}.$$

Absence of the internal electrodes inside the ceramic composite doesn't change the resulting parameters because of dielectric origin of the PZT composite. So, multilayer composite approach can provide a fine adjustment of piezoelectric and dielectric constants of the resulting composite by choosing appropriate PZT piezoceramic compositions, as well as a number and thickness of PZT layers.

For verification of the developed approach a line of 3-layer composites were fabricated and tested (Figure 2). As one example, center layer of the composite was composed from PZT1 material with d_{33} = 350 and $\varepsilon^T{}_{33}$ = 2000, whereas upper and bottom layers were fabricated from PZT2 material with d_{33} = 250 and $\varepsilon^T{}_{33}$ = 800. Multilayer structure were pressed and sintered in one cycle at 1220 °C during 2 hours. Thickness of the center layer was 2 mm while upper and bottom layers have 1 mm thickness. The diameter and the total thickness of composite samples were 10 and 4 mm correspondingly.

Figure 2. Multilayer ceramic composite elements comprising co-sintered layers of soft and hard piezoceramics.

The multilayer composite samples demonstrate the following parameters: d_{33} = 275 and $\varepsilon^T{}_{33}$ = 600 that agree well with the theoretical predictions. To satisfy the specific technical requirements for piezoceramic materials used in energy harvesting devices, technologically and chemically compatible PZT compositions with appropriate parameters can be chosen and sintering or hot pressing regimes providing reproducible composite parameters should be adjusted.

3.2.2. Porous Piezoceramics

Porous ceramic technology permits changing of main parameters of initial piezoceramic composition in a wide range. Low porosity piezoceramics with fine closed pores are characterized by high piezoelectric properties along with high mechanical stability and durability. Besides, porous piezoceramic elements are easily processed, are more stable against thermal and mechanical shocks and provide higher adhesion of electrodes in comparison with standard dense piezoelements.

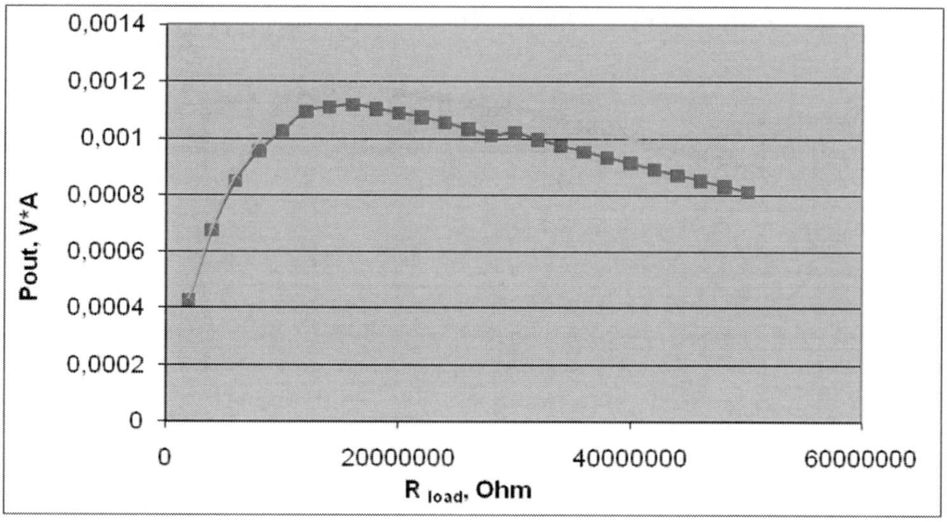

Figure 3. Output power P_{out} *vs* load resistance R_{load} for R-1119P porous ceramics (relative porosity 20%, $d_{33} = 410 \cdot 10^{-12}$ C/N, $\varepsilon_{33}{}^{T} = 1100$) at 2000 and 5000 cycles.

The dielectric constant $\varepsilon_{33}{}^{T}/\varepsilon_{0}{}^{*}$ of porous ceramics decreases practically linearly with increase in relative porosity p, that is caused by essential difference of dielectric constants of piezoceramics and air. The piezoelectric modulus d_{33}* for porous PZT piezoceramics in porosity range 0 – 70% has a minor alteration caused by the continuity of the rigid "quasi-rod" ceramic skeleton in the polarization direction (sample thickness). Reduction of the relative surface area of the piezoceramic phase in this case is compensated by an increase in relative pressure applied to the ceramic skeleton; however it can restrict maximum mechanical stress acceptable for porous ceramic elements. So, porous piezoceramics at appropriate porosity can have high piezoelectric voltage constant (*g*) along with high $(d \cdot g)$ product and mechanical strength.

Different types of proprietary porous piezoelectric ceramics with 3-0 connectivity (closed porosity) based on PZT compositions with intermediate ferroelectric "hardness" were fabricated and tested. Experimental samples of porous piezoceramics in porosity range 0–20% were fabricated using specially developed and approved technology that provides the controllable formation of micro porous structures with reproducible parameters.

The porous PZT ceramics were prepared by burning out of organic powders. Mixtures of organic crystal powders with different particle shapes, decomposition temperatures and grain-size distribution (from 5 to 10 μm) were used for producing the fine-pore volumes. The porous ceramics samples with vacuum deposited Cr/Ni and fired silver electrodes were poled in air at cooling from 380°C to 90°C for 1 min by applying a *dc* electric field of 1 – 1.5 kV/mm.

The following PZT porous ceramics were fabricated and tested at constant and cyclic mechanical loadings:

(i) R-1119P (relative porosity 20%, $d_{33} = 400$ and $\varepsilon^{T}{}_{33} = 1130$);
(ii) R-1101P (relative porosity 20%, $d_{33} = 280$ and $\varepsilon^{T}{}_{33} = 530$);
(iii) R-1102P (relative porosity 8%, $d_{33} = 250$ and $\varepsilon^{T}{}_{33} = 630$).

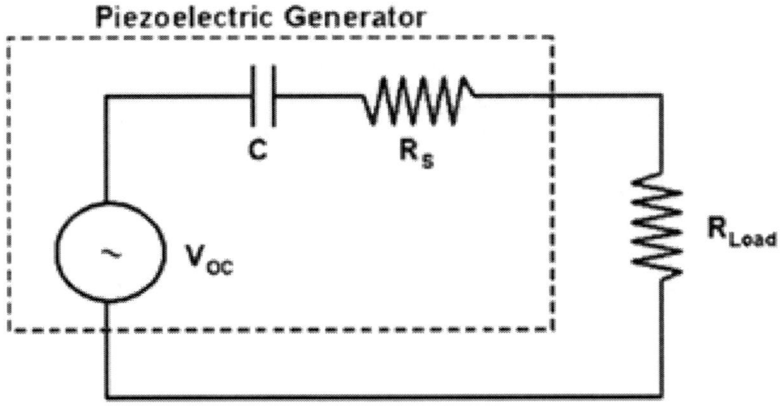

Figure 4. The equivalent circuit of the piezoelectric generator for the frequency range far from the resonance.

The results of cyclic loading of R-1119P piezoelement (disk 10×4 mm^2) by mechanical stress 38.2 MPa at a frequency of 5 Hz at different load resistances are shown in Figure 3.

It is obvious from Figure that in spite of very high power efficiency of porous ceramics R-1119P (g_{33} = 42.14 mV· m/N and $(d_{33}{\cdot}g_{33}) = 17.27 \cdot 10^{-15}$ m^2/N) its P_{out} values at cyclic loading are far from those for dense piezoceramics. Possible reasons for this discrepancy are:

- low internal impedance $|Z_{int}| = (R_s^2+X^2)^{1/2}$ that can lead to saturation of P_{out} *vs* R_{load} dependence;
- piezoceramic depolarization induced by cyclic stresses;
- mechanical destruction of porous piezoceramic skeleton.

The load dependence of the piezoelectric generator in the frequency range far from the resonance can be calculated by using the equivalent circuit representation shown in Figure 4.

In this circuit the voltage source is taken to be the open circuit voltage across the piezoelectric generator ($V_{OC} = Q/C = d{\cdot}F/C$, where F is the force, C is the damped capacity, d is the piezoelectric coefficient).

The voltage across the load can then be expressed as

$$V_{load} = V_{OC}\left|\frac{R_{load}}{R_{load} + \dfrac{1}{j\omega C} + R_s}\right|,$$

where R_s is the series resistance ($R_s = \dfrac{\tan\delta}{2\pi f C}$), $X = 1/jwC$ is the reactance, and C is the damped capacitance of the piezoelectric generator. The average power delivered to the load can then be found using the expression:

$$P = \frac{V_{load}^2}{2R_{load}}.$$

The power P reaches a maximum at an optimum load value (internal impedance should be equal to load resistance):

$$R_{load}^{opt} = \left| \frac{1}{j\omega C} + R_s \right|$$

and can be expressed as:

$$P_{\max} = \frac{V_{OC}^2 R_{load}}{(R_{load} + |Z_{\text{int}}|)^2},$$

where $|Z_{int}| = \sqrt{(R_s^2 + X^2)}$.

The graphs of output power P_{out} and voltage V_{load} as a function of resistive load R_{load} have the typical view shown in Figure 5, and $|Z_{int}|$ of piezoelectric generator can be estimated from the graphs as R_{load} at which P reaches maximum and starts to decay.

Measurements of electric impedance of porous ceramics have shown that at 5 Hz $|Z_{int}|$ for porous ceramics reaches 200 MΩ that exceeds corresponding values for dense ceramics and $|Z_{int}|$ can't be responsible for obtained dependencies.

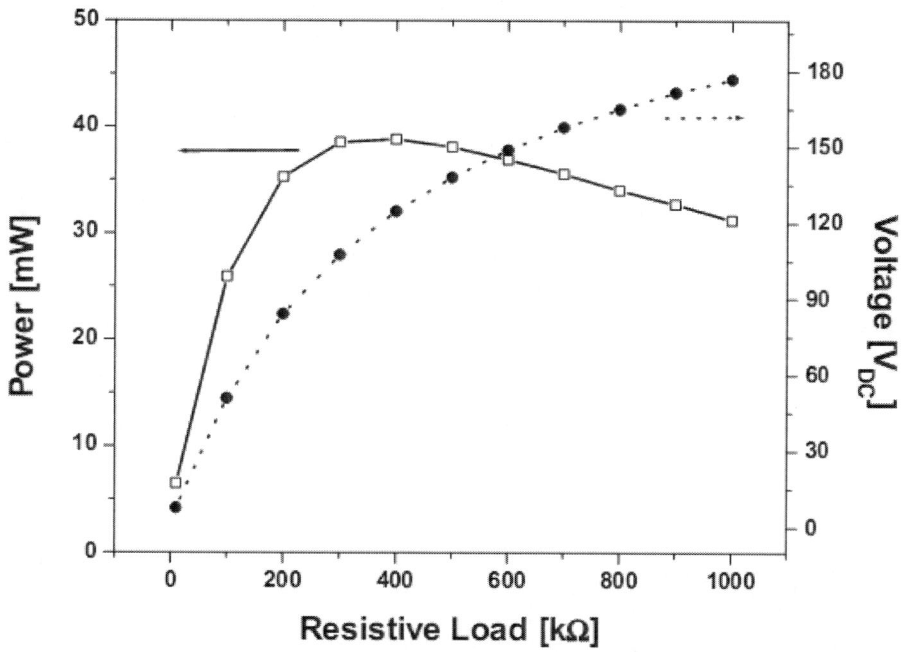

Figure 5. Output power P_{out} and voltage V_{load} as a function of resistive load R_{load} for piezoelectric generator with $|Z_{int}|$ = 300 kΩ.

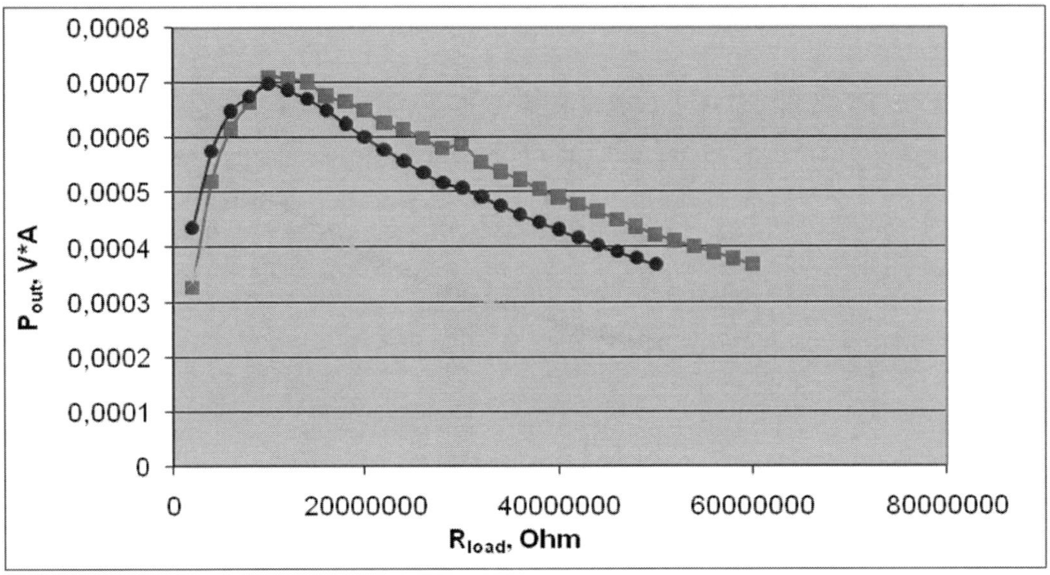

Figure 6. Output power P_{out} *vs* load resistance R_{load} for R-1101P porous ceramics (relative porosity 20%, $d_{33} = 280{\cdot}10^{-12}$ C/N, $\varepsilon_{33}{}^{T} = 530$) at 2000, 5000 and 50000 cycles.

Additional experiments have been carried out for porous piezoceramics R-1101P and R-1102P with low and fine porosity. For comparison dense PZT-based ceramics R-1111 ($d_{33} = 500\times10^{-12}$ C/N, $e_{33}{}^{T} = 2400$) was tested also.

It was found that at the same loading conditions porous ceramics provide 1.5–2 times higher power efficiency then dense hot-pressed ceramics. At any loading forces up to 4 kN deformation of porous ceramics was 1.3 times larger than that for dense ceramics but this pre-eminence decreased as loading force grew.

Cyclic loading tests revealed a notable decrease in output power and piezoelectric modulus d_{33} versus the number of loading cycles for the porous piezoceramics R-1101P, R-1102P and R-1119P. This decrease seems to be caused by high deformations inducing depolarization of piezoceramic. The results of cyclic loading of R-1101P piezoelements (diameter of 10 mm and height of 4 mm) by mechanical stress 38.2 MPa at a frequency of 5 Hz at different load resistances are shown in Figure 6.

The advantages of porous ceramics at low loading stresses are more obvious and can be used for energy harvesting devices with restricted force loadings. It's also clear that porous ceramics approach can be used for fine adjustment of piezoceramic parameters at low and fine porosity keeping mechanical durability of piezoelements for energy harvesting device applications.

3.3. PZT Piezoceramics

3.3.1. Selection of Prospective Chemical Compositions

Hot-pressed ceramics of multicomponent PZT-based solid solutions were chosen as prospective piezomaterials. These materials are characterized by higher diversity of allowable chemical compositions and electromechanical parameters.

Several PZT-based compositions were fabricated and tested using a step compression mechanical loading:

(i) $PbTiO_3 - PbZrO_3 - Pb(W_{1/2}Cd_{1/2})O_3$
(ii) $(Pb, Sr)TiO_3 - (Pb, Sr)ZrO_3 - (Pb, Sr)(Na_{1/2}Bi_{1/2})O_3$
(iii) $(Pb, Sr)TiO_3 - (Pb, Sr)ZrO_3 - (Pb, Sr)(W_{1/2}Mg_{1/2})O_3 - (Pb, Sr)(Sb_{2/3}Li_{1/3})O_3$
(iv) $PbTiO_3 - PbZrO_3 - PbNb_{2/3}Zn_{1/3}O_3 - PbW_{1/2}Mn_{1/2}O_3$
(v) $PbTiO_3 - PbZrO_3 - PbNb_{2/3}Mg_{1/3}O_3 - PbW_{1/2}Zn_{1/2}O_3$

These piezoceramics were prepared by either hot-pressing or conventional sintering methods. Preliminary studies have shown that for all ceramic compositions used, the output voltage decreases with compressive stress. This decrease seems to be caused by depolarization of piezoceramics and depolarization factor was found to depend on ferroelectric hardness of piezoceramics.

The results have shown that the decay of piezoelectric constant d_{33} at compressive stress increase is more pronounced for ferroelectrically "soft" ceramics and caused by ferroelectric domains re-orientations. Depolarization factor calculated as a ratio of d_{33} values before and after loading at compressive stress $\sigma_c = 2.8\cdot10^8$ Pa is equal to 6 – 10 % for "hard" piezoceramics compared to 40 – 50 % for "soft" piezoceramic compositions (compressive strength limit for all types of piezoceramics was $\sim 3\cdot10^8$ Pa).

Comparison of experimental data for piezoceramic samples fabricated by hot-pressing and conventional sintering methods shows that hot-pressed ceramics are much more stable to compressive stress σ_c in the range from $1.25\cdot10^8$ to $2.8\cdot10^8$ Pa (depolarization factors for hot-pressed and conventional sintered "hard" ceramics were 6.6 % и 22.0 % correspondingly).

In order to better demonstrate this fact the changes in the output power and piezoelectric modulus d_{33} from the number of load cycles for the soft PZT piezoelectric material has been measured.

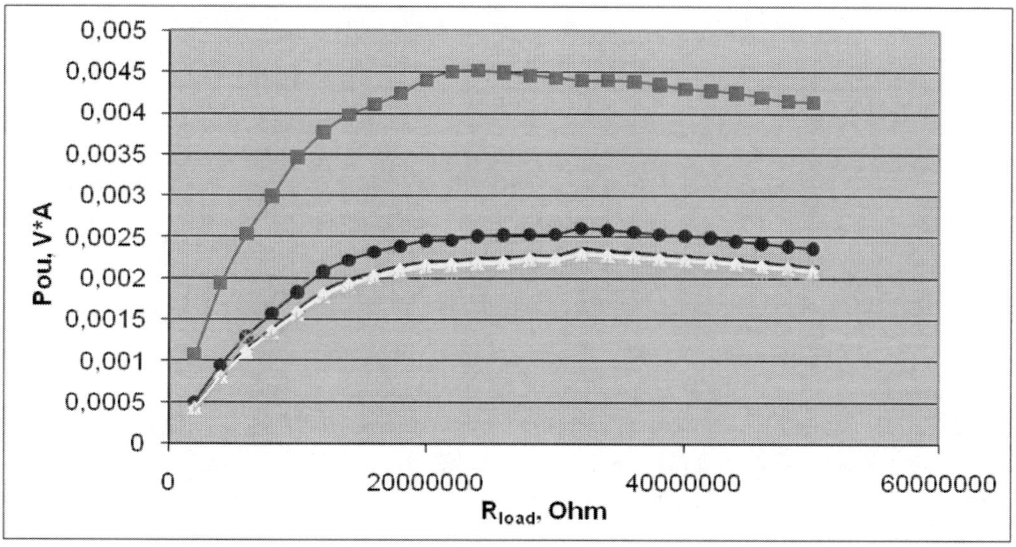

Figure 7. Output power P_{out} *vs* load resistance R_{load} for the "soft" PZT piezoceramics ($d_{33} = 650\cdot10^{-12}$ C/N, $\varepsilon_{33}{}^T$=2930) at 2000 up to 50000 load cycles.

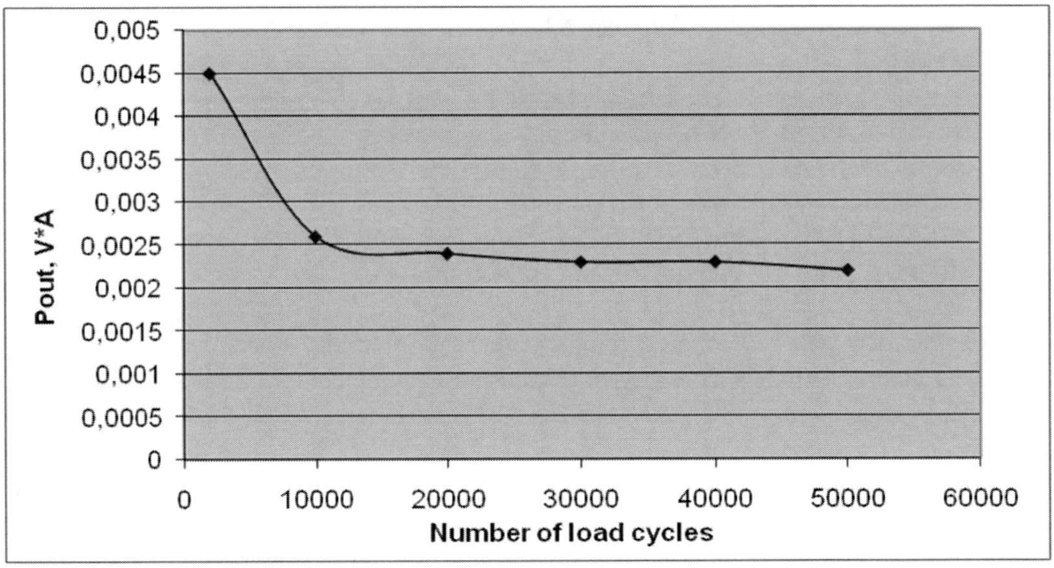

Figure 8. Output power P_{out} *vs* number of load cycles for the "soft" PZT piezoceramics ($d_{33} = 650 \cdot 10^{-12}$ C/N, $\varepsilon_{33}{}^{T} = 2930$).

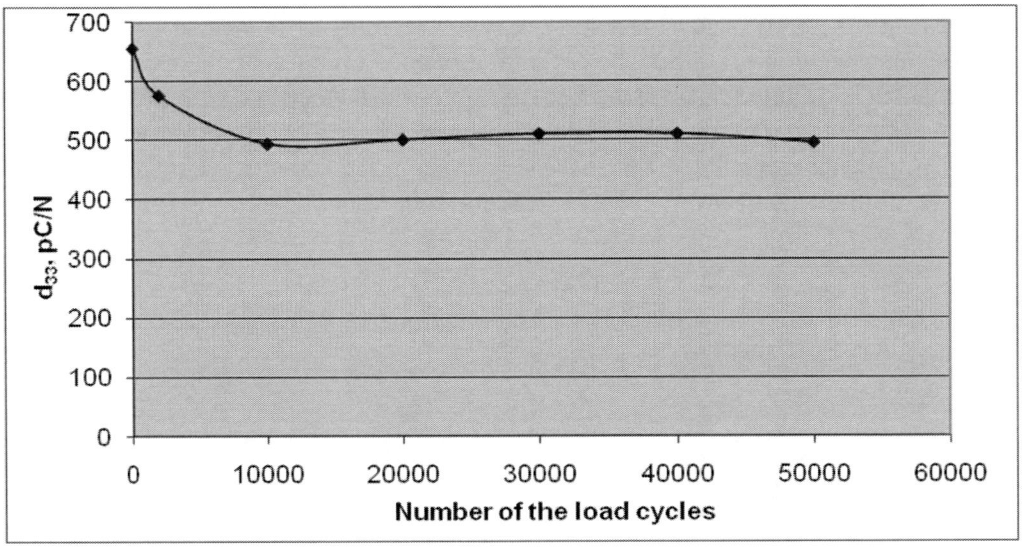

Figure 9. Piezoelectric modulus d_{33} *vs* number of load cycles for the "soft" PZT-based piezoceramics ($d_{33} = 650 \cdot 10^{-12}$ C/N, $\varepsilon_{33}{}^{T} = 2930$).

The results of cyclic loading of "soft" PZT composition belonging to the system 5 (diameter of 10 mm and height of 4 mm) by mechanical stress 38.2 MPa with the frequency 5 Hz at different numbers of load cycles are shown on Figures 7–9.

For comparison, the results of cyclic loading of intermediately "hard" PZT composition belonging to the system 2 (diameter of 10 mm and height of 4 mm) by mechanical stress 38.2 MPa with the frequency 5 Hz at different numbers of load cycles are shown in Figure 10.

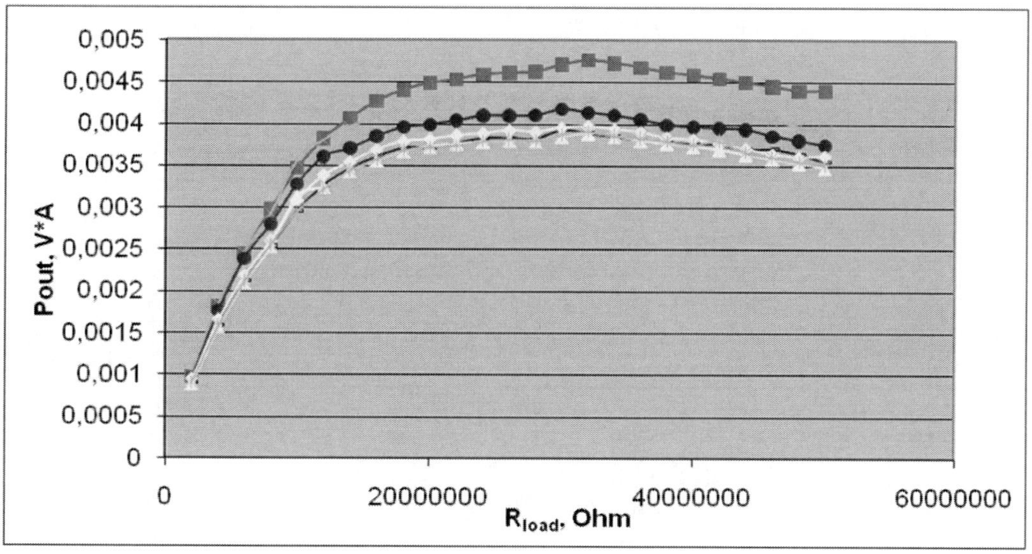

Figure 10. Output power P_{out} *vs* load resistance R_{load} for the PZT-based piezoceramics of intermediate hardness ($d_{33} = 470 \cdot 10^{-12}$ C/N, $\varepsilon_{33}{}^{T} = 1785$) at 2000 up to 50000 load cycles.

Figures 7–10 clearly show that 'soft' piezoelectric materials are unstable under the action of cyclic mechanical stress and that this instability is caused by partial depolarization of piezoceramics.

Thereby, despite of high piezoelectric efficiency "soft" piezoceramic compositions are unsuitable for commercial piezoelectric generator applications because of low mechanical, time and temperature stability reasons. "Hard" or intermediately "hard" hot-pressed piezoceramics are preferable as piezoelectric generators for commercial energy harvesting applications with long life-time at cyclic loadings.

3.3.2. Correction and Doping of Compositions

The following PZT compositions were chosen for further chemical composition correction and doping to increase a power efficiency and stability of energy harvesting generators:

(i) $PbTiO_3 - PbZrO_3 - PbNb_{2/3}Mg_{1/3}O_3 - PbW_{1/2}Zn_{1/2}O_3$ ("soft");
(ii) $PbTiO_3 - PbZrO_3 - Pb(W_{1/2}Cd_{1/2})O_3$ (intermediately "hard").

It was found that the variation in the magnitude of piezoelectric stress constant d_{33} and dielectric constant $\varepsilon_{33}{}^{T}$ is similar and they are increasing or decreasing simultaneously upon stoichiometric modification of the PZT composition resulting in minor changes in the magnitude of constant g.

Thus to obtain a large magnitude of the product ($d \cdot g$), the piezoelectric compositions can be tailored by either over-stoichiometric doping or processing technique such that the change in piezoelectric and dielectric constant leads to a decrease in the magnitude of constant n and to a rise in ($d \cdot g$) product.

Table 7. Parameters of piezoceramic samples

Parameter	R12-1	R12-2	R12-3
d_{33}, pC/N	620	520	460
$\varepsilon_{33}^T/\varepsilon_0$	2850	2350	1750
$d_{33}^2/\varepsilon_{33}^T$ (10 – 15 m²/N)	15240	13000	13660

Table 8. Parameters of piezoceramic samples

Parameter	R12-4	R12-5	R12-6
d_{33}, pC/N	250	260	420
$\varepsilon_{33}^T/\varepsilon_0$	650	630	1650
$d_{33}^2/\varepsilon_{33}^T$ (10 – 15 m²/N)	10860	12120	12080

Piezoceramics of system (1) depending on component content have the values of d_{33}= 350–600 and ε_{33}^T=1500–3200 [3]. A set of piezoceramic compositions belonging to this system with additional doping by Mn was prepared and tested:

$Pb(Ti_xZr_{1-x-y-z})(Nb_{1/2}Zn_{1/2})_y(W_{1/2}Mg_{1/2})_zO_3$ + (0 – 1)wt.% MnO_2, where (x = 0.42 – 0.46, y = 0.02 – 0.04, z = 0.01 – 0.03);

R12-1: + 0 wt.% MnO_2; R12-2: + 0.1 wt.% MnO_2; R12-3: + 0.5 wt.% MnO_2.

Experimental samples were fabricated using either standard technological process or hot-pressing [3]. Parameters of the experimental samples are listed in Table 7.

Modification of the components content for 3-component solid solutions #2 $PbTiO_3$ – $PbZrO_3$–$Pb(W_{1/2}Cd_{1/2})O_3$ and adding (0 – 1) wt.% Ta_2O_5 give us the compositions $Pb(Ti_x,Zr_{1-x-y})(W_{1/2}Cd_{1/2})_yO_3$ + 1 wt.% Ta_2O_5, where (x = 0.44 – 0.45, y = 0.02) with the parameters listed in Table 8.

4.1. Experimental Results and Discussion

The results of piezoelectric and dielectric parameters measurements for experimental batches of piezoelements R12-1 – R12-6 are listed in Table 9.

The results of mechanical loading of R12-1 – R-12-6 piezoelements (5 samples with diameter of 10 mm and height of 4 mm of each composition) by mechanical stress 38.2 MPa with the frequency 5 Hz at different load resistances are shown in Figure 11–16.

Table 9. Piezoelectric and dielectric parameters of experimental batches of piezoelements

Composition	# of samples, pcs.	d_{33}, pC/N	C, pF	$\varepsilon_{33}^T/\varepsilon_0$
R12-1	5	630 ± 2	505 ± 6	2900
R12-2	5	530 ± 2	416 ± 6	2395
R12-3	5	470 ± 4	310 ± 5	1785
R12-4	5	250 ± 4	113 ± 5	650
R12-5	5	267 ± 4	115 ± 10	660
R12-6	5	415 ± 5	295 ± 5	1698

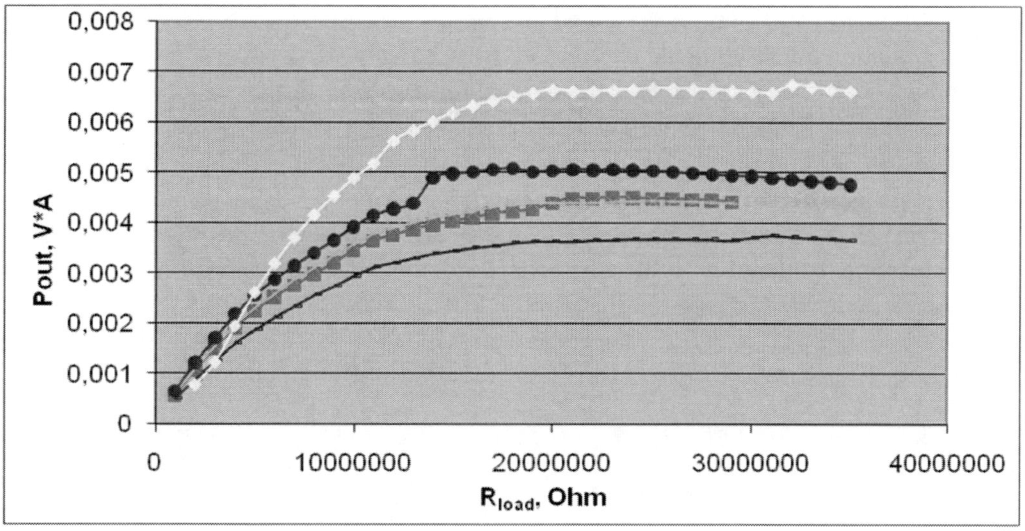

Figure 11. Output power P_{out} *vs* load resistance R_{load} for 5 samples of R12-1 piezoceramics.

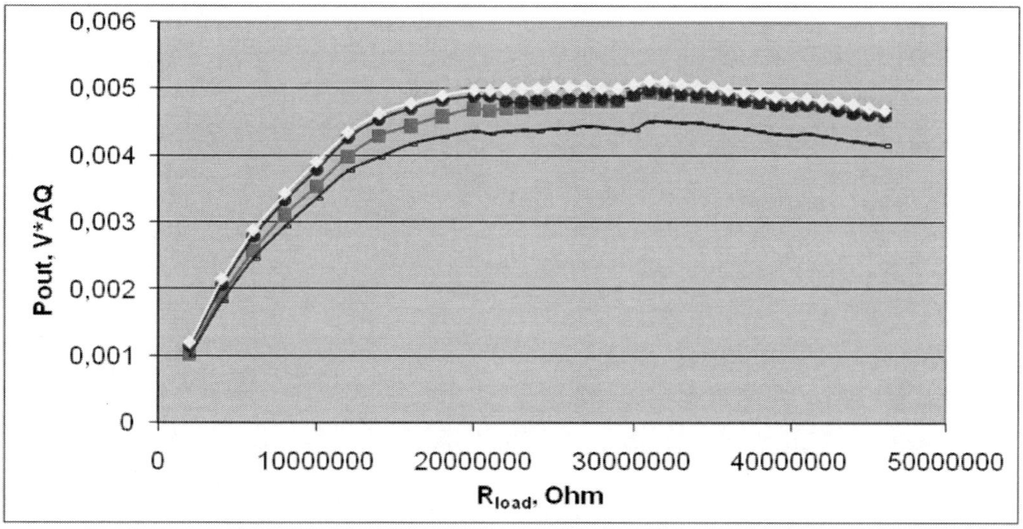

Figure 12. Output power P_{out} *vs* load resistance R_{load} for R12-2 piezoceramic samples.

Electric impedance characteristics (modulus of complex impedance $|Z|$, active resistance R_s, imaginary resistance X, capacitance C_s and loss tangent tan δ) measured for R12-1 – R12-6 piezoceramics samples at 5 Hz frequency are listed in Table 10.

The results of cyclic loading of R12-1 – R12-6 piezoceramic elements have shown that compositions R12-3 – R12-6 can withstand mechanical stresses up to 50 MPa and demonstrate stable power output during 50000 load cycles at a mechanical stress of 38.2 MPa with the frequency 5 Hz.

However more "soft" composition R12-1, despite of higher piezoelectric activity, are starting degradation after 2000 load cycles, like was shown in Figures 7–9 for typically "soft" piezoceramics. It's obvious, that R12-1 composition requires further modification, namely more precise doping by MnO_2 as well as additional adjustment of technological regimes.

Comparison of the results shown on Figures 11–16 and listed in Tables 3 and 4 confirms conclusion that piezoceramics with higher ($d{\cdot}g$) product demonstrate higher power output.

The influence of complex impedance of piezoceramic elements on power efficiency of energy harvesting generators is closely related to the electric load type, matching and measurement circuits and requires further study.

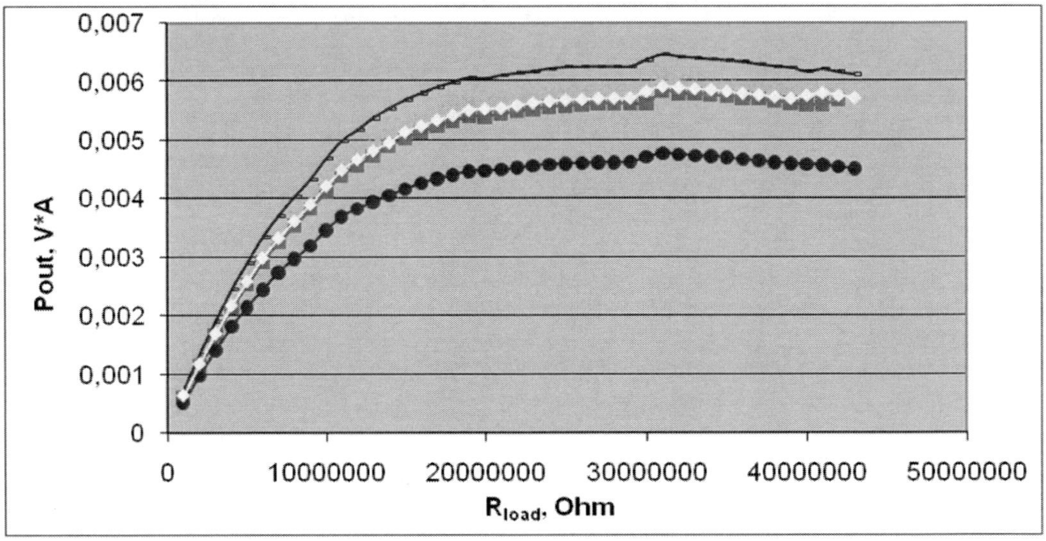

Figure 13. Output power P_{out} *vs* load resistance R_{load} for R12-3 piezoceramic samples.

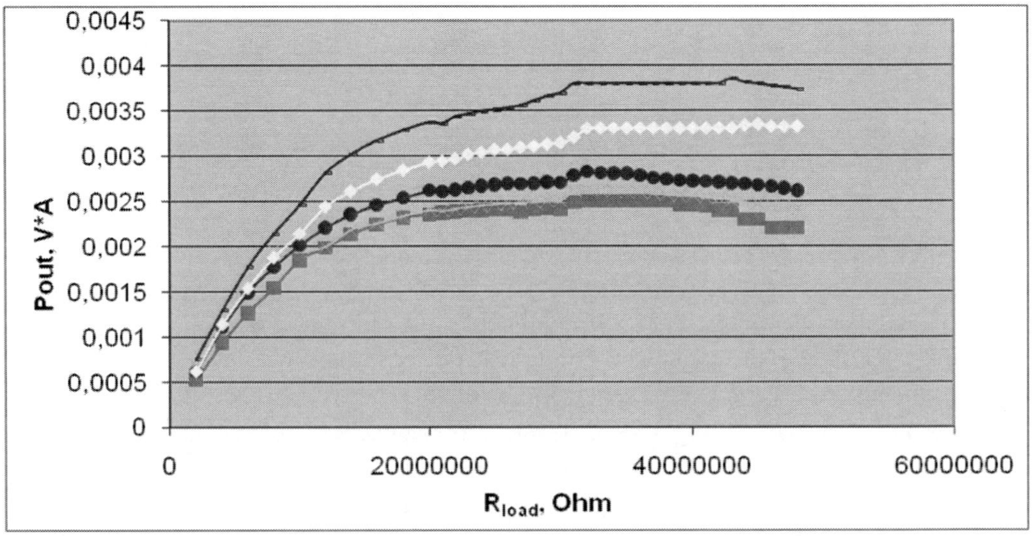

Figure 14. Output power P_{out} *vs* load resistance R_{load} for R12-4 piezoceramic samples.

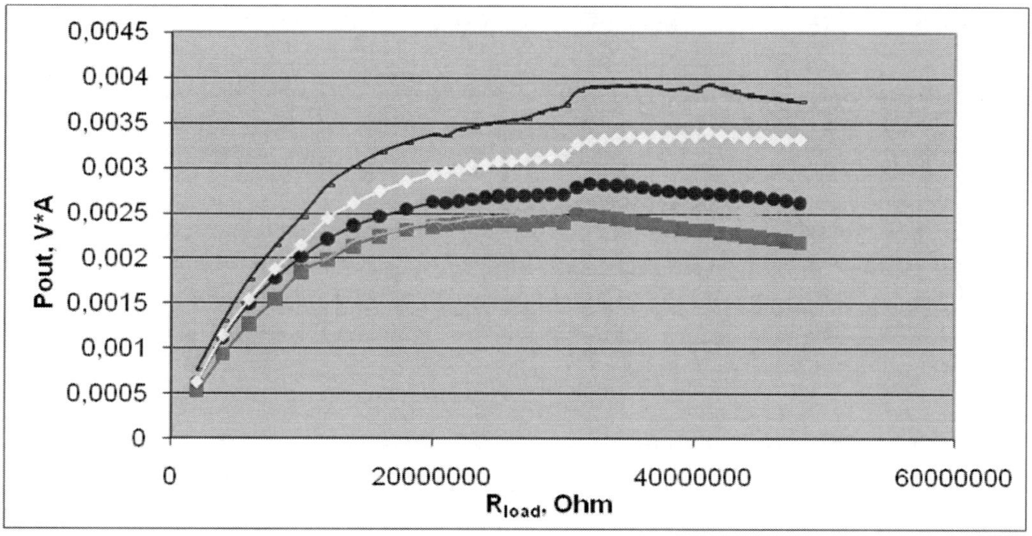

Figure 15. Output power P_{out} *vs* load resistance R_{load} for R12-5 piezoceramics samples.

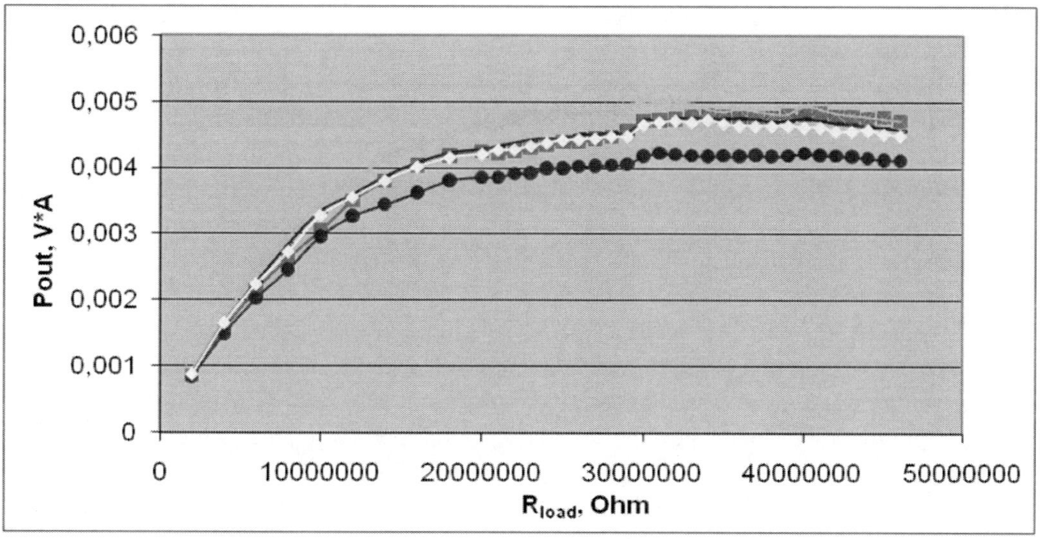

Figure 16. Output power P_{out} *vs* load resistance R_{load} for R12-6 piezoceramics samples.

Table 10. Impedance characteristics of R12-1 – R12-6 piezoceramics samples

PZT composition	$\|Z\|$, MΩ	R_s, MΩ	X, MΩ	C_s, nF	tan δ
R12-1	57	0.68	57	0.560	0.012
R12-2	69	0.69	69	0.460	0.01
R12-3	88	1.5	88	0.359	0.017
R12-4	209	4.0	209	0.152	0.02
R12-5	214	2.14	214	0.149	0.01
R12-6	95	0.95	95	0.333	0.01

Preliminary experiments with complex shaped piezoelements showed that the giant increase in power efficiency is provided by flex-tensional elements like membranes, shells or flat-concave elements. As an example, flat concave element with diameter 10 mm and thickness difference 1/2 mm made from R12-3 piezoceramics can withstand to 20 MPa pressure and demonstrates d_{33} value equal to 2500 pC/N with corresponding increase in power efficiency.

The results of development and detailed studies of energy harvesting devices with higher overall efficiency, stability, life-time and low self-cost, based on both new materials (piezoelectric, pyroelectric and electret ceramics and composites) and new physical principles (giant field-induced piezoelectric and pyroelectric responses, multimode vibrations and complex deformations) will be published elsewhere.

CONCLUSIONS

1. The piezoelectric materials properties are the main limiting factor in piezoelectric transducer efficiency. The modifications of these properties along with the development of new transducers designs will yield dramatic improvements in performance and efficiency of energy harvesting devices.
2. The variation in the magnitude of piezoelectric stress constant d and dielectric constant ε is similar and they are increasing or decreasing simultaneously upon stoichiometric modification of the PZT composition resulting in minor changes in the magnitude of constant g. Thus to obtain a large magnitude of the product $(d \cdot g)$, the piezoelectric compositions can be tailored by modifying with an over stoichiometric doping or processing technique such that the change in piezoelectric and dielectric constant leads to a rise in $(d \cdot g)$ product.
3. A high-energy density material must be characterized by a large magnitude of a product of piezoelectric voltage constant (g) and piezoelectric strain constant (d) given as $(d \cdot g)$. Absolute values of d_{33} and $\varepsilon_{33}/\varepsilon_0$ have not a sense for piezoelectric generator power efficiency, but can be important for particular generator design and electric load resistance.
4. The high-energy density piezoelectric ceramic compositions were realized in the system: $Pb(Ti_xZr_{1-x-y-z})(Nb_{1/2}Zn_{1/2})_y(W_{1/2}Mg_{1/2})_zO_3 + (0-1)$wt.% MnO_2, where (x = 0.42 – 0.46, y = 0.02 – 0.04, z = 0.01 – 0.03). The composition R12-1 with 0 wt.% MnO_2, has d_{33} = 620 pC/N, $\varepsilon_{33}^T/\varepsilon_0$ = 2850, and a large product $(d_{33} \cdot g_{33})$ of 15240×10^{-15} m^2/N . The composition R12-3 with 0.5 wt.% MnO_2 has d_{33} = 460 pC/N, $\varepsilon_{33}^T/\varepsilon_0$ = 1750, and a large product $(d_{33} \cdot g_{33})$ of 13660×10^{-15} m^2/N . The hot-pressed piezoceramic elements can withstand to cyclic load up to 40 MPa with 5 Hz frequency during 50000 load cycles and sustain compression stress up to 100 MPa.
5. Ferroelectrically "soft" piezoceramics are characterized in general by higher energy density $(d \cdot g)$ then a "hard" ones, but "hard" piezoceramics can tolerate higher stress levels without depolarization or piezoelectric coefficient decrease. So, "hard" piezoceramics or piezoceramics with intermediate hardness are preferable for commercial piezoelectric generators with long life-time at cyclic loadings.

6. The decay of piezoelectric constant d_{33} at compressive stress increase is more pronounced for ferroelectrically "soft" ceramics and caused by ferroelectric domains re-orientations. Depolarization factor calculated as ratio of d_{33} before and after loading at compressive stress σ_c = 2.8·10^8 Pa is equal to 6 – 10 % for "hard" piezoceramics compared to 40 – 50 % for "soft" piezoceramic compositions (compressive strength limit for all types of hot-pressed piezoceramics was ~ 3·10^8 Pa).
7. Comparison of experimental data for piezoceramic samples fabricated by hot-pressing and conventional sintering methods shows that hot-pressed ceramics much more stable to compressive stress σ_c in the range from 1.25·10^8 to 2.8·10^8 Pa (depolarization factors for hot-pressed and conventional sintered "hard" PZT ceramics were 6.6 % и 22.0 %, respectively).
8. Porous ceramics are characterized by very high power efficiency. Porous PZT piezoceramics R-1119P at relative porosity 20% has d_{33} = 410×10^{-12} C/N, $\varepsilon_{33}{}^T$ = 1100, g_{33} = 42.14 mV·m/N and (d_{33}·g_{33}) = 17270 × 10^{-15} m^2/N. Porous ceramics with fine and low relative porosities provide 1.5–2 times higher power efficiency at cyclic loadings then dense hot-pressed ceramics because of higher deformation of porous skeleton and can be used as piezogenerators at low loading pressures (up to 10 MPa). Porous ceramics approach can be used for fine adjustment of piezoceramic parameters at low and fine porosity keeping high mechanical durability of piezoelements for energy harvesting device applications.
9. The multilayer composite approach is prospective for fine adjustment of piezoceramic parameters. 3-layers composite samples, fabricated in one co-sintering cycle demonstrate following parameters: d_{33} = 275, $\varepsilon^T{}_{33}$ = 600, g_{33} = 51.8 mV·m/N and (d_{33}·g_{33}) = 14240 × 10^{-15} m^2/N.
10. The giant increase in power efficiency can be provided by flex-tensional elements like membranes, shells or flat-concave piezoelements at relatively low loading forces.

This work has been supported by Russian Foundation for Basic Research under Grant No 12-08-01137 a.

REFERENCES

[1] Kim, H.; Tadesse, H.; Priya, S. In *Energy Harvesting Technologies*, S. Priya, D.J. Inman (Eds.). Springer Science + Business Media, LLC: 2009, Chapter 1, pp. 3–39.

[2] Anton, S. R.; Sodano, H. A. *Smart Mater. Struct*. 2007, *vol. 16*, R1–R21.

[3] Cook-Chennault, K. A.; Thambi, N.; Sastry, A. M. *Smart Materials and Structures*, 2008, *vol. 17*, 043001.

[4] Roundy, S.; Wright, P.; Rabaey, J. *Energy Scavenging for Wireless Sensor Networks*. Kluwer Academic, Boston, MA. 2003, pp. 1–167.

[5] Erturk, A.; Inman D.J. *Piezoelectric Energy Harvesting*. John Wiley and Sons, 2011, pp. 1-276.

[6] Priya, S. *J. Electroceram*. 2007, *vol. 19*, pp. 165–182.

[7] Sodano, H. A.; Inman, D. J.; Park, G. *Shock and Vibration Digest.* 2004, *vol. 36*, 197–205.

[8] Keawboonchuay, C. *IEEE Transactions on Plasma Science.* 2003, *vol. 31*, 123–128.

[9] Keawboonchuay, C.; Engel, T. G. *IEEE Transactions on Ultrasonics, Ferroelectrics, and Frequency Control.* 2003, *vol. 50*, 1377–1382.

[10] Poulin, E. S. G.; Costa, F. *Sensors and Actuators.* 2004, *vol. A 116*, 461–471.

[11] Abramovich, H.; Harash, E.; Milgrom, C.; Amit, U; Azulay, L. E. Power harvesting from railway; apparatus system and method. US Pat. 7812508 B2. 12.10.2010.

[12] Abramovich, H.; Harash, E.; Milgrom, C.; Azulay, L.E.; Amit, U. Piezoelectric stack compression generator. US Patent Application 20110291526. 01.12.2011.

[13] Ren, B.; Zhang, Y.; Zhang, Q.; Li, X.; Di, W.; Zhao, X.; Luo, H.; Or, S. W. *Appl. Phys. A.* 2010, *vol. 100*, 125–128.

[14] Platt, S. R.; Farritor, S.; Haider, H. *IEEE/ASME Trans. Mechatronics*. 2005, *vol. 10*, 240–252.

[15] Baker, J.; Roundy, S.; Wright, P. *Proc. 3rd Int. Energy Conversion Engineering Conf.* (San Francisco, CA.). 2005, pp. 959–970.

[16] Roundy, S.; Wright, P. K.; Rabaey, J. *Comput. Commun*. 2003, *vol. 26*, 1131–1144.

[17] Ng, T. H.; Liao, W. H. *J. Intell. Mater. Syst. Struct*. 2005, *vol.16*, 785–797.

[18] Roundy, S.; Sundararajan, V.; Baker, J.; Carleton, E.; Reilly, E.; Otis, B.; Rabaey, J. ; Wright, P. *Energy scavenging in support of ambient intelligence: techniques, challenges, and future directions*, Mukherjee, S.; et al. (Ed.). Amiware Hardware Technology Drivers of Ambient Intelligence: 2006, Chapter 5.1, pp. 265–284.

[19] Wood, R. J.; Steltz, E.; Fearing, R. S. *J. of Sensors and Actuators A: Physical.* 2005, *vol. 119(2)*, 476–488.

[20] Xu, Jia Wen; Shao, Wei Wei; Kong, Fan Rang; Feng, Zhi Hua. *Appl. Phys. Lett.* 2010, *vol. 96*, 152904.

[21] Mossi, K.; Ounaies, Z.; Smith, R.; Ball, B. *Proc. SPIE: Smart Structures and Materials 2003: Active Materials: Behavior and Mechanics*, San Diego, CA, USA. 2003, *vol. 5053*, 423–435.

[22] Yoon, K. J.; Park, K. H.; Park, H. C.; Perreux, D. *Composites Science and Technology.* 2003, *vol. 63(3–4)*, 501–506.

[23] Kim, H. W.; Batra, A.; Priya, S.; Uchino, K.; Markley, D.; Newnham, R. E.; Hofmann, H. F. *J. Appl. Phys.* 2004, *vol. 43*, 6178–6183.

[24] Dogan, A.; Fernandez, J. F.; Uchino, K.; Newnham, R. E. *IEEE Trans. on Ultrasonics, Ferroelectrics and Frequency Control.* 1997, *vol. 44*, 597 – 605.

[25] Kim, H. W.; Priya, S.; Uchino, K.; Newnham, R. E. *J. Electroceram.* 2005, *vol. 15*, 27–34.

[26] Priya, S.; Taneja, R.; Myers, R.; Islam, R. In: *Piezoelectric and Acoustic Materials for Transducer Applications.* A. Safari, E. K. Akdogan (Eds.). Springer Science + Business Media. LLC: 2008, Chapter 18, pp. 373–388.

[27] Tadesse, Y.; Zhang, S.; Priya, S. *J. Intelligent Material Systems and Structures,* 2009, *vol. 20*, 625–632

[28] Islam, R. A.; Priya, S. *Appl. Phys. Lett.* 2006, *vol. 88*, 032903.

[29] Islam, R A.; Priya, S. *J. Amer. Ceram. Soc*. 2006, *vol. 89*, 3147–3156.

[30] Kim, H.; Tadesse, Y.; Priya, S. In: *Energy Harvesting Technologies*, S. Priya, D. J. Inman (Eds.). Springer Science + Business Media, LLC: 2009. Chapter 1, pp. 19–49.
[31] Richards, C. D.; Anderson, M. J.; Bahr, D. F.; Richards, R. F. *J. Micromech. Microeng.* 2004, *vol. 14*, 717–721.
[32] Lefeuvre, E.; Sebald, G.; Guyomar, D.; Lallart, M.; Richard, C. *J. Electroceram.* 2009, *vol. 22*, 171–179.
[33] Trollier-Mckinstry, S.; Muralt, P. *J. Electroceram.* 2004, *vol. 12*, 7–17.
[34] Badel, A.; Benayad, A.; Lefeuvre, E.; Lebrun, L.; Richard, C.; Guyomar, D. *IEEE Trans. Ultrason. Ferroelectr. Freq. Control.* 2006, *vol. 53*, 673–684.
[35] 35. Taylor, G. W.; Burns, J. R.; Kammann, S. A.; Powers, W. B.; Welsh, T. R. *IEEE J. Oceanic Eng.* 2001, *vol. 26*, 539–547.
[36] Shrout, T. R.; Zhang, S. J. *J. Electroceram.* 2007, *vol. 19*, 111–124.
[37] Guo, Y.; Kakimoto, K.; Ohsato, H. *Appl. Phys. Lett.* 2004, *vol. 85*, 4121–4123.
[38] Yuan, Y.; Zhang, S.; Zhou, X.; Liu, J. *Jpn. J. Appl. Phys.* 2006, *vol. 45*, 831–834.
[39] Takenaka, T.; Nagata, H. *J. Eur. Ceram. Soc.* 2005, *vol. 25*, 2693–2700.
[40] Zhao, P.; Zhang, B.P.; Li, J.F. *Appl. Phys. Lett.* 2007, *vol. 91*, 172901.
[41] Zang, G. Z; Wang, J. F.; et al. *Appl. Phys. Lett.* 2006, *vol. 88*, 212908.
[42] Ming, B. Q.; Wang, J. F.; Qi, P.; Zang, G. Z. *J. Appl. Phys.* 2007, *vol. 101*, 054103.
[43] Park, H. Y.; Ahn, C. W.; Song, H. C.; Lee, J. H.; Nahm, S.; et al. *Appl. Phys. Lett.* 2006, *vol. 89*, 062906.
[44] Ahn, C. W, Park, H. Y, Nahm, S, et al. *Sens. Actuators A.* 2007, *vol. 136*, 255.
[45] Cho, K. H.; Park, H. Y.; Ahn, C. W.; Nahm, S.; Lee, H. G.; Lee, H. J. *J. Am. Ceram. Soc.* 2007, *vol. 90*, 1946–1949.
[46] Oh, Y.; Noh, J.; Yoo, J.; Kang, J.; Hwang, L.; Hong, *J. IEEE Trans. Ultrason. Ferroelectr. Freq. Control.* 2011, *vol. 58*, 1860–1866.
[47] Simon-Seveyrat, L.; Hajjaji, A.; Emziane, Y.; Guiffard, B.; Guyomar, D. *Ceram. Internat.* 2007, *vol. 33*, 35–40.
[48] Liu, W.; Ren, X. *Phys. Rev. Lett.* 2009, *vol. 103*, 257602.
[49] Xue, D.; Zhou, Y.; Bao, H.; Gao, J.; Zhou, C.; Ren, X. *Appl. Phys. Lett.* 2011, *vol. 99*, 122901.
[50] Liu, Y.; Ren, K. L.; Hofmann, H. F.; Zhang, Q. *IEEE Trans. Ultrason. Ferroelec. Freq. Control*. 2005, *vol. 52*, 2411–2417.
[51] Priya, S.; Taneja, R.; Myers, R.; Islam R. In: *Piezoelectric and Acoustic Materials for Transducer Applications*. A. Safari, E. K. Akdogan (Eds.). Springer Science + Business Media, LLC: 2008, Chapter 18, pp. 124–164.
[52] Daue T. Overview Using the MFC for Energy Harvesting and other Commercial Applications. UTA Workshop on Energy Harvesting, Jan 27, 2006.
[53] Bryant, R. G.; Effinger, I. V.; Aranda, Jr. I.; Copeland, Jr.; Covington, E. W. *Proc. SPIE, Smart Struc. and Materials – Active Materials: Behavior and Mechanics.* San Diego, CA. 2002, Paper 4699, pp. 40.
[54] Ren, K.; Liu, Y.; Geng, X.; Hofmann, H. F.; Zhang, Q. M. *IEEE Trans. Ultrason. Ferroelec. Freq. Control.* 2006, *vol. 53*, 631–638.
[55] Peng, Hu.; Zhaoxian, Xiong; Hao, Xue; Jie, Pan; Qingjun, Lu; Xiaopeng, Xiao. *Proc. 2011 International Symposium on Advanced Packaging Materials (APM 2011).* 2011, pp.43–46.

[56] Rybyanets, A. N.; Rybyanets, A. A. *IEEE Trans. UFFC.* 2011, vol. 58, 1757–1774.
[57] Rybyanets, A. N. *IEEE Trans. UFFC.* 2011, *vol. 58*, 1492–1507.
[58] Dantziger A. J.; et al. *Multicomponent Systems of Ferroelectric Solid Solutions: Physics, Crystallochemistry, Technology. Design Aspects of Piezoelectric Materials.* Rostov State University Press, Rostov-on-Don, Russia; 2001, vol. 1–2.
[59] Rybyanets, A. N. In *Piezoceramic Materials and Devices*, Ivan A. Parinov (Ed.). Nova Science Publishers Inc. NY: 2010, Chapter 3, pp. 113 – 174.
[60] *IEEE Standard on Piezoelectricity*. ANSI/IEEE Std. 1987, pp. 1–176.
[61] PRAP (Piezoelectric Resonance Analysis Program). TASI Technical Software Inc. www.tasitechnical.com
[62] Rybyanets, A. N.; Tasker R. *Ferroelectrics.* 2007, *vol. 360*, 90 – 95.

In: Physics and Mechanics of New Materials …
Editors: Ivan A. Parinov and Shun-Hsyung Chang
ISBN: 978-1-62618-535-7

Chapter 23

THEORETICAL AND EXPERIMENTAL INVESTIGATIONS OF PIEZOELECTRIC GENERATORS OF VARIOUS TYPES

V. A. Akopyan[1,2], Yu. N. Zakharov[3], I. A. Parinov[2], E. V. Rozhkov[2], S. N. Shevtsov[4], P. C. Wu[5] and J. K. Wu[6]

[2]Vorovich Mechanics and Applied Mathematics Research Institute, Southern Federal University, Russia

[3]Institute of Physics, Southern Federal University, Rostov-on-Don, Russia

[4]South Scientific Center of Russian Academy of Science, Rostov-on-Don, Russia

[5]Nanoscience and Nanotechnology Center, National Chung Hsing University, Taichung, Taiwan

[6]Department of Marine Engineering, National Kaohsiung Marine University, Taiwan

ABSTRACT

This paper presents a brief overview of modern research directions, methods and results, obtained at designing and processing devices for energy harvesting applications. The presented work allowed one to propose original models of piezoelectric generators (PEGs). Various types of energy harvesting PEGs, including cantilever and multilayer piezo-stack samples were examined. Different electrical schemes for the developed energy harvesting devices are discussed in detail and stability investigation of electro-physical parameters of piezoceramic materials used for sensitive elements of PEGs is presented. Optimal characteristics of sensitive elements of the cantilever type PEGs are defined by using finite-element method. In order to study spectral and physical characteristics of signals of the considered PEGs, we considered the created experimental set-up and numerical data obtained with its use. Finally, it is presented critical analysis of various physical parameters, ensured by the treated types of PEGs.

[1] akop@math.rsu.ru.

1. Introduction

Today, the problem of creation of piezoelectric generators (PEGs) for different applications is not solved in a whole due to low-energy efficiency of PEGs and low output power, developed by test constructions. Among different problems, arising at the PEG processing, the most significant problems are connected with: (i) selection of energy-efficient piezoceramic materials (PMs), (ii) development of electric schemes of accumulation of the electric energy with minimal charge leakage, (iii) search of geometric configurations and technologies of assembling sensitive element of PEG, which can provide maximal output power. The research approaches, used in solution of these problems, differ significantly in dependence on application region and specific use of PEG.

2. Overview of Modern Methods and Approaches

The problems of energy harvesting have been presented in References [1-4]. The following studies in this region concentrated on studies of characteristics of small-power PEGs, for which the considerable practical results have been attained.

There is a region of small energetics, where power of supplied energy is no great. This direction covers small home appliance, cellular phones, I-phones, notebooks, wireless sensor systems for control and diagnostics of technical state of the different objects, etc. These techniques do not require powerful sources of energy. Therefore, autonomous electrochemical sources of electric energy (small accumulators, and batteries) are very often applied for them.

In the last years, the priority for feeding sources of cellular phones and notebooks is also given to small lithium-ion accumulators with working voltage of 3 – 5 V and energy density of 100 – 180 W·hr/kg. The main shortcoming of the accumulators is connected with that what they do not give a great discharge current. In difference of the lithium-ion accumulators, the nickel-cadmium ones with working voltage of 1.2 V ensure fast discharge and allow their additional charging by current impulses.

However, as rule, serve duration of these types of sources is limited. Prolonged use of the devices, especially for home appliance leads to necessity of often replacement of the electrochemical sources. In autonomous sensor modules of devices applied, for example in the systems of technical diagnostics and monitoring of high (and not only) blocks, the replacement of feeding sources is enough labor-consuming and expended operation. Inopportune fulfillment of this operation even in home appliance (no speaking on other systems) leads to unexpected switching off of the system. It is very often these devices refuse in most critical moment that may lead to supernumerary situations. The existing attempts in the last time to use electric double-layer capacitors as autonomous sources (with low energy density equals to only 30 – 32 W·hr/kg), now to be at experimental stage and require following investigations.

In the last years, above-mentioned factors compel scientists to search more effective self-renewal autonomous sources of electric energy based on novel principles. Among them, the most attention is directed to research and development of self-renewal sources of energy, in which piezoceramic and thin-film elements (PE) are used, which transform environment energy (vibrations, wind, solar energy, heat, radiation, etc.) into electric one with the following its accumulation and transfer to receiver. Analysis of requirements, produced to the sources of energy (energy harvesting devices), showed that they significantly depend on that for which types of devices they are used. Nevertheless, majority piezoelectric systems of energy harvesting for low-power consumers of electric energy, obtained from environment, consist of functionally-homogeneous components, the structure of which could be presented by the scheme of Figure 1.

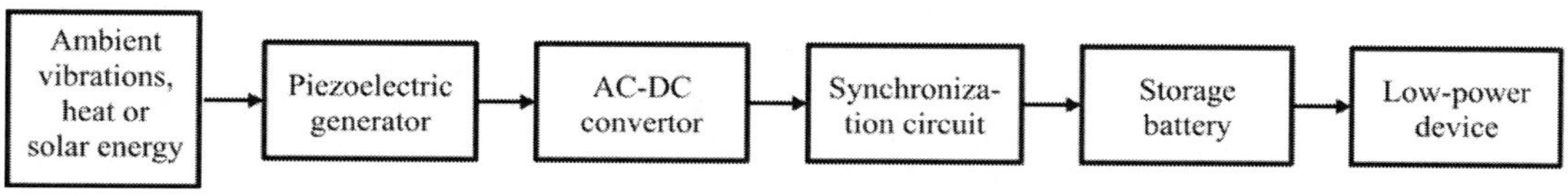

Figure 1. Structural scheme of piezoelectric system of the energy harvesting.

The structural schemes of systems of the energy harvesting for devices of high and low power in a whole differ weakly one of other.

2.1. Transducers of Mechanical, Chemical and Heat Energy into Electrical One and Their Applications

2.1.1. Known Schemes of Devices

Today, the piezoelectric transducers (energy sources) obtained most application, which could be used in several types of devices. Among them, it is necessary to note: (i) small devices of wireless electronics with prolonged service time, (ii) low-power built-in and wireless communication devices (for example, for cellular phones and smart-phones), (iii) home appliance electric mechanics and electronics (for example, electronic watch), (iv) piezoelectric generators for enough capacitive local systems of lighting and signalization on distant infrastructure objects (for example, on rail-roads and automobile roads), etc.

In the last years, the test samples of piezoelectric harvesting devices have been researched and developed for various applications, that could be divided into two various classes differing on type of the used oscillation modes (the transverse piezomodulus and the longitudinal piezomodulus on the longitudinal oscillation modes).

The principal schemes of PEGs, working on the longitudinal (relatively a direction of polarization vector) and bending oscillation modes, are very well-known [1 – 4]. In the last years, test samples of PEGs have been developed and studied, in which applied not only above-mentioned schemes in pure form, but their modifications [5]. In particular, both the solid piezoceramic elements and thin-film (mono- and multilayered) piezoelements have been used in the PEGs as sensitive elements. Moreover, there are also research results, devoted to estimation of output characteristics of PEGs, which differ by piezoplate topology on substrate of its sensitive element (SE) [6]. Then, it has been developed the hybride type PEG, in which SE presented itself a commbination of piezoelectric, light- and thermo-sensitive layers.

It should be noted, that the test samples of one type of low-power PEG were known long ago and applied as piezoactuators or antenna-driving mechanisms of satellite guidance [7]. Later, the test samples of multilayer PEG of high output power have been developed, which worked on longitudinal oscillation modes and were designed as sources of autonomous electric feeding for lighting and signalization systems of automobile roads. Today, these devices are subjected to tests [8].

Moreover, the piezoelectric energy harvesting devices are perspective, based on PEGs of cantilever type, working due to bending oscillations (Figure 2). They differs one of other both as quantity of bimorph piezoelements as their disposition on substrate, for instance steel plate. In order to ensure higher output power the plate PE could be made in multilayer form. However, the processing technology of these PE is sufficiently complex and now be at the test stage [9].

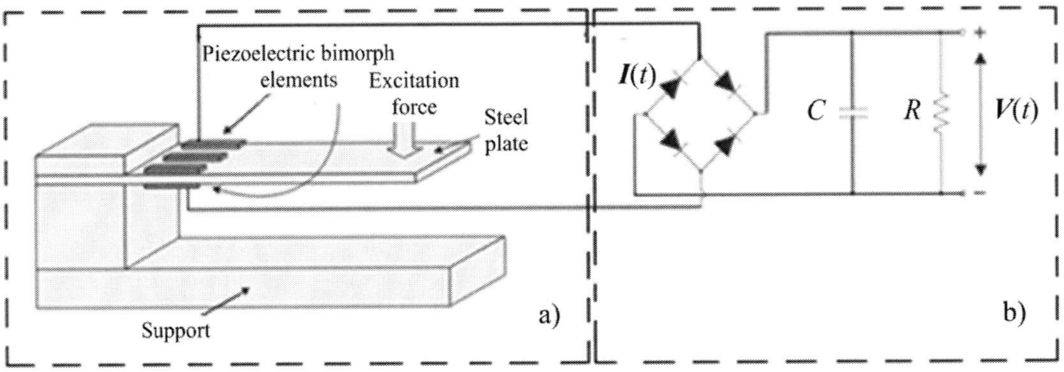

Figure 2. Standard scheme of piezoelectric energy harvesting device of the cantilever type: (a) piezoelectric generator scheme, and (b) scheme of energy transformation and storage.

In dependence on applications, the piezoelectric energy harvesting devices, which include the cantilever type PEGs, could be divided into two groups, namely micropiezogenerators (μPEG) and macropiezogenerators (MPEG). Among early constructive solutions of the first group devices, it has been proposed and described scheme developed in 2004 by Lu F., Lee H. P. et al. Principle of work of this storage could be easily understood from its scheme presented below (see Figure 3). The investigations showed that this μPEG, including cantilever with thickness of 0.1 mm and transverse sizes 5 × 1 mm^2, at the oscillation amplitude on free edge of the plate equal to 0.1 mm generated output power of 1.6 mW.

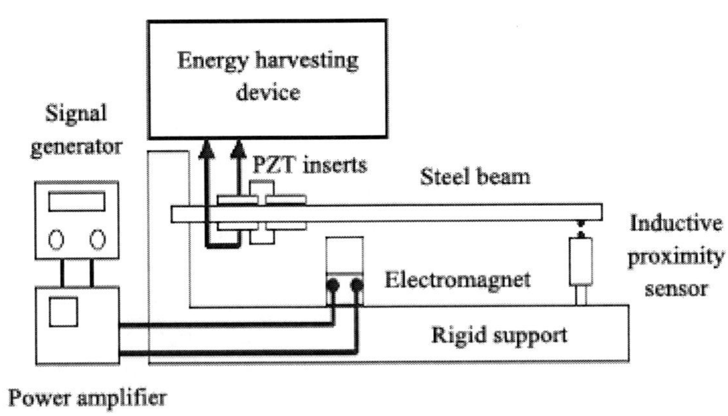

Figure 3. Principle scheme of piezoelectric energy harvesting device [5].

Then Lefeuvre E. et al. created and tested experimental sample of similar PEG with other principle scheme, which proposed to state several plate-type bimorph piezoelements on steel plate [5]. Output power of MPEG (piezoelectric energy harvesting device) of the last type was equal to 4.6 mW at the plate oscillations with the frequency of 60 Hz. After modifications, connected with application of new electric scheme, in which synchronization (on phase) of voltage signals and transverse displacements of piezocantilever has been carried out, the output power of MPEG equal to 12.3 mW was attained.

Moreover, there are known μPEGs which differ by topology (or geometry of PE disposition on substrate), but their output power is slightly upper than in above-mentioned MPEGs. The presented schemes of MPEG and μPEG are intended for small devices of home appliance, cellular communication and sensor systems of signalization and control. An attempt to use MPEG as feeding source for booster charge of storages of the cellular communication has been also made.

In other work, it has been proposed μPEG acting on the longitudinal oscillation mode. In this device, single piezoelement of μPEG was introduced into foot-wear and generated a small power at fast movement (run) of man. This μPEG was described by Ville Kaajakan from Technical University of Louisiana (USA) [19] and processed in the form of spiral platelet piezoelement. Longevity and reliability of this device are doubtful because of brittleness of piezoceramic material. It is possible this idea could be productive in the case of application of flexible piezopolimer plates. To regret, these materials now only be at the research stage. Other device of the same type has been developed in Wisconsin University of USA. Miniature piezogenerator built-in into usual foot-wear allows one to harvest energy which maybe enough for some booster charge (measured by several mW) of smart-phone, but it could be enough only for "mixing" audio-tracks in portative player (for comparison, cellular device Apple iPhone consumes about 5 mW of energy). Now researchers develop a device generating 10 W of energy. Moreover, the test samples of piezoelectric energy harvesting device (MPEG) with relatively great power of varios constructions are known. Well-known experimental system of energy harvesting, based on piezogenerators built-in floor covering at entrance in Marunouchi station of Tokyo metro (see Figure 4), relates to most interesting developments of these devices [5].

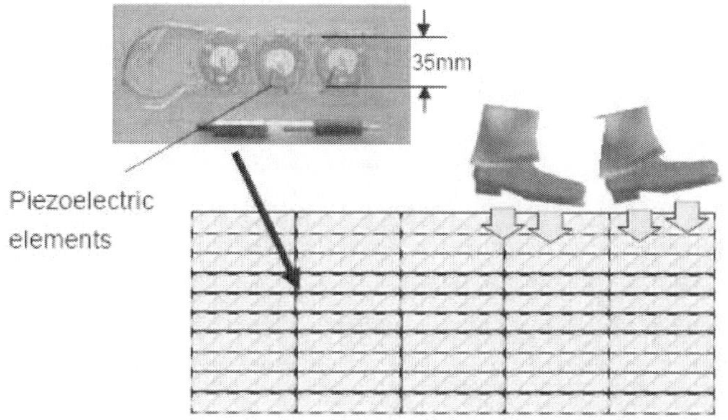

Figure 4. Scheme of energy harvesting into floor covering at entrance in Marunouchi station of Tokyo metro [5].

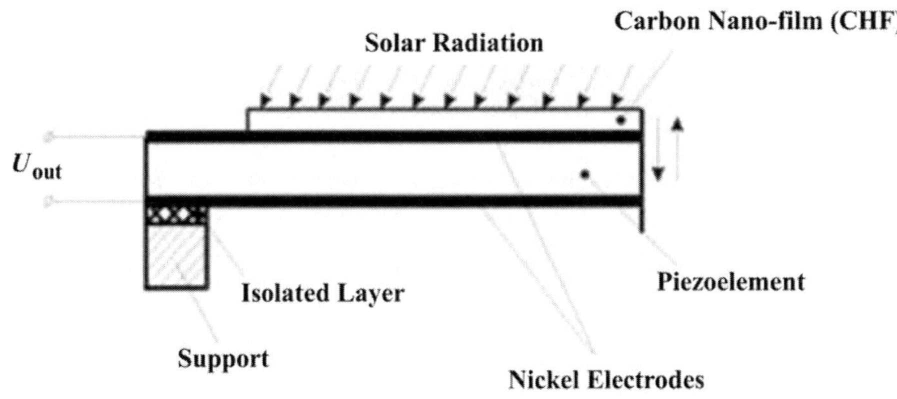

Figure 5. Principle scheme of hybrid microgenerator [22].

First tests have shown that output power of this small system of MPEG provided feeding 100 W-lamp of lighting during 100 s. Further tests have demonstrated prospective of this development. Other known application of enough powerful MPEG has been found for feeding bycicle headlight. In this case, the power of MPEGs, stated on bycicle pedal, was enough for periodical feeding electrical headlight during some time.

It is known another persective research direction connected with PEG application as energy source, caused by strain of road surface due to moving transport and proposed by Israel company Innowatech [10]. Essence of the development consists in that the PEGs are stated under asphalt covering of auto-road at the definite distance one of other, transforming the strain energy of the covering, caused by moving transport, into electric energy, stored in original energy harvesting devices and transferred to consumers, disposed near the road. This technology has been transferred for approbation in 2011 on test site of *highway Venice – Trieste* (Italy).

Among perspective developments of new μPEG schemes, we note hybrid schemes of energy harvesting including sensitive elements consisting not only of piezoelements, but also thermo-sensitive films. One of these schemes is presented in Figure 5. These μPEGs be at stage of test models and they will be possibly developed for special devices in future.

As it follows from brief analysis of characteristics of the piezoelectric energy harvesting devices, developed today and described above, their output power is unsuffitient for wide applications. Therefore, the search of possibilities to increase output power is carried out by many scientists. In particular, main trends in researches are directed to increasing efficiency of the piezoelectric energy harvesting devices with help of selection of the optimal physical parameters and geometrical configurations of their sensitive elements. These approaches together with original adaptive electrical schemes of the energy harvesting have been discussed in References [11, 12]. Analysis of PEG electric schmes is presented in [13]. Some known constructive schemes of PEGs, including most effective configuraions of SE with viewpoint of increasing output power have been briefly described above. The set of problems, connected with development of effective electric schemes of the energy harvesting, is discussed in [14]. For example, the standard electric scheme of piezoelectric energy harvesting device with several bimorh piezo-elements, used in its SE, is shown in Figure 2.

Several works have been devoted to descrition of modernized electric schemes of the energy harvesting, for example two-stage scheme. In one of the works, it is made an attempt to increase PEG output power for account of matching levels of oscillating mechanical load and load resistance (impedance) [14]. Erturk A. and Inman D. J. published overview [15] on the problem of piezoelectric energy harvesting, in which various known schemes of the energy harvesting by using AC-AC and AC-DC converters have been presented. It has been made a conclusion on that the method of synchronous switch harvesting on inductor (SSHI), realized in above-mentioned scheme of the energy harvesting (see Figure 11.8 from [15]), could be considered as effective way to increase output power of PEG, especially for weakly connected systems of the energy harvesting. However, this conclusion has not been confirmed by calculations of power and output voltage in the case of using above-mentioned scheme and method. The dependences of these PEG parameters on frequency and oscillation shapes of its SE have been presented in [15] for no all significant oscillation modes. At the same time, it is well-known that in dependence on eigenshapes and oscillation frequency of SE, the value of electric charge, generated by PEG of cantilever type, could be significantly changed. This important factor has been studied in work of Lefeuvre E., Badel A., Richard C. et al. [16] and then in our paper [17]. These papers have studied output parameters of PEG for several oscillation modes in the case of using different electric schemes of energy harvesting. In particular, Reference [16] studied the presented below schemes, including AC-DC (Figure 6) and DC-DC power converters (Figure 7). The synchronous electric charge extraction circuit is presented in Figure 8.

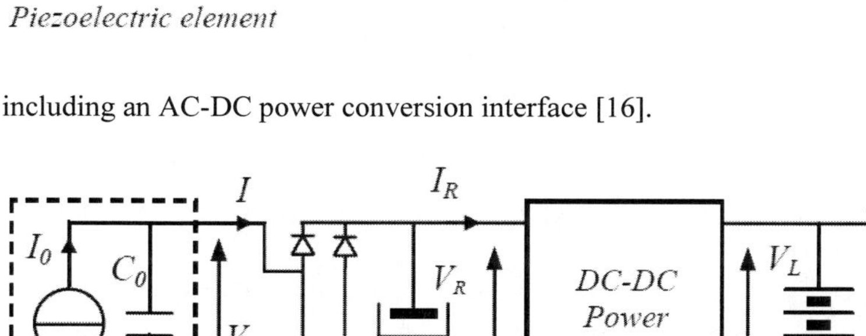

Figure 6. PEG including an AC-DC power conversion interface [16].

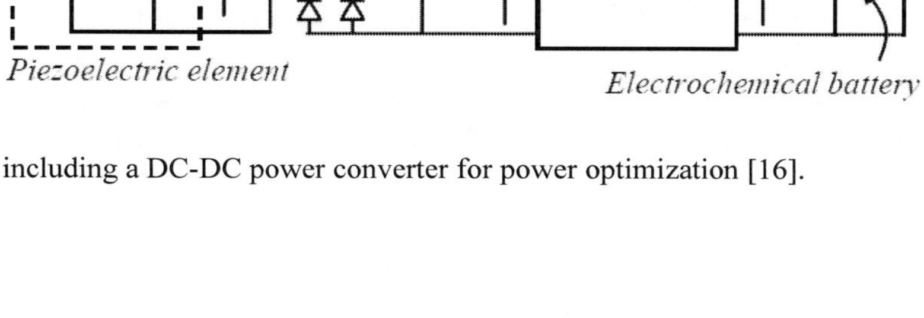

Figure 7. PEG including a DC-DC power converter for power optimization [16].

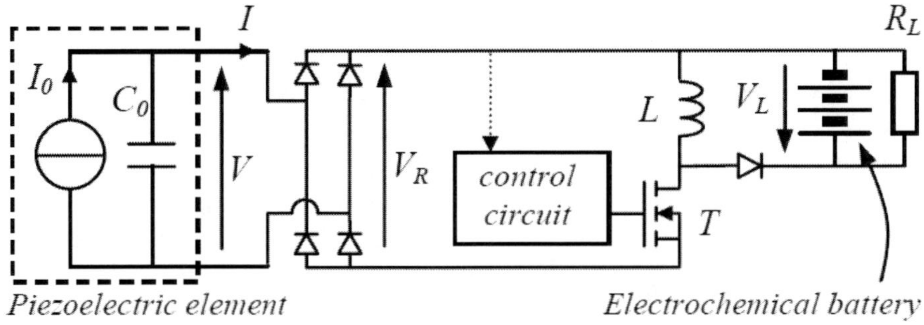

Figure 8. Synchronous electric charge extraction circuit [16].

Reference [16] also presents plots of dependences for PEG output power on load resistance for several oscillation modes. These plots for 1st and 2nd Modes are presented in Figures 9 and 10. Comparative analysis of the dependences of output power on the load resistance for 1st (Figure 9) and 2nd (Figure 10) Modes of oscillations showed that the output power of the studied type PEG at oscillations on frequency of 334 Hz equals to 2.5 mW, that is, greater on 20% than power, obtained at oscillations on frequency of the 1st Mode (56 Hz). This result takes place at the same load resistance (100 kHz) and at application of so-called method of synchronous electric charge extraction (SECE). From these plots, obtained in [16], it is obvious, that application of the SECE method allows one to increase output power of PEG as minimum on 15 %. In the case of application of standard method the power diminishes as minimum in two times. The electric scheme, in which this method has been realized, is shown in Figure 8. Analysis of these dependences also shows that maximal output power of PEG is reached (in the case of the 1st Mode) at different levels of load resistance, namely: it is equal to 100 kΩ in standard method, and already beginning from 10 kΩ in the SECE method. These values of parameters of the measurement schemes should be taken into account at develoment of actual PEG models. The calculations of values of the output power and voltage have been conducted by using the next relations [16]:

$$P_{\max} = \frac{2C_0\omega}{\pi} V_{Ropt}^2 ; \; V_{Ropt} = \frac{\alpha}{2C_0} U_m , \tag{1}$$

where P_{max} is the maximal value of PEG output power, C_0 is the capacity of smoothing capacitor (see Figure 8), ω is the circular frequency, V_{Ropt} is the optimal value of output voltage, U_m is the maximal output voltage.

The brief analysis of validity of these schemes showed that the SECE method relates to most productive with viewpoint of suplying maximal energetic effeciency of the studied type PEG. It is truthful only at use of the selected optimal regimes (that is, selected oscillation mode). The results, obtained in works [15, 16], witness on prospective of this approach. This conclusion is confirmed also by the results of our investigations, relating to the connection of mechanical stresses, arising in SE of generator, with oscillation forms and hence with value of the generated piezoelectric charge [17]. Partially, these data are presented in the next chapters of our paper.

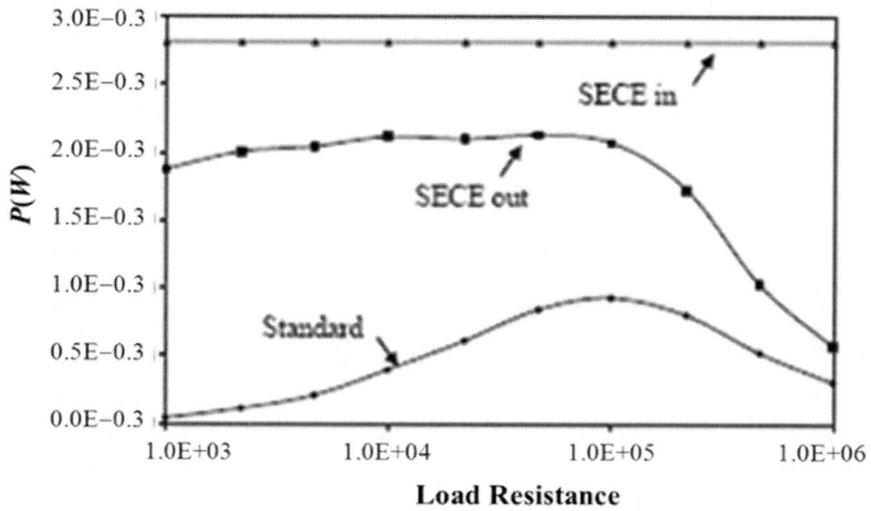

Figure 9. Plots of dependence for PEG output power on load resistance for 1st Mode (56 Hz) [16].

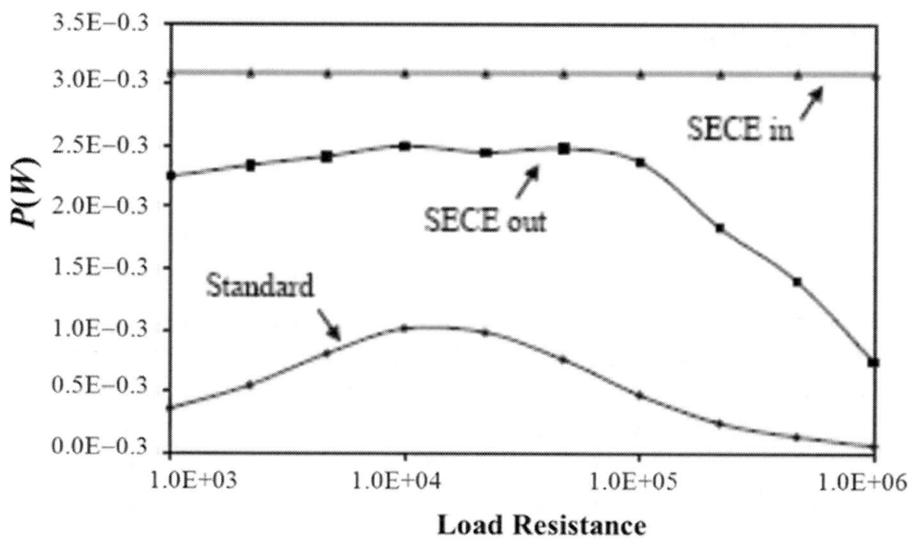

Figure 10. Plots of dependence for PEG output power on load resistance for 2nd Mode (334 Hz) [16].

2.2. Stability Investigation of Electro-Physical Characteristics of Piezoceramic Materials Being Sensitive Elements of Piezogenerators

The problem of ensuring reliable prolonged exploitation of piezogenerator devices had insufficient attention, while it has most value for statement of optimal characteristics of the devices proposed for applications, especially of increased power, at work in conditions of great mechanical loads. Before, the problem of PM strength has been considered in detail into framework of studying piezostackers for truss structures [16]. The strength of PEG elements is connected with features, differing from working conditions of piezostacks with standard scheme. In this direction, the works of Anton S. R., Erturk A., and Inman D. J. [14, 15] present the research results of static strength of several lead zirconate titanate (PZT) compositions, used as sensitive elements of PEGs. The problems of cyclic strength of PMs are connected directly with stability of electro-physical characteristics of the piezoceramics. The numerous studies of aging processes of the PMs and non-stability of values of their piezoelectric and elastic constants have been devoted to solution of this problem. There have been obtained interesting results, having important value at designing constructions of PEG sensitive element. For example, with growth of mechanical stress and temperature, piezoelectric constants (and, hence piezoelectric charge) of widely used piezoceramic materials as PZT-19 and PCR-1[2] diminish in two and more times [12]. In particular, the authors have studied an influence of mechanical stresses up to 2×10^8 Pa on magnitudes of piezomoduli and piezocharge of the SE plate from different PMs [17]. It has been stated, that only PCR-78 (composition of PZT) provides acceptable for PEG stability of piezocharge into ranges up to 10%. Other considered PM compositions demonstrated decreasing piezocharge down to 50%. It is obvious, that absence of account for aging and re-construction of domain structure under mechanical stresses decreases reliability of above-discussed results.

Other important factor, having especial value for any PEG, is scale factor. Transition from microscale μPEG (10^{-6} m), on the base of which the microelectromechanical systems (MEMS) have been processed with enough high factor of useful action (caused by high value of electromechanical couple factor for PM), to nanoscale (10^{-9} m) systems (NEMS) allows one to increase factor of useful action (basing on some references [17]) at least on two and even three orders of magnitude.

This direction began to develop owing to possibility to apply nanoscale SEs of piezogenerators for cellular communication, smart-phones and players [15]. Our recent investigations [17] of characteristics of three types of cantilever PEGs on bending oscillations (see Figure 2), which had sizes of PEs and substrates, differing in 5 times, showed fruitfulness of the proposed approach to development of the piezoelectric energy harvesting devices. Output voltage of the small-size PEG (length of composite substrate was 0.035 m, wide was 0.005 m and thickness was 0.0004 m), developed by us, exceeded 1.4 V. Other perspective scheme of hybride μPEG has been proposed by Long Que from Louisiana Technical University [19]. Piezocharge in this μPEG is harvested for account of both as PE strain, as heating nano-film by solar rays (see Figure 5). In [19], the technical characteristics of this device are absent, but the development could be very perspective.

3. Calculation of Optimal Characteristics of SE for Cantilever Type PEG by Finite-Element Method

The finite-element models of PEG have been proposed by De Marqui C. Jr., A. Erturk, D. J. Inman [20] and Elvin N. G., Elvin A. A. [21]. In [21, 22], it has been shown a convergence of these models to analytical solution for corresponding PEG models, presented in [15]. It has been also investigated a possibility to use the finite-element model of PEG for development

---

[2] PCRs are the piezoceramics processed in Rostov-on-Don, Russia.

of variants and optimization of piezoelement topology on substrate of PEG sensitive element [6]. Basing on part of these results, we have carried out numerical and experimental analysis of electric and physical parameters of cantilever type piezogenerator [17]. Here, optimal characteristics of original construction PEG have been calculated and confirmed, allowing obtainment of its output power sufficient for charging storages of low-power devices of cellular and radio communications, disposed, in first order, in hard-to-rich places for replacement of the storages. The problem has been solved by using finite-element modeling and natural physical experiment, describing the process of generating electrical charge at different levels and types of load resistance and transverse displacements of substrate. In order to state correlation relationships and couples between electrophysical, mechanical and geometrical parameters of the model PEG and its output electric power into wide range of eigenfrequencies of the PEG oscillations and impedance of its sensitive element, it has been conducted computer simulation by using Comsol Multiphysics software. At first stage of the research, the modal analysis of the system has been carried out, which allowed one to define eigenfrequencies of first three modes of the bending oscillations for consol beam in plane of lowest stiffness. It has been developed the finite-element method for calculation of characteristics of the cantilever type PEG sensitive element.

In natural physical experiments, conducted by using original experimental set-up [7], it has been stated significant dependence of electric power given off by the piezotransducer to load in dependence on the load. The calculation results of optimal impedance of the load for first four eigenfrequencies of bending oscillations of the cantilever showed that with growth of the eigenfrequency the optimal impedance of the load decreased significantly. The causes of a such dependence have been stated in [24], where considered a dependence of frequency replies of electric potential and power, created by harmonically-deformed single piezoplate due to its tension-compression, which worked in regime of d_{31}. It has been shown, that the electric voltage, generated by piezoplate, had the complex form:

$$V = \frac{i\omega e_{31}\delta b}{Y_L + iC_{pz}} = \frac{i\omega e_{31}\delta b}{Y_L + Y_{Cpz}}, \tag{2}$$

where e_{31} is the electric constant, Y_L, Y_{Cpz} are the conductivities of load and piezoplate of capacity type, respectively, δ is the amplitude of dynamic displacement of the piezoplate edge, b is the piezoplate wide, ω is the circular frequency, C_{pz} is the capacity. At the purely active load $(Y_L = 1/R)$, the absolute value of expression (2) reduces to the form:

$$|V| = \frac{\omega e_{31}\delta bR}{\sqrt{1+(\omega RC_{pz})^2}}. \tag{3}$$

From here, it is followed that with growth of the load impedance, R, voltage on its aspires to the finite limit, $e_{31}\delta b / C_{pz}$.

The total electric power in the complex form at the ideal active load finds as

$$W_{el} = \frac{1}{2} V \cdot I^* = \frac{1}{2} V \cdot V^* \cdot Y_L^* = \frac{1}{2} \frac{(e_{31}\omega\delta b)^2}{1+(R\omega C_{pz})^2} R, \tag{4}$$

has purely active character and reaches its maximal value:

$$W_{el}^m = \frac{(e_{31}\delta b)^2}{2C_{pz}} \omega, \tag{5}$$

when the value of load impedance is equal to

$$R_m = (\omega C_{pz})^{-1}, \tag{6}$$

that is, at equality of the load impedance to internal impedance of piezoplate, demonstrating purely capacitive character. The calculation results of the optimal load impedance for first three eigenfrequencies of bending oscillations of the considered cantilever are present in Figure 11. The plots show that with growth of the eigenfrequency, the value of optimal impedance decreases very significantly. For instance, at the oscillations on 1st Mode frequency of ~ 15 Hz, the value of optimal impedance equals to units of MΩ, at the same time, the otimal impedance equals ~ 500 – 600 kΩ already on 2nd Mode at the frequency of ~ 80 Hz.

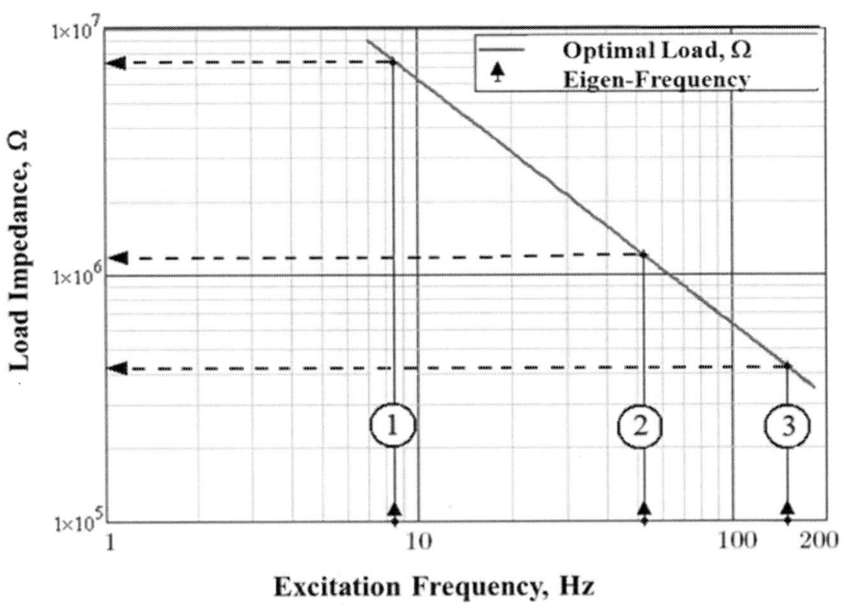

Figure 11. Dependence of optimal load impedance (corresponding to maximal power given off by piezoelement to the load) on eigenfrequencies of first four modes of cantilever bending oscillations.

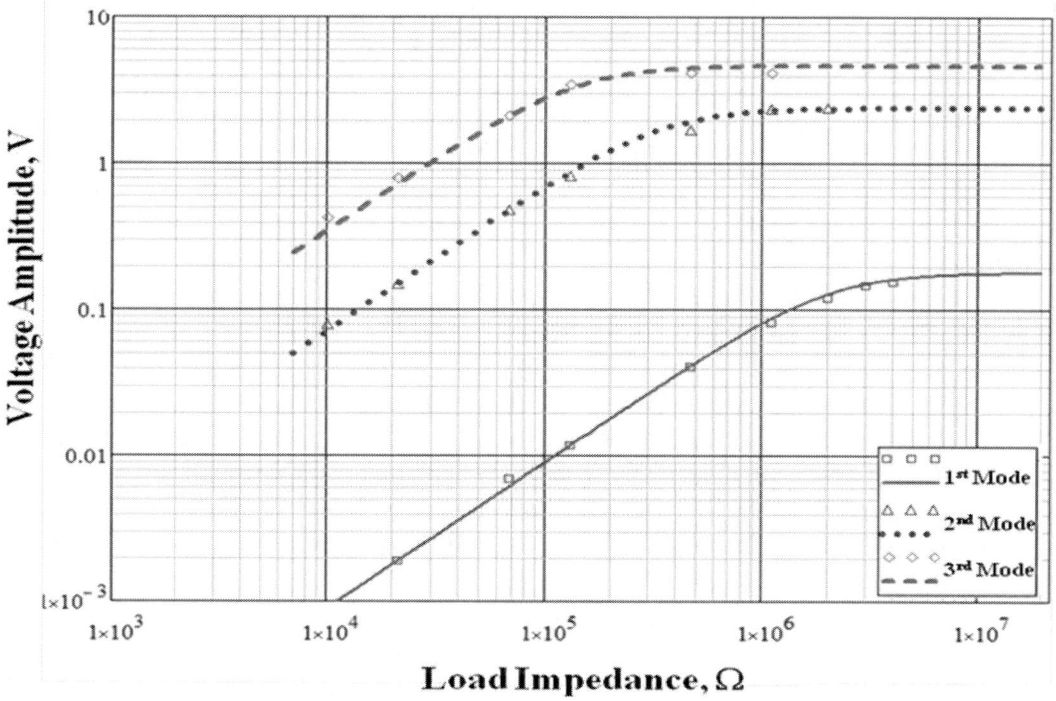

Figure 12. Dependence of voltage amplitude on load on its impedance for first three modes of bending oscillations of the model beam.

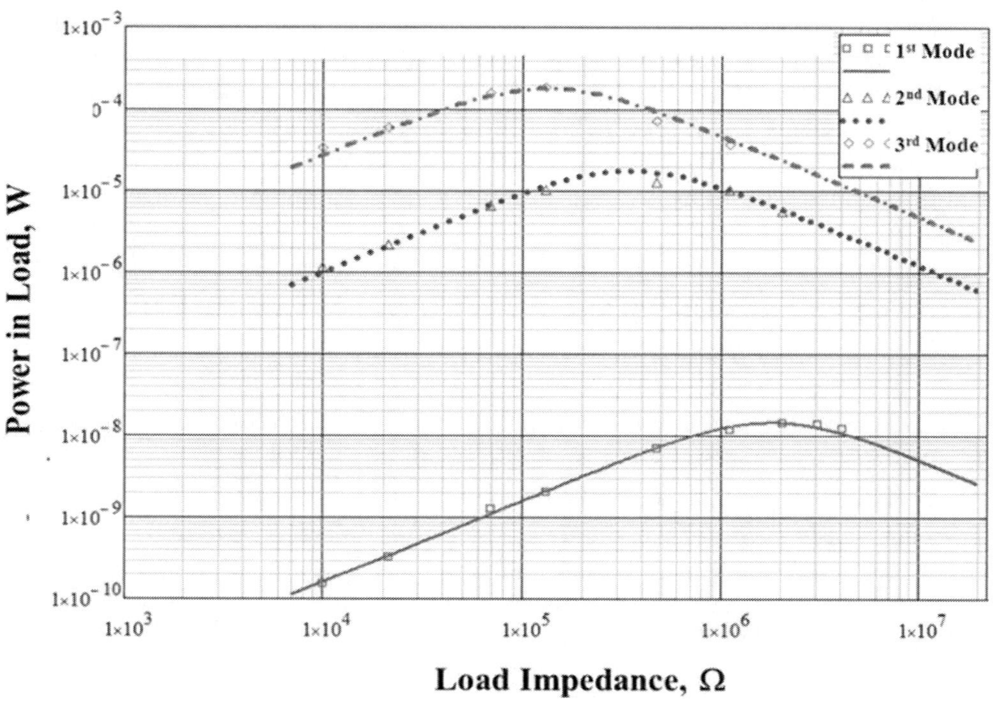

Figure 13. Dependence of electrical power, given off to load, on the load impedance for first three modes of bending oscillations of the model beam.

The considered relationships have been obtained at the piezoplate study under deformation of pure tension-compression and therefore the conclusions, made on the base of their analysis, had qualitative character. The plates, stated on surface of the bended beam, are subjected to more complex strains (non-uniform tension-compression). By this, the thinner thickness of piezoplates, compared to the beam thickness, and their shorter length lead to closer character of deformation to the ideal one, considered in [24]. These conclusions have been confirmed after working out the numerical results of experiments, presented in Figures 12 and 13.

The dependencies of electric potential, presented in Figure 12, converge to different values of limit at growth of the load impedance that could seem contrasting to the conclusion on independence of this limit on frequency (see formula (3)). However this difference is caused by that what oscillations in the model beam are excited while with the same amplitude of displacement of the beam free edge, but on different oscillation modes. The different curvatures of the axis near rigid fixing, oscillation amplitude and strains in the zone of the piezoplates disposition are the sequences of this situation.

Analysis of the dependencies of power, given off to load, presented in Figure 13, confirm their extreme character. Similar to the result, obtained analytically on the base of the simplified model with only one piezoelement, we make the following conclusion: with growth of frequency, the power extreme displaces to smaller values of the load impedance. In particular, the optimal impedance, corresponding to power extreme on first oscillation mode, is equal to ~ 2 MΩ, one is equal to ~ 1 MΩ on second oscillation mode, and be near to 200 – 400 kΩ on third oscillation mode.

Since the finite-element results are obtained at small oscillations of beam, when there is linear approximation, then small changes of amplitudes of the applied loads and corresponding oscillations cannot lead to change of the dependence character, presented in Figure 13. Test points in Figures 12 and 13 are approximated by curves, constructed in accordance with relations (3) and (4). Their ideal coincidence with the test points confirms additionally an adequacy of the conclusions, obtained in Reference [24], that allows their use for semi-quantitative estimation of working efficiency of the developed piezogenerator.

4. Test Samples of Small Piezogenerators of Different Types with Original Construction

Basing on the results of above-mentioned studies of the PEG characteristics, we developed at first stage the sample of small PEG with original construction. Below, the test sample of cantilever type PEG (the hull sizes are 45 × 30 mm^2) with maximal output voltage of 5.8 V and electric power of 0.55 mW is presented in Figure 14. At the next stage, it has been developed a new model of the cantilever type piezogenerator with a proof mass at the free edge of cantilever, depicted in Figure 15. As, it is well-known, the proof mass changes eigenfrequency of the cantilever oscillations, that may increase the sensitivity range of piezotransducer [23]. Test studies of this model have demonstrated perspectiveness of this direction to increase the output characteristics of the considered type piezogenerator.

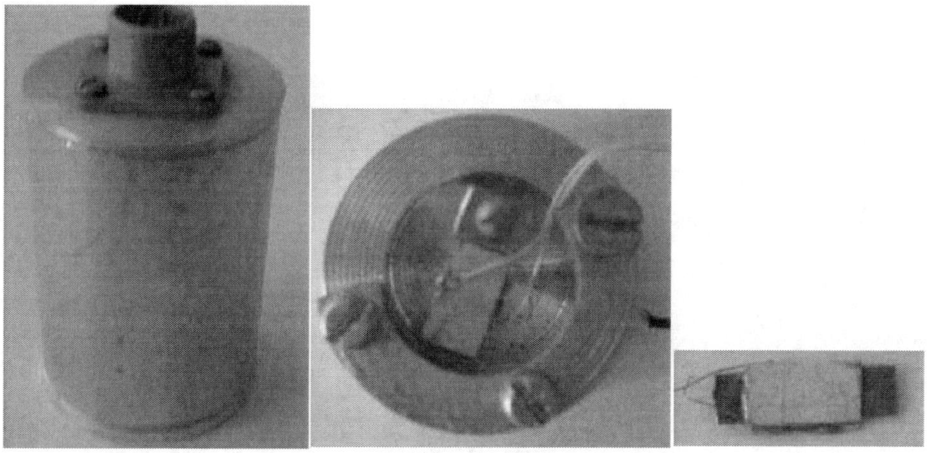

Figure 14. Cantilever type piezogenerator, working on bending oscillations, from left to right: assembled cantilever module, its form with deleted cup of hull, and PEG piezoelement.

The block of improved PEG model (see Figure 15) was mounted in piezoelectric energy harvesting device with original construction (see Figure 16), that allowed us to obtain prospective results. In any case, a power generated by this PEG was enough for feeding two light-emitting diodes (LEDs) during some time. In Figure 16, presenting the piezoelectric energy harvesting device, the lighting LEDs (5) of green color are well seen.

The modernized model of PEG was also mounted in the piezoelectric energy harvesting device with impact impulse exciter of constrained oscillations (see Figure 17). In the figure, it is also seen lighting LED (6). Some physical parameters for the cantilever type PEG are shown in Figures 18 – 21. We are planning more intensive investigations of this model piezoelectric energy harvesting device.

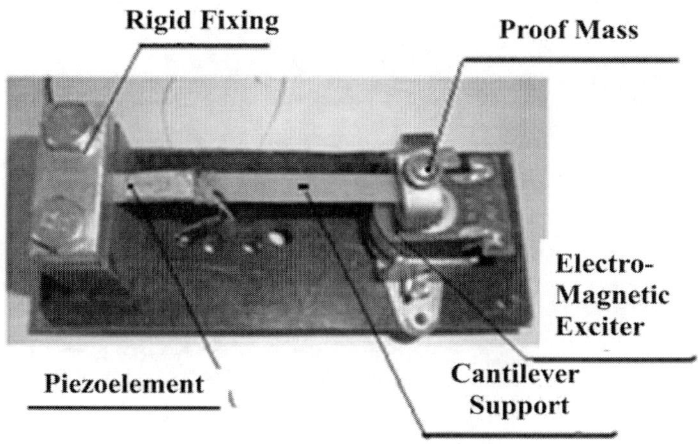

Figure 15. Block of cantilever type PEG with a proof mass (piezoelement from PZT-19 with sizes of 15 × 7 × 0.5 mm^3).

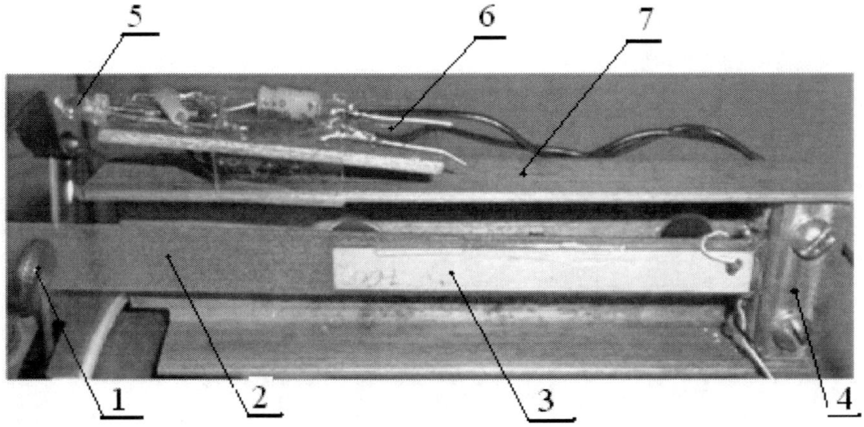

Figure 16. Cantilever type PEG with mechanical excitation. Digits design: 1 – mechanical plunger, 2 – support of bimorph, 3 – bimorph plate of generator, processed from PZT-19 with sizes of 60 × 10 × 0.5 mm^3 and d_{33} = 690 × 10^{-12} pC/N, 4 – bimorph rigid fixing, 5 – LED, 6 – standard piezoharvesting device, 7 – generator hull.

Besides above cantilever type PEG (see Figure 15), we R&D several laboratory samples of PEGs, working on longitudinal mode of exciting the mechanical oscillations (see Figures 22). The piezogenerator, processed on the base of high-sensitive multilayered PE, differs by high values of fàctors on charge or voltage. Increasing the PEG power could be reached owing to increasing the additive mass (inertial element), selection of PM with high values of corresponding constants and growth of number of the parallel-switched on layers of piezoelements. The PEG consists of 11-layered piezoelements, two metalized insulators and inertial element with proof mass of 200 g. The construction of the PEG is presented in Figure 22.

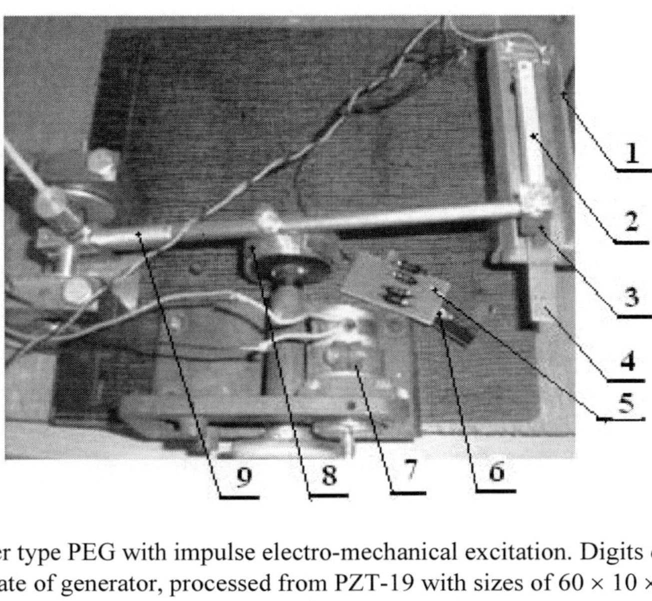

Figure 17. Cantilever type PEG with impulse electro-mechanical excitation. Digits design: 1 – PEG hull, 2 – bimorph plate of generator, processed from PZT-19 with sizes of 60 × 10 × 0.5 mm^3 and d_{33} = 690 × 10^{-12} pC/N, 3 – mechanical plunger, 4 – woven-glass plate of bimorph, 5 – standard piezo-storage device, 6 – LED, 7 – electromotor, 8 – ratchet-wheel, 9 – mechanical plunger bar.

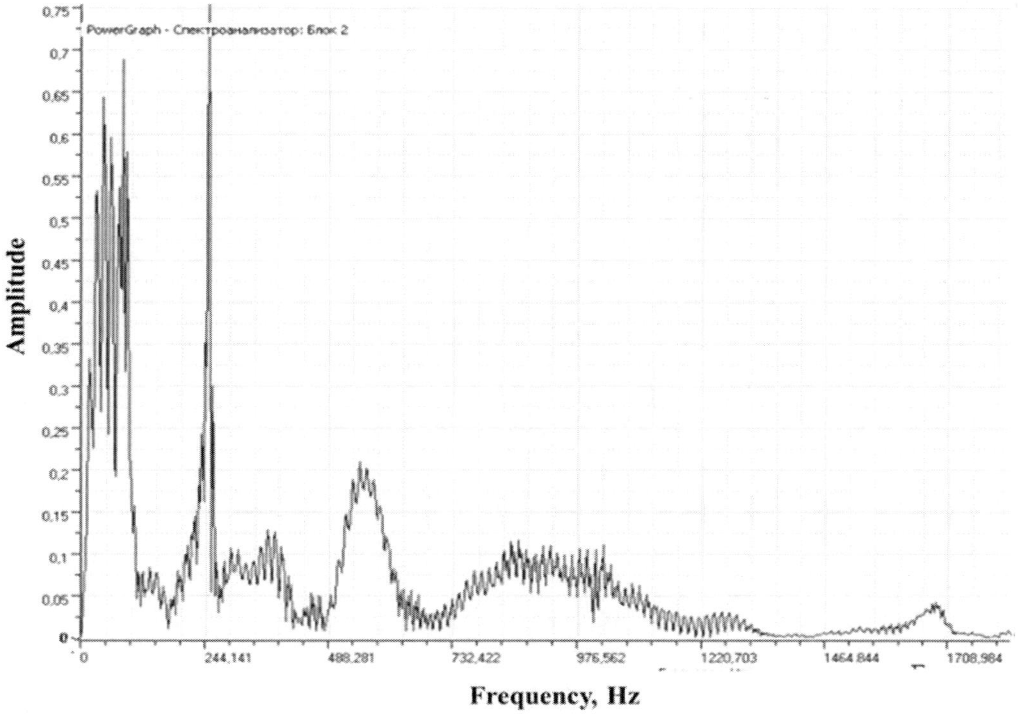

Figure 18. Signal spectrum at excitation of cantilever type PEG at active load of storage in 1 MΩ.

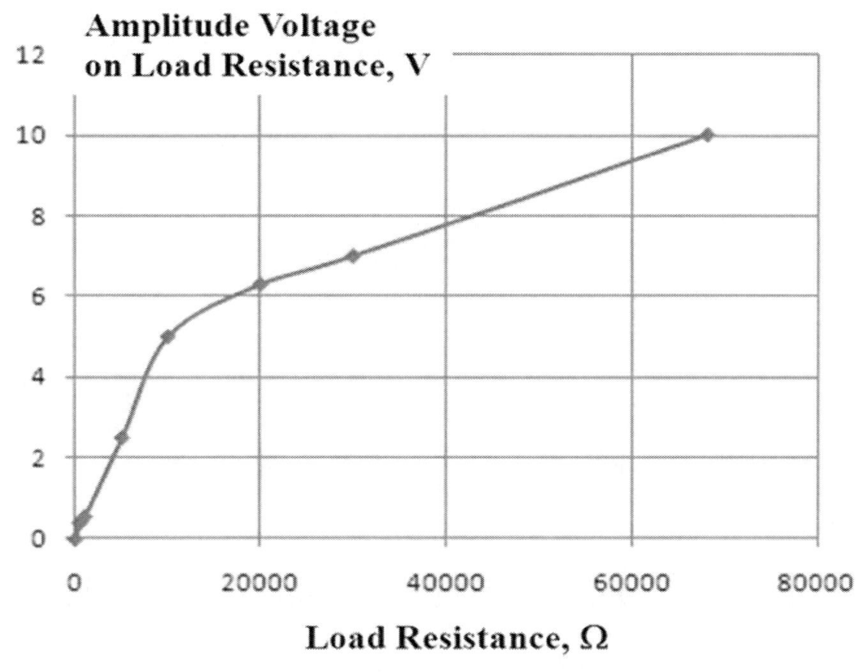

Figure 19. Dependence of amplitude voltage of cantilever type PEG on active load at impact loading.

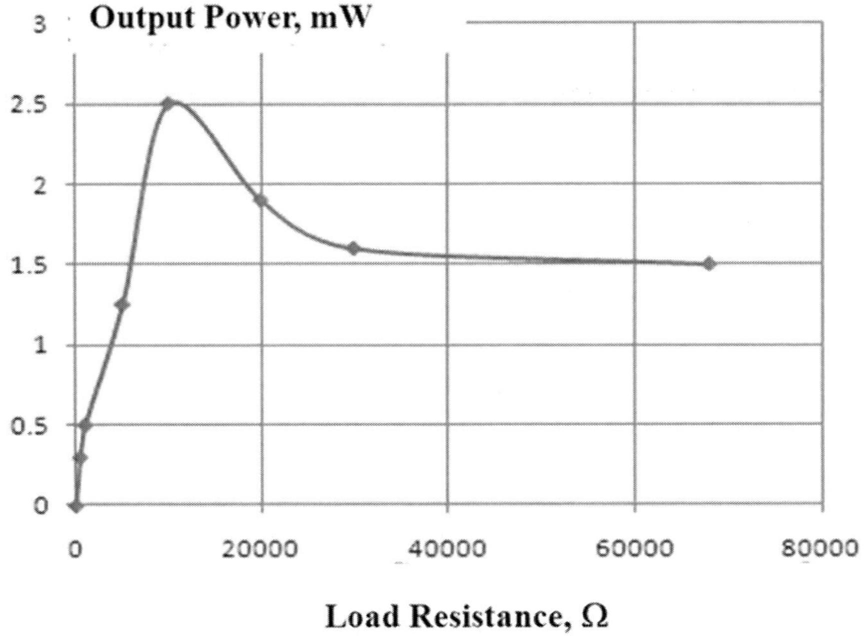

Figure 20. Dependence of output power for cantilever type PEG on active load.

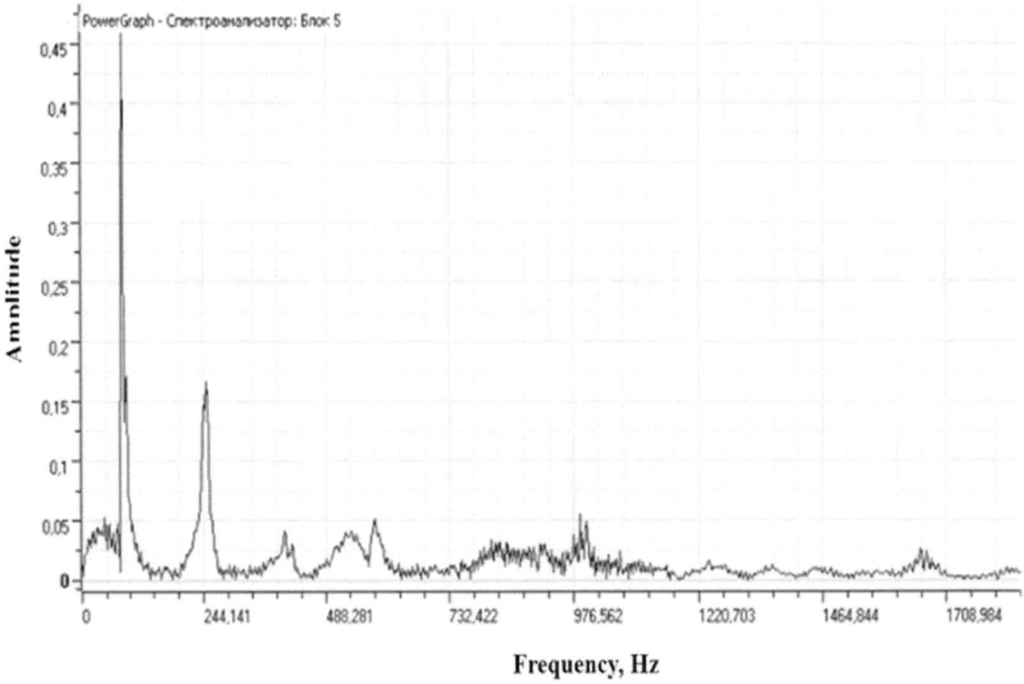

Figure 21. Signal spectrum at excitation of cantilever type PEG by mechanical plunger at active load of storage by capacity of 1.8 F.

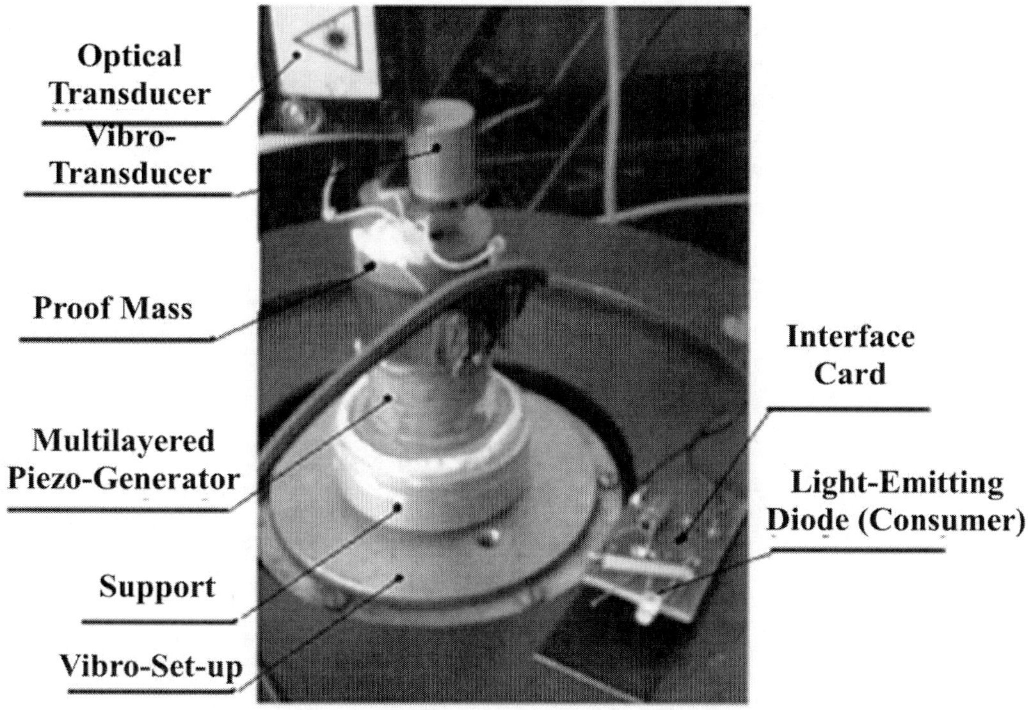

Figure 22. Multilayered PEG, working on longitudinal oscillations (includes 11 piezoelements from PZT-19); output power is equal to 0.34 mW.

Figure 23. Experimental set-up: 1 – multilayered sensitive PE of generator, working on longitudinal oscillations, 2 – vibro-set-up "Robotron-11077", 3 – acceleration transducer "ADXL-103", 4 – optical transducer of displacements "optoNCDT-2200", 5 – optical microscope for measurement of displacements, 6 – ADP module.

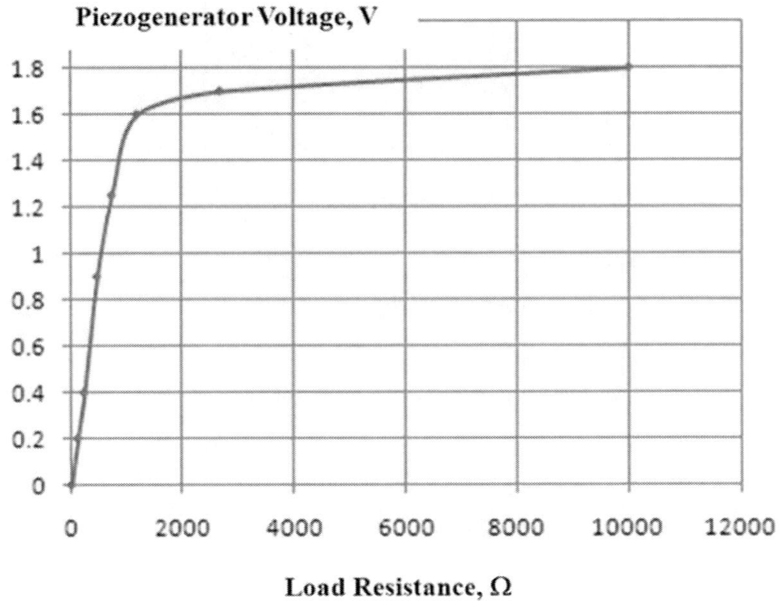

Figure 24. Dependence of voltage of the multilayered PEG, working on longitudinal oscillations, on active load.

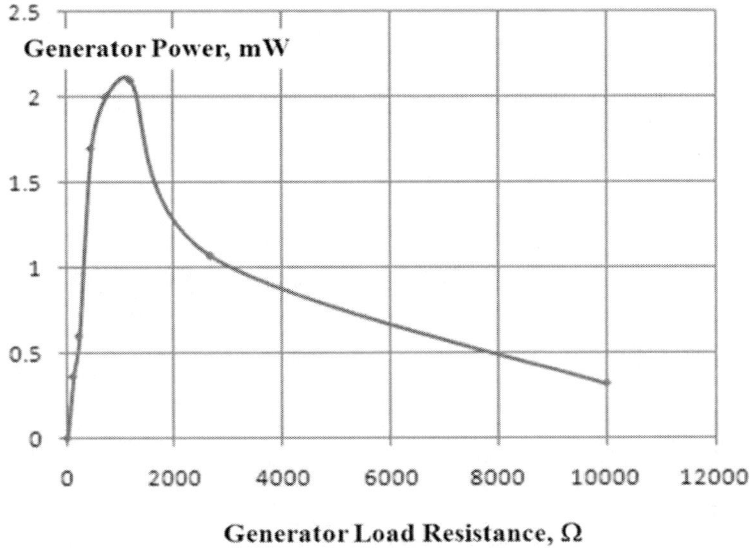

Figure 25. Dependence of output power of the multilayered PEG on active load.

In order to conduct studies of output characteristics of the different type PEGs, it has been created original experimental set-up (see Figure 23). As exciter of mechanical oscillations, it has been used the well-known vibro-set-up "Robotron-11077". Current values of accelerations, and axis displacements were registered by using sensors, depicted in the Figure, and measurement microscope. In measurement tract of the set-up, it has been realized the standard electric scheme of the piezoelectric energy harvesting device (see Figure 2). In detail, the scheme and measurement algorithm of output characteristics have been presented in our paper [17]. The set-up structure and components of measurement tract are understood from Figure 23.

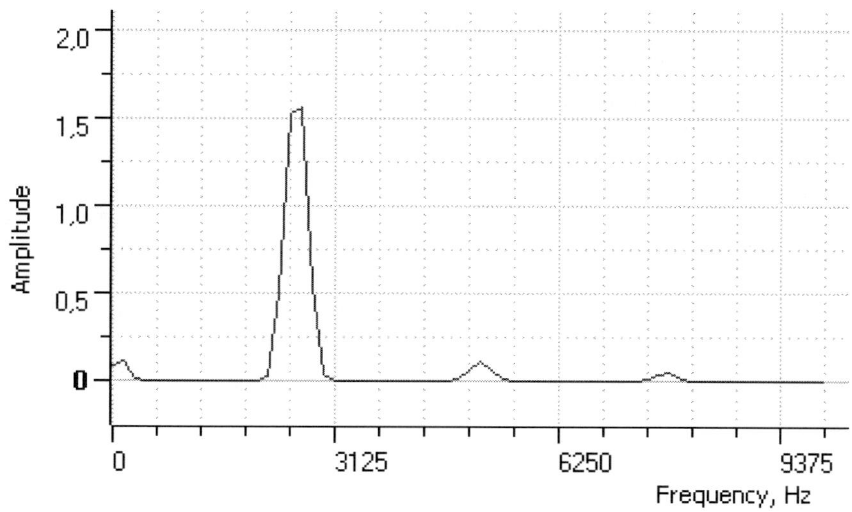

Figure 26. Spectrum of output signal of the multilayered generator.

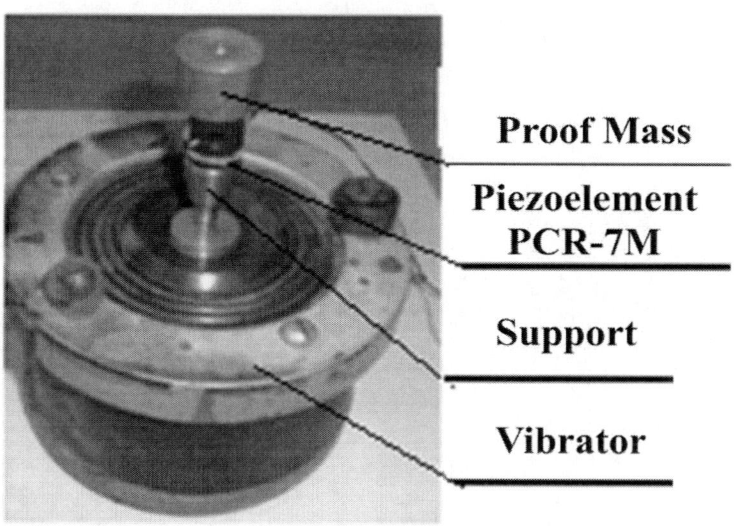

Figure 27. Piezogenerator, working on longitudinal oscillations (single-layered PE from PCR-7M).

By using the set-up, it has been measured output voltage and corresponding load resistance of test PEG sample, depicted in Figure 22. The results of the measurements have been obtained and the dependencies of output voltage on electric load resistance (Figure 24), and also output power of the generator (Figure 25) constructed on their base. Spectrum of output signal is shown in Figure 26. From analysis of the output signal spectrum, it is followed that most output power of the generator is reached at the frequency of 2655 Hz. At this frequency, the dependencies of voltage and electric power on load have been obtained.

Analysis of dependence of the output power on electric load showed that the test sample of the generator also allowed one to obtain energy corresponding to power of 2.25 mW at the load resistance about 1 kΩ. The power level of the generator was no worse than for the known analogue [15]. It is possible that at transition to other electric scheme of the piezoelectric energy harvesting device, the output characteristics of the multilayered PEG could be significantly improved.

During research, we developed test sample of PEG (see Figure 27) with single PE of the same type that in Figure 22, but in difference on it the piezoplate in the new piezotransducer has been processed from PCR-7M with piezomodulus d_{33}, which in two times exceeds the piezomodulus of PZT-19 [16].

By using the set-up, output characteristics of the model PEG have been measured. The measurements allowed us to construct spectrum of output signal (Figure 28), dependencies of output voltage (Figure 29), and also output power of generator (Figure 30) on electrical resistance of load.

From the spectrum of output signal, it is followed that most output power of signal be at frequency of 1718 Hz. Therefore, the plots of voltage and power were constructed at this frequency. Analysis of the dependencies showed that this model of PEG allows one to obtain output power which is equal to 5.7 μW, that is, lower than previous PEG model on three orders of magnitude. This shows that material of PEG sensitive element influences output power comparatively in insignificant degree. Significantly greater role for reaching higher output power of PEG plays number of piezoelements in multilayered transducer. It has been shown by comparative tests of the developed PEG models (see Figures 22 and 27).

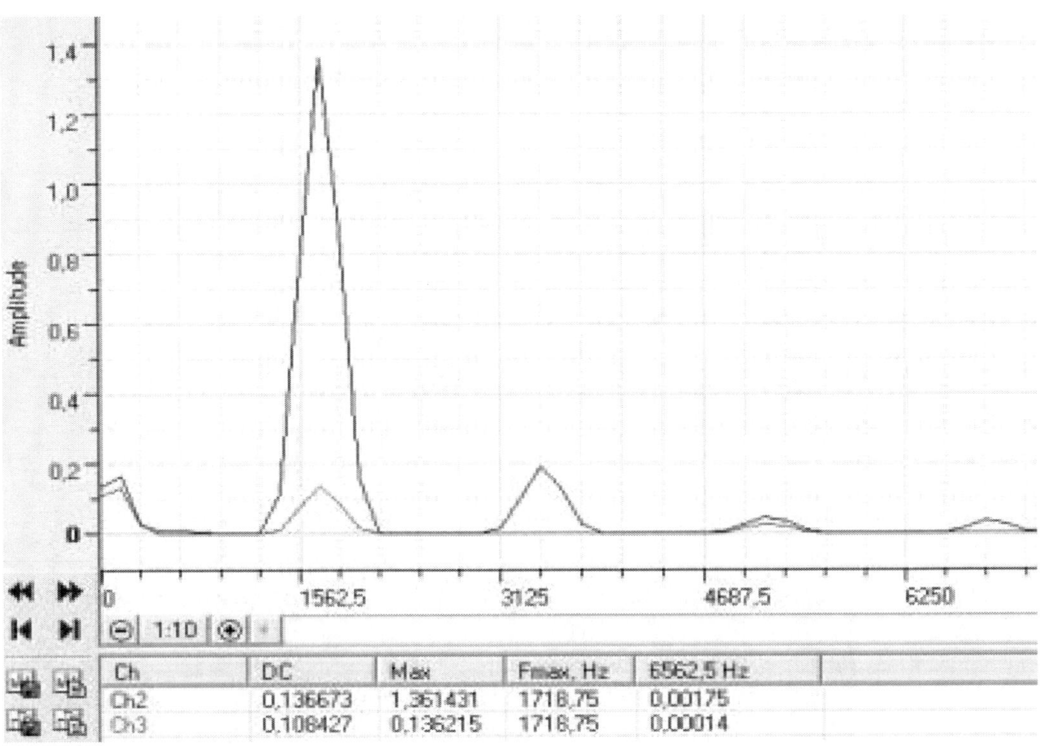

Figure 28. Spectrum of PEG output signal.

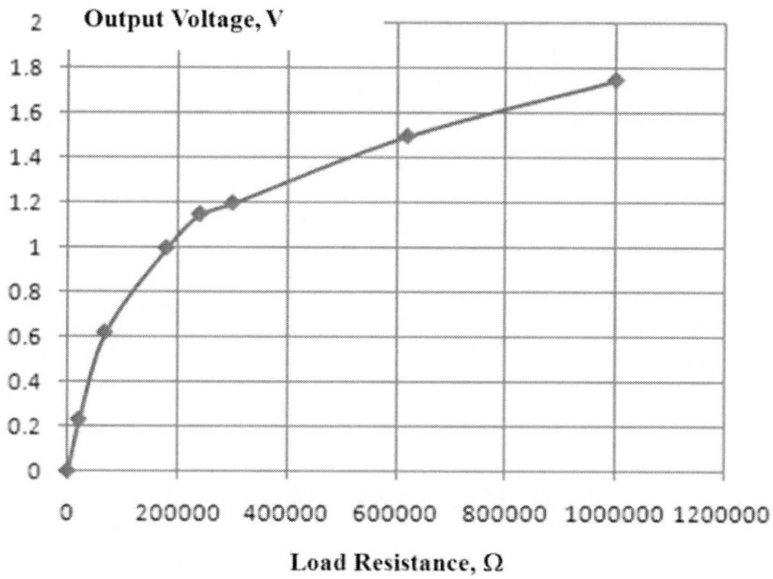

Figure 29. Dependence of PEG output voltage on active load.

Finally, it should be noted, that earlier difference of one of the electric schemes of SSHI realization, described in [24], consisted in inclusion in it in series of switch, inductivity and resistance (see [24], p. 896). Due to, the maximal amplitudes of the voltage and displacement have been combined in on-line regime. This allowed an increasing output power of piezoelectric energy harvesting device. Therefore, the next investigations require not only this approach to solution of the problem of increasing PEG output power, but other ones, discussed in [15].

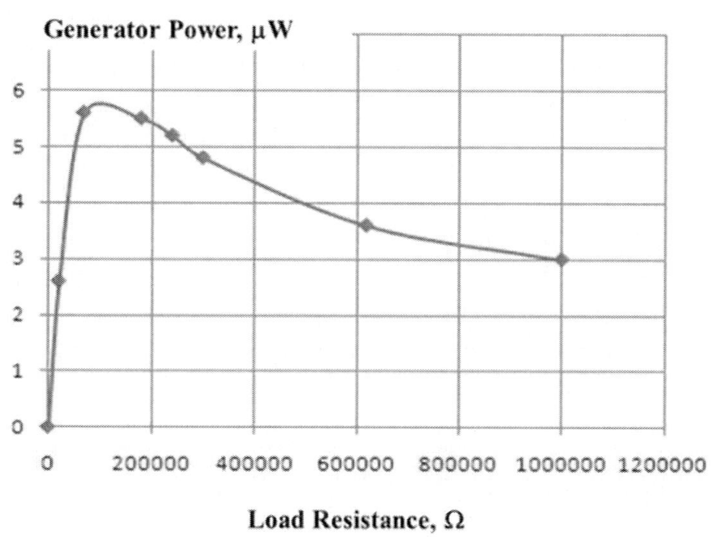

Figure 30. Dependence of PEG output power on active load.

The above brief analysis of research results of the piezoelectric energy harvesting devices does not pretend on final presentation of total results in this area. One only discussed the characteristics of some original PEG models and moreover presented introductory in some problems, which in our opinion should be interesting for specialists in this region.

ACKNOWLEDGEMENT

This work has been carried out at partial support of Russian Foundation for Basic Research (Projects Nos. 10-08-00093-a, 10-08-00136-a, 10-08-13300 RT_omi, 12-08-01137-a).

REFERENCES

[1] Sodano, H. A.; Inman, D. J.; Park, G. *Shock and Vibration Digest*. 2004, *vol. 36*, 197–205.
[2] Anton, S. R.; Sodano, H. A. *Smart Mater. Struct*. 2007, *vol. 16*, R1–R21.
[3] Priya, S. J. *Electroceram*. 2007, *vol. 19*, 165–182.
[4] Cook-Chennault, K. A.; Thambi, N.; Sastry, A. M. *Smart Mater. Struct*., 2008, *vol. 17*, 043001.
[5] Proc. 7th Brazilian Conf. on Dynamics, Control and Applications. "Dincon, 2008". www.dem.feis.unesp.br.
[6] Rupp, C. J.; Evgrafov, A.; Maute, K.; et al. *J. Intel. Mater. Syst. Struct.* 2009, *vol. 20*, 1923-1939.
[7] Panich, A. E.; Sokallo, A. I.; Grishin, E. A.; et al. *Microsystem Technique*, 2002. *No. 4*, 60-64.
[8] Howells, C. A. *Energy Conversion and Management*. 2009, *vol. 50*, 1847–1850.
[9] Grohmann, B.; Maucher, C.; Prunhuber, T.; et al. In *Proc. First Int. Symp. Demass 1. Turin, Italy*, 2006.
[10] Abramovich, H.; Harash, E.; Milgrom, C.; Azulay, L.E.; Amit, U. Piezoelectric stack compression generator. US Patent Application 20110291526. 01.12.2011.
[11] Sadano, H. A. *Power Harvesting Using Piezoelectric Materials*. http://cap.ee.ic.ac.uk/pdm 97/2007.
[12] Jeon, Y. B.; Sood, R.; Jeong, J.-H.; et al. *Sensors and Actuators. A. Physical*. www.elsever.com/locate/sna
[13] Shu, Y. C.; Lien, I. C.; Wu, W. J. *Smart Mater. Struct*. 2007, *vol. 16*, 2253-2264.
[14] Anton, S. R.; Erturk, A.; Inman, D. J. In: *Proc. of SPIE*. 2010. p. 7643.
[15] Erturk, A.; Inman D. J. *Piezoelectric Energy Harvesting*. John Wiley & Sons, Ney Jersey, 2011, pp. 1–276.
[16] Lefeuvre, E.; Badel, A; Richard, C.; et al. *DTIP of MEMS and MOEMS*, Stresa, Italy, 26-28 April, 2006, 6 p.
[17] Akopyan, V. A.; Zakharov, Yu. N.; Parinov, I. A.; et al. *Nano- and Microsystem Technique*. 2012, *No. 3*, 42–47.

[18] Akopyan, V. A.; Zakharov, Yu. N.; Matveev, S. Yu.; et al. *Nano- and Microsystem Technique*. 2012, *No. 1*, 37–41.
[19] PHYSORG. Louisiana University http://phys.org/news191502442.html.
[20] De Marqui, C. Jr.; Erturk, A.; Inman, D. J. *J. Sound and Vibrations*. 2009, *vol. 327*, 9–25.
[21] Elvin, N. G.; Elvin, A. A. *J. Intel. Mater. Syst. Struct*. 2009, *vol. 20*, 587–595.
[22] Yang, Y.; Tang, L. J. *Intel. Mater. Syst. Struct*. 2009, *vol. 20*, 2223 – 2235.
[23] Akopyan, V. A.; Soloviev, A. N.; Parinov, I. A.; Shevtsov, S. N. *Definition of Constants for Piezoceramic Materials*. Nova Science Publishers Inc., N.-Y., 2010, pp. 1 – 205.
[24] Shevtsov, S. N.; Akopyan, V. A. In: *Piezoceramic Materials and Devices*, Ivan A. Parinov (Ed.). Nova Science Publishers Inc., N.-Y., 2010, pp. 286 – 323.
[25] Badel, A.; Guyomar, D.; Lefeuvre, E.; et al. *J. Intel. Mater. Syst. Struct*. 2005, *vol. 16*, 889–901.

In: Physics and Mechanics of New Materials …
Editors: Ivan A. Parinov and Shun-Hsyung Chang
ISBN: 978-1-62618-535-7

Chapter 24

ANALYSIS OF FINITE ELEMENT MODELS FOR PIEZOELECTRIC DEVICES OF ENERGY HARVESTING

A. N. Soloviev*[1,2,3,4,5]*, I. A. Parinov*[3]*, L. V. Duong*[2,6]*,
***C. C. Yang*[7]*, S. H. Chang*[7] *and J. C. Y. Lee*[7]**

[2]Department of Strength of Materials, Don State Technical University, Rostov-on-Don, Russia
[3]Vorovich Mechanics and Applied Mathematics Research Institute, Southern Federal University, Rostov-on-Don, Russia
[4]Department of Mathematics, Mechanics and Computer Sciences, Southern Federal University, Rostov-on-Don, Russia
[5]Department of Mechanics of Active Materials, Southern Scientific Centre of Russian Academy of Sciences, Rostov-on-Don, Russia
[6]Department of Mechanical Engineering, Le Quy Don Technical University, Hanoi, Vietnam
[7]Department of Microelectronics Engineering, National Kaohsiung Marine University, Kaohsiung City, Taiwan

ABSTRACT

The paper presents finite element modeling (FEM) of axial-type piezoelectric device for energy harvesting. The considered device has axial construction, where an active element consists of one or several piezoelectric layers. It is used FEM in the Lagrange's classical formulation to solve dynamic problems of acoustic electric elasticity. Numerically, the paper studies an effect of ratio of the heights of piezoceramic layer and inertial mass at different ratios of material densities of the piezoceramics and inertial mass, and also compares the models with and without account of strain of the inertial mass. The modeling is carried out by using the finite element software ANSYS and ACELAN.

---

[1] solovievarc@gmail.com.

1. INTRODUCTION

Today the requirements of mobility and energetic independence become to be first-ordered in creation of sensor and measurement systems. There are numerous ways to solve this actual problem. They covers a decreasing energetic consumption of sensors and transducers included into a measurement system, manufacture of accumulators and batteries with super-prolonged service time and transfer, for example super-high-frequency magnetic or light energy by using air channels. The autonomy could be also reached owing to statement on sensor device of the element, which can excite and accumulate external energy. The wide-known devices of this type present solar cell panels, which use a light energy, electromagnetic transducers of mechanical energy, etc. The piezoelectric devices of energy harvesting are alternative energy sources, which allow us to collect energy of small-power sources from environment, for instance due to a vibration of oscillating bridges, moving mechanisms, etc. In the last time, the energy harvesting devices, in particular piezoelectric ones, have been intensively studied [1-22].

Typical structures of the piezoelectric devices of energy harvesting use a deformation of piezoelement, caused by the main regimes of energy transformation, namely due to

(i) a tension-compression along a polarization axis,
(ii) a shear along the same axis,
(iii) a tension-compression in the direction perpendicular to the polarization axis, caused by a bending strain of the device.

The inertial mass is one of important elements of similar devices; it is fixed directly on piezoelement or on some elastic element, a destination of which is to ensure given amplitudes of deformations of active element. The devices of the third type are most studied ones, using as constructive element the cantilever beams or plates.

An equivalent mechanical model of this construction is shown in Figure 1 [3]; the model is reduced to the ordinary differential equation relatively of vertical coordinate of the inertial mass, $u(t)$:

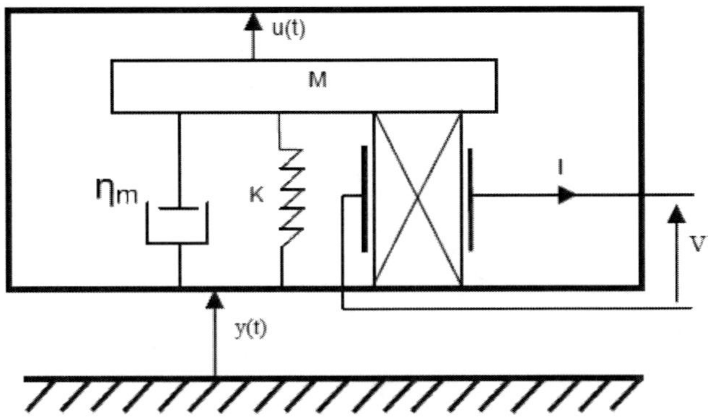

Figure 1. Equivalent model of piezoelectric devices of energy harvesting.

$$M\ddot{u} + \eta_m \dot{u} + Ku + K_p u + \alpha V = -M\ddot{y} \,. \tag{1}$$

The voltage, V, at the piezoelement electrodes and power of energy harvesting, P, are defined by using the vertical coordinate as [3]

$$\frac{V}{y} = \frac{jM\alpha R\omega^3}{(-M\omega^2 + j\eta_m\omega + K + K_p)(1 + jRC\omega) + j\alpha^2 R\omega};$$
$$P = \frac{VV^*}{2R} = \frac{y^2}{2} \frac{M^2\alpha^2 R\omega^4}{((\alpha^2 R + \eta_m)^2 + (RC\eta_m\omega)^2)}. \tag{2}$$

In this paper, it is studied the first type device, having an axial construction, in which an active element consists of one or several piezoelectric layers. Scheme and photo of this test device of low-frequency oscillations (the elaboration of E. V. Rozhkov) are depicted in Figure 2.

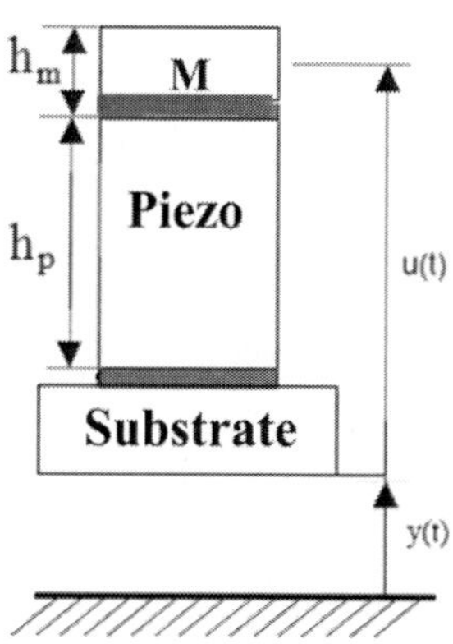

Figure 2. Energy storage piezoelectric device of axial construction.

A designing of high-tech devices for the energy harvesting is aimed to optimization of geometry and selection of materials for construction elements. However, this optimization is impossible without development of adequate models, namely: (i) analytical ones, considering one-axial movements, (ii) finite-element ones, when external influences have complex character, and (iii) simplified ones, when deformations of separate elements could be neglected. The paper studies an effect of ratio of the heights h_m, h_p (see Figure 2, a) at different ratios of material densities of the piezoceramics and inertial mass, and also compares the models with and without account of strain of the inertial mass. The modeling is carried out by using the finite element software ANSYS and ACELAN.

2. Continual and Finite-Element Statements of Problem

2.1. Continual Statement of Problem of the Acoustic Electric Elasticity

The axial-type piezoelectric device of energy harvesting (see Figure 2) presents itself a combined elastic and electroelastic solid, making small oscillations in mobile coordinate system, the movement of which is given by surface, to which the device base (for example $y(t)$) is fixed. In these conditions, enough adequate mathematical model of work of the device presents itself an initial-boundary problem into framework of linear theory of electroelasticity [23].

Let us consider a piezoelectric transducer Ω, presented by a set of areas $\Omega_j = \Omega_{pk}$; $k = 1, 2,..., N_p$; $j = k$ with the properties of piezoelectric materials, and a set of areas $\Omega_j = \Omega_{em}$; $m = 1,2,...,N_e$; $j = N_p + m$ with the properties of elastic materials. It is appropriate to describe the physical and mechanical processes taking place in the media Ω_{pk} and Ω_{em} in the framework of piezoelectricity (electric elasticity) and elasticity theory, respectively.

We assume that the following constitutive equations are satisfied for the piezoelectric media $\Omega_j = \Omega_{pk}$:

$$\rho_{pk}\ddot{\mathbf{u}} + \alpha_{dj}\rho_j\dot{\mathbf{u}} - \nabla\cdot\boldsymbol{\sigma} = \mathbf{f}_j;\ \nabla\cdot\mathbf{D} = 0; \tag{3}$$

$$\boldsymbol{\sigma} = \mathbf{c}_j^E\cdot\cdot(\boldsymbol{\varepsilon} + \beta_{dj}\dot{\boldsymbol{\varepsilon}}) - \mathbf{e}_j^T\cdot\mathbf{E};\ \mathbf{D} + \varsigma_d\dot{\mathbf{D}} = \mathbf{e}_j\cdot\cdot(\boldsymbol{\varepsilon} + \varsigma_d\dot{\boldsymbol{\varepsilon}}) + \mathbf{э}_j^S\cdot\mathbf{E}; \tag{4}$$

$$\boldsymbol{\varepsilon} = (\nabla\mathbf{u} + \nabla\mathbf{u}^T)/2;\ \mathbf{E} = -\nabla\varphi, \tag{5}$$

where $\rho(x,t)$ is the continuous function of coordinates (material density), $\mathbf{u}$(x) is the vector-function of displacements, $\boldsymbol{\sigma}$ is the stress tensor, $\mathbf{f}$ are the mass forces, $\mathbf{D}$ is the electric induction vector, $\mathbf{c}_j^E$ are the components of the elastic constant tensor, $\boldsymbol{\varepsilon}$ is the strain tensor, $\mathbf{E}$ is the electric field vector, $\mathbf{э}_j^S$ are the components of the dielectric permittivity tensor, $\varphi(x)$ is the electric potential function, α_{dj}, β_{dj}, ζ_d are the non-negative coefficients of damping.

For the media $\Omega_j = \Omega_{em}$ with pure elastic properties, only stress fields would be considered. Similar equations (3) – (5) and constitutive relationships are used with neglected electric fields and piezoelectric connectivity effects.

Finally, the piezoelectric device can be loaded by using the acoustic operating media $\Omega_j = \Omega_{al}$; $l = 1, 2,..., N_a$; $j = l + N_p + N_e$. For these media, we use acoustic equations with linear dissipative effects taken into account as

$$\frac{1}{\rho_j c_j^2}\dot{p} + \nabla\cdot\mathbf{v} = 0\quad \mathbf{v} = \nabla\psi; \tag{6}$$

$$\rho_j \dot{\mathbf{v}} = \nabla \cdot \boldsymbol{\sigma}\,; \quad \boldsymbol{\sigma} = -p\mathbf{I} + b\nabla\mathbf{v}\,, \tag{7}$$

where ρ_j are the equilibrium values of material densities, c_j are the sound velocities, b_j are the dissipative coefficients for the media $\Omega_j = \Omega_{em}$, p is the sound pressure, $\mathbf{v}$ is the vector of velocity, ψ is the velocity potential, $\boldsymbol{\sigma}$ is the stress tensor, $\mathbf{I}$ is the identity tensor.

It should be noted that Equations (3) – (7) define specific models in order to take into account a damping factor. The Rayleigh model of the damping is used for the elastic media. The model (3) – (5) is generalized to the piezoelectric media.

In addition to Equations (1) – (5), it is necessary to define appropriate principal and natural boundary conditions, the conditions of field matching along the contact boundaries of different media, the impedance conditions for the 'truncation' of acoustic areas and initial conditions of non-stationary problems [24, 25].

In particular, for an electrode (S_e), included in external circuit, besides the condition of constancy of the electric potential, which presents itself unknown function in this case, it is necessary to add the condition, defining electric current passes through this electrode:

$$\int_{S_e} \dot{D}_m ds = I\,. \tag{8}$$

2.2. FEM Formulation

We use FEM in the Lagrange's classical formulation to solve dynamic problems of acoustic electric elasticity. We choose coherent finite-element mesh specified into areas Ω_{hj}, which approach the areas Ω_j. There are unknown field functions $\mathbf{u}$, φ and ψ in this mesh. We define them as

$$\mathbf{u}(\mathbf{x},t) = \mathbf{N}_u^T(\mathbf{x}) \cdot \mathbf{U}(t)\,; \;\; \varphi(\mathbf{x},t) = \mathbf{N}_\varphi^T \cdot \boldsymbol{\Phi}(t)\,; \;\; \psi(\mathbf{x},t) = \mathbf{N}_\psi^T(\mathbf{x}) \cdot \boldsymbol{\Psi}(t)\,, \tag{9}$$

where $\mathbf{N}_u$ is the shape function matrix for the displacement field $\mathbf{u}$; $\mathbf{N}_\varphi$ and $\mathbf{N}_\psi$ are the shape vector-functions for the electric potential fields φ and speed potential in acoustic medium ψ, respectively; $\mathbf{U}(t)$, $\boldsymbol{\Phi}(t)$, $\boldsymbol{\Psi}(t)$ are the global vectors of the corresponding nodal degrees of freedom.

FEM approximations (9) of the generalized formulations for dynamic problems (3) – (7), including principal and natural boundary conditions, reduce to the following system of differential equations:

$$\mathbf{M} \cdot \ddot{\mathbf{a}} + \mathbf{C} \cdot \dot{\mathbf{a}} + \mathbf{K} \cdot \mathbf{a} = \mathbf{F}\,, \tag{10}$$

$$\mathbf{M}=\begin{pmatrix}\mathbf{M}_{uu} & 0 & \widetilde{\mathbf{R}}_{u\psi}\\ 0 & 0 & 0\\ \widetilde{\mathbf{R}}^{T}_{u\psi} & 0 & -\mathbf{M}_{\psi\psi}\end{pmatrix};\qquad \mathbf{C}=\begin{pmatrix}\mathbf{C}_{uu} & 0 & \mathbf{R}_{u\psi}\\ \varsigma_{\mathrm{d}}\mathbf{K}^{T}_{u\varphi} & 0 & 0\\ \mathbf{R}^{T}_{u\psi} & 0 & -\mathbf{C}_{\psi\psi}\end{pmatrix};$$

$$\mathbf{K}=\begin{pmatrix}\mathbf{K}_{uu} & \mathbf{K}_{u\varphi} & 0\\ \mathbf{K}^{T}_{u\varphi} & -\mathbf{K}_{\varphi\varphi} & 0\\ 0 & 0 & -\mathbf{K}_{\psi\psi}\end{pmatrix};\quad \mathbf{F}=\begin{Bmatrix}\mathbf{F}_{u}\\ \mathbf{F}_{\varphi}+\varsigma_{\mathrm{d}}\dot{\mathbf{F}}_{\varphi}\\ 0\end{Bmatrix}, \tag{11}$$

relatively of the vector of unknown $\mathbf{a}=[\mathbf{U},\mathbf{\Phi},\mathbf{\Psi}]^{T}$. Here $\mathbf{C}_{uu}=\sum_{j}(\alpha_{dj}\mathbf{M}_{uuj}+\beta_{dj}\mathbf{K}_{uuj})$, where $\mathbf{M}_{uuj}$ and $\mathbf{K}_{uuj}$ are the structural finite elements of mass and stiffness matrices. Other elements of the sub-matrices (10), (11) are described in Reference [25].

The FEM equation system for *eigenvalue (eigenfrequency) problems* has the following form:

$$-\omega^{2}\mathbf{M}\cdot a+\mathbf{K}\cdot a=0, \tag{12}$$

$$\mathbf{M}=\begin{pmatrix}\mathbf{M}_{\mathrm{uu}} & 0\\ 0 & 0\end{pmatrix},\ \mathbf{K}=\begin{pmatrix}\mathbf{K}_{uu} & \mathbf{K}_{u\varphi}\\ \mathbf{K}^{T}_{u\varphi} & -\mathbf{K}_{\varphi\varphi}\end{pmatrix}. \tag{13}$$

Steady-state vibration problems arise, when $\mathbf{F}_{u}=\widetilde{\mathbf{F}}_{u}(\mathbf{x})\exp(j\omega t)$, $\mathbf{F}_{\varphi}=\widetilde{\mathbf{F}}_{\varphi}(\mathbf{x})\exp(j\omega t)$, $\mathbf{a}=\widetilde{\mathbf{a}}(\mathbf{x})\exp(j\omega t)$. It is easy to derive a system of linear algebraic equations from (10), (11) relatively of the amplitude vector $\widetilde{\mathbf{a}}$ in the form:

$$\mathbf{K}_{c}\cdot\widetilde{\mathbf{a}}=\widetilde{\mathbf{F}}_{c};\ \widetilde{\mathbf{F}}_{c}=[\widetilde{\mathbf{F}}_{u},\widetilde{\mathbf{F}}_{\varphi},0]^{T}, \tag{14}$$

$$\mathbf{K}_{c}=\begin{pmatrix}\mathbf{K}_{uuc} & \mathbf{K}_{u\varphi} & \mathbf{K}_{u\psi c}\\ \mathbf{K}^{T}_{u\varphi} & -\mathbf{K}_{\varphi\varphi c} & 0\\ \mathbf{K}^{T}_{u\psi c} & 0 & -\mathbf{K}_{\psi\psi c}\end{pmatrix}, \tag{15}$$

$$\mathbf{K}_{\eta\eta c}=-\omega^{2}\mathbf{M}_{\eta\eta}+i\omega\mathbf{C}_{\eta\eta}+\mathbf{K}_{\eta\eta};\ \eta=u,\psi;$$

$$\mathbf{K}_{u\psi c}=-\omega^{2}\widetilde{\mathbf{R}}_{u\psi}+i\omega\mathbf{R}_{u\psi};\ \mathbf{K}_{\varphi\varphi c}=\frac{1}{(1+i\omega\zeta_{d})}\mathbf{K}_{\varphi\varphi}.$$

3. Computer Simulation

3.1. Axi-Symmetric Model

Let's consider a cylindrical structure of energy storage piezoelectric device combined from piezoelement in the form of short cylinder and two transit elastic layers of the same radius. The piezoelement consists of piezoceramic polarized on thickness. The device sizes are presented in Figure 3. Its total height, by taking into account insulator, piezoelectric and inertial elements is fixed and equal to 7.5 mm.

So, an optimization of stress value is carried out by changing ratio of heights of the inertial h_m and piezoelectric h_p elements at the condition keeping a whole height of the energy storage piezoelectric device.

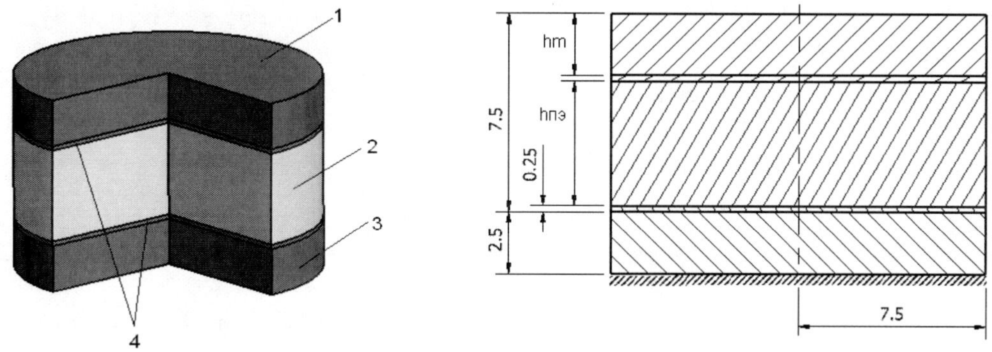

(*a*) (*b*)

Figure 3. Structure of energy harvesting piezoelectric device with cylindrical form: 1 – inertial element, 2 – piezoelectric element, 3 – substrate, 4 – insulators

Below, we give the material constants used in calculations.

Piezoceramic PZT-4:

$$[C]=\begin{bmatrix} 13.9 & 7.78 & 7.43 & 0 & 0 & 0 \\ 7.78 & 13.9 & 7.43 & 0 & 0 & 0 \\ 7.43 & 7.43 & 11.5 & 0 & 0 & 0 \\ 0 & 0 & 0 & 3.06 & 0 & 0 \\ 0 & 0 & 0 & 0 & 2.56 & 0 \\ 0 & 0 & 0 & 0 & 0 & 2.56 \end{bmatrix}\times 10^{10}\ \text{Pa};$$

$$[e]=\begin{bmatrix} 0 & 0 & -5.2 \\ 0 & 0 & -5.2 \\ 0 & 0 & 15.1 \\ 0 & 0 & 0 \\ 0 & 12.7 & 0 \\ 12.7 & 0 & 0 \end{bmatrix}\ \text{C/m}^2;$$

$$[\varepsilon] = \varepsilon_0 \begin{bmatrix} 730 & 0 & 0 \\ 0 & 730 & 0 \\ 0 & 0 & 635 \end{bmatrix}$$

$\varepsilon_0 = 8.85 \times 10^{-12}$ F/m ; $\rho = 7.5 \times 10^3$ kg/m^3 ; $Q = 200$ is the quality factor.

The elastic properties of isotropic materials are found by Young's modulus, E, and Poisson's ratio, ν.

Brass LS59:

$\rho = 8.8 \times 10^3$ kg/m^3; $E = 10.5 \times 10^{10}$ Pa; $\nu = 0.17$.

Some averaged characteristics have been selected for the connective layer, namely:

$\rho = 7.5.10^3$ kg/m^3; $E = 0.4.10^{10}$ Pa; $\nu = 0.3$.

External excitations have the form:

$$y = y_0 h_m e^{-i(2\pi f)t}, \tag{16}$$

where y_0 is the oscillation amplitude (we select $y_0 = 0.2$), f is the oscillation frequency.

The considered piezoelement consists of one layer of polarized piezoceramic material. The finite element model, taking into account axial symmetry of the construction, is demonstrated in Figure 4.

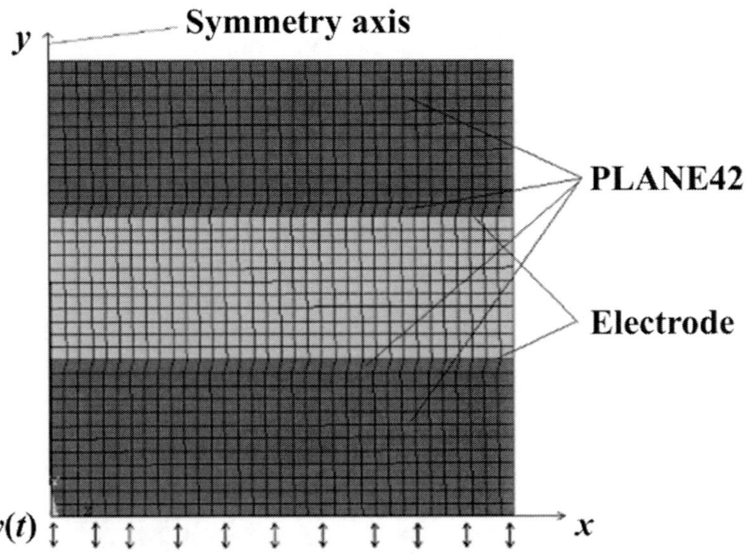

Figure 4. Axisymmetric finite element model of energy harvesting piezoelectric device.

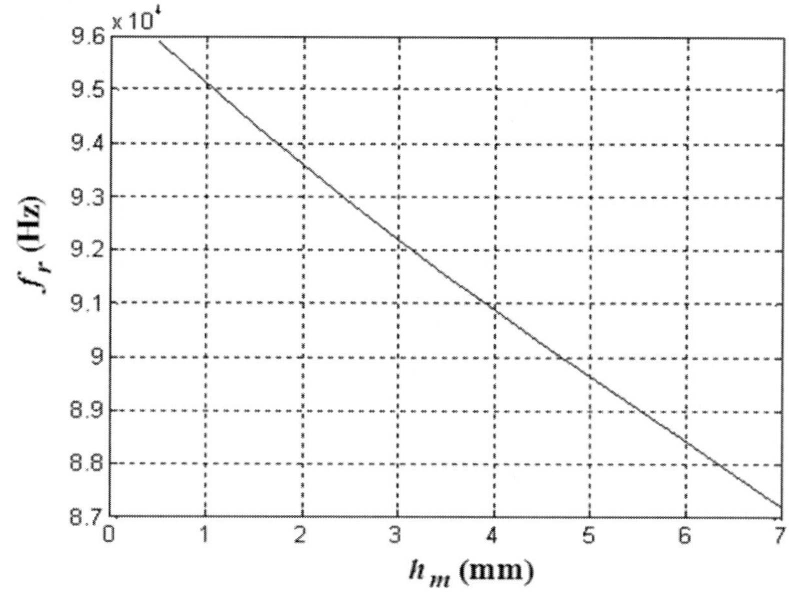

Figure 5. Dependence of eigenfrequency on height of inertial element.

Lower surface of substrate is fixed on radius (on x-axis), and vibro-accelerations act along y-axis with given frequency. The calculations are conducted by using finite element method into framework of the program software ANSYS. The study defines the values of h_m at which voltage V attains a maximal value.

Firstly, we consider influence of height h_m of the inertial element on eigenfrequency at condition of keeping a whole height of the piezoelectric device. The obtained dependence is shown in Figure 5. From this figure, it is followed that eigenfrequencies locate far higher than working frequencies, disposed into range of 87– 96 kHz at any acceptable ratio of heights of the inertial and piezoelectric elements.

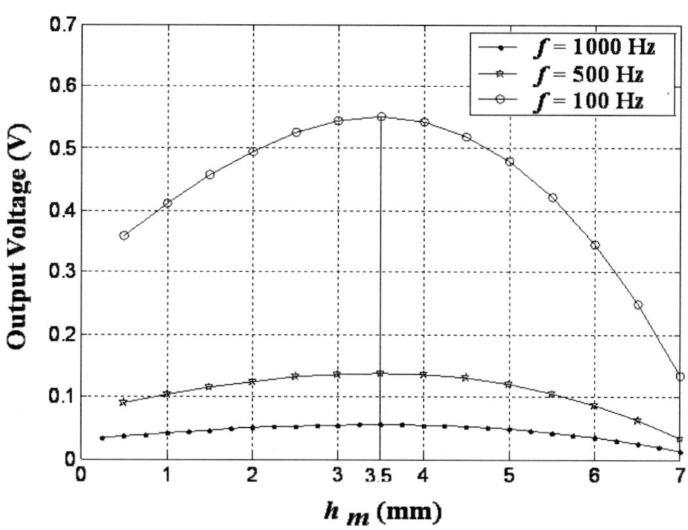

Figure 6. Dependences of voltage on height of inertial element at various eigenfrequencies.

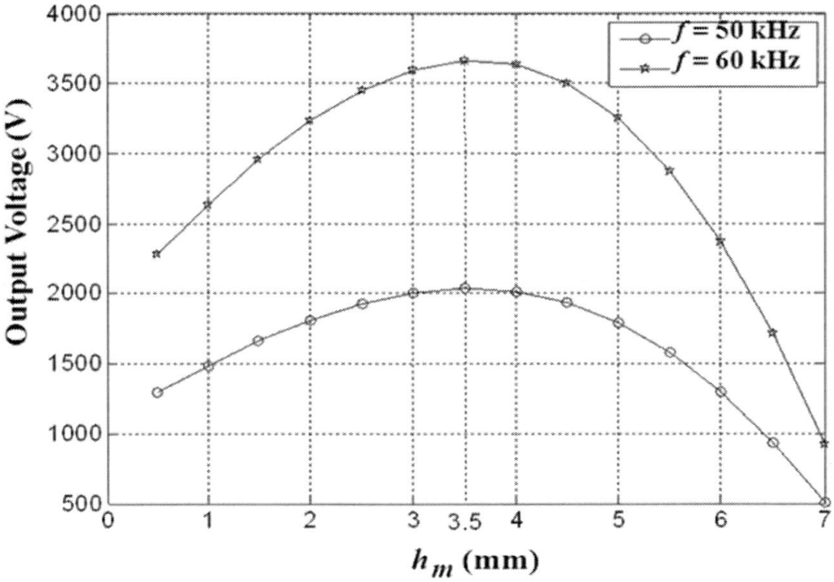

Figure 7. Dependences of voltage on height of inertial element.

Secondly, we find dependence of voltage V on the height h_m. This problem is solved for different excitation frequencies and the obtained results are depicted in Figures 6 – 9. The numerical results for the case, when the excitation frequency of oscillations disposes lower than eigenfrequency of the energy harvesting device, are shown in Figures 6, 7. From these figures we conclude that the values of voltage are maximal at the value of the height h_m = 3.5 mm.

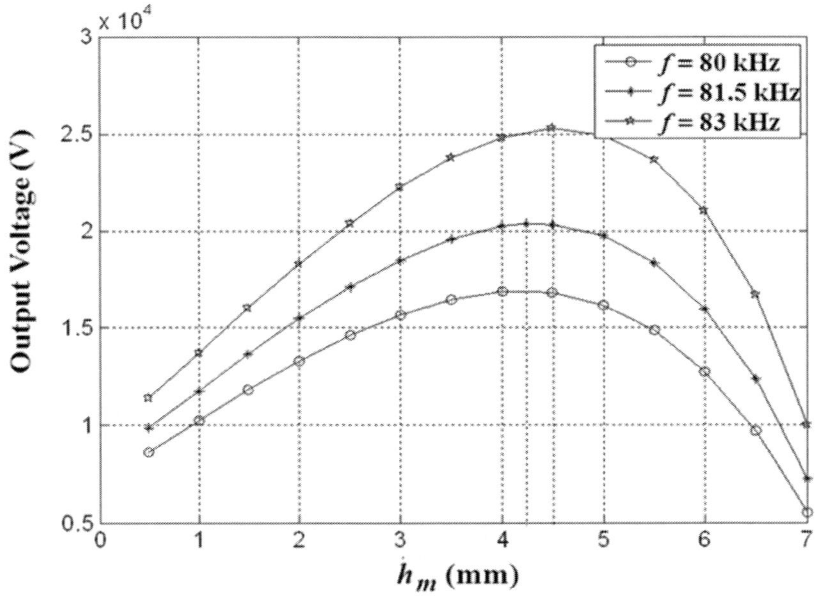

Figure 8. Dependences of voltage on height of inertial element.

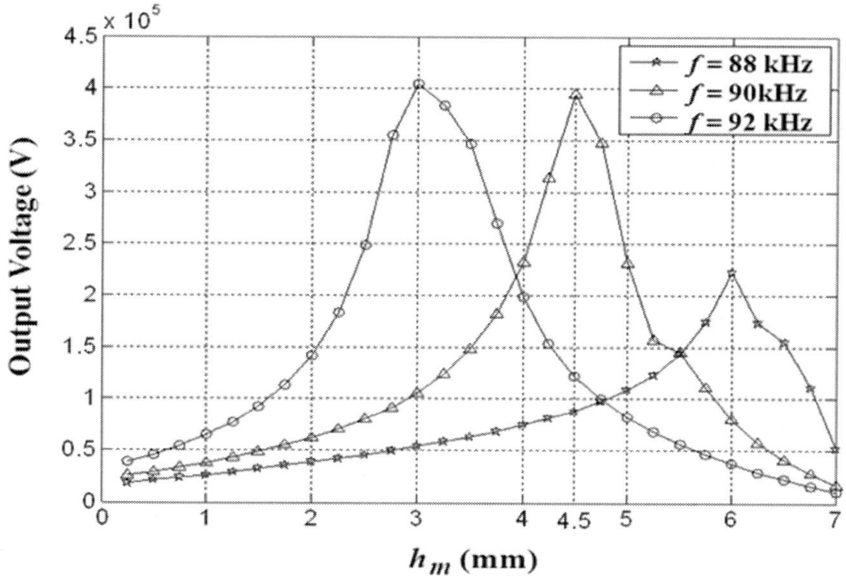

Figure 9. Dependences of voltage on height of inertial element.

The numerical results for the case, when the excitation frequency of oscillations disposes near of eigenfrequency of the energy storage device and lower its, are shown in Figure 8. From Figure 8, it is followed that the values of the height h_m, at which the voltage reaches maxima, depend on the value of excitation frequency. If the value of the excitation frequency increases then the height h_m also increases.

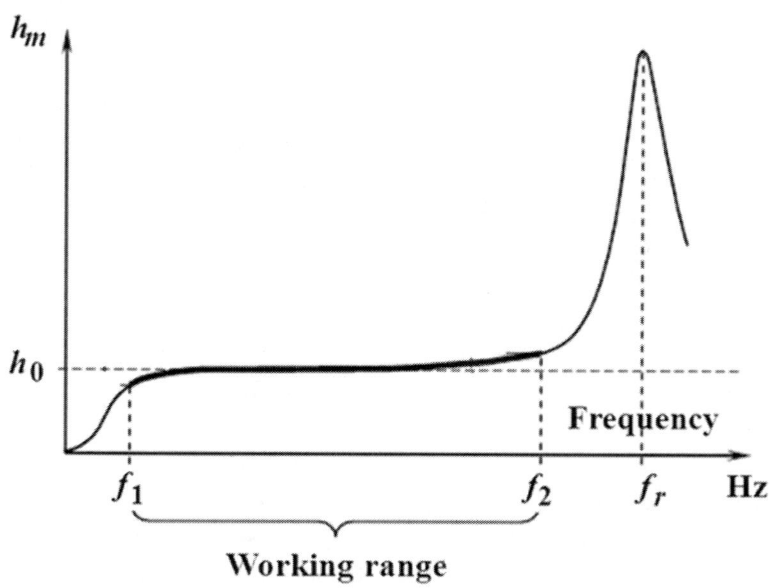

Figure 10. Dependence of height of the inertial element on excitation frequency; f_r is the resonance eigenfrequency.

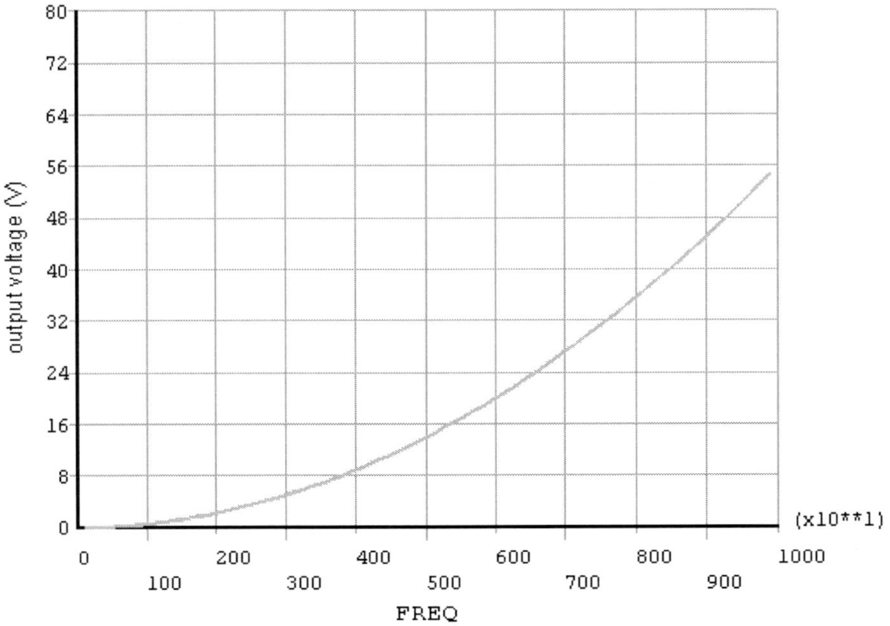

Figure 11. Dependence of voltage on vibration frequency.

The numerical results for the case, when the excitation frequency of oscillations disposes higher than eigenfrequency of the energy storage device, are shown in Figure 9. From Figure 9, it is followed that the values of the height h_m, at which the voltage reaches maxima, depend on the value of excitation frequency. If the value of the excitation frequency increases then the height h_m also increases.

By analyzing above results we can obtain the dependence of height h_m on the excitation frequency, which is presented in Figure 10.

Analysis of the results shows that the energy harvesting piezoelectric device with the height of inertial element equals to 3.5 mm demonstrates optimal parameters at the selected constraints for sizes. At the height h_m = 3.5 mm and external excitation of the form, $y = y_0 e^{-i(2\pi f)t}$, where y_0 = 1 mm is the oscillation amplitude, the frequency f disposes into range of $10^2 - 10^4$ Hz. The obtained results of dependence of stress on vibration frequency are presented in Figure 11.

3.2. Model of Plane Strain

Then on example of piezoelectric transducer, consisting of two layers (piezoceramic material and inertial mass, see Figure 3, b without lower element), it is considered an issue of applicability of simplified finite element model in plane statement, when a deformation of inertial element is neglected. In this case, the mechanical boundary condition at upper surface of the piezoelement has the form, taking into account this inertial mass:

$$\sigma_z = M\omega^2 (u_z + y_0)/S\,;\ \tau_{zx} = 0\,,\ \text{at}\ z = h_p\,, \tag{17}$$

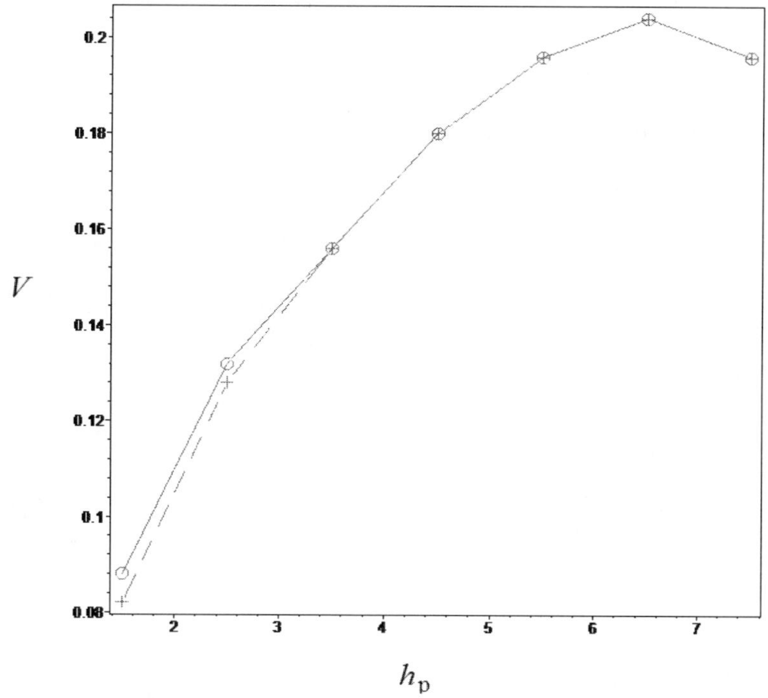

Figure 12. Dependence of voltage on height h_p for two considered models.

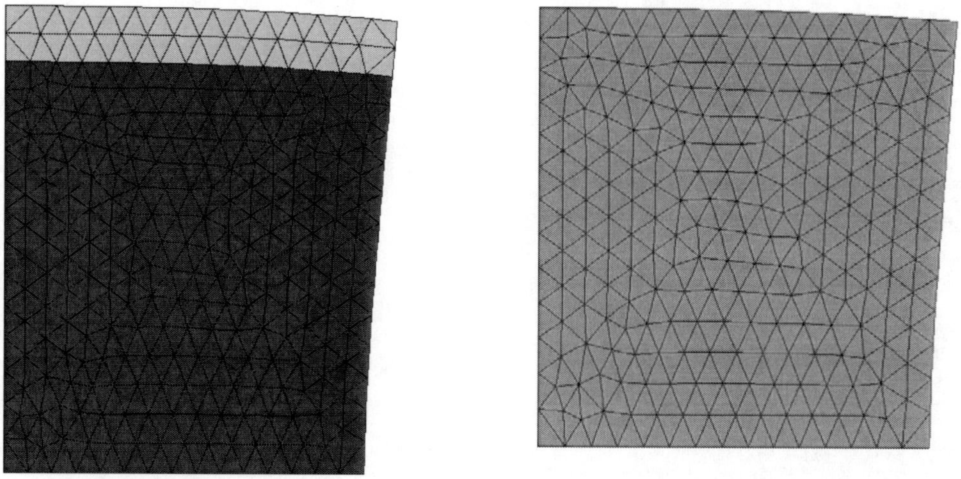

Figure 13. Deformed state of region for two models.

where M is the inertial mass, ω is the circular frequency of oscillations, S is the square of upper site of the piezoelement, to which the inertial mass is fixed.

Figure 12 presents dependencies of voltage on upper free electrode for frequency of constrained oscillations of 100 Hz (the calculations were carried out by using the program software ACELAN) on height of the inertial layer. The solid curve with light circulars corresponds to the finite element model with two layers; the dashed line with crosses corresponds to the model for piezoelectric layer and boundary conditions (17).

Figures 13 – 19 present the components of stress-strain and electric states of the piezoelement at $h_p = 6.5$ mm. In the all cases, the distributions of some parameters for the piezoelement are shown at the left sides of the figures, taking into account inertial mass in the finite element model, and the results for the model with boundary conditions (17) are presented at the right sides.

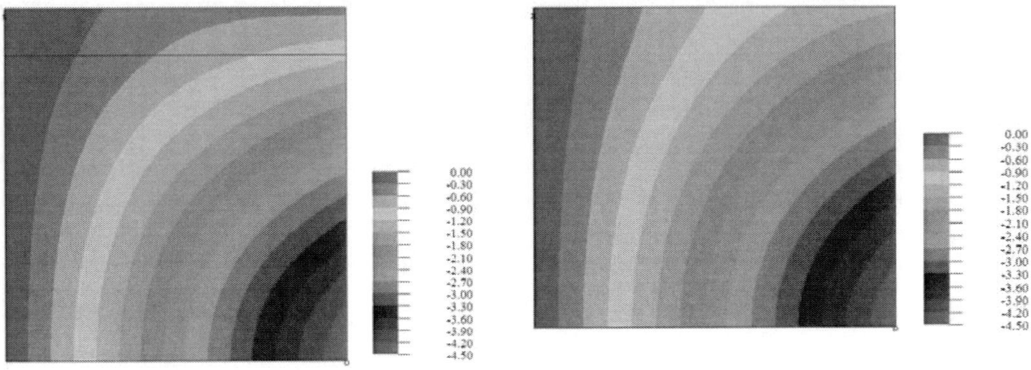

Figure 14. Distribution of horizontal displacement.

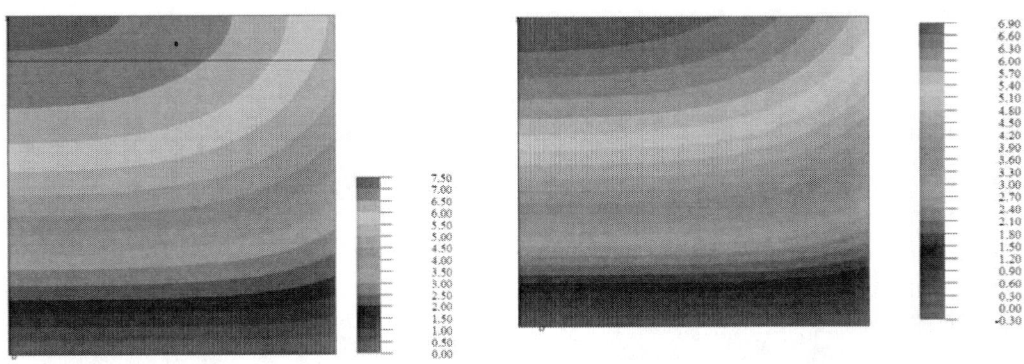

Figure 15. Distribution of vertical displacement.

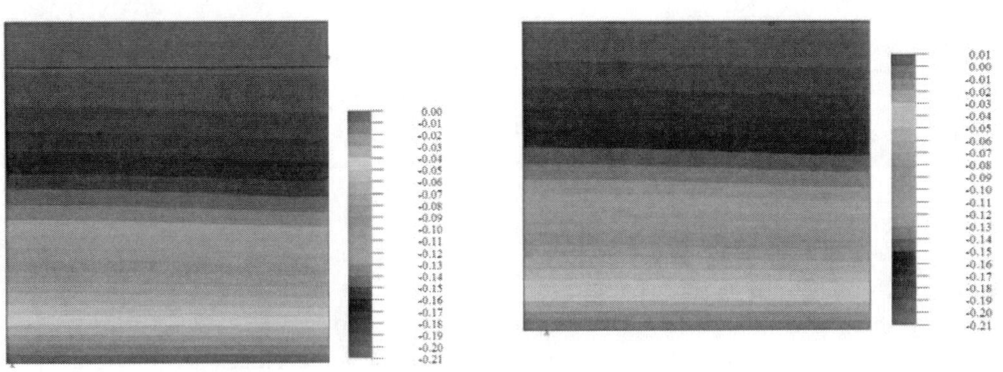

Figure 16. Distribution of voltage.

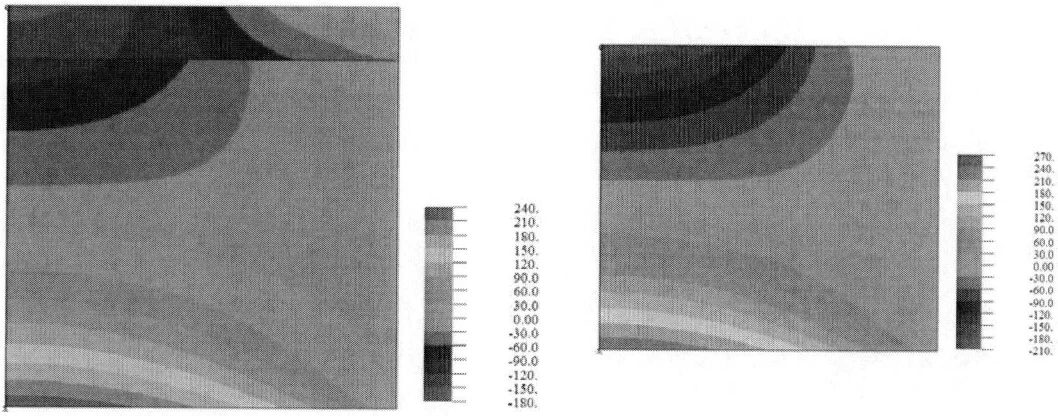

Figure 17. Distribution of mechanical stress σ_{11}.

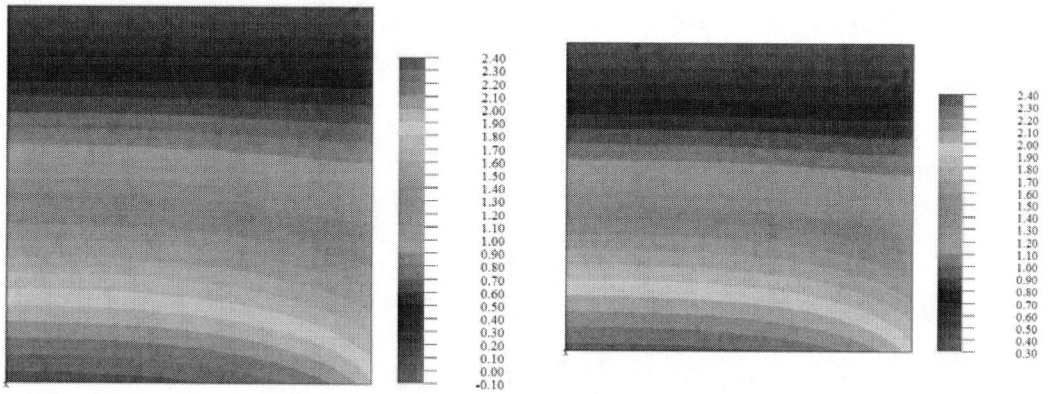

Figure 18. Distribution of mechanical stress σ_{33}.

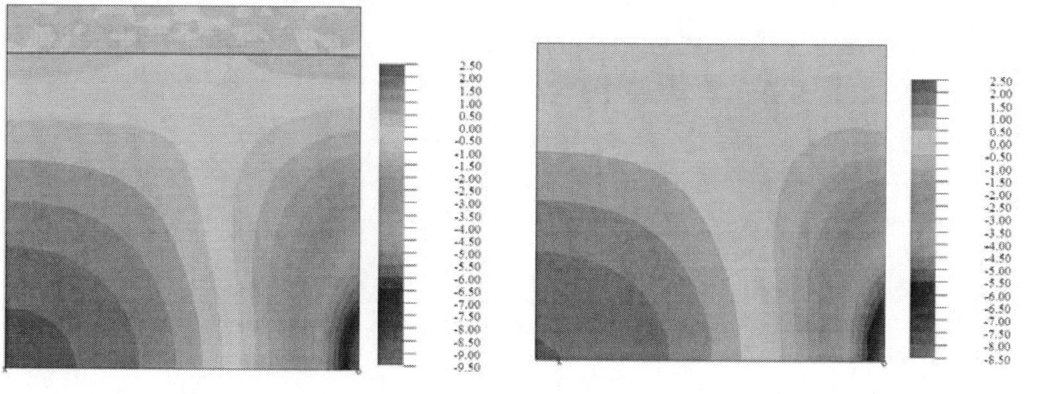

Figure 19. Distribution of vertical component of electric displacement vector.

The results of these models show enough well coincidence, especially in the value range, where the relative height of inertial mass is no great.

ACKNOWLEDGEMENT

This work has been carried out at partial support of Russian Foundation for Basic Research (Grant numbers 10-08-01296-a, 10-01-00194-a, 10-08-13300-RT_omi, 12-08-91165-GFEN_a).

REFERENCES

[1] *Energy Harvesting Technologies* S. Priya, D. J. Inman (Eds.). Springer Science+Business Media, LLC, Berlin, 2009.

[2] Erturk, A.; Inman D. J. *Piezoelectric Energy Harvesting*. John Wiley & Sons, New Jersey, 2011, pp. 1–276.

[3] Minazara, E.; Vasic, D.; Costa, F. *Smart Mater. Struct*. 2008, *vol. 15*.

[4] Anton, S. R.; Sodano, H. A. *Smart Mater. Struct*. 2007, *vol. 16*, R1–R21.

[5] Beeby, S. P.; Tudor, M. J.; White, N. M. *Meas. Sci. Technol*. 2006, *vol. 17*, R175–R195.

[6] Lefeuvre, E.; Badel, A.; Benayad, A.; Lebrun, L.; Richard, C.; Guyomar, D. *J. Physique IV*, 2005, *vol. 128*, 177–186.

[7] Lefeuvre, E.; Badel, A.; Richard, C.; Petit. L.; Guyomar, D. *Sensors Actuators A*, 2006, *vol. 126*, 405–416.

[8] Priya, S. J. Electroceram. 2007, *vol. 19*, 165–182.

[9] Sodano, H. A.; Inman, D. J.; Park, G. *Shock and Vibration Digest*. 2004, *vol. 36*, 197-205.

[10] Sodano, H.; Inman, D.; Park, G. *J. Intell. Mater. Syst. Struct*. 2005, *vol. 16*, 67–75.

[11] Tanner, T.; Inman, D. J. In: *Proc. 73rd Shock and Vibration Symp.*, Newport Beach, RI, 2002.

[12] Adhikari, S.; Friswell, M. I.; Inman, D. J. *Smart Mater. Struct*. 2009, *vol. 18*, 115005.

[13] Litak, G.; Friswell, M. I.; Adhikari, S. *Appl. Phys. Lett*. 2010, *vol. 96*, 214103.

[14] Jyh-Cheng, Yu; Chin-Bing, Lan. *Sensors and Actuators A*, 2001, *vol. 88*, 178-186.

[15] Ottman, G. K.; Hofmann, H. F.; Bhatt, A. C.; Lesieutre, G. A. *IEEE Transactions on Power Electronics*, 2002, *vol. 17(5)*, 669-676.

[16] Richard, C. D.; Anderson, M. J.; Bahr, D. F.; Richards, R. F. *Journal of Micromechanics and Microengineering*, 2004, *vol. 14(5)*, 717–721.

[17] Lefeuvre, E.; Badel, A.; Richard, C.; Guyomar, D. *Journal of Intelligent Material Systems and Structures*, 2005, *vol. 16(10)*, 865–876.

[18] Guyomar, D.; Badel, A.; Lefeuvre, E.; Richard, C. *IEEE Transactions on Ultrasonics, Ferroelectrics, and Frequency Control*, 2005, *vol. 52(4)*, 584–595.

[19] duToit, N. E.; Wardle, B. L.; Kim, S. G. *Integrated Ferroelectrics*, 2005, *vol. 71*, 121–160.

[20] Sodano, H. A.; Park, G.; Inman D. J. *Strain*, 2004, *vol. 40(2)*, 49–58.

[21] Adhikari, S.; Friswell, M. I.; Inman, D. J. *Smart Mater. Struct*. 2009, *vol.18*, 115005.

[22] Parton, V. Z.; Kudryavtsev, B. A. *Electromagnetoelasticity of Piezoelectrics and Electrically Conductive Solids*; Nauka, Moscow, 1988; pp 1 – 472.

[23] Belokon, A. V.; Nasedkin, A. V.; Soloviev, A. N. In: *Theor. Appl. Mech.*; Donetsk National University Press: Kharkov, 2001; *No 33*, 45-51.

[24] Nasedkin, A. V. *High School Proceedings. Northern Caucasus Region. Natural Science*. 1999, *No 1*, pp 48-51.

[25] Belokon A.V., Nasedkin A.V., Soloviev: A.N. *J. Applied Math. Mech*. 2002, *vol. 66*(*3*), 481–490.

In: Physics and Mechanics of New Materials … ISBN: 978-1-62618-535-7
Editors: Ivan A. Parinov and Shun-Hsyung Chang

Chapter 25

MULTI-SECTION SUPERCONDUCTING MAGNETIC ENERGY STORAGE FOR RAILROAD TRACTION APPLICATIONS

***V. N. Noskov*[1,*] *and M. Yu. Pustovetov*[1,2,†]**

[1]Rostov-on-Don State University of Transport Communications, Rostov-on-Don, Russia
[2]Department of Machine-Tool Systems Automation and Electric Drives, Power Engineering and Machine-Building Institute, Don State Technical University, Rostov-on-Don, Russia

ABSTRACT

In this paper we explain why SMES systems have a good prospective for use in railroad transport. We suggest a research method for effective and flexible operation of the SMES systems in the railroad traction applications, such as electric power supply system and electric power transmission of autonomous railroad locomotive. This method is based on sectionalizing of SMES system and DC-DC converter technology.

1. INTRODUCTION

It is well-known a non-stationarity of graphic of the power consumption for railroad traction applications, such as electric power supply systems and autonomous locomotives with electric power transmission. At the same time, these systems are the important technical areas of applications for the energy storage devices, now rapidly developed. Among them, the superconducting magnetic energy storage (SMES) is one of the best candidates. In this case, the main goal is a recuperation of braking power in energy-supplying of locomotive's

* E-mail: nvn_nis@sci.rgups.ru, 1, Rostovskogo Polka Narodnogo Opolchenia Square, 344038 Rostov-on-Don, Russia.
† E-mail: mgsn2006@rambler.ru, 6, Studencheskaya Street, 344023 Rostov-on-Don, Russia.

traction motors by using SMES. In SMES systems, this energy can be stored during time. Moreover, when a power of railway substation or traction generator is insufficient for use as traction force or auxiliary needs, it is possible to use additionally a power, stored in SMES, or it could be only used for supplying of locomotive.

We developed a scheme for SMES system, integrated in traction system of electric power supplying or electric power transmission of autonomous railroad locomotive. This scheme is based on DC-DC converter technology [1]. The converter can be used not only with DC-DC electric systems, but also with AC-DC (in presence of input rectifier) or with AC-AC (in presence of input rectifier and output autonomous inverter) electric systems.

Most effective and flexible functioning scheme can be achieved by sectionalizing SMES. The possible variants are the set of SMES devices with parallel junctions and the sectionalized coil in single SMES system. The technical solution with multi-sectional SMES devices permits to organize several power supply channels with separate or mutual-dependent control. For example, it is possible to charge first section of SMES system from energy source, discharge the second section due to the load resistance and keep the third one in a mode of current storage. Moreover, a work synchronization of the SMES sections is also possible.

2. Is SMES for Railroad Applications Necessary?

One of the actual goals of railroad transport in Russia is the service organization of freight trains with high weight. The solution of this problem is connected with readiness of power supply systems of the electric traction. Figures 1 and 2 show the measurement results (catenary voltage and train velocity versus distance) for ride tests of freight train. Weight of this train was 10442 tons. Traction was carried out by two electric locomotives of VL10u type in a head of train.

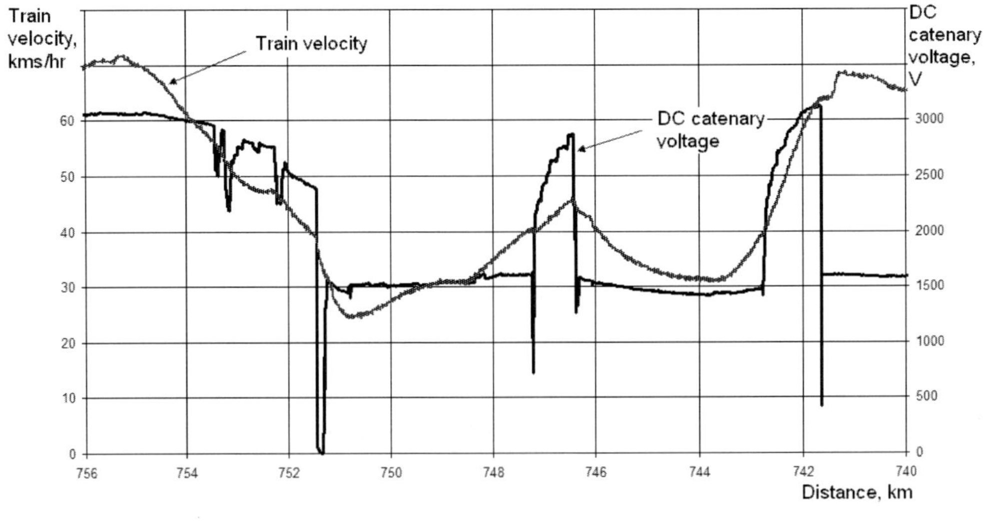

Figure 1. Measurement results (fragment 1) for ride tests of freight train at average traction power equal to 8.4 MW.

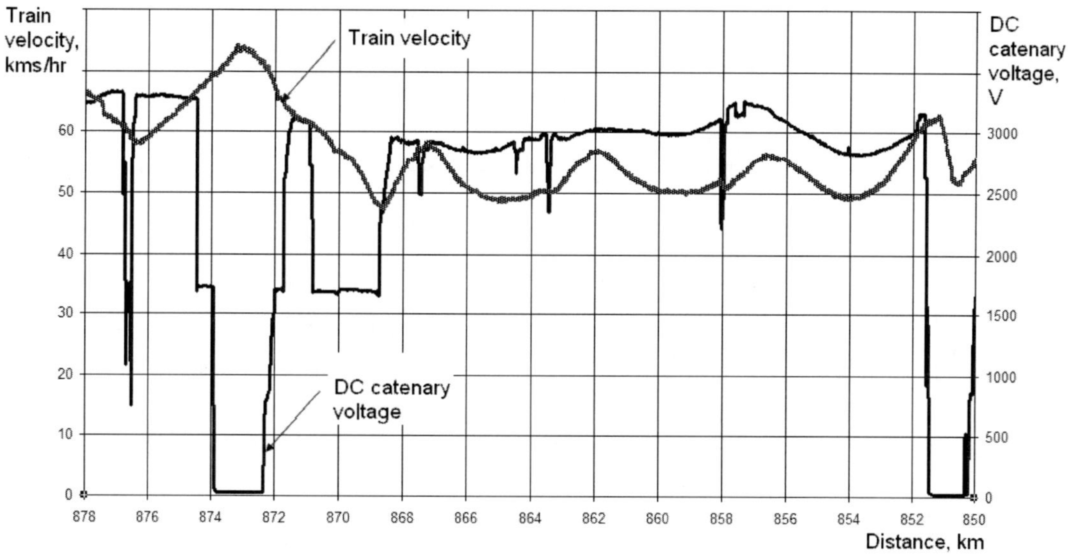

Figure 2. Measurement results (fragment 2) for ride tests of freight train at average traction power equal to 6 MW.

Analysis of the curves shows that power of traction energy of the supplying system is not sufficient for that heavy train. We can see that the catenary voltage decreases to 1.5 kV from nominal 3 kV level during 2 – 3 minutes and the voltage disappears for 1.1 - 1.7 minutes. Normal level of DC catenary voltage should be into range 2 – 4 kV. It is well-known from practice, that 6 – 7 freight trains could be between two neighboring traction DC substations, disposed on distance of 20 – 30 km. Taking into account the ride test results (catenary voltage decrease and duration of voltage disappearance), it is no difficult to calculate a quantity of electric energy, required to eliminate such deleterious effects: one equals approximately to 3.0 – 5.5 GJ. These results well agree with estimations scientists of Kurchatov Institute (Moscow) and TRINITI (Troitsk), which confirm 5 GJ-capacitance of energy storage per DC railroad traction substation to fully remove the effects of voltage sags and surges from the external power supply. It means that such energy storage device permits to solve the problem of recuperation of the braking power for energy-supplying of train, caused by the work of the feeding power system. In above-mentioned both cases of electric energy consumption of the traction substation, a whole smoothing out of the plots will not be observed in any case.

In Figure 3, we can see measurement plots of active power consumption for AC traction substation "Vostochnaya" (Rostov-on-Don), which demonstrate extremely irregular mode of consumption different for each day.

So, a profitability of energy storage devices for railroad traction substations is caused by

(i) a decrease of payment for electric energy;
(ii) a minimization of effects of the voltage disappearances, sags and surges;
(iii) a decrease of expanses of the traction substations at diminishing of the electric energy, transferred by the substation, due to elimination of consumption peaks.

Therefore, the energy storage device provides fulfillment of the following functions:

(i) a stabilization of catenary voltage (voltage on a load);

(ii) a stabilization of electric power (current) consumption from feeding power system;
(iii) an energy recuperation by using electric locomotive via catenary to energy storage device.

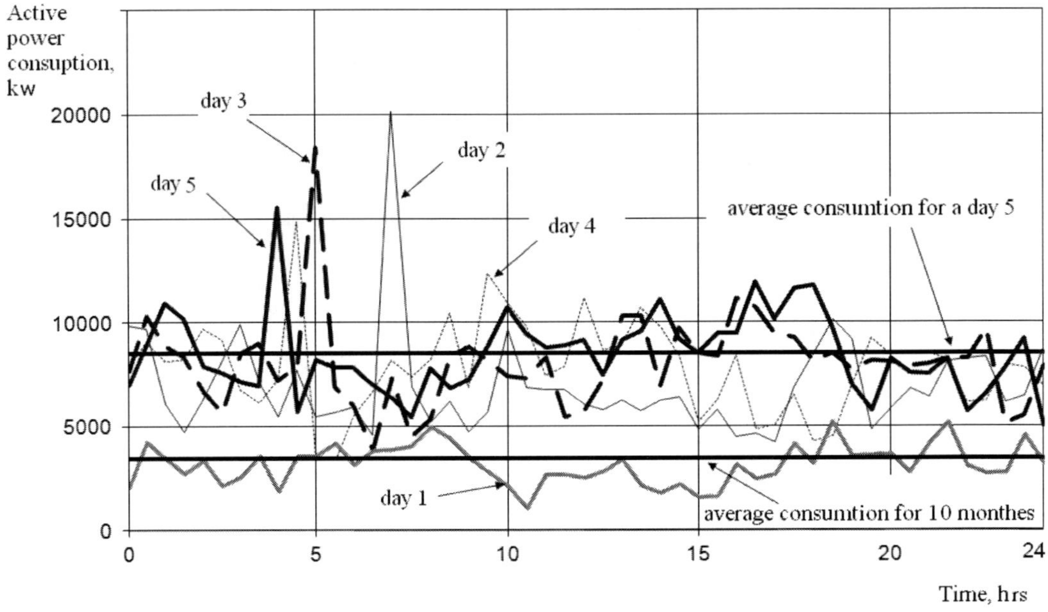

Figure 3. Measurement plots of active power consumption by AC traction substation “Vostochnaya”.

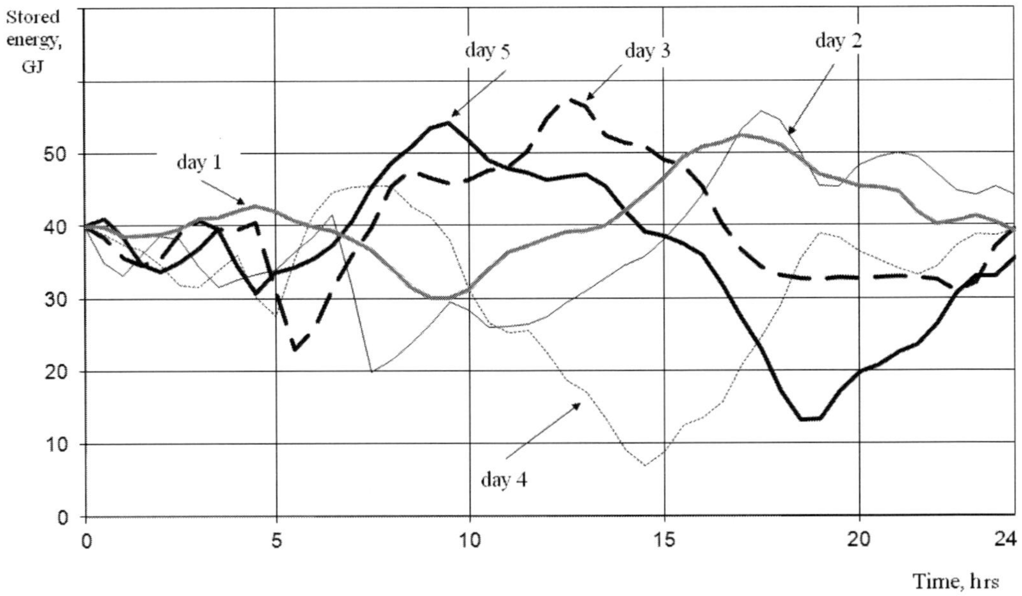

Figure 4. Calculated plots of stored energy for AC traction substation “Vostochnaya”, corresponding to data of Figure 3 in the case of constant everyday active power consumption from feeding power system.

It is very often a more convenient variant of fulfillment of the second function is to provide a necessary law of power (current) consumption.

Our calculations show, that if all of above-mentioned functions will be fulfilled, then the capacitance of energy storage per railway traction substation must be 4 – 20 GJ (1.11 – 5.55 MW·hr) for DC conversion and 20 – 80 GJ (5.55 – 22.22 MW·hr) in the case of AC conversion [2] (see Figure 4).

Expression (1), presented below, allows one to calculate with reasonable accuracy a sufficient energy capacitance of energy storage for set of power consumption plots:

$$EC_{energy\ storage} = EC_{dead\ amount} + EC_{discharge}^{\max} + EC_{charge}^{\max} = EC_{dead\ amount} + EC_{calculated}. \quad (1)$$

Here $EC_{dead\ amount}$ is the minimal amount of energy that keeps in some types of energy storage devices (e.g., SMES) at most discharge, which may be during the operation, usually equal to 10 – 15% of the calculated energy, $EC_{calculated}$, of SMES; $EC_{discharge}^{\max}$ is the maximal amount of energy, lost by energy storage, during set of monotonous discharging; $EC_{charge}^{\max}$ is the maximal amount of energy, accumulated by the energy storage, during set of monotonous charging.

During movement of autonomous locomotive the traction power of motors is changed in wide limits. Due to, it is necessary to recuperate a braking power of the motor traction by using energy storage. Here this energy could be stored during time. Moreover, while the power, generated by traction of generator is insufficient for creation of a traction force, we can use the stored power for additional simultaneous with generator supplying of the traction motors [3, 4]. Such a way, it is possible to lead to energy storage and provides the diesel-generator work with constant power at most high efficiency. During some time, it is preferential to supply the traction motors from SMES only, for example from ecological viewpoint, when a locomotive starts its movement from station with turned off diesel.

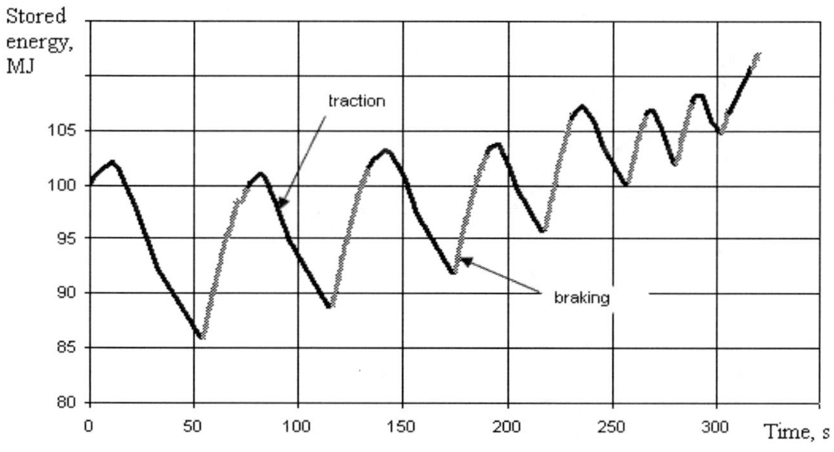

Figure 5. Calculated energy, stored in energy storage system on board of shunted locomotive in working mode “shunt by push”.

Capacitance of on-board energy storage system for diesel-electric shunt locomotive was discussed in [4, 5] and approximately equal to 120 MJ. This value is suitable for diesel-trains, too [6, 7]. Examples of calculated curves of energy stored in energy storage system on board of Soviet shunt diesel-electric locomotive TEM2 in presence of 400 kW-electric generator and peak power 650 kW of traction of the electric motors for four different working modes are shown in Figures 5 - 8 [4, 5].

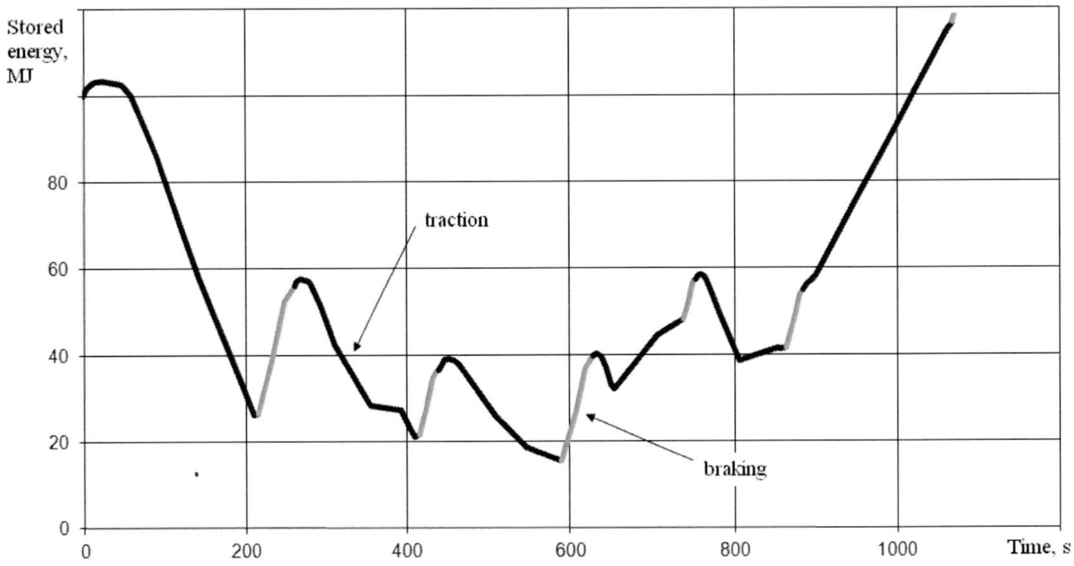

Figure 6. Calculated energy, stored in energy storage system on board of shunted locomotive in working mode “freight yard”.

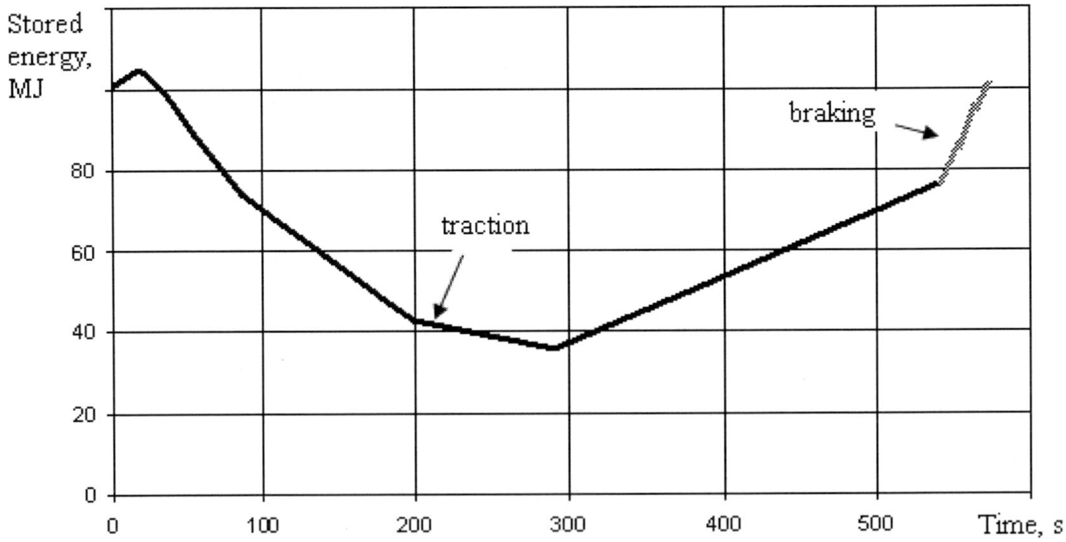

Figure 7. Calculated energy, stored in energy storage system on board of shunted locomotive in working mode “pull from park to park”.

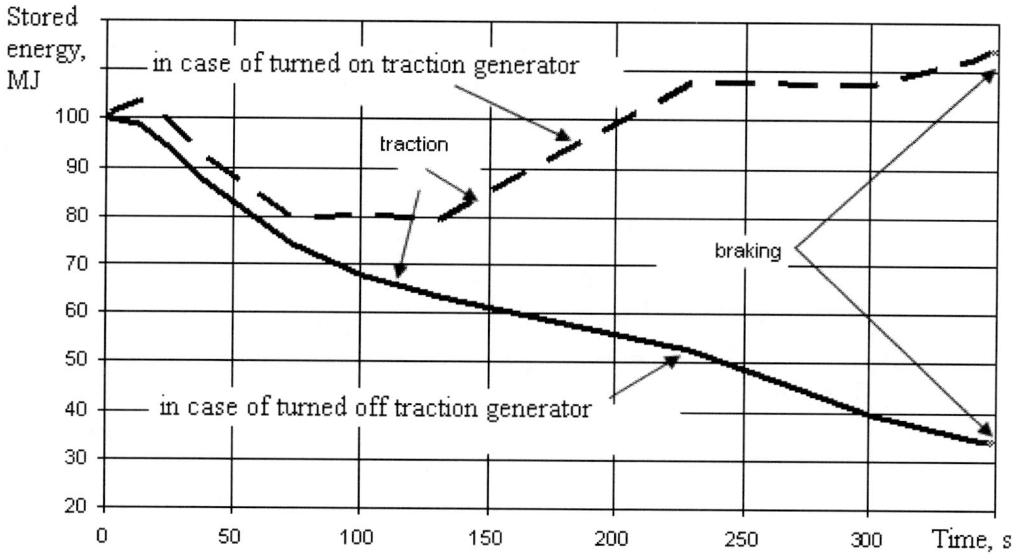

Figure 8. Calculated energy, stored in energy storage system on board of shunted locomotive in working mode "slide".

Table 1 contains summary data for three main perspective types of energy storage systems, which could be applied for railway needs: flywheels, supercapacitive and SMES [8 - 11]. Latter two types, in our opinion, are more preferable for exploitation in electric systems because of static energy conversion principle, absence of mutual conversion electrical and mechanical energy, high efficiency and reliability. Moreover, Russian scientists have an experience in design of SMES and power systems with SMES including railroad applications.

Table 1. Summary data for different types of energy storage

Characteristics	Units	Type of energy storage system		
		Flywheel	Supercapacitive	SMES
Optimal energy capacitance	MJ (kW·hr)	up to 10^3 (up to 280)	1 – 37 (0.28 – 10.2)	10 – 10^7 (2.8 – 2.8^7)
Optimal power	MW	up to 50	up to 500	> 500
Efficiency of charge-discharge cycle	%	70 – 92	88 - 90	85 – 99
Specific energy capacitance	MJ/m^3 (kW·hr/m^3)	10 (2.8)	50 (14)	100 (28)
Specific power	kW/m^3	-	up to 10^5	up to 60
Functioning time (up to)	s	1 – 60	10 - 10^4	10^2 - 10^4
Energy reverse time	s	0.1	0.01	0.01
Duration of energy storage	s	Limited (up to 10^3 hrs)	Limited (up to 100 hrs)	Unlimited
Number of working cycles	-	10^5 - 10^7	10^3 - 10^8	10^6 - 10^7

Figure 9 demonstrates schematic diagram of the physical model for the shunt locomotive electric power transmission with SMES. In 1998, the experimental studies have been carried out by specialists of Kurchatov Institute (Moscow) and Rostov-on-Don State University of Transport Communications (RSTU, Rostov-on-Don) by using this physical model.

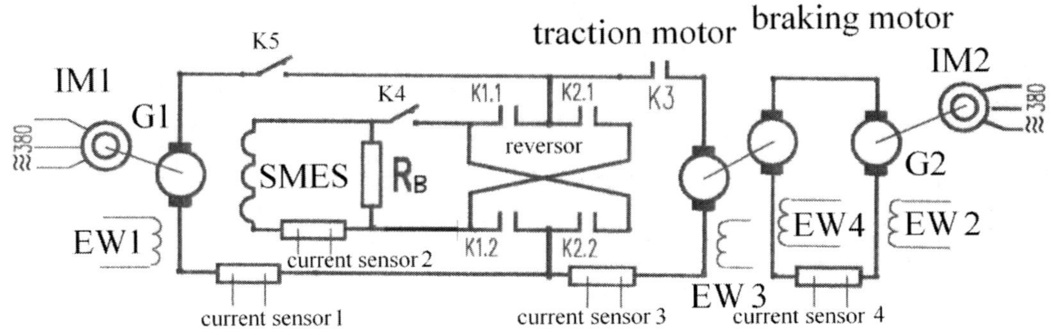

Figure 9. Schematic diagram of the physical model of the shunt locomotive electric power transmission with SMES.

Figure 10 demonstrates experimental results (the scaled down physical modeling) for the processes in shunt locomotive electric power transmission in the working mode "shunt by push". For this experiment, SMES with maximal energy capacitance equal to 1 MJ was used.

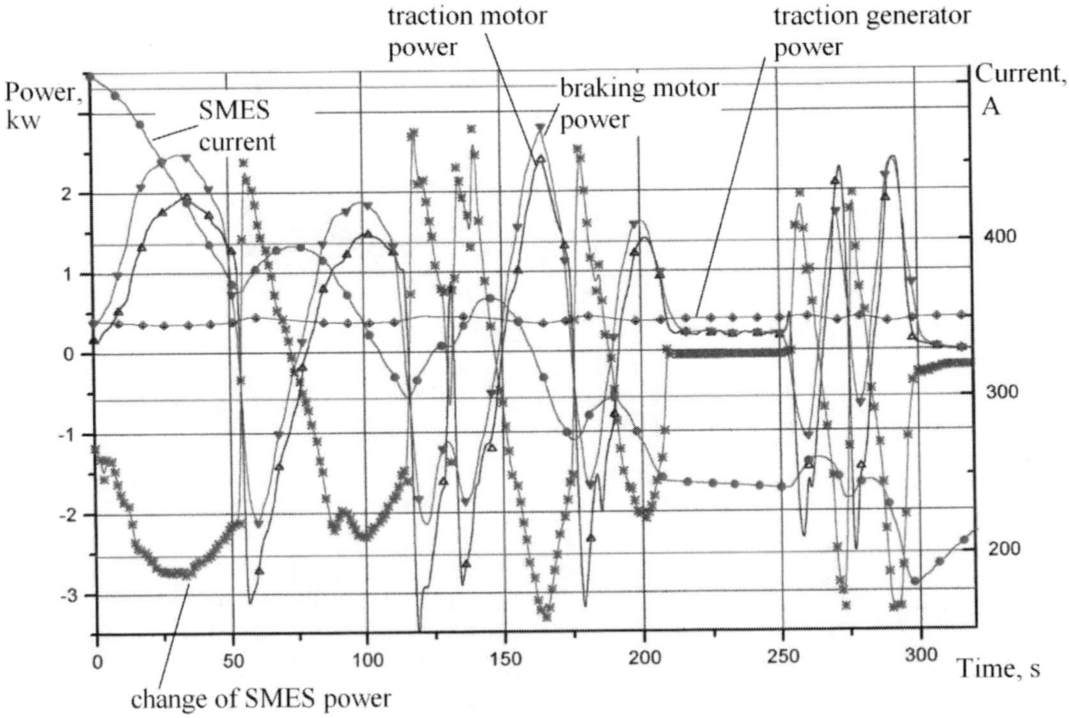

Figure 10. Experimental results (the scaled down physical modeling) for processes in shunt locomotive electric power transmission in the working mode "shunt by push".

3. Perspective Construction of SMES System

The central element of the SMES is the coil of superconducting wire. For SMES, the technically acceptable form is winding configuration with closed magnetic flux. Its stray magnetic field decreases rapidly with distance from the coil and, therefore, does not violate the performance of ambient electrical and electronic equipment, and also ensures compliance with sanitary standards. In the case of the same energy capacitance, the superconducting wire for windings without stray magnetic field in 2 - 4 times more powerful than short solenoid, "optimal" for the energy density of the magnetic field. However, the complexity of SMES with large stray magnetic field, embedded in the existing electric power system, makes questionable the feasibility of designing such devices. Among variety of possible options for design of the windings without stray magnetic field, the greatest interest presents winding with uniform density of the accumulated energy in the energy storage volume, as well as the most technological advanced winding, presenting a set of identical sections. The stray magnetic field is imperceptible at a distance of half a meter from the coil of such a SMES system.

In 2000s years, the RSTU (Rostov-on-Don), Kurchatov Institute (Moscow) and TRINITI (Troitsk) have designed and manufactured an experimental SMES equipment for railroad traction power supplying systems with energy capacitance of 1.5 MJ or 0.42 kW·hr (due to non-optimal properties of a single sector of torus) for DC current with value of 10 kA at voltage up to 5 kV (see Figure 11) [12, 13]. The main result of this work defined by the developed technology for original SMES equipment.

Figure 11. Experimental SMES for railroad traction power supplying systems (author: Vladimir N. Noskov, RSTU).

The device, shown in Figure 11, presented one of 12 sectors of the "compact torus" with a height of 1100 mm and a diameter of 1150 mm. Each sector was made from 20 panels of equal height and width, decreasing from center to periphery of the sector. The main element of the SMES structure was a system of superconducting electrical current coils, which had a complex spatial structure close to the torus shape with a frame structure, made of dielectric composite materials on the base of polymer matrix. Winding of each of the sectors was performed in the form of continuous segment of the superconducting wire with the niobium-titanium filaments into copper matrix. It developed fields of 5 - 6 T at the rates of change in the fields equal to 0.2 - 0.5 T/s.

Due to lack of specialized polymer materials, which are only developed for the panels, we used glass-textolite plates, including the foiled ones. The coils of superconducting wire were attached by using special glue on the perimeter of the panels. The foil on the glass-textolite plates were used for the formation of electric heating elements, which were necessary to safely transfer the SMES system into non-superconducting (normal) state without destroying of the structure. When a signal on the arising of the normal state is fed, a current passed in the heaters, distributed over the volume of the magnetic system, and transferred the entire winding in the normal state. It provided a uniform distribution of stored energy in the winding, and its heating and also transferred structure to thermal state no exceeding 100 – 150 K. The current passed in the heater through group of pre-charged capacitors.

Photo-etching pattern tracks of the heating elements, processed on the base of glass-textolite foil, are shown in Figure 12. This technology was used to create printed circuit boards of large size.

The "compact torus" shape of the magnet system SMES30 with 12 sectors (see Figure 13) could provide the energy up to 30 MJ with maximum load power of 6 MW. It is selected a distributed toroidal winding with the calculated inductance of 1.35 H. It is advisable to use the current leads for superconducting magnet coils due to non-superconducting parts of the device, installed on the current of 5 – 20 kA (the estimated current is equal to 10 kA) at the DC voltage of 10 kV.

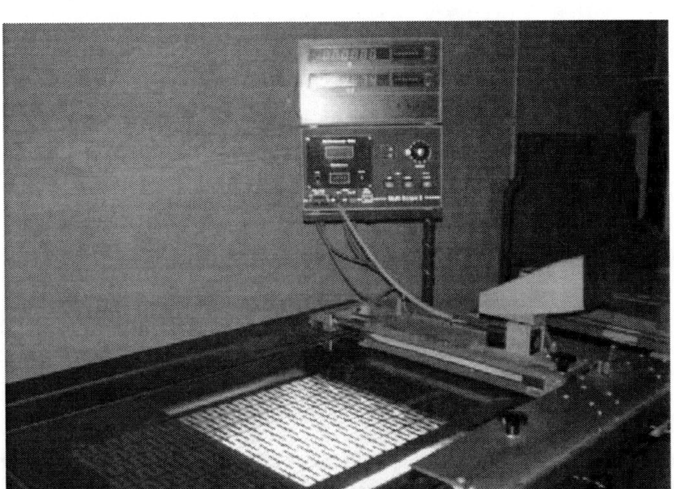

Figure 12. Photo-etching tracks pattern for heating elements on a foiled glass-textolite.

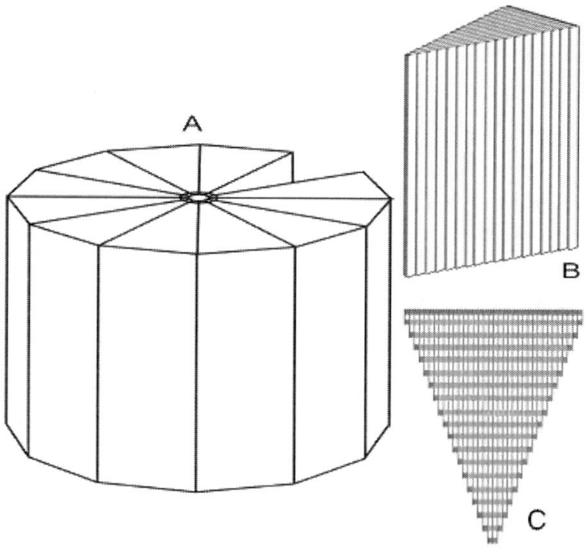

Figure 13. Scheme of the SMES magnetic system of toroidal structure with uniform distribution of energy in the volume: (a) shape of the magnetic system SMES30, (b) shape of one sector of the magnetic system, (c) cross-section of the sector.

Note, that for the SMES system, working at the cryogenic temperature of liquid helium 4.2 K, the corresponding cooling system (no shown in Figures 4, 6) is required.

Design and manufacture of SMES30 require approximately 60 million rubles of investment. Modern possibilities of SMES systems, processed in Russia, are currently limited by energy capacitance of about 100 MJ.

Table 2. Characteristics of perspective SMES devices for traction and no traction needs

Parameter name	Units	Value	Value
Type of SMES	for traction power supplying system		for no traction needs
Winding type	Toroid with distributed winding		
Energy capacitance	MW·hrs (GJ)	1.0 – 1.5 (3.6 – 5.4)	0.0277 (0.1)
Maximum power	MW	300	8.3
Number of sectors	–	48 – 60	12
Outer diameter of winding at nominal magnetic field induction of 5 T	m	6	2.1
Height of winding	m	6 – 9	1.2
Diameter of vacuum envelope	m	7.5	2.6
Height of vacuum envelope	m	10 – 15	2.6
Cold mass of the winding	tones	250 – 400	4.3

The characteristics of perspective SMES devices for traction power supplying systems and for no traction needs in the case of stationary location or for on-board disposition are presented in Table. 2. The SMES device for no traction needs could be used to arisen electric critical loads, namely: in data centers, in centers for traffic management of alarm systems, in automation and communication systems by ensuring qualitative and uninterrupted power supplying.

4. Scheme of Power Static Converter and Its Modes

There is a fundamental difference between the inductive storage as a current source, and the source of DC voltage. An important task is the introduction of SMES device into electric circuit. The suggested scheme of power static converter is presented in Figure 14 [4, 12]. This converter converts input DC-voltage, V_{IN}, into output DC-voltage, V_{OUT}, with other characteristics. Here, L is the inductance of SMES coil, and Z_{LOAD} is the complexor of load resistance.

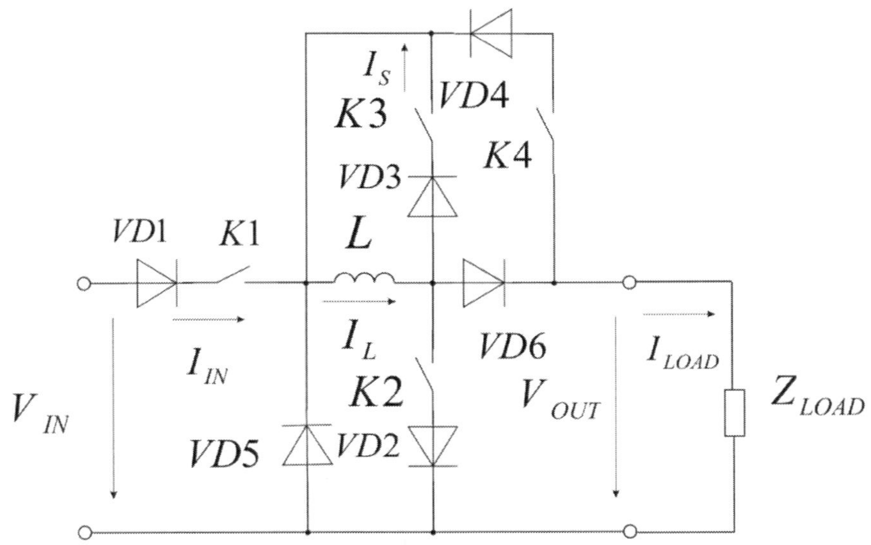

Figure 14. Electric scheme of power static converter with SMES coil.

In Figure 14, the following designations are used: $VD5$, $VD6$ are the diodes; $VD1$ and $K1$, $VD2$ and $K2$, $VD3$ and $K3$, $VD4$ and $K4$ present the combinations of diode and key full-controlled semiconductor triode such as GTO.

This converter can fulfill several operations, namely: (i) to charge SMES from generator, (ii) to discharge SMES to load, (iii) to store DC-current in SMES coil, and (iv) to recuperate energy from load to SMES.

In Tables 3 – 8 the states of the keys are shown for different modes of converter. The state "1" means the key is in position "ON". The state "0" means the key is in position "OFF".

Mode 1: SMES coil is charged from voltage source, V_{IN} (circuit $V_{IN}+ \rightarrow VD1 \rightarrow L \rightarrow VD2 \rightarrow V_{IN}$-).

Table 3. States of keys in Mode 1

Key name	K1	K2	K3	K4
Key state	1	1	0	0

The equation for circuit has the form:

$$v_{IN}(t) = 2R_{VD} \cdot i_L(t) + L\frac{di_L(t)}{dt}, \tag{2}$$

where $i_L(t)$ is the current passes through SMES coil, R_{VD} is the diode forward resistance, when conducting.

Mode 2: SMES is discharged to load (circuit $L \rightarrow VD6 \rightarrow Z_{LOAD} \rightarrow VD5$).

Table 4. States of keys in Mode 2

Key name	K1	K2	K3	K4
Key state	0	0	0	0

The equation for circuit has the form:

$$0 = 2R_{VD} \cdot i_L(t) + L\frac{di_L(t)}{dt} + v_{OUT}(t). \tag{3}$$

Mode 3: It is the mode of current storage in the SMES coil with partition discharge of SMES coil to load (circuits $L \rightarrow VD3$ and $L \rightarrow VD6 \rightarrow Z_{LOAD} \rightarrow VD5$).

Table 5. States of keys in Mode 3

Key name	K1	K2	K3	K4
Key state	0	0	1	0

The equations for circuits are written as

$$\begin{cases} 0 = R_{VD} \cdot i_S(t) + L\dfrac{di_S(t)}{dt}; \\ 0 = 2R_{VD} \cdot i_{LOAD}(t) + L\dfrac{di_{LOAD}(t)}{dt} + v_{OUT}(t); \\ i_L(t) = i_{LOAD}(t) + i_S(t), \end{cases} \tag{4}$$

where $i_S(t)$ is the current, stored in SMES coil, $i_{LOAD}(t)$ is the current passes through the load.

Mode 4: SMES is charged from the load, being in the recuperating mode (voltage source, V_{IN}, is turned off, circuit $Z_{LOAD} \rightarrow VD4 \rightarrow L \rightarrow VD2$).

Table 6. States of keys in Mode 4

Key name	K1	K2	K3	K4
Key state	0	1	0	1

The equation for circuit has the form:

$$v_{OUT}(t) = 2R_{VD} \cdot i_L(t) + L\frac{di_L(t)}{dt}. \quad (5)$$

Mode 5: SMES is charged from the recuperating load and voltage source, V_{IN}, at the same time (circuits $Z_{LOAD} \rightarrow VD4 \rightarrow L \rightarrow VD2$ and $U_{IN}+ \rightarrow VD1 \rightarrow L \rightarrow VD2 \rightarrow U_{IN}-$).

Table 7. States of keys in Mode 5

Key name	K1	K2	K3	K4
Key state	1	1	0	1

The equations for circuits are written as

$$\begin{cases} v_{OUT}(t) = R_{VD} \cdot i_{LOAD}(t) + L\dfrac{di_{LOAD}(t)}{dt}; \\ v_{IN}(t) = 2R_{VD} \cdot i_{IN}(t) + L\dfrac{di_{IN}(t)}{dt}; \\ i_L(t) = i_{LOAD}(t) + i_{IN}(t), \end{cases} \quad (6)$$

where $i_{IN}(t)$ is the current from the voltage source, V_{IN}.

Mode 6: it is the mode of current storage in the SMES with the load supplying from voltage source V_{IN}, at the same time (circuits $L \rightarrow VD3$ and $U_{IN}+ \rightarrow VD1 \rightarrow L \rightarrow VD6 \rightarrow Z_{LOAD} \rightarrow U_{IN}-$).

Table 8. States of keys in Mode 6

Key name	K1	K2	K3	K4
Key state	1	0	1	0

The equations for circuits are written as

$$\begin{cases} 0 = R_{VD} \cdot i_S(t) + L\dfrac{di_S(t)}{dt}; \\ v_{IN}(t) = 2R_{VD} \cdot i_{LOAD}(t) + L\dfrac{di_{LOAD}(t)}{dt} + v_{OUT}(t); \\ i_L(t) = i_{LOAD}(t) + i_S(t). \end{cases} \quad (7)$$

It should be noted that in scheme of Figure 14, the absence of key $K4$ is possible because of the current, passing through diode $VD4$, will exist only in the case of $V_{OUT} > V_{IN}$. The modes 1 – 6 can operate in arbitrary order.

Some results of computer modeling for scheme of working modes in Figure 14 are given in [4, 14] at assumption of identical parameters and modes for all traction motors.

4. The Reasons for Sectionalizing of SMES Device

In practice, the working modes of locomotive's traction motors are no identical because of the spread of parameters and different adhesion conditions of wheelsets with rails. It is maybe that one motor works in the traction mode but another in the braking one. Due to this, it will be better to feed each motor from a separate SMES system.

Traction substation can feed several, often two sites of feeding. However, it is the most rational to organize the work of each SMES system for single site of the feeding. This way reaches the maximal "targeted" compensation of power fluctuations that can give a gain in necessary SMES energy capacitance and cost of SMES systems.

Thus, we come to the desirability of sectioalizing of SMES system. There are two possible variants, namely: several SMES devices with parallel junctions or a sectionalized coil in the separate SMES. The last case is acceptable because of modularity of the "compact torus" type winding. In any case, the scheme of DC-DC converter [1] can provide the work with several SMES coils or sections. An important distinctive feature of this scheme is the ability to combine the accumulation time of energy in the inductive energy storages, and transmission time to the load. This result is achieved by using sequential switching circuits with inductive storages. Technical solution, based on the multi-section SMES devices, permits to organize several power supplying channels with separate or mutual-dependent control (it is assumed, that work of cooling fans for electric drives on locomotive depends on work of traction motors). For example, it is possible to charge first section of SMES system from energy source, discharge the second section to the load and keep the third one in the mode of current storage. Moreover, a synchronization of work of the SMES sections is also possible.

5. Description of Converter for Use with Multi-Section SMES System

Figure 15 presents the variant of electric scheme of the power static converter with multi-section SMES system. The following designations are used: 1-1, ... ,1-n are the working control gates, such as GTO; 2-1, ... ,2-n are the inductive storage elements (SMES sections); 3-1, ... ,3-n are the main valves, for example, semiconductor diodes; 4-1, ... ,4-n are the reverse control gates; 5 is the DC power supply; 6 is the load; 7-1, ... ,7-n are the bypass valves; 8 or 8-1, ... ,8-n are the control entrance gates; 9 or 9-1, ... ,9-n are the additional control gates; 10 is the discharge control gate; 11 is the discharge resistance. The arrows indicate the direction of the current in inductive storage elements.

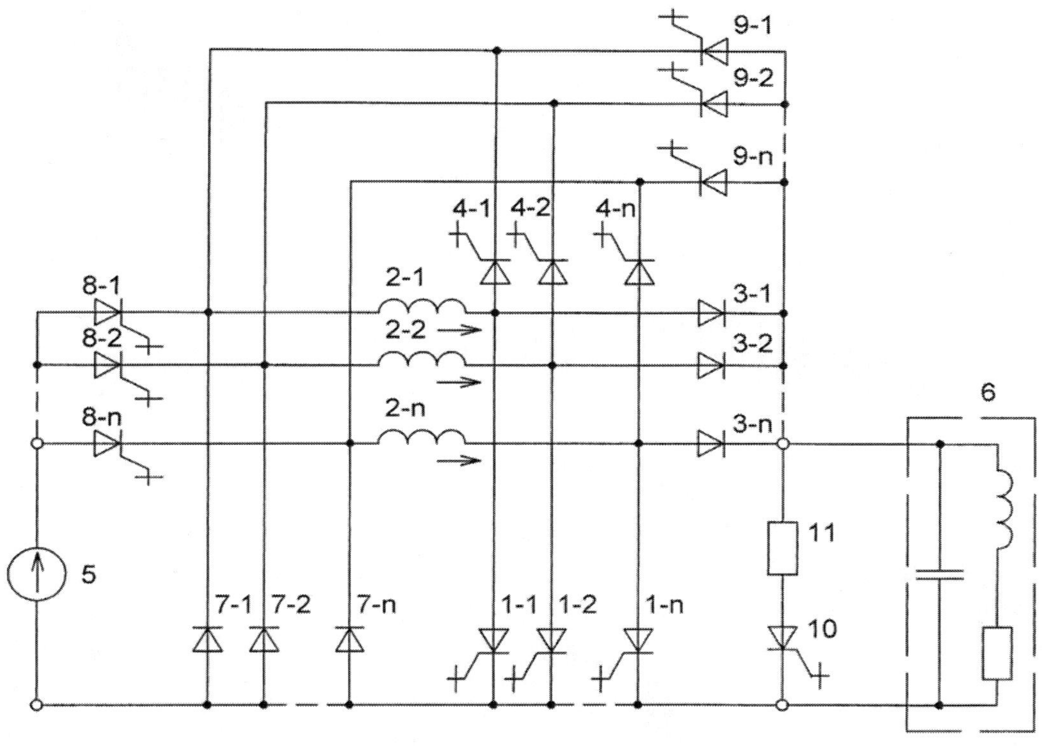

Figure 15. Electric scheme of power static converter with multi-section SMES system.

This DC-DC voltage converter operates as following. In the process of work, the inductive storage elements 2-1, ... , 2-n may serve as the main, reserve and additional sources of electrical energy. In the initial state, the inductive storage elements are discharged. For the charging of inductive storage elements, the working control gates 1-1, ... ,1-n are turnd on in series or in groups (depending on DC voltage source). By reaching the desired current value, being significantly higher than the rated load 6, the current, corresponding reverse control gates 4-1, ... , 4-n are turned on, at the same time, appropriate working control gates are switched off. In this time, the current inductive storage element is closed through the reverse control gates 4-1, ... , 4-n. The inductive storage elements pass to DC current store mode and can be in this mode for a long time. The load 6 in this mode, is governed by the DC voltage source 5 by using the control entrance gate 8 and 3-1, ... , 3-n main valves. When you turn off (e.g. emergency) the DC voltage source 5, the load 6 passes to the power supplying from inductive storage elements. In this case, the reverse control gates 4-1, ... , 4-n are turned on in series or in groups. The inductive storage elements is partially discharged to the load 6 through the bypass valve 7, at the same time the current, transferred to the load, is determined by duration of turn off state of the reverse control gate. After selection of the required amount of energy to the load, the control reverse gates are turned on. The choice of duration of turn off state of the reverse gate determines the effective voltage across the load with periodic shutdown and turn on of reverse gate. The value of the average voltage across the load in this mode can be maintained in the range below the nominal voltage of the DC source 5. The presence of the input control gate 8 allows one to implement a mode of operation of the converter, in which periodic shutdown of the DC input voltage 5, controlled by the gate 8, allows one to adjust the average voltage across the load 6 in range of narrower than the input voltage. In this mode, when the controlled entrance gate is turned on, the working control gates 1-1, ... ,1-n are turned on and the charging of inductive storage elements 2-1, ... , 2-n is carried out. The reverse control gates 4-1, ... , 4-n are turned off, for example, by the voltage source 5.

The turn off of the input control gate 8 and the working control gates 1-1, ... , 1-n allows one to pass energy to the load 6 from inductive storage elements. In this case the current, which passes through the inductive storage elements 2-1, ... , 2-n, working controlled gates 1-1, ... , 1-n and the load 6, is closed by the bypass valve 7. It should be noted, that turn off of working control gates can be conducted at the same time for all groups, and one by one. Duration of power output to the load from each *i*-th inductive storage element, and hence the value of the average voltage across the load, is determined by duration of turn off state of *i*-th reverse control gate. When the turn off operations for different control gates 1-1, ... , 1-n are conducted asynchronously, the current, which passes through control working gate 1, is closed by bypass valve 7 and the corresponding inductive storage element. In this mode, the load 6 is governed by the energy, stored in the inductive storage elements 2-1, ... , 2-n. For this scheme of the inductive storage elements a pass from the mode of accumulation to the saving mode is possible, when synchronously working control gates are turned off and the appropriate reverse control gates are also turned off. It is caused, for example, by alternate energy output to the load at a short-time turning off of reverse control gates. Alternate mode of the power output from several inductive storages in the form of pulses with variable duration allows one to obtain mono-impulses of desired shape on the load.

6. Computer Modeling Results for Converter with Two-Section SMES System

Figure 16 presents the computer modeling results for current of two-section SMES system. Both sections operate in different modes. Section 1 demonstrates alternate charge – discharge processes. Section 2 is in DC-current store mode during first 7 seconds and then demonstrates alternate processes of charging and storage asynchronously with section 1. In this case we have *R-L* load. V_{OUT} is stabilized by parallel to the load capacity. Figure 17 presents model results for converter with two-section SMES system V_{IN} and V_{OUT}, corresponding to the currents of the SMES system, depicted in Figure 16. It is clear, that by combining the working modes of two SMES sections, we can get different shapes of V_{OUT} pulses and additional way to regulate value of average voltage across the load.

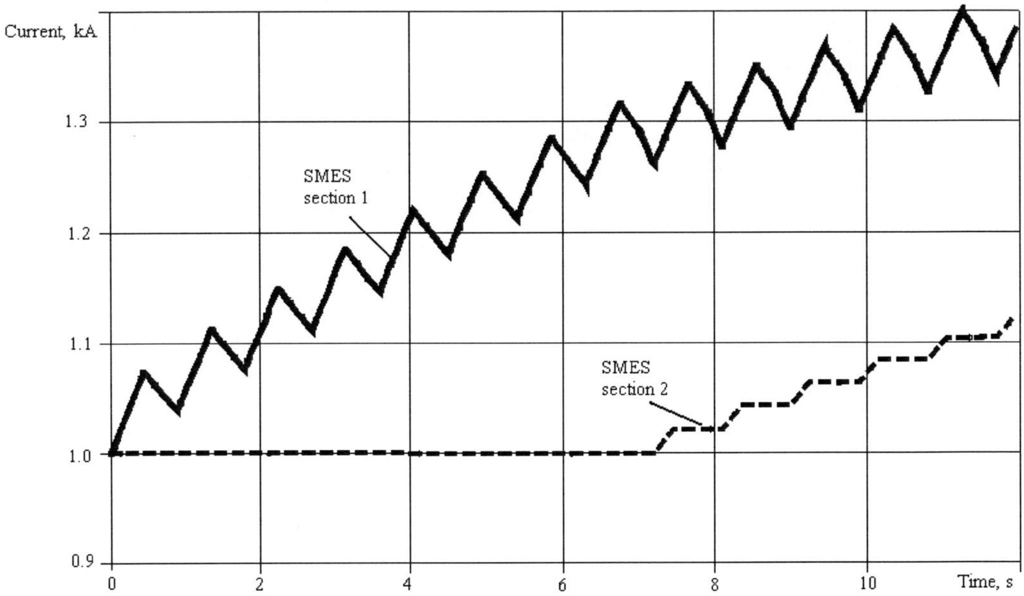

Figure 16. Computer modeling results for current of two-section SMES system.

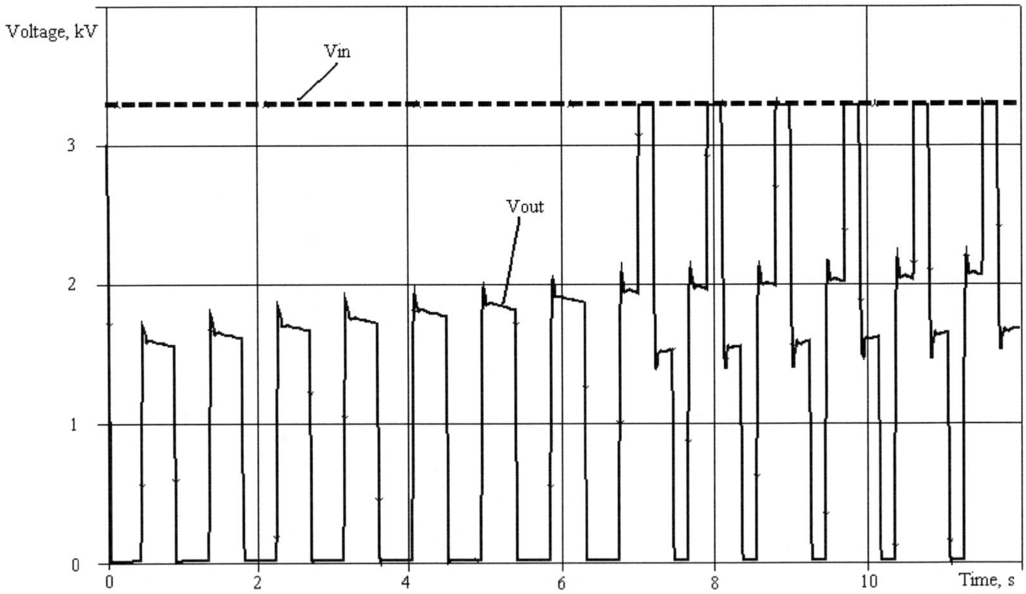

Figure 17. Computer modeling results of V_{IN} and V_{OUT} for converter with two-section SMES system.

Conclusion

The scheme of DC-DC voltage converter with multi-section SMES system for integration of energy storage device in traction electric power supplying system or electric power transmission of autonomous railroad locomotive is suggested. The converter can be used not only with the DC-DC electric systems but also with AC-DC (in presence of input rectifier) or with AC-AC (in presence of input rectifier and output autonomous inverter) electric systems. This device provides most effective and flexible operation of SMES system. The technical solution with multi-section SMES system permits to organize several power supplying channels with separate or mutual-dependent control. Equations of main modes of this scheme are presented. Energy capacitance of the SMES system in the cases of traction substations and shunt locomotive are calculated. The scheme capability to work is confirmed by results of computer modeling.

References

[1] Klimenko, E.Yu.; Noskov, V. N.; Poltanov, A. E.; Polulyakh, E. P.; Pustovetov, M. Yu.; Rindin, V. N.; Flegontov, N. S. *DC-Voltage Converter.* Russian Federation Patent No. 2370875, April 17, 2008.

[2] Noskov, V. N.; Pustovetov, M. Yu.; Chirkov V. C. *Vestnik of Rostov-on-Don State University of Transport Communications*, 2008, *No. 1*, 120 – 127.

[3] Johnson, B.; Low, J.; Saw, G. *IEEE Trans. Appl. Supercond.* 2001, *vol. 11*, 1900-1903.

[4] Noskov, V. N.; Pustovetov, M. Yu. In: *Ferroelectrics and Superconductors: Properties and Applications*. Ivan A. Parinov (Ed.); Nova Science Publishers: New York, 2012; pp 273 – 280

[5] Noskov, V. N.; Pustovetov, M. Yu. *Vestnik of All-Russian Science and Research Railway Transport Institute*, 2008, *No. 5*, 42 – 44.

[6] Kossov, E. E.; Azarenko, V. A.; Komaritzky, M. M. *Transport of Russian Federation. Special Issue. Science and Transport. of Russian Railways*, 2007, pp 20 – 21.

[7] Destraz, B.; Barrade, P.; Rufer, A. *Supercapacitive Energy Storage for Diesel-Electric Locomotives*. SPEEDAM 2004 June 16th – 18th, Capri (Italy).

[8] Traction substations load smoothing out by energy storage devices. *Rail International*, 1997, *No. 1*, 48 – 50.

[9] Astahov, Yu. N.; Venikov, V. A.; Ter-Gazarian, A. G. *Energy Storage in Electrical Systems;* High School: Moscow, 1989; pp 1 – 159.

[10] But, D. A.; Alievsky, B. L.; Miziurin, S. R.; Vasiukevich, P. V. *Energy Storage;* Energoatomizdat: Moscow, 1991; pp 1 – 400.

[11] Krutikov, A. *Power Electronics*, 2005, *No. 2*, 28 – 29.

[12] Pustovetov, M. Yu.; Noskov, V. N. In: *Trans. of 4th Int. Symp. "Electrification and the Organization of High-Speed and Heavy Rail Corridors"*; Saint-Petersburg State University of Transport Communications: Saint-Petersburg, 2007, p. 82.

[13] Semionov, V. I.; Shakimov, L. A.; Abramova, N. G.; Klimenko, E.Yu.; Polulyakh, E. P.. *Structures Made of Composite Materials*, 2009, No. 3, 29 – 34.
[14] Noskov, V. N. *Vestnik of Rostov-on-Don State University of Transport Communications*, 2008, *No. 4*, 41 – 47.

In: Physics and Mechanics of New Materials ...
Editors: Ivan A. Parinov and Shun-Hsyung Chang
ISBN: 978-1-62618-535-7

Chapter 26

THE FABRICATION OF HYDROPHONE BASED ON EPITAXIAL PZT FILM FOR ACOUSTIC DEVICE APPLICATIONS

Yuang-Tung Cheng[1], Jiun-Chen Liou[2], Kuo-Chang Hou[3], Shun-Hsyung Chang[*3], David. R. C. Chu[4], I. A. Parinov[5], C. C. Yang[3], Ian Y. Y. Bu[3], M. S. Shevtsova[6], I. V. Zhilyaev[6] and V. A. Akopyan[5]

[1]LS Center, Chung-Shan Institute of Science and Technology. Lung-Tan, Tao-Yuan, Taiwan
[2]Dept. of Electrical Eng., National Taiwan Ocean University, Keelung, Taiwan
[3]Dept. of Microelectronic Eng., National Kaohsiung Marine University, Kaohsiung, Taiwan
[4]College of Aviation, China University of Science and Technology, Henshan Shiang, Hsinchu, Taiwan
[5]Vorovich Mechanics and Applied Mathematics Research Institute, Southern Federal University, Rostov-on-Don, Russia
[6]Southern Center of Russian Academy of Sciences, Rostov-on-Don, Russia

ABSTRACT

This paper presents the design, fabrication, and characterization a center-supported circular diaphragm sensor with a figure of merit described. By using a circular diaphragm the sensing leaf can be made more flexible, having larger deformations as a result from sound pressure. The proto-type hydrophones of the electrical impedance and receiving sensitivity of the hydrophones in water were evaluated in the frequency range of 10-200 kHz. The finite element model results were compared with the experimental results. The deposited PZT thin film micro electro-mechanical systems (MEMS) fabrication processes

* E-mail: stephenshchang@me.com

were used to fabricate the acoustic devices. We report the preparation of epitaxial multi-layer structures that are suitable for sensor fabrication on Si substrates, and we evaluate the sensitivity of these ultrasonic sensors to provide details of the first reported use of epitaxial PZT films on Pt/SiO_2/Si structures. The frequency is governed by the dimensions and the layered materials of the diaphragm and can be controlled independent of layer thickness, offering more design flexibility and control over the frequency. The sharp resonance peaks observed for the typical hydrophone were broadened to a large extent for porous piezocomposite hydrophones due to a weaker coupling of the structure. The highest levels of transmission sensitivity and receiving sensitivity which could be obtained were 70 dB and –65 dB, respectively. The acoustic device has better performance when poled PZT thin film is used.

1. INTRODUCTION

The accurate measurement of sound pressure in the region of hydrophone devices is very important from the viewpoint of underwater acoustic multimedia equipment calibrations [1-4]. The important parameters will include the material density, piezoelectricity, dielectric constant and high electro-mechanical coupling factor. The advantage of using the piezoelectric principle is that it can be used as both a transmitter and a receiver of information [5-7]. In 1880, Jacques and Pierre Curie first discovered the piezoelectric effect. This provides a convenient transducer effect between electrical and mechanical oscillations. The property of piezoelectricity is dictated by both the atoms in the crystal and the particular way in which that crystal formed materials such as reochelle salt ($NaKC_2H_4O_6{\bullet}4H_2O$), barium titanate ($BaTiO_3$), lead titanate ($PbTiO_3$), lead zirconate titanate ($PbZr_{0.52}Ti_{0.48}O_3$), zinc oxide (ZnO), aluminum nitride (AlN), polyvinyliden fluoride (PVDF), etc. [8-14]. Since lead zirconate titanate (PZT) has excellent piezoelectric properties, a high Curies temperature, high spontaneous polarization and high electromechanical coupling coefficient, it has become an important industrial product. PZT is the main material used to fabricate our acoustic devices [15-17]. However, piezoelectric materials such as this have insufficient sensitivity when the sensing area of their active elements becomes smaller in size. The advantage of our silicon-based microelectromechanical system (MEMS) hydrophone is an easier integration with the eventual IC and signal processing [18-20]. These devices are fabricated on Si wafers of either (100) or (111) orientations [21, 22]. Silicon has controllable electrical properties, and economical and desirable mechanical properties. In addition, the fabrication of transmission devices on Si substrates permits integration with signal processing circuits such as preamplifiers and impedance matching circuits. Hydrophones detect the pressure variation of acoustic signals and noise in the water and produce an output voltage proportional to the pressure. The pressure vibrating element is formed by multi-layered membrane of the diaphragm actuated by the active layer, usually a thin film. Due to the nature of the diaphragm which has low acoustic impedance, it can be made to match with fluid impedance more easily. The performance of the hydrophone often is affected by noise voltages generated due to the thermal agitation in any internal resistance. The most important parameter is the open-output-voltage and signal-to-noise ratio. The new design was initially incorporated with traditional PZT piezoceramic materials. This design was also computer simulated by Russian co-authors. The simulation helped to modify the PZT in order for optimum performance. Another parameter of interest was the various thicknesses of the layers in the MEMS structure. We evaluate the sensitivity of these devices to provide details by using epitaxial PZT films on $Pt/SiO_2/Si$ structures.

2. Hydrohone Physical Characterization and Design

In our numerical investigation we have modeled a frequency response of a bimorph PZT disc placed in water at depths from 1 to 100 meters. Axisymmetric coupled-field elements were used for the active material. The model includes the PZT layer, two SiO_2 layers, and a Pt film electrode surrounding water medium. It is considered an axial symmetric domain with a 0.5m diameter and 1m length. For the disk behavior, the equations of linear elasticity in the cylindrical coordinates are used, and a finite-element mesh inside the disk has nearly 100 elements. In order to obtain good numerical accuracy, the size of the mesh has been kept very small compared to the dominant wavelength of the sound waves in PZT and water. Because acoustic wave action causes very small displacements of the disk surface, these displacements were neglected. All finite element models were built in Comsol Multiphysic's commercial soft package. The model and the finite-element mesh are shown in Figure 1.

For acoustic devices fabrication, a 1 Ωcm n-type CZ (111)-oriented silicon wafer sized 12.5×12.5 cm^2 was used. The wafer thickness was about 250 μm. The silicon substrates are cut into pieces approximately 5×5 cm^2 size, taken from the manufacturer without any pretreatment; different layers are heterojunctions on the substrate surface.

The three main elements in conventional silicon processing are deposition, lithography, and etching. In this research, the Si substrates are cleaned by de-ionized water.

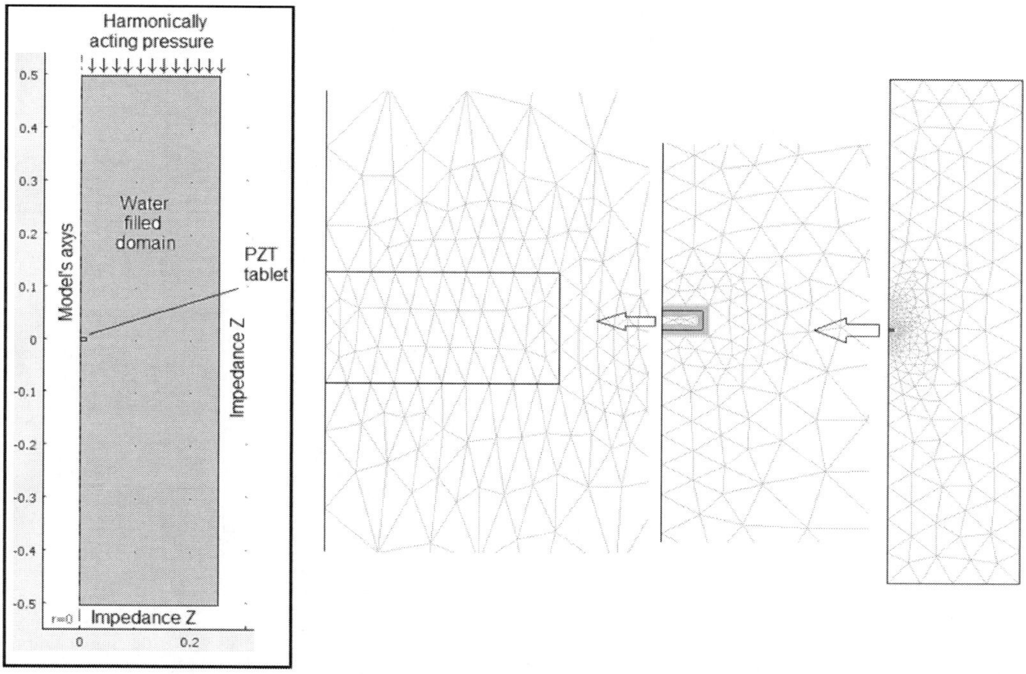

Figure 1. The finite-element model of a dense PZT hydrophone.

Figure 2 summarizes the fabrication process of the acoustic devices. Basically, the diaphragm is made by anisotropically etching from the back side of Si substrates. The process PZT used to fabricate thin film was as follows: After the surface of the *n*-type (111) silicon was indented by dry etching, a 40 μm oxide layer was produced using a wet oxidation process. A 300 nm Pt electrode on the oxide layer was evaporated. Since the piezoelectric coefficient tends to increase with increasing film thickness due to a reduction in substrate constraints [12], we prepared a relatively thick PZT film using the sol-gel method. PZT sol was prepared using three precursors, $Pb(CH_3COO)_2$, $Zr(C_3H_7O)_4$ and $Ti((CH_3)_2CHO)_4$. The process PZT used to fabricate thin film was as follows: The sol-gel PZT solutions were prepared using 2-methoxyethanol (2-MOE) as the solvent. The sol was spin-coated onto the substrate at 1,500 rpm for 40 seconds. After spin-coated, the sample was dried at 650°C for 1 hour to evaporate the solvent. Then, the sample was crystallized at 1,000°C for 2 hours in a rapid thermal annealing furnace. The thickness of a PZT layer was about 4 μm. No polish or any special pre-treatment was done to the PZT layer before electrode deposition. A 300 nm-thick Pt/Ti layer was deposited on the PZT film as the top electrode. The SiO_2 layer thickness in the top layer was about 30 μm thick. To make via holes, the SiO_2/Pt/Ti/PZT/Pt/SiO_2/Si structure was etched by DRIE to reveal the structure. For miniaturized mechanical structures, viscous damping dominated the dissipation mechanism. As a result, we have employed a wafer-level polymer bonding technology to directly integrate the bulk PZT to the Si wafer. Major analyses can be conducted to determine the various layers thickness. The sound frequency was governed by the dimensions and the layer materials of the diaphragm, and can be controlled independent of layer thickness offering more design flexibility and control over the frequency.

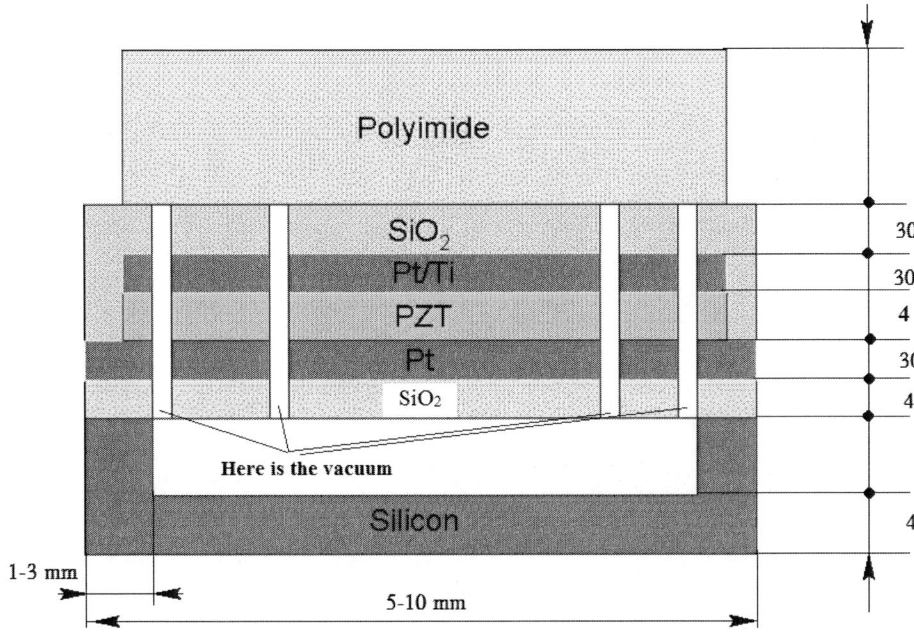

Figure 2. Schematic illustration of an acoustic device structure for underwater hydrophones.

3. Results and Discussion

A crucial issue in PZT deposition is crystallinity. In general, the deposited PZT film has a near-amorphous structure. In order to study the effect of PZT layer properties, each sample was XRD measured for the inspection of its crystal structure. The structure of the thin films was examined using a Siemens (Model D5000) X-ray diffraction (XRD) with CuKα radiation. Patterns were recorded at a rate of 1°/minute in the 2θ range of 20° to 60°. Based on the pattern of the XRD spectra presented in Figure 3, the phase intensity of PZT was found to be a pure perovskite phase with a strong peak. The film was randomly oriented because the SiO_2 film was randomly oriented. It is likely that the crystallization and growth of the PZT film was significantly influenced by the SiO_2 film structure due to the close matching of lattice parameters between SiO_2 and PZT [23].

Figure 4 shows a SEM cross-sectional image of various thick films. The PZT layer is fabricated by sol-gel processing. The 4 μm-thick of a PZT layer is fabricated with SiO_2 and Pt coating layers. The acoustic device microstructure clearly shows the well developed grain structure of the bulk PZT, which guarantees the excellent piezoelectric properties of the PZT layer.

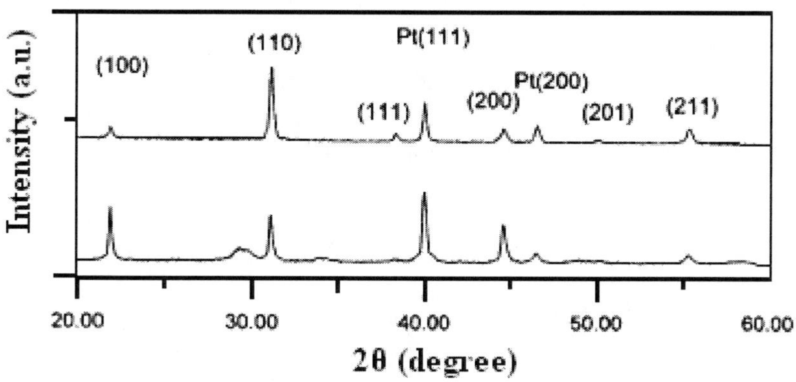

Figure 3. XRD pattern of 4 μm-thick PZT film grown on a SiO_2 layer.

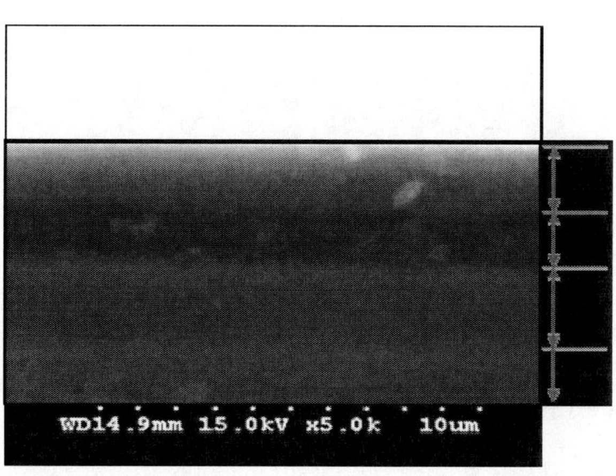

Figure 4. Cross-section image of various thick films.

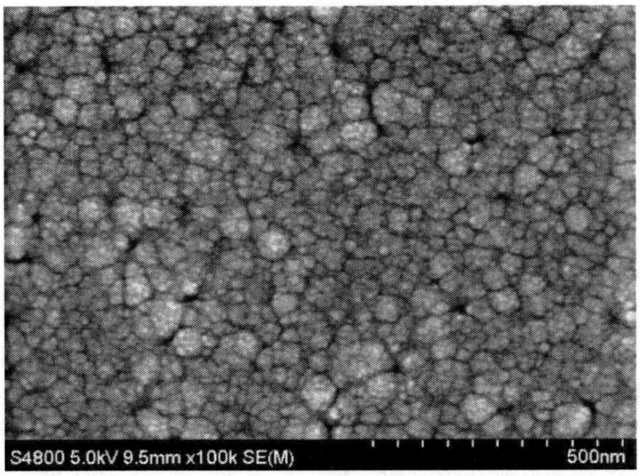

Figure 5. SEM image of the PZT surface which is grown on Pt surface.

Thin and thick layers grown on the Si wafer could not obtain such a well developed grain microstructure. As can be seen from Figure 5, the grain size of the PZT thin layer grown under the above conditions is about 100 nm. No defects were found on the coatings, and the surface was smooth on this length scale. Therefore, it can be expected that these PZT crystal structures have piezoelectricity. These micrographs were obtained using scanning electron microscopy (SEM, Hitachi Field-Emission S-4800). The SEM is used to characterize the crystal structure of the PZT layers. In the next stage, the properties of porous ceramics for the hydrophone's applications in water will be investigated; to achieve this objective, the sensitivity of porous ceramics measured and the hydrostatic characteristics were calculated.

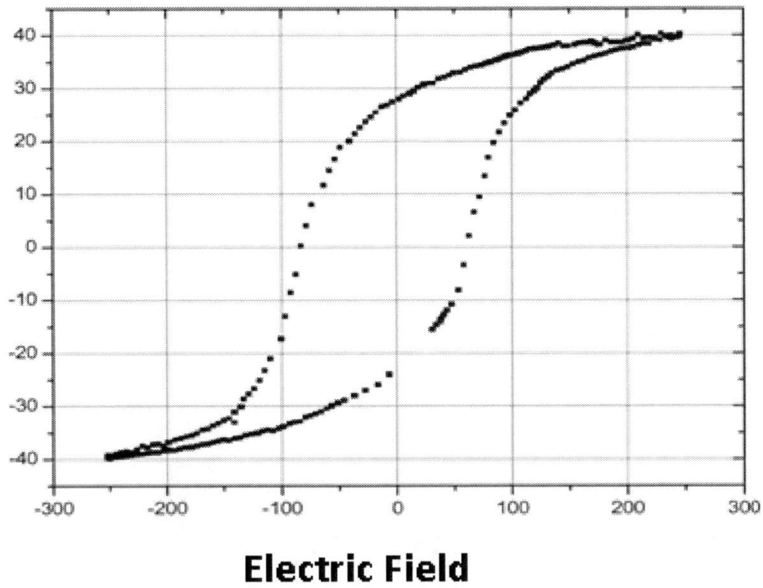

Figure 6. P-E hysteresis loop of the PZT film.

After fabrication, an acoustic device with a diameter of 10mm was employed to measure the ferroelectric properties of the lapped PZT layer. Figure 6 shows the polarization-electric field (P-E) hysteresis loops of the PZT films. The bias field is proportional to the poling effects of the ferroelectrics, indicating that the PZT film exhibits a piezoelectric effect before polling. The thicker film has higher remnant polarization (P_r) and coercive field (E_c) values. These experimental measurements reveal that a thicker film has better ferroelectric characteristics.

CONCLUSION

This research investigated the properties of PZT ceramics for hydrophones application in water. To achieve this objective, the sensitivity of PZT ceramics was measured and the hydrostatic characteristics were calculated. A SiO_2/Pt/Ti/PZT/Pt/SiO_2/Si structure was used to fabricate the acoustic devices. A PZT film with 4 μm in thickness was prepared by the sol-gel method. The characteristics of the PZT thin film were also analyzed. This acoustic device structure is very stable during the fabrication process after preparation of the PZT. No defects were found on the coatings, and the surface was smooth on SEM morphology. The electrical characteristics were evaluated in terms of dielectric constants and hysteresis curves. The thicker PZT film had higher remnant polarization and coercive field values. Finally, MEMS have a cost advantage due to their mass fabrication by the matured IC fabrication techniques. These results suggest that the fabricated acoustic devices can be used as high sensitivity hydrophones.

ACKNOWLEDGMENTS

This research was supported in part by the Russian-Taiwan Goal Oriented Research on Interdisciplinary Collaborative Research Program (Grant No. 10-08-13300). The authors would also like to thank the National Science Council for financial support under the grants NSC 99-2923-E-022-001-MY3.

REFERENCES

[1] Bernstein, J. J.; Finberg, S. L.; Houston, K.; Niles, L. C.; Chen, H.; Cross, L. E.; Li, K.; Udayakumar, K. Micromachined high frequency ferroelectric sonar transducers. *IEEE Trans. Ultrason., Ferroelect., Freq. Contr.* 1997, *vol. 44(5)*, 960–969.

[2] Young, J. W. Optimization of acoustic receiver noise performance. *J. Acoust. Soc. Am.* 1977, *vol. 61*, 1471-4.

[3] Moffett, M. B.; Powers, J. M.; Clay, W. L., Jr. Ultrasonic microprobe hydrophones. *J. Acoust. Soc. Am.* 1988, *vol. 84*,.1186-5.

[4] Aravamudhan, S.; Bhansali, S. Reinforced piezoresistive pressure sensor for ocean depth measurements. *Sensors Actuators A*. 2008, *vol.142*, 111–117.

[5] Yoo, K.; Yeh, J.-L. A.; Tien, N. C.; Gibbons, C.; Su, Q.; Cui, W.; Miles, R. N. Fabrication of a Biomimetic Corrugated Polysilicon Diaphragm with Attached Single

Crystal Silicon Proof Masses. *Solid-State Sensors and Actuators, Transducers'01*, June, 2001, pp. 130-133.

[6] Chan, H. L. W.; Lau, S. T.; Kwok, K. W.; Zhang, Q. Q.; Zhou, Q. F.; Choy, C. L. Nanocomposite ultrasonic hydrophones. *Sens. Actuators A*, 1999, *vol. 75*, 252-256.

[7] Akasheh, F.; Myers, T.; Fraser, J. D.; Bose, S.; Bandyopadhyay, A. Development of piezoelectric micromachined ultrasonic transducers. *Sens. Actuators A*. 2004, *vol. 111*, 275–287.

[8] Amanuma, K.; Hase, T.; Miyasaka, Y. Crystallization behavior of Sol-Gel derived $Pb(Zr,Ti)O_3$ thin films and the polarization switching effect on film microstructure. *Appl. Phys. Lett.* 1993, *vol. 65(24)*, 3140-3142.

[9] Auciello, O. Synthesis and characterization of $Pb(Zr_xTi_{1-x})O_3$ thin films produced by an automated laser ablation deposition technique. *J. Appl. Phys*. 1993, *vol. 73(10)*, 5197-5207.

[10] Akai, D.; Ohba, Y.; Okada, N.; Ito, M.; Sawada, K.; Takao, H.; Ishida, M. Fabrication of Ultrasonic Transducers using Epitaxial $Pb(Zr, Ti)O_3$ Thin Films on Epitaxial γ-Al_2O_3/Si Substrates for Smart Sensors. *Sens. Mater.* 2006, *vol. 18*, 161-169.

[11] Ito, M.; Okada, N.; Takabe, M.; Otonari, M.; Akai, D.; Sawada, K.; Ishida, M. High sensitivity ultrasonic sensor for hydrophone applications, using an epitaxial PZT film grown on $SrRuO_3$/Pt/γ-Al_2O_3/Si," *Sensors and Actuators A*. 2008, *vol. 145-146*, 278-282.

[12] Kim, D. M.; Eom, C. B.; Nagarajan, V.; Ouyang, J.; Ramesh, R.; Vaithyanathan, V.; Schlom, D. G. Thickness dependence of structural and piezoelectric properties of epitaxial Pb $(Zr_{0.52}, Ti_{0.48})O_3$ films on Si and $SrTiO_3$ substrates,. *Appl. Phys. Lett.* 2006 *vol. 88(142904)*, 1-3.

[13] Wado, H.; Shimizu, T.; Ishida, M. Epitaxial growth of γ-Al_2O_3 layers on Si (111) using Al solid source and N_2O gas molecular beam epitaxy. *Appl. Phys. Lett.* 1995, *vol. 67*, 2200-2202.

[14] Braun, B.; Kwak, S.; Erbil, A. Exitaxial lead zirconate-titanatethin films on sapphire. *Appl. Phys. Lett.* 1993, *vol. 63(4)*, 467-469.

[15] Dey, K.; Budd, K. D.; Payne, D. A. Thin-film ferroelectrics of PZT by Sol-Gel processing. *IEEE Trans. Ultrason. Ferroelectr. Freq. Contr.* 1988, *vol. 35(1),* 80-81.

[16] Wang, Z.; Zhu, W.; Zhu, H.; Miao, J.; Chao, C.; Zhao, C.; Tan, O. K. Fabrication and characterization of piezoelectric micromachined ultrasonic transducers with thick composite PZT films. *IEEE Trans. Ultrasonics Ferroelectrics Freq. Control* 2005, *vol. 52*, 2289-2297.

[17] Chen, H. D.; Udayakumar, K. R.; Gaskey, C. J.; Cross, L. E. Fabrication and electrical properties of lead zirconate titanate thick films. *J. Amer. Ceram. Soc.* 1996, *vol. 79(8)*, 2189–2192.

[18] Mescher, M.; Houston, K.; Bernstein, J. Novel MEMS micro-shell transducer arrays for high-resolution underwater acoustic imaging applications. *Proc. of the IEEE Sensors*. 2002. *vol. 1*, 541-546.

[19] Oralkan, O.; Sanli Ergun, A.; Khuri-Yakuh, T. Capacitive micromachined ultrasonic transducers: Next generation arrays for acoustic imaging. *IEEE Trans. on Ultrasonics Ferroelectrics and Frequency Control*. 2002, *vol. 49*, 11-16.

[20] Cheng, Y. T.; Ho., J. J.; Wang, C. K.; Lee, W.; Lu, C. C.; Yau, B. S.; Nain, J. L.; Chang, S. H.; Chang, C. C.; Wang, K. L. Improvement of organic solar cells by flexible substrate and ITO surface treatments. *Applied Surface Science*, 2010, *vol. 256*, 7606–7611.

[21] Finberg, S. L.; Houston, K.; Niles, L. C.; Chen, H. D.; Cross, L. E.; Li, K. K.; Udayakumar, K. Micromachined high frequency ferroelectric sonar transducers. *IEEE Tran. on Ultrasonics, Ferroelectrics and Frequency Control,* 1997. vol. *44 (5)*, 960-969.

[22] Isarakorn, D.; Sambri, A.; Janphuang, P.; Briand, D.; Gariglio, S.; Triscone, J. M.; Guy, F.; Reiner, J. W.; Ahn, C. H., de Rooij, N. F. Epitaxial piezoelectric MEMS on silicon. *J. Micromech. Microeng.* 2010, *vol.20 (055008)*, 1-9.

[23] Chen, S. Y.; Chen, I. W. Texture development, microstructure evolution, and crystallization of chemically derived PZT thin films. *J. Amer. Ceram. Soc.* 1998, *vol. 81(1)*, 97–105.

In: Physics and Mechanics of New Materials …
Editors: Ivan A. Parinov and Shun-Hsyung Chang

ISBN: 978-1-62618-535-7

Chapter 27

A MIMO-OFDM UNDERWATER ACOUSTIC MULTIMEDIA COMMUNICATION

C.F. Lin*[*a], *S. H. Chang*[b], *C. C. Lee*[a], *W. C. Wu*[a], *W. H. Chen*[a], *K. H. Chang*[a], *J. C. Y. Lee*[b] and *I. A. Parinov*[c]

[a] Department of Electrical Engineering,
National Taiwan Ocean University, Taiwan, R.O.C.
[b] Department of Microelectronics Engineering,
National Taiwan Ocean University, Taiwan, R.O.C.
[c]Vorovich Mechanics and Applied Mathematics Research Instiate,
Southern Federal University, Russia

ABSTRACT

In the paper, a transmission scheme based on multi-input multi-output (MIMO) orthogonal frequency-division multiplexing (OFDM) was proposed for underwater acoustic multimedia (UWAM) communication. The direct mapping (DM) and space-time block code (STBC) strategies, OFDM, adaptive modulation, and unequal error protection (UEP) was used in the proposed UWAM system. Simulation results show that the proposed scheme not only fulfils the quality of services (QoS) required of a UWAM system, but also achieves maximum transmission bit rates or minimum transmission power.

1. INTRODUCTION

Underwater acoustic (UWA) transmission scheme is an interesting field of communications research. Several reviews have discussed the research and development of UWA communication systems carried out in the past 10 to 30 years [1-12]. The areas of discussion have included channel physics, receiver algorithms, time or spatial diversity, source and channel coding, and UWA networks. This and related advanced underwater

* e-mail:lcf1024@mail.ntou.edu.tw.

communication schemes can be applied in underwater acoustic telemetry, robot communication and control, vehicles, sensor networks, and image/audio transmission. The major problems of underwater communication are deep multipath fading, highly time-varying channels, large transmission delays, and power consumptions. The underwater channels are severely band-limited, bandwidth-efficient transmission schemes play a central role in current research. Several papers have discussed the application of spread spectrum technologies to the UWA communication scheme. We have previously proposed signal-input signal-output (SISO), and an underwater acoustic multimedia (UWAM) transmission scheme based on direct mapping (DM) with orthogonal variable spreading factor (OVSF) codes [13-14]. In the present paper, we extend upon a previous study [15] to propose a MIMO-OFDM UWAM communication scheme. In the proposed scheme, DM and STBC strategies, a power assignment mechanism, OFDM, adaptive modulation, block-by-block basis transmission strategy, and unequal error protection (UEP) are integrated in the transmission of audio, image, and sensor signals over an UWA channel.

2. The Proposed MIMO-OFDM UWAM Scheme

Figure 1 shows the proposed MIMO OFDM-based UWAM communication scheme. Sensor, audio, and image signals are transmitted using the block transmission mode in UWAM applications. A G.729 encoder, a Joint Photographic Experts Group 2000 (JPEG2000) encoder, a UEP scheme, an object-composition petri-net (OCPN) model, blocks by blocks transmission strategies, DM or STBC MIMO transmission technology, adaptive modulation, serial-to-parallel and parallel-to-serial schemes, OFDM modulation, and a power assignment mechanism are all integrated. Digital sensor data bit streams are inputted to the UEP encoder, and the UEP sensor data bit streams are extracted as output. Audio and image signals are inputted to G.729 and JPEG2000 encoders, respectively, and the G.729 audio, and JPEG2000 image bit streams are extracted as output, are then inputted to the UEP encoder to extract UEP audio and video bit streams as output. The digital sensor data, G.729 audio, and JPEG2000 images UEP bit streams are inputted to the OCPN model, and blocks of the data, audio, and image UEP bit streams are decomposed into several data, audio, and image packets, respectively, to extract as output. The OCPN model [18], can describe the temporal relationships for various components of UWAM information including type, size, throughput requirements, and duration of presentation. The transmission time and amounts of sensor data, audio, and image packets can be computed by OCPN models for multimedia player in the receiver. The quality of service (QoS) requirements for different messages in UWAM applications is different. In the proposed system, the acceptable transmission BERs for data, audio, and image packets are assumed to be 10^{-6}, 10^{-3}, and 10^{-4}, respectively. Assume that in the UWAM communication system, there are P users, M OFDM symbols in an OFDM transmission packet, N sub-carriers in an OFDM symbol. Furthermore, assume that each packet can be modulated independently. Let $S_{p,m,k,l}$ denote the adaptive complex modulated symbol transmitted in the m-th OFDM symbol over the k-th sub-carrier using the l-th transmission antenna before the Inverse Fast Fourier Transform (IFFT) for the considered OFDM block in the p-th user. The transmitted signal after the IFFT can then be represented as

$$s_{p,m,k,l} = \frac{1}{\sqrt{N}} \sum_{k=0}^{N-1} S_{p,m,k,l} u_{pl} e^{\frac{j2\pi lk}{N}}, \quad (1)$$

where μ_{pl} is the transmission power weighting factor of the l-th antenna of the p-th user, and $\frac{1}{30} \le \mu_{pl} \le 1$. The received signal of the k-th subcarrier of the m-th OFDM symbol using the z-th received antenna for the p-th user can be expressed as

$$y_{p,m,k,z} = \sum_{l=1}^{L} s_{p,m,k,l,z} * h_{p,m,k,l,z} + e_{p,m,k,l,z}, \quad (2)$$

where $h_{p,m,k,l,z}$ is the impulse response of the underwater channel of the k-th subcarrier of the m-th OFDM symbol using the l-th transmission antenna and z-th received antenna for the p-th user. $e_{p,m,k,l,z}$ is the additive white Gaussian noise of the k-th subcarrier of the m-th OFDM symbol using the l-th transmission antenna and z-th received antenna for the p-th user. The received signal of the k-th sub-carrier of the m-th OFDM symbol for the p-th user can then be expressed as

$$y_{p,m,k} = \sum_{z=1}^{Z} y_{p,m,k,z}. \quad (3)$$

The double window detection algorithm [17] is used to detect the signal-to-noise ratios (SNRs) of sensor data, image, and audio packets. The energies of windows $q-1$ and q are $c_{p,q-1}$ and $c_{p,q}$, respectively, for the p-th user.

$$c_{p,q-1} = \sum_{k=1}^{N} y_{p,q-1} y^{*}_{p,q-1,k} = \sum_{k=1}^{N} \left| y_{p,q-1,k} \right|^2; \quad (4)$$

$$c_{p,q} = \sum_{k=1}^{N} y_{p,q} y^{*}_{p,q,k} = \sum_{k=1}^{N} \left| y_{p,q,k} \right|^2.$$

The decision parameter d_p of the p-th user is defined as

$$d_p = \frac{c_{p,s-1}}{c_{p,s}} = \frac{R_p + N_p}{N_p} = SNR_p + 1, \quad (5)$$

where R and N are the sums of the signal energies and noise energies, respectively, across the two windows for the p-th user. SNR_p is SNR of the p-th user. The receiver makes a decision according to the specified thresholds.

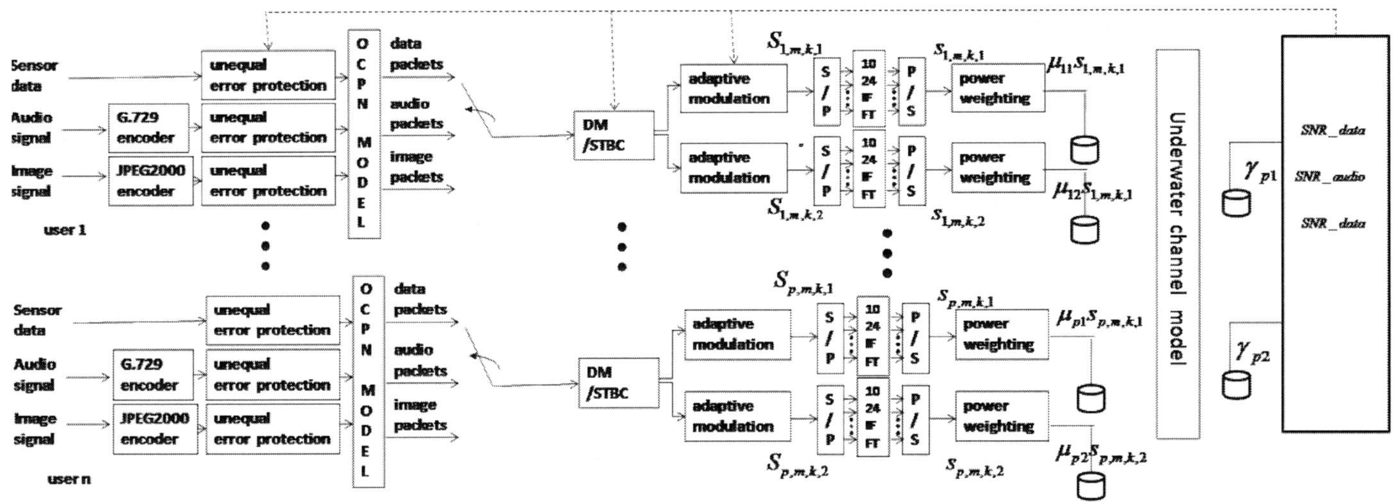

Figure 1. The proposed MIMO OFDM-based UWAM communication scheme.

3. Simulation Results

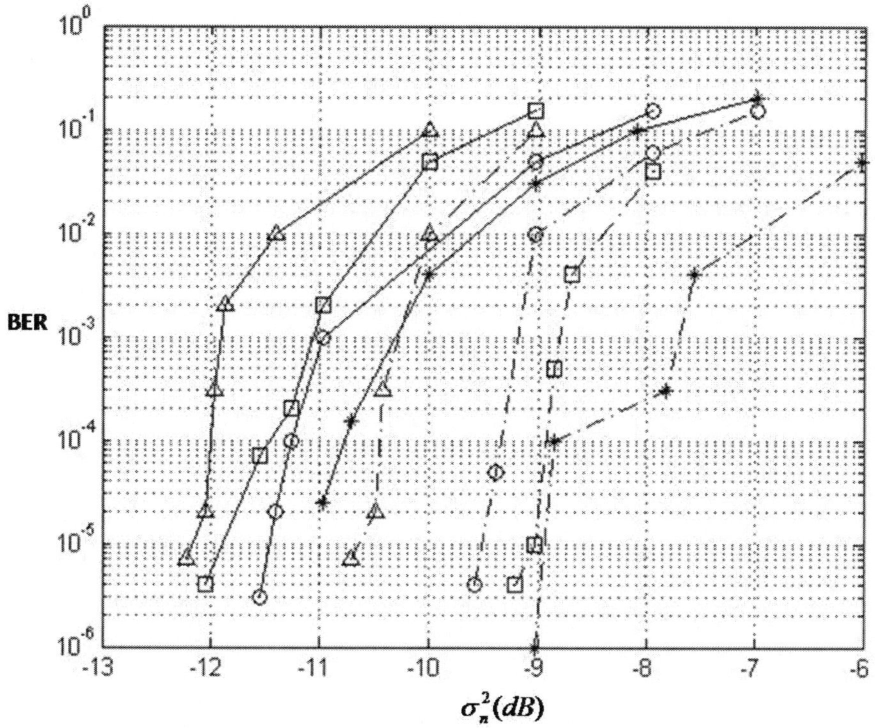

Figure 2. The BERs performance of the UWAM communication system. (Δ__:DM, QPSK, K = 91/2 (561,753) convolution code, Δ--:STBC, QPSK, K = 9 1/2 (561,753) convolution code, ○__: DM, BPSK, K = 9 1/2 (561,753) convolution code, ○--: STBC, BPSK, K = 9 1/2 (561,753) convolution code, □__: DM, QPSK, K = 9 1/3 (557, 663, 771) convolution code, □--: STBC, QPSK, K = 9 1/3 (557, 663, 771) convolution code, *__: DM, BPSK, K = 9 1/3 (557, 663, 771) convolution code, *--: STBC, BPSK, K = 9 1/3 (557, 663, 771) convolution code).

We carried out simulations to demonstrate proper functionality of the proposed UWAM communication system. In the proposed system, we use BPSK or QPSK modulations, K = 9 1/2 (561,753) or 1/3 (557, 663, 771) convolution codes with soft decoding, and 2 × 2 DM or STBC strategies for the transmission of audio, image and sensor data packets. We adopt an underwater channel model described in a reference [16]. The transmission distance, carrier central frequency, and underwater channel bandwidth are 1 km, 11.5 kHz, and 4 kHz, respectively. The maximum data transmission rate is 6 kbps using QPSK modulation, 1/2 channel coding, and a 2 × 2 DM transmission strategy.

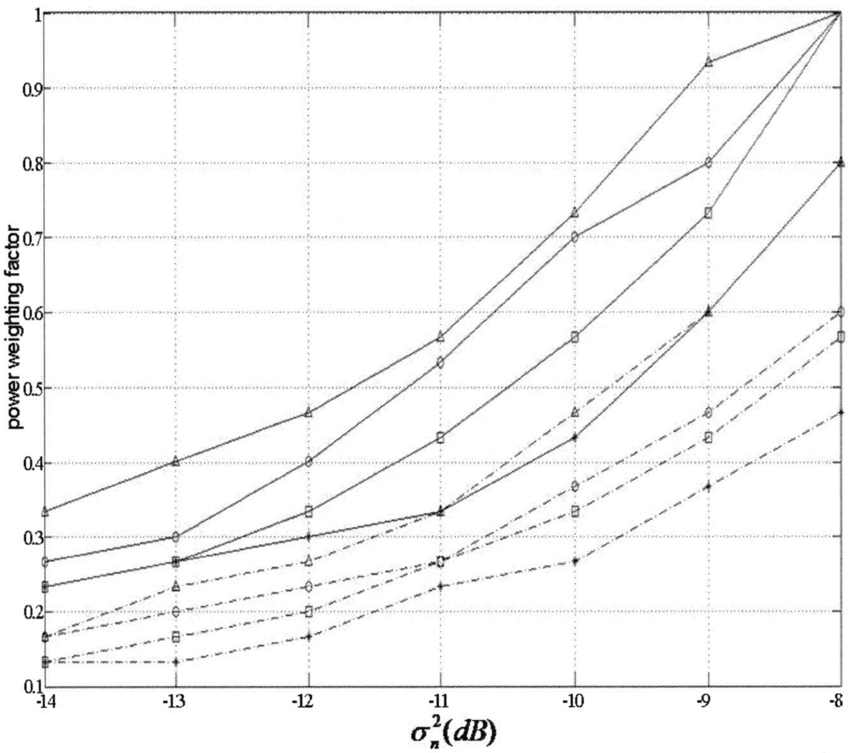

Figure 3. The performance of the power weighting factor of the UWAM system with a transmission BER of 10^{-3}.(∆__:DM, QPSK, K = 9 1/2 (561,753) convolution code, ∆--:STBC, QPSK, K = 9 1/2 (561,753) convolution code, o__: DM, BPSK, K = 9 1/2 (561,753) convolution code, o--: STBC, BPSK, K = 9 1/2 (561,753) convolution code, □__: DM, QPSK, K = 9 1/3 (557, 663, 771) convolution code, □--: STBC, QPSK, K = 9 1/3 (557, 663, 771) convolution code, *__: DM, BPSK, K = 9 1/3 (557, 663, 771) convolution code, *--: STBC, BPSK, K = 9 1/3 (557, 663, 771) convolution code).

Figures 2 and 3 respectively show the performance of the BER of the UWAM system and that of the power weighting factor with a transmission BER of 10^{-3}. Symbols "∆—" , "∆--", "o—", "o--", "□—", "□--", "*—", and "*--" denote transmission modes of "DM, QPSK, K = 9 1/2 (561,753) convolution code", "STBC, QPSK, K = 9 1/2 (561,753) convolution code", "DM, BPSK, K = 9 1/2 (561,753) convolution code", "STBC, BPSK, K = 9 1/2 (561,753) convolution code", "DM, QPSK, K = 9 1/3 (557, 663, 771) convolution code", "STBC, QPSK, K = 9 1/3 (557, 663, 771) convolution code", "DM, BPSK, K = 9 1/3 (557, 663, 771) convolution code", and "STBC, BPSK, K = 9 1/3 (557, 663, 771) convolution code", respectively. The transmission mode is lowest when we obtain the BER of the transmission scheme and the power weighting factor using the STBC strategy, BPSK modulation, and K = 9 1/3 (557, 663, 771) convolution code; it is highest when we obtain the BER and power weighting factor using the DM strategy, QPSK modulation, and K = 9 1/2 (561,753) convolution code. From these simulation results, we see that the proposed UWAM system is suitable for transmission of audio, image, and sensor signals to achieve maximum data transmission rates or minimum power consumption levels.

CONCLUSION

In this paper, we proposed a transmission scheme to apply to UWAM communication. Advanced communication schemes including the DM or STBC MIMO transmission strategies, UEP, adaptive modulation, OFDM modulation, and power assignment mechanism were integrated to transmit audio, image, and sensor signals over UWA communication channels. Simulation results showed that the proposed UWAM communication system could meet the QoS of underwater multimedia application and achieve maximum transmission data rates or minimum power consumption level.

ACKNOWLEDGMENTS

The authors acknowledge the support of the grant from The National Science Council of Taiwan NSC 98-2221-E-022-018, NSC 99-2221-E-022-019, NSC 99-2923-E-022-001-MY3, and the valuable comments of the reviewers.

REFERENCES

[1] Quazi, A. H.; William, L.-K. Underwater Acoustic Communications. *IEEE Communications Magazine*. 1982, vol.20, 24-30

[2] Baggeroer, A. B. Acoustic Telemetry-An Overview. *IEEE J. Oceanic Eng.* 1984, vol.9, 229-235

[3] Stojanovic, M. Recent Advances in High-Speed Underwater Acoustic Communications. *IEEE J. Oceanic Eng*. 1996, vol.21, 125-136.

[4] Kilfoyle, D. B.; A. B. Baggeroer, A. –B. The State of the Art in Underwater Acoustic Telemetry. *IEEE J. Oceanic Eng*. 2000, vol.25, 4-27

[5] Sozer, E. M.; Stojanovic, M.; Proakis, J. –G. Underwater Acoustic Networks. *IEEE J. Oceanic Eng*. 2000, vol.25, 72-83

[6] Proakis, J. G.; Sozer, E. –M.; Rice, J. -A.; Stojanovic, M. Shallow Water Acoustic Networks. *IEEE Communications Magazine*. 2001, vol.39, 114-119

[7] Cui, J. H.; Kong, J.;Gerla, M.; Zhou, S. The Challenges of Building Scalable Mobile Underwater Wireless Sensor Networks for Aquatic Applications. *IEEE Network*. 2006, vol.20, 12-18

[8] Stojanovic, M. Underwater Acoustic Communication: Design Considerations on the Physical Layer. *MTS/IEEE Ocean'08*.

[9] Chitre, M.; Shahabudeen, S.; Freitag, L.; Stojanovic, M.Recent Advances in Underwater Acoustic Communications & Networking. *MTS/IEEE Ocean'08.*

[10] Rutgers, D. P.; Akyildiz, I. –F. Overview of Networking Protocols for Underwater Wireless Communications. *IEEE Communications Magazine*. 2009, vol.47, 97–102

[11] Singer, A. C.; Nelson, J. –K.; Kozat, S. S. Signal Processing for Underwater Acoustic Communications. *IEEE Communications Magazine*. 2010, vol.47, 90-96

[12] Stojanovic, M.; Preisig, J. Underwater Acoustic Communication Channels: Propagation Models and Statistical Characterization. *IEEE Communications Magazine*. 2009, vol.47, 84-89

[13] Lin, C. F.; Chang, S. –H.; Chen, J. Y.; Yan, J. –T.; A Power Assignment Mechanism for Underwater Wireless Multimedia. *MTS/IEEE Ocean'08.*

[14] Lin, C. F., Chen, J. –Y.; Yu, Y. J.; Yan, J. –T.; Chang, S. H. Direct Mapping OVSF-based Transmission Scheme for Underwater Acoustic Multimedia Communication. *Journal of Marine Science and Technology*. 2010, vol.18, 413-418

[15] Lin, C. F.; Shih, C. -H.; Chen, C. P.; Leu, S. –W.; Wu, J. K.; Tseng, C. -H.; Hung, H. S.;. Lu, F. –S.; Parinov, I. A.; Chang, S. –H. An OFDM-based Transmission Scheme for Underwater Acoustic Multimedia. *WSEAS Transactions on Communications*. 2009, vol.8, 343-352

[16] Tao, J.; Zheng, Y. -R.; Xiao, C.; Yang, T. C.; Yang, W. -B. Time-Domain Receiver Design for MIMO Underwater Acoustic Communications. *MTS/IEEE Ocean'08*.

[17] Terry, J.; Heiskala, J.; *OFDM Wireless LANs: A Theoretical and Practical Guide*, Indianapolis,Sams Publishing: Indiana, 2002, 51-53

[18] Woo, M.; Prabhu, -N.; Ghafoor, A.Dynamic resource allocation for multimedia services in mobile communication environments. *IEEE J Sel Areas Commun*. 1995, vol.13, 913–922

In: Physics and Mechanics of New Materials …
Editors: Ivan A. Parinov and Shun-Hsyung Chang
ISBN: 978-1-62618-535-7

Chapter 28

The Process of Silicon-Based Resonant Tunneling Diodes and Their Application for Underwater Acoustic Multimedia Communication

Yuang-Tung Cheng[1], Jiun-Chen Liou[2], Kuo-Chang Hou[3], David. R. C. Chu[4], Ping-Chen Wu[5] and Shun-Hsyung Chang*

[1]ILS Center, Chung-Shan Institute of Science & Technology, Lung-Tan, Tao-Yuan, Taiwan
[2]Department of Electrical Engineering, National Taiwan Ocean University, Keelung, Taiwan
[3]Department of Microelectronic Engineering, National Kaohsiung Marine University, Kaohsiung, Taiwan
[4]College of Aviation, China University of Science and Technology, Henshan Shiang, Hsinchu, Taiwan
[5]Department of Materials Science and Engineering, National Chung Hsing University, Taichung, Taiwan

Abstract

A structure of microwave components is the Al/SiO2/n-Si/SiO2/n-Si/Al hetero-structure designed and fabricated with a DC/RF sputter deposition system. In this paper, we demonstrate that the quantum well structure of silicon material is deposited at room temperature. NDR effects are directly observed when voltage peaks and conductance peaks appear at the same voltage positions in the I–V characteristics. The effect of barrier thickness on the peak-to-valley current ratio, peak voltage Jp and peak is studied. From experiments, we get the best double-barrier thickness, equal to 0.6 μm, Jp = 184 nA/cm2, and Jv = 110 nA/cm2, and an associated PVCR = 1.67, suitable for low-power

* E-mail: stephenshchang@me.com.

microwave device applications. Compared with past papers, our research has many advantages. The PVCR of our designed structure is lower, Si-based material is affordable, the process can be practiced at room temperature, and the size of device is small. In addition, the result demonstrates the designed device can use little power, switch quickly, and be produced easily.

1. INTRODUCTION

The increasing demand for faster switching, smaller size and lower power consumption are keys for the next generation of microwave device applications [1–3]. A new technology corresponding to the requirement is based on the incorporation of resonant tunneling diodes (RTD) in the circuits. Several microwave circuits using RTD have been demonstrated as in the following: a low-noise amplifier [4], mixer [5], frequency multiplier [6] and oscillator [7]. The RTD that exhibit negative differential effects (NDR) in their current-voltage (I-V) characteristics are very useful for circuit applications because they can increase circuit speed and reduce power consumption [8, 9].

NDRs were first observed in 1973 [10] after the demonstration of GaAs/AlGaAs [11, 12]. The most representative NDR material is GaAs; however, it is expensive. Because of the nano-structure and chemistry of the interfaces, lattice mismatch, defect density, and interface roughness are critical in improving the performance of the RTD structures. Moreover, it is difficult to fabricate high-quality silicon (Si) on a GaAs substrate. All of these above-mentioned fabrications of RTDs utilize III-V materials, which are not compatible with main and affordable Si device technology. Thus, it is of practical importantance and necessity to study the resonant tunneling and NDR effects in microwave devices. In this paper, we provide a good basis for the observation of resonant tunneling and NDR effects in the Si-based material, which is suitable for microwave oscillating applications [13-15].

The simplest Si-based structure for an RTD with multiple NDR is a double-barrier heterostructure, and resonant levels are created in one quantum well. Hetero-junction tunneling diodes allow the highest peak-valley current ratio (PVCR) and an excellent cut-off frequency. Therefore, in order to utilize their unique transport properties in microwave devices, studying the resonant tunneling and NDR effects in an Si substrate is necessary. We use Wolfram Mathematica 6 to simulate I-V characteristics, negative differential conductance, and power consumption. The purpose is to propose a novel RTD-based microwave circuit and to estimate the performance. In contrast to conventional approaches where the oscillation is based on a small-signal behavior around the RTD negative differential resistance region, the present circuit exploits a large-signal behavior of the RTD (i.e., a switching from the peak state to the valley state in the I-V characteristics). We successfully simulate NDR phenomenon in SiO_2/Si/SiO_2 hetero-structures. In this paper, SiO_2 deposited under room temperature is chosen as a barrier. Since SiO_2 made using the ion reactive DC/RF magnetron sputtering system has a relatively high potential height of 3.2 eV, it can be an ideal barrier for quantum confinement in Si. The characteristics of RTD devices are low power, low leakage current, and large PVCR, and are suitable for being applied to an acoustic multimedia communication component.

2. Theoretical Analysis

At first, we propose a parameter extraction method for paralleling with three RTD devices and analyze their RTD's DC characteristics. Theoltage peak is directly observed in the I-V characteristics while the conductance peak is identified at the same voltage positions. This phenomenon is the NDR effect. In this paper, we simulate the quantum well structure under a room-temperature deposition. All results are compared with corresponding experimental data on the extracted different parameters. In past, the material used for NDR was mostly GaAs, but this paper provided a good basis for the observation of resonant tunneling and NDR effects in the Si material. The different I-V curves of RTDs models have been developed. The current equations of both Gaussian function and exponential function are used in two NDR devices in parallel and one NDR device. A simple model for negative resistance in the room temperature is expressed below in Equations (1). The dc characteristics for RTD (J_{RTD}) current density can be described by $J_{RTD} = J_1 + J_2 + J_3$. Where the J_1 is the resonant tunneling current, J_2 is the resonant tunneling parallel current and J_3 is the saturation current. The parameter extraction method for our simulations is as follows [16-19]:

$$
\begin{aligned}
J_1 &= A \ln\left[\frac{1+e^{(B-C+n_1V)e/KT}}{1+e^{(B-C-n_1V)e/KT}}\right]\left[\frac{\pi}{2}+\tan^{-1}\left(\frac{C-n_1V}{D}\right)\right] \\
J_2 &\propto \left[\frac{\pi}{2}+\tan^{-1}\left(\frac{F-\frac{n_3eV}{2}}{S}\right)\right] \\
J_3 &= H\left(\exp\left(\frac{n_2V}{V_T}\right)-1\right)
\end{aligned}
\tag{1}
$$

where A is effective mass and temperature, B is the material coefficient of diode, C is the energy of quantum well, D is one half of the width of the resonance layer and on the order of one meV or less for even quite thin barrier widths [16], H is the diode saturation current, n_1 and n_2 are the resonance tunneling diode material coefficients, and n_3 is the divergence coefficient [20].

The I-V model of a double-well resonant-tunnel diode with a Si-based material, which is suitable for Wolfram Mathematica 6, is presented in this paper. Characteristics of I-V and conductance-voltage (Cond-V) are identified under different conditions. Figure 1 shows the simulated I-V curve of the double-barrier single quantum well RTD all at room-temperature operations. The material coefficient (B) in Equations (1) of p-type Si (P-Si), Si, and n-type Si (N-Si) is equal to – 0.358, 0, and 0.358 eV, respectively. Parameters extracted by the method described above are: A = 1.554 (room temperature), B = 0.0465, C = 0.0689, D = 0.0052, n_1 = – 0.358 eV to 0.358 eV, n_2 = – 0.358 eV to 0.358 eV, n_3 = 0.9, S = 0.07 to 0.1, F = 0.07 to 0.1, and $H = 1.43\times10^{-8}$ in the simulation. However, the material coefficient is bound up with peak voltage. The change of threshold voltage is not much. We can clearly see that the material coefficient increases as the peak tunneling current increases. Apparently, all the features mentioned above are quite different from the NDR effects reported in the quantum well structure. We can obtain the good results under different conditions. It is obviously found that the threshold voltage shifts with the change of the n_1, n_2 parameter. In Equation 1, explication, J_1, which impacts the tunneling current, is the material coefficient. The material coefficient of the threshold voltage is 0.9 V. The material coefficient of the peak current density is 20, 40, 480 (μA/cm^2) respectively. Figure1 shows that a satisfactory result is obtained.

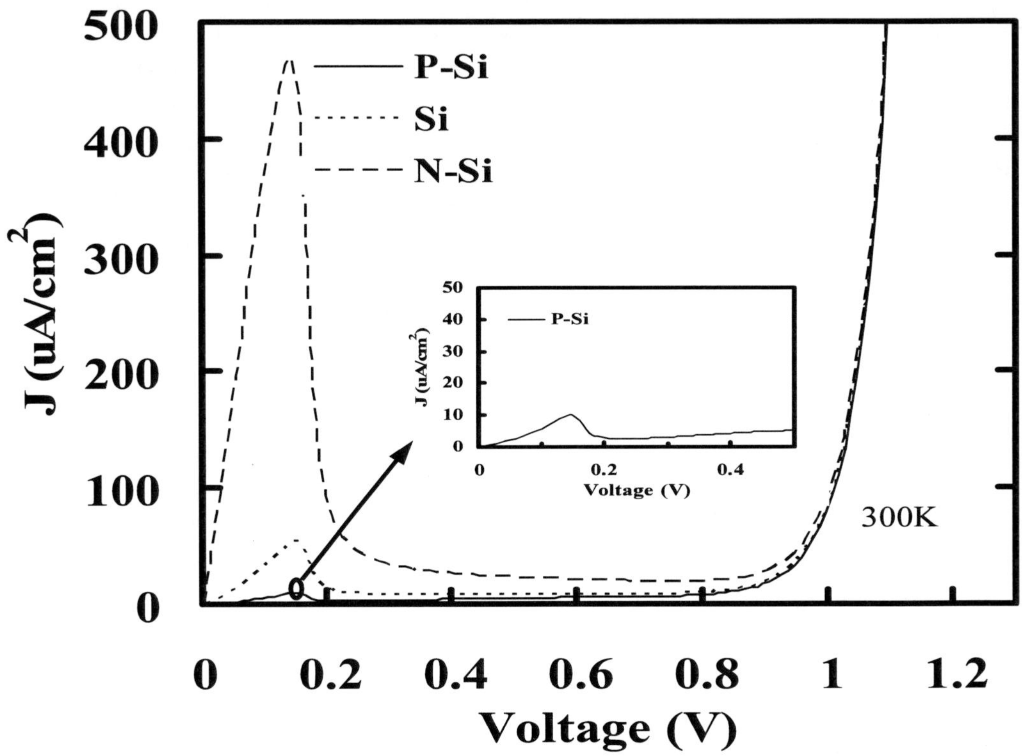

Figure 1. Numerical simulate results for the *I-V* characteristics of the double-barrier single quantum well RTD with P-Si, Si, and N-Si at room temperature; solid, dotted, and dashed curve, respectively. The inset plots the normalized *I-V* characteristics of RTD with P-Si.

The changing point of the material coefficient affects the negative resistance, which is very useful for the high-frequency circuit applications. Figure 2 shows that connecting a negative-resistance diode with a resonant cavity forms an equivalent circuit. As two ends are connected, the DC voltage can be generated in the resonant cavity of microwave oscillations. Resonant cavity results from the equivalent lumped circuit of R_o, L_o, C_o parallel resonance circuits; R_d is the negative resistance in RTD; C_d is the capacitance parameter of the charge region in NDR diodes. The distribution parameters (C and L) are the diode lead shell of the left-side dash line in Figure 2. The diodes have negative resistance, $-R_d$, and load resistor, R_0, $-R_d$ and R_0 are in parallel. Their resistance is given as:

$$R' = \frac{R_d R_o}{R_d + R_o}. \tag{2}$$

When DC power is turned on, R_d is smaller than R_o, R' is negative, and $-R_d$ is nonlinear. As the amplitude $|-R_d|$ increases, the value gradually decreases. In Equation (2), R' is infinite when $|-R_d|$ is equal to R_o when R', L', and C' are connected in parallel and R' is an open-circuit, thus generating persistent oscillation.

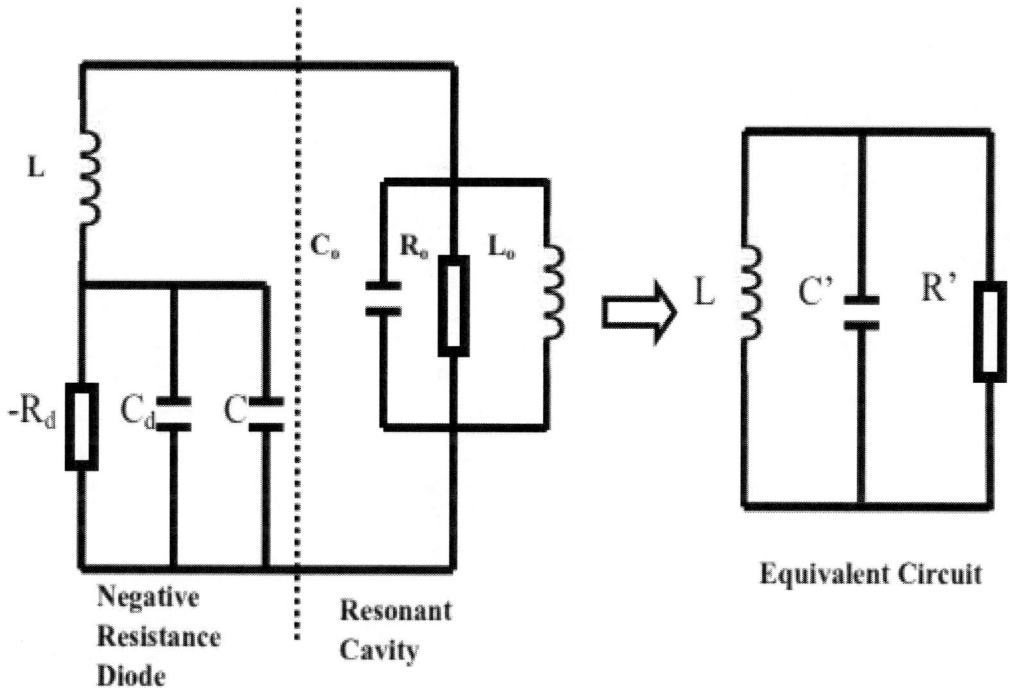

Figure 2. A circuit schematic of the negative-resistance equivalent, consisting of a negative-resistance diode and resonant cavity. The R', L', and C belong to the simplified, equivalent circuit with connected parallel, R' is open-circuit, generating persistent oscillation.

Electrical conductance is used to measure how easily electricity flows along a certain path through an electrical element. The conductance is calculated as:

$$G = \frac{1}{R} = \frac{I}{V}, \tag{3}$$

where G is the electrical conductance, R is the electrical resistance, I is the electric current, and V is the voltage. Figure 3 shows the simulated conductance–V curve of the double-barrier single quantum well RTD and set parameters at room temperature. The material coefficient of the resonance tunneling diode is − 0.358, 0, and 0.358 eV, respectively. The negative conductance ranges from 0.15 to 0.35 volts. In the conductance–V characteristics, we also observe a similar peak structure with their peak position being the the same as that observed in I–V characteristics. It has been obviously found that all of peak conductance shifts with the change of the material coefficient. Thus, Equation (3) can demonstrate that the negative differential resistance area is suitable for the negative conductance.

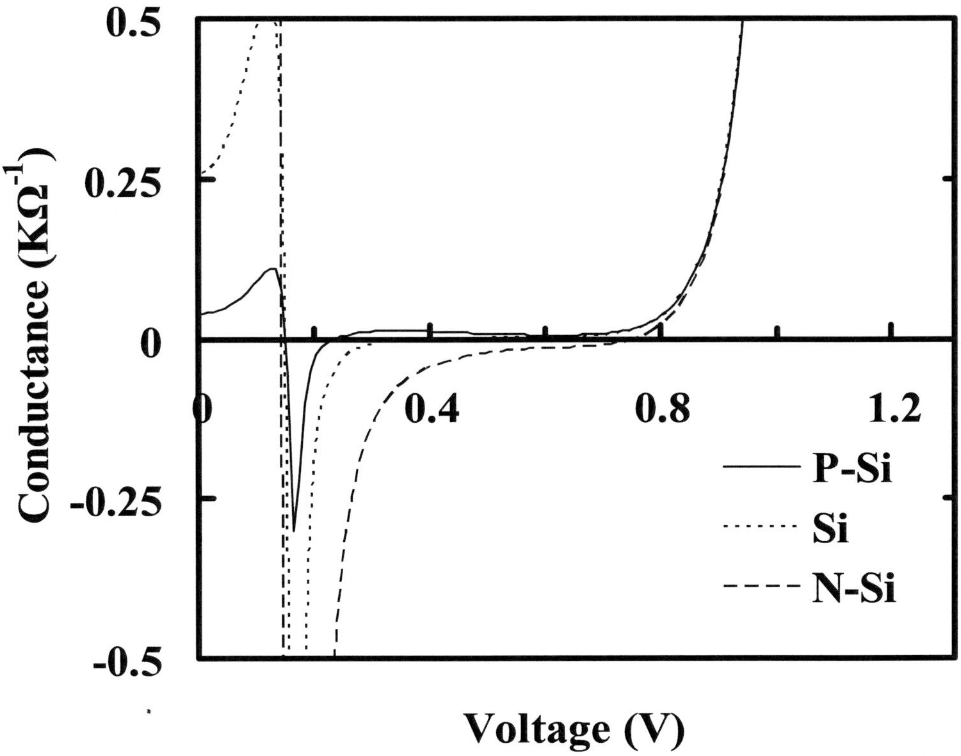

Figure 3. Numerical simulation results for the conductance-V curves of the double-barrier single quantum well RTD with P-Si, N-Si, and Si at room temperature; solid, dotted, and dashed curve, respectively.

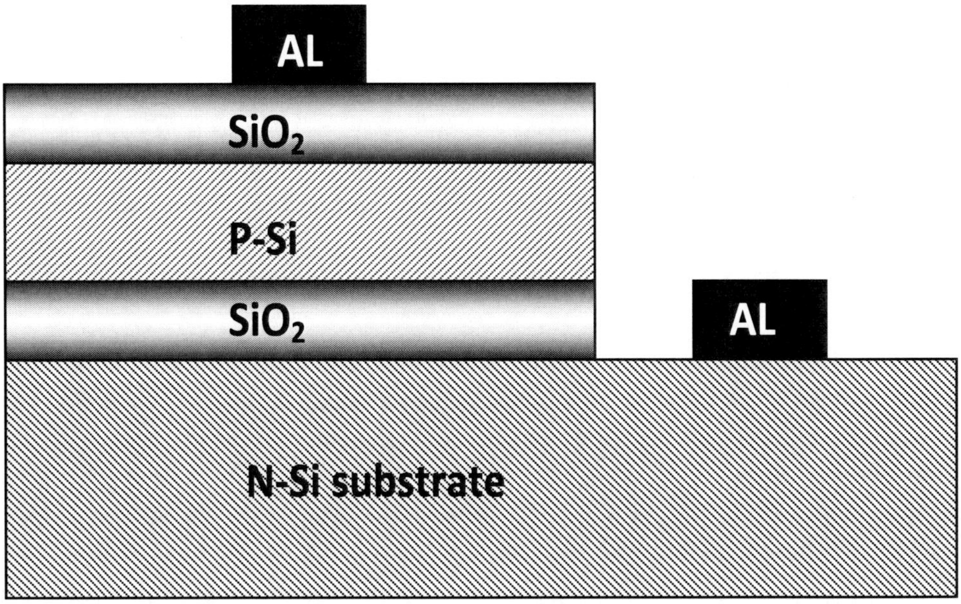

Figure 4. A schematic of the basic the Al/SiO_2/P-Si/SiO_2/N-Si device structure designed in this paper. The different SiO_2 thicknesses of the double barrier is obtained by DC/RF sputtering.

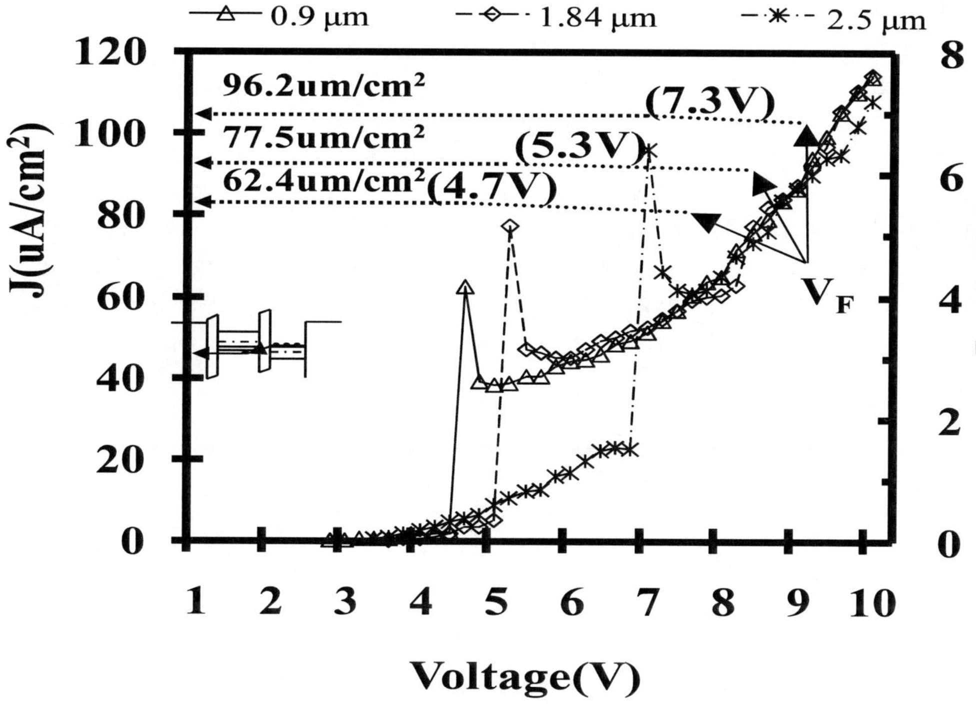

Figure 5. The effect of SiO_2 thickness on NDR peaks of current and voltage. The schematic band diagram under forward bias is shown in the bottom-left inset.

3. Experimental

The basic structure of the Si-based double-barrier single quantum well RTDs, shown in Figure 4, was deposited by conventional DC/RF magnetron sputtering. The *p*-Si single quantum well was sandwiched between the SiO_2 barriers. The structure of the double-barrier, silicon-based RTD was the Al/SiO_2 barrier/*p*-Si quantum well/SiO_2 barrier/*n*-Si (100) substrate. The RTD studied here varied with the overall double-barrier thickness. The RTD fabrication process was described as follows.

Prior to depositing SiO_2, we used buffered oxide etch (BOE) to remove the native oxide layers. The SiO_2 thin film was deposited onto the Si substrate using DC sputtering. The diameter associated with the target of SiO_2 was 2 inches. The DC power applied to the target was fixed at 50 W, and the films were deposited in the presence of a mixture of Ar (20 sccm) and O_2 (3 sccm) at a deposition pressure of 5×10^{-3} torr. This double-barrier layer had been grown by employing an ion-reactive, DC-magnetron sputtering process to deposit SiO_2 at 400°C. The *p*-Si single quantum well with a thin film of 1.5 μm was deposited on the SiO_2/Si substrate using DC sputtering. The diameter associated with the target of *p*-Si is 2 inches. The sputtering power was 50 W, and the thin film was deposited in the presence of argon (20 sccm) at a working pressure of 5×10^{-3} torr.

To get the best efficiency, we also employed the three-target alternation RF magnetron sputtering to deposit SiO_2 thin films onto a Si substrate at room temperature. We used BOE to remove the native oxide layers before depositing the SiO_2. The RF power applied to the target was fixed to 150 W, and the flow rate of the argon gas was 10 sccm during deposition. The background pressure in this chamber was kept below 5×10^{-6} torr, and sputtering was carried out at an argon pressure of 4×10^{-3} torr. The *p*-Si single quantum well's thin film, at a thickness of 1.5 μm, was deposited on the SiO_2/Si substrate by RF sputtering. The diameter associated with the target of SiO_2 was 2 inches. The sputtering power was 100 W, and the flow rate of argon gas during deposition was 10 sccm. The background pressure in this chamber kept below 5×10^{-6} torr, and sputtering was carried out at an argon pressure of 4×10^{-3} torr. Al electrodes on the top and bottom were evaporated through the photoresister mask which had a diameter of 200 μm. All the electrodes were the same size, $(100)^2\times\pi$ (μm^2). We can observe the size of RTD device as 1×1 mm^2.

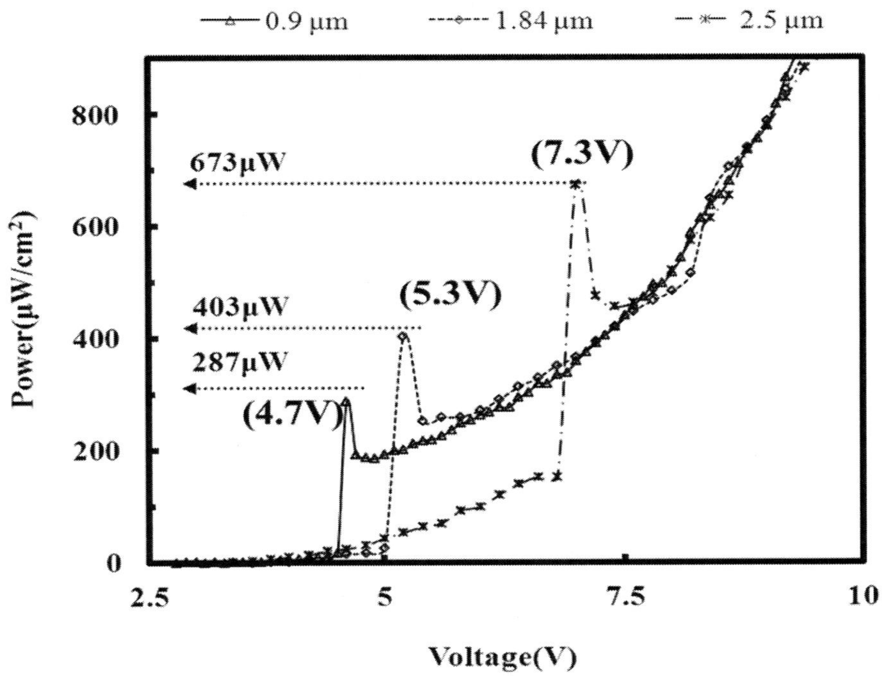

Figure 6. A typical power-V curve with obvious NDR peaks for different SiO_2 thicknesses.

4. Results and Discussion

PVCR is an important parameter for RTD, such that the competition for higher PVCR never stops. For RTDs, the PVCR is determined by the ratio between the desired tunneling current and the unwanted excess current through the energy states in a bandgap. Therefore, the goal to obtain a higher PVCR is to reduce the excess current component while maintaining the desired tunneling current. An increase of the barrier thickness results in a shift of the current peak location and PVCR to higher values. Peak current density (J_p) is another important parameter in addition to PVCR. For high speed applications, a large J_p is required to quickly charge or discharge the junction capacitance, but for low power applications, a small J_p is desired to lower the power consumption. The direct pathway to vary J_p is simply to adjust the SiO_2 thickness, since the J_p of the RTD is exponentially dependent on the thickness of the tunneling barrier.

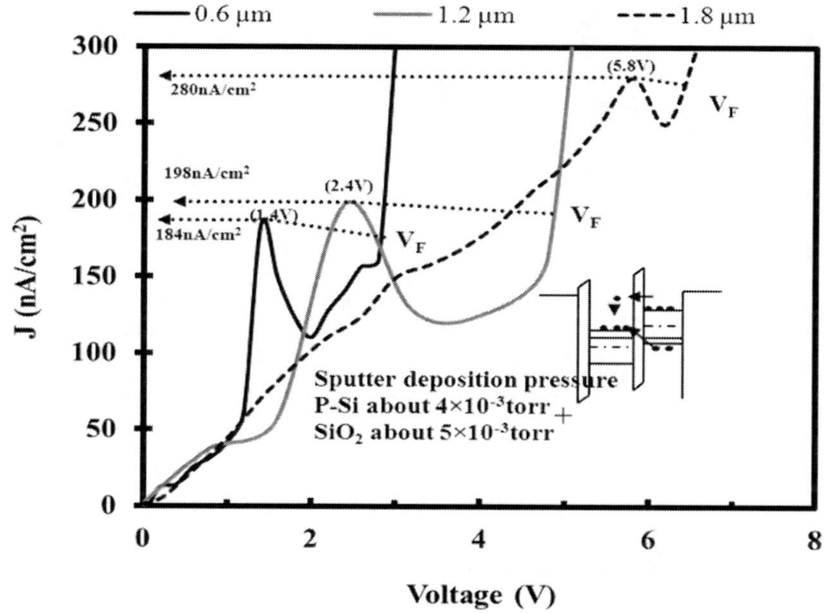

Figure 7. Measurement of room-temperature I–V characteristics for devices showing obvious NDR peaks for these three different SiO_2 thicknesses. An electrostatic band diagram under forward-biasing conditions is shown on the bottom-right inset.

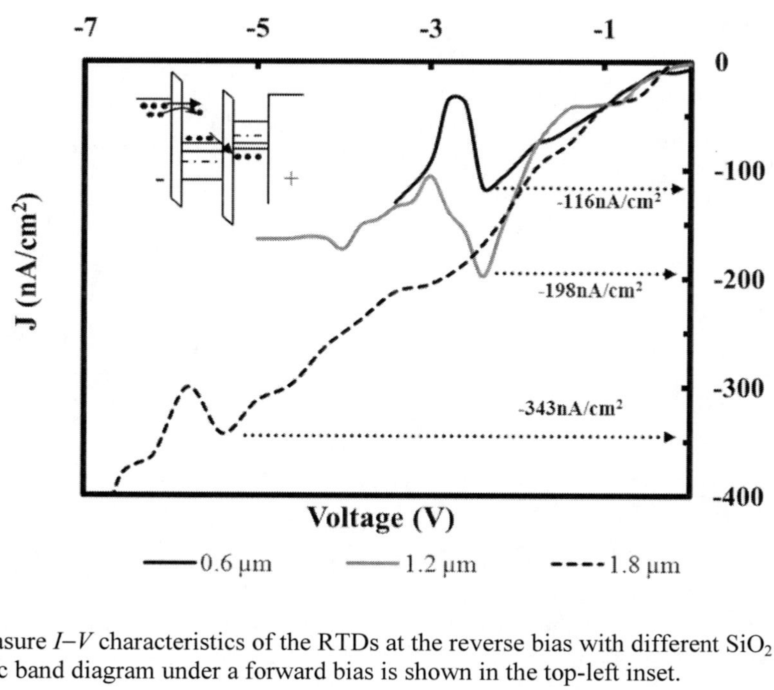

Figure 8. Measure I–V characteristics of the RTDs at the reverse bias with different SiO_2 thicknesses. The schematic band diagram under a forward bias is shown in the top-left inset.

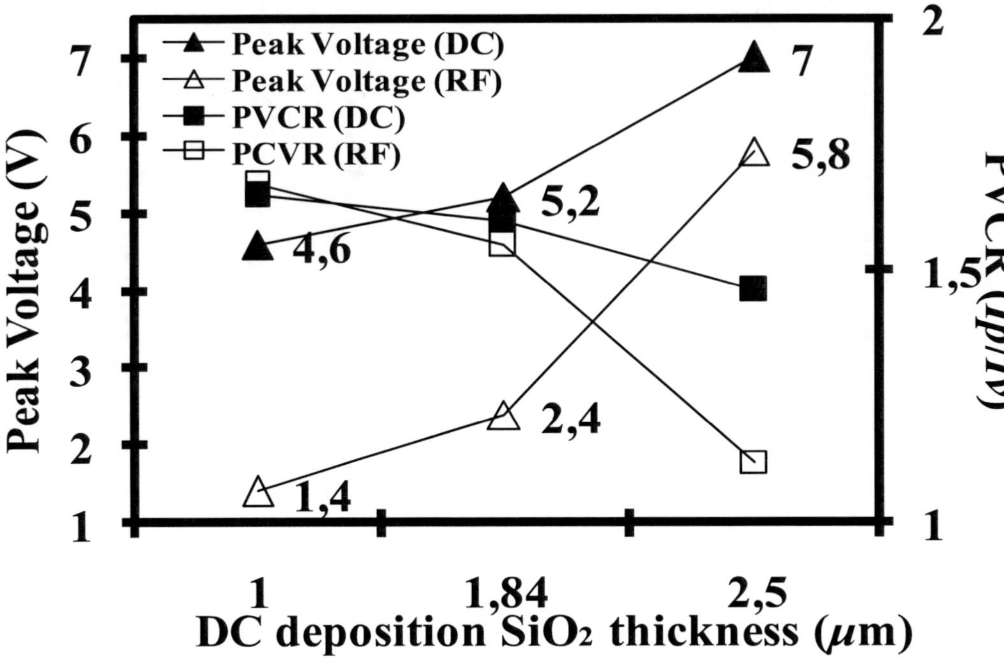

Figure 9. The PVCR compared to DC magnetron sputtering with RF magnetron sputtering.

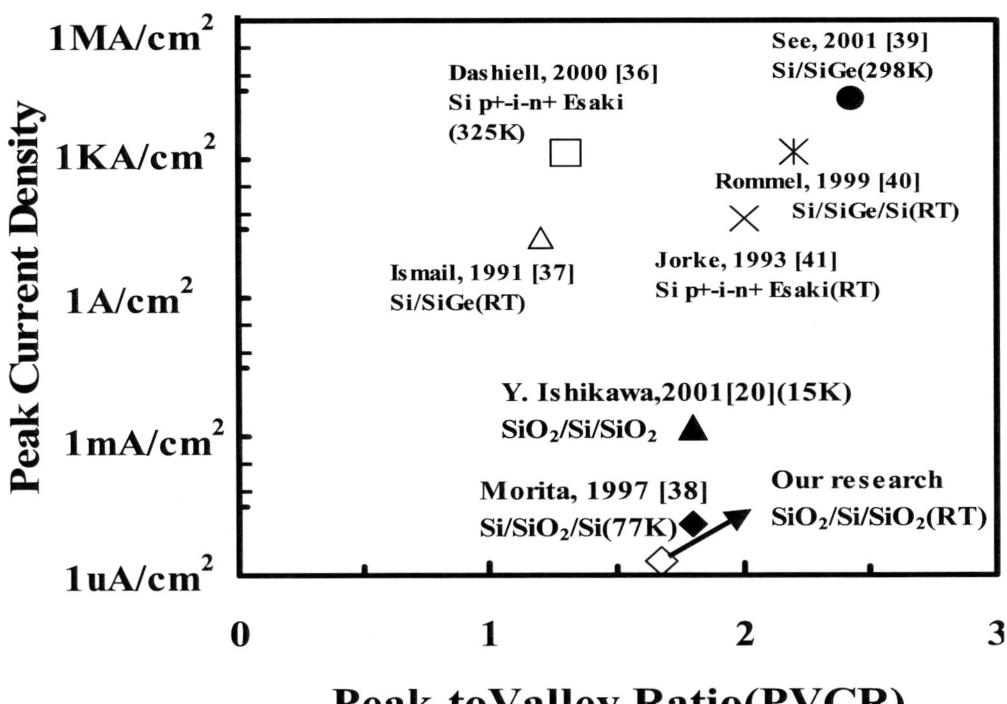

Figure 10. A graph of the peak-to-valley current ratio plotted against the peak current density of a number of Si-based tunnel diodes including those presented in this study.

We fabricate with different thicknesses by using DC magnetron sputtering to deposit SiO_2 according to the structure shown in Figure 4. The thicknesses of SiO_2 are 0.9, 1.84, and 2.5 μm, respectively. At each sample, characteristics of PVCR exit on the microwave device. Figure 5 shows typical *I*–*V* characteristics. RTD parameters include J_p, the peak voltage, (V_p), the valley current, and PVCR. The direct pathway to vary J_p is simply to adjust the thickness of SiO_2 since the J_p of the RTD is exponentially dependent of the thickness of the tunneling barrier. Eventually, the effects of the spacer SiO_2 thickness on PVCR, V_v (Valley voltage), and switching speed are also discussed.

In Figure 5, obvious NDR-affected region (PVCR = 1.6, J_p is 62.4 μA/cm^2, J_v is 38.9 μA/cm^2) can be identified in the voltage range from 4.4 V to 5 V when SiO_2 is 0.9 μm. When SiO_2 is 1.84 μm and the voltage ranging from 5 V to 5.5 V, the PVCR, J_p, and J_v are 1.65, 77.5 μA/cm^2, and 46.9 μA/cm^2, respectively. When SiO_2 is 2.5 μm and voltage range is between 7 V and 7.5 V, the PVCR, J_p and J_v, are 1.46, 96.2 μA/cm^2, and 66.1 μA/cm^2, respectively. From the parameters of the experiment, the minimum peak voltage is 4.7 Volts when SiO_2 is 0.9 μm. Therefore, it can be known that when the SiO_2/*p*-Si/SiO_2/*n*-Si structure increases, the tunneling decreases.

In order to reduce the power consumption, a Si-based RTD is used to show the negative resistance characteristics at a low voltage. This paper presents that a voltage-controlled oscillator (VCO) has extremely low power consumption based on the SiO_2/*p*-Si/SiO_2/*n*-Si structure. In Figure 6, when SiO_2 is 0.9 μm, the bias current is 62.4 μA/cm^2, supply voltage is 4.7 V, and the corresponding DC power in the VCO core is 287 μW, the oscillation frequency is 4.7619 kHz. When SiO_2 is 1.84 μm and the corresponding DC power in the VCO core is 287 μW, the oscillation frequency is 33.932 kHz. When SiO_2 is 2.5 μm and corresponding DC power is 673 μW, oscillation frequency is 26.88 kHz. Therefore, in this work, the VCO has extremely low power consumption characteristics.

Considerable amounts of data can be acquired to analyze or improve resonant tunneling diodes, and we fabricate three samples with different thicknesses by using RF magnetron sputtering to deposit SiO_2 according to the structure shown in Figure 4. The thickness of SiO_2 is 0.6, 1.2, and 1.8 μm, respectively. Figure 7 shows *I*–*V* characteristics of RTD. When SiO_2 is 0.6 μm and the voltage range is between 1 V to 2 V, the NDR effect region (PVCR = 1.67, J_p is 184 nA/cm^2, J_v is 110 nA/cm^2) can be identified. When SiO_2 is 1.2 μm and the voltage range is about 1.6 V and 3.2 V, the PVCR, J_p, and J_v are 1.55, 198 nA/cm^2, and 128 nA/cm^2. When SiO_2 is 1.8 μm and the voltage range is about 5.8 V to 6.2 V, the PVCR, J_p, and J_v are 1.12, 280 nA/cm^2, and 250 nA/cm^2. From the parameters of the experiment, the minimum peak voltage is 1.4 Volt when SiO_2 is 0.6 μm. Therefore, it can be known that when the SiO_2/*p*-Si/SiO_2/*n*-Si structure increases, the tunneling decreases.

In Figure 8, when the SiO_2 is 0.6 μm, and SiO_2/*p*-Si/SiO_2/*n*-Si diodes are under a reverse bias, PVCR, J_p, and J_v are 3.28, – 116 nA/cm^2, and – 35.3 nA/cm^2. When the SiO_2 is 1.26 μm, PVCR, J_p, and J_v are 1.88, – 198 nA/cm^2, and – 198 nA/cm^2. When SiO_2 is 1.8 μm, PVCR, J_p, and J_v are 1.14, – 343 nA/cm^2, and – 300 nA/cm^2. The RF sputtering system of the deposited ratio is quite small, cause by the tunneling distance that is also quite small. Therefore, the result of the PVCR increases as tunneling distance is decreased. Figure 9 shows the value of the PVCR and peak voltage of SiO_2 deposited by the DC magnetron sputtering and RF magnetron sputtering. The deposited ratio of the RF sputtering system is slower, thus double-barrier quantum-wells are thinner. The peak voltage increases while

PVCR decreases. Therefore, the DC peak voltage is higher than RF by 3.2 volts. The longer deposition times, the thicker SiO_2, and the smaller the peak-to-valley current ratio. In conclusion, the thin SiO_2 deposition can cause low power, low leakage currents, and large peak-to-valley-current-ratios; therefore, the thickness of SiO_2 is one important parameter of microwave components. We successfully use RF sputtering to deposit SiO_2 under room temperature.

CONCLUSION

PVCR is a very important parameter for tunnel diodes, and one of the primary tasks in this work is to improve the PVCR of Si-based RTDs. Therefore, instead of maintaining the desired tunneling current, the key to obtain a higher PVCR is to reduce the excess current component. Si is an affordable material, thus it is suitable for being applied on acoustic multimedia communication devices. The double-barrier, quantum-well structure has apparent negative differential resistance effects. Figure 10 summarizes the past status of Si-based tunnel diodes. The highest PVCR, 1.67 (J_p = 184 nA/cm^2, J_v = 110 nA/cm^2), is obtained by using RF sputtering to deposit a 0.6 μm tunneling film at room temperature. The report of 2001 indicates a PVCR at 1.8 for $SiO_2/Si/SiO_2$ (J_p = 2mA/cm^2) at 15 K. Compared with past reports, our research has many advantages. The PVCR of a $SiO_2/Si/SiO_2$ structure is lower, Si based material is affordable, and the process can be practiced at room temperature.

ACKNOWLEDGEMENTS

This research was supported in part by the Russian-Taiwan Goal Oriented Research on Interdisciplinary Collaborative Research Program (Grant NSC 99-2923-E-022-001-MY3).

REFERENCES

[1] van der Wagt, J. P. A.; Seabaugh, A. C.; Beam, E. A. III, RTD/HFET Low Standby Power SRAM Gain Cell, *IEDM Tech. Dig*., pp. 425–428, 1996.

[2] Abidi, A. A. Low-power Radio-frequency ICs for Portable Communications, *Proc. IEEE,* Apr. 1995, pp. 544–569.

[3] de Los Santos, H. J.; Chui, K. K.; Chow, D. H.; Dunlap, H. L. IEEE Microw. Wireless Compon. *Lett.*, 2001, vol. 11(5), 193–195.

[4] Yuji, Ando; Alain, Cappy, IEEE Trans. *Electron Devices*, 1998, vol. 45(1).

[5] Cidronali, A.; Collodi, G.; Deshpande, M. R.; El-Zein, N.; Nair, V.; Manes, G.; Goronkin, H. IEEE Trans. *Microwave Theory and Techniques,* 2001, vol. 49(12).

[6] Liou, W.-R. *Solid-Srate Electronics*, 1997, vol. 41(6). 865-869.

[7] Nair, V.; El-Zein, N.; Lewis, J.; Deshpande, M.; Kramer, G.; Kyler, M.; Maracas, G.; Goronkin, H. *IEEE GaAs IC Technical Digest*, 1998, pp. 191-194.

[8] Cidronali, A.; Collodi, G.; Camprini, M.; Nair, V.; Manes, G.; Lewis, J.; Goronkin, H. *IEEE Radio Frequency Integrated Circuits Symposium*, 2002, pp. 297–300.

[9] Kidner, C.; Mehdi, I.; East, J. R.; Haddad, G. I. *IEEE Trans. Microwave Theory and Techniques,* 1990, vol. 38(7), 864 - 872.

[10] Chang, L. L.; Esaki, L.; Howard, W. E.; Ludeke, R. J. *Vac. Sci. Technol*, 1973, vol. 10(1), 11-16.

[11] Rudra, A.; Pelucchi, E.; Oberli, D. Y.; Moret, N.; Dwir, B.; Kapon, E. *J. Crystal Growth,* 2004, vol. 272 , 615–620.

[12] Vickerst, A. J.; Strawt, A.; Roberts, J. S. Semicond. Sci. Technol, , 1989, vol. 4, 743-746.

[13] Wu, K. H.; Fang, Y. K.; Ho, Jyh-Jier; Chen, T. J. *IEE Electronics Letters,* 1997, vol. 33(21), 1824-1825.

[14] Wu, K. H.; Fang, Y. K.; Ho, Jyh-Jier; Hsieh, W. T.; Chen, T. J. *Appl. Phys. Lett*., 1998, vol. 72(23), 3017-3019.

[15] Wei, T. Y.; Yang, C. C.; Ho, J. J.; Chang, S. H. *The Study of Negative Differential Resistance Effect on Si-based RTID with Double Wells*, OPT'08, Sat-S8-04, 2008.

[16] Schulman, J. N.; De Los Santos, H. J.; Chow, D. H. *IEEE Electron Device Letters,* 1996, vol. 17(5), 220-222.

[17] Yan, Z.; Deen, M. J. *IEEE Transactions on Computer-AIDED Design of Integrated Circuits and System*, vol. 14(2), 1995.

[18] Zhang, B.; Chen, K. J.; Gang, R.; Chen, R. M. M. *IEEE Elecrron Device Letters*, 1999, vol. 1(2), 178 -181.

[19] Sun, J. P.; Haddad, G. I.; Mazumder, P; Schulman, J. N. *Proceedings of the IEEE*, 1998, vol. 86(4).

[20] Mizuta, H.; Tanoue, T.; Takahashi, S. IEEE Trans. *Electron Devices*, 1988, vol. 35(11).

In: Physics and Mechanics of New Materials …
Editors: Ivan A. Parinov and Shun-Hsyung Chang
ISBN: 978-1-62618-535-7

Chapter 29

RESEARCH OF OPPORTUNITY TO USE MISM STRUCTURES FOR COOLING OF LIGHT-EMITTING DIODES

G. Ya. Karapetyan* and V. G. Dneprovski
Vorovich Mechanics and Applied Mathematics Research Institute,
Southern Federal University, Rostov-on-Don, Russia

ABSTRACT

The thermoelectric refrigerator, based on MISM structures for cooling of light-emitting diode (LED) matrices, is proposed. In difference of known coolers on the basis of the Peltier effect, the proposed cooler has not hot side that allows one to transform an absorbed heat into electric energy. The series connection of light-emitting diode matrices with the proposed cooler leads to decrease in power consumption, and cooling of the matrices leads to significant increase of their longevity.

1. INTRODUCTION

Despite of known advantages of light-emitting diodes (LEDs), application of LED lamps is very insignificant (less than 10 % of all electric devices and goods). Very important reason of this problem is insufficient quality of LEDs. Due to no thermal nature of the LED radiation, absence of a glower should theoretically cause a fantastic service duration. In reality, as well as in all other light sources, parameters of LEDs worse eventually, and this degradation is coupled with such factors, as value of direct current, heat-conducting path, type and quality of used chips. In particular, the light flow can be decreased down to 50 % and less from original value.

One of the reasons of the LED degradation is connected with its overheating. In this paper, we propose to use for decrease of overheating the specific cooling devices, based on

* jorichkaka@rambler.ru.

the planar structures, namely metal-insulator (dielectric)-semiconductor-metal (MISM) systems. The feature of these cooling devices is defined by possibility to transform a heat on input of LED into electric energy. Thus, the proposed MISM-structures not only decrease a degradation of LEDs, but also diminish a power consumption, because the thermal energy, emitted by LED in its working cycle, is converted into the electric power in this case.

The microelectronic devices, for example, Peltier elements maybe used enough long time and they are cooled by circulating electric current (see Figure 1) [1].

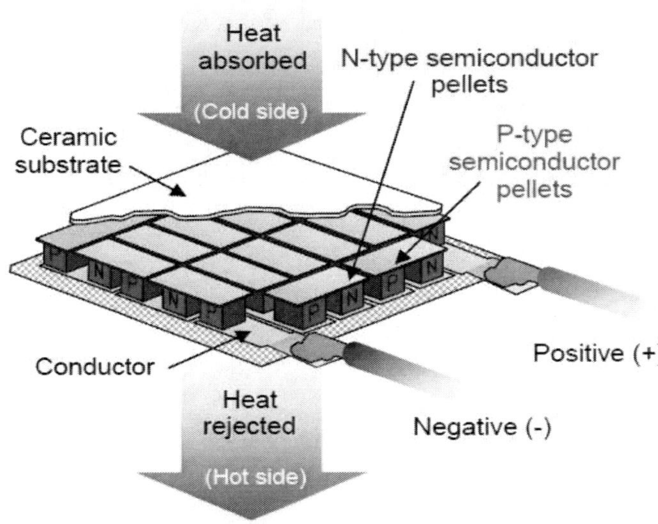

Figure 1. The semiconductor thermoelectric cooler of old type with hot and cold sides.

The purpose of the present work is to develop a novel design of the cooler, based on the MISM-structure, without hot side (see Figure 2), being the thermoelectric capacitor.

Such a cooler is very well suitable for a flat LED matrix (Figure 3), which generates heat during its work. In this case, the cooler, based on the MISM structure, disposes under the light-emitting diode matrix, as it is shown in Figure 4. Since such a cooler only absorbs heat, then accordingly to the law of energy conservation, it will transform the absorbed heat into electric energy. In particular, it is very actually for the converter of environment heat into electric energy. Therefore, at its connection in series with LED matrix and electric power source, the energy consumption from the power supply should be decreased due to heat absorption from environment by the proposed cooler.

The cooler basis is the thermoelectric capacitor (see Figure 5), which consists of a semiconductor substrate of *n*-type, on lower surface of which the ground electrode 2 with ohmic contact to the substrate is located. On upper side of the substrate, the metallic electrodes 3 dispose, which form the impoverished layer 8 in the contact with a semi-conductor substrate. The electrodes 3 locate in deepening of the substrate. On the substrate surface, the dielectric layer 4 disposes between profound electrodes 3, which present themselves small blocks on the surface. The electrode 5, called a gate, disposes on this dielectric layer, and the electrodes 6 and 7, placed on the metal electrodes 3, are called a drain and source, respectively. This structure is similar to the field transistor [3] with the same designations (i.e. drain, source, and gate) of electrodes, but the drain and source are joined together in the last case.

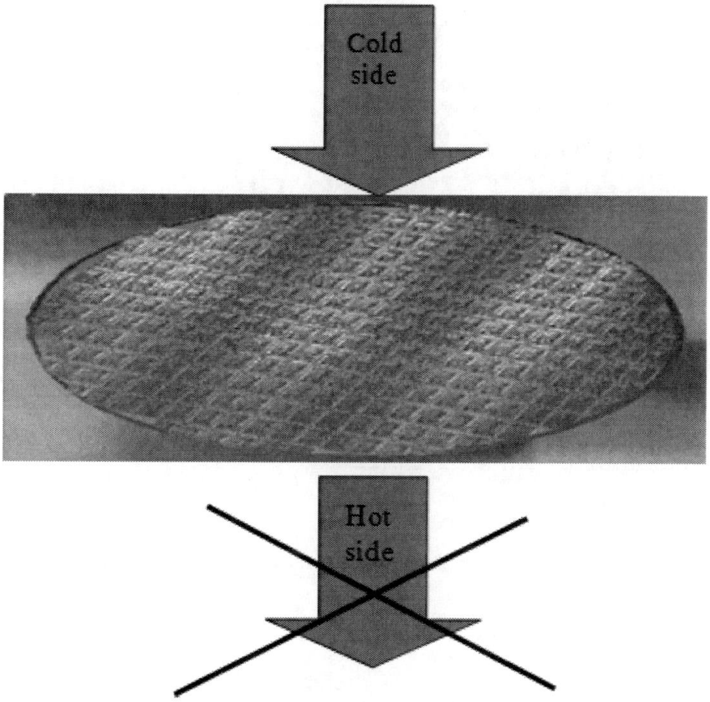

Figure 2. The semiconductor thermoelectric cooler of novel type without hot side.

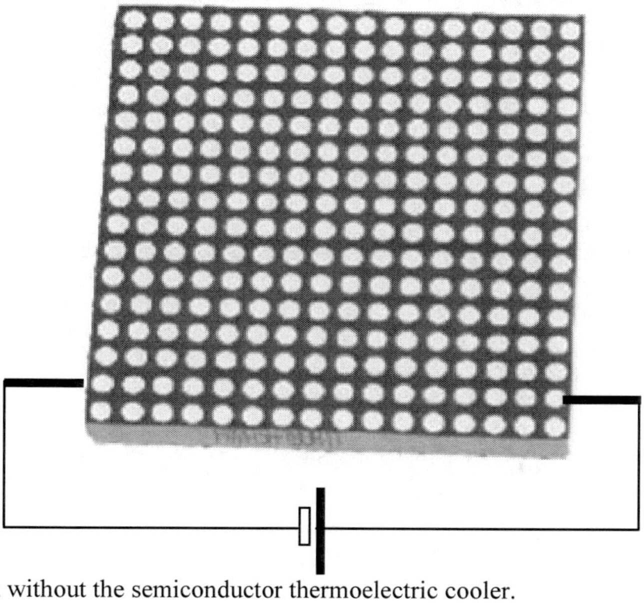

Figure 3. LED matrix without the semiconductor thermoelectric cooler.

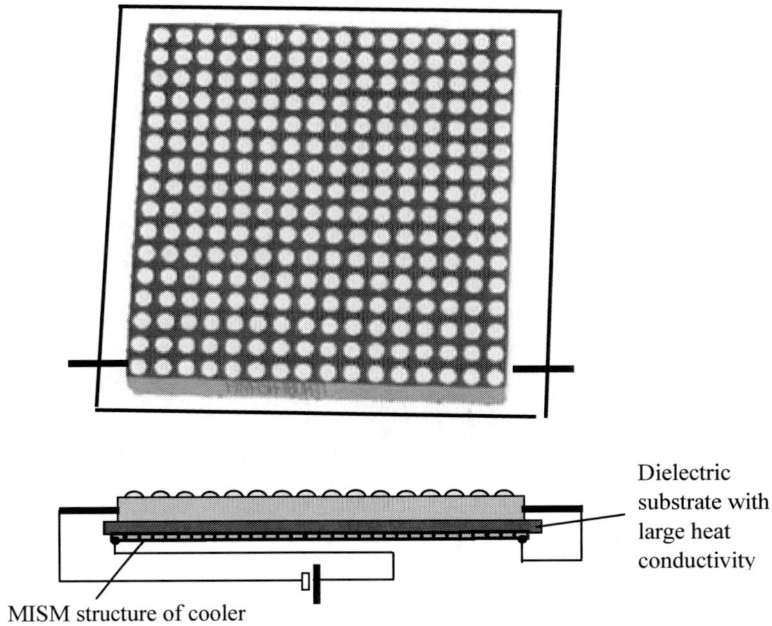

Figure 4. LED matrix with new type of semiconductor thermoelectric cooler and decreased energy consumption.

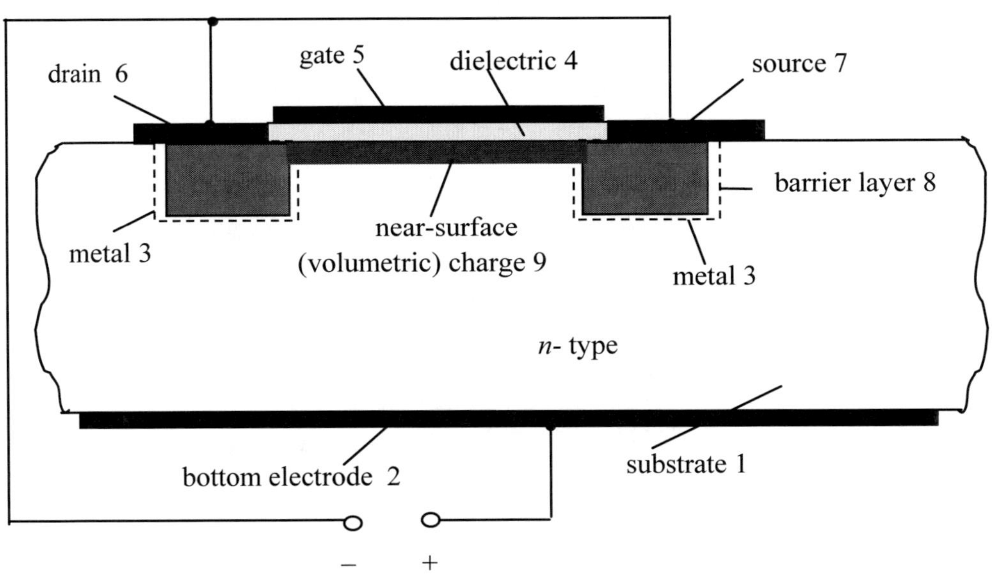

Figure 5. Metal-insulator-semiconductor-metal (MISM) structure for thermoelectric cooler of new type.

2. CONCEPT

Let us consider a charging of such thermoelectric capacitor (see Figure 6, in this case, the key K1 is close, and key K2 is open) from voltage source through resistor *R*. At the closing of key K1, the capacity between the gate 5 and the substrate is charged to voltage U_0 through resistor R. The polarity of voltage is selected in such a manner that power zones in the semiconductor bend downwards (see Figure 6) and in near-surface layer under dielectric is formed the enriched by electrons the volume charge 9, in which concentration of electrons is much more, than in the substrate. The voltage U_0 is selected so that electrons in the volume charge 9 would be in degenerate condition because of strongly raised concentration of the electrons. In this case the barrier layer 8 between metallic electrodes 3 and near-surface charge 9 disappears (because barrier layers are characteristic only for contact of metal with non-degenerative semiconductor), i.e. drain and source are attached to area with the raised concentration of electrons.

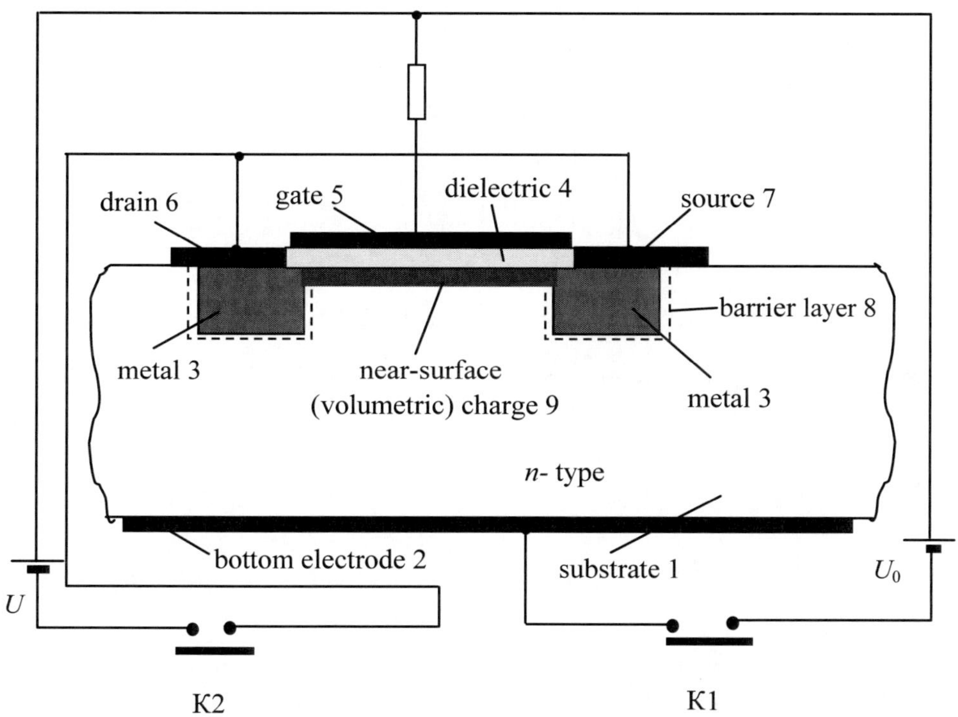

Figure 6. Design of the thermoelectric capacitor (MISM-structure).

The equation, describing this process has the following form (see Figure 7):

$$RC\frac{du}{dt}+u=U_k+U_0, \tag{1}$$

where u is the voltage on capacitor plates, C is the capacity of the thermoelectric capacitor, U_k is the junction potential difference between metals, U_0 is the voltage of the power supply (Figure 7).

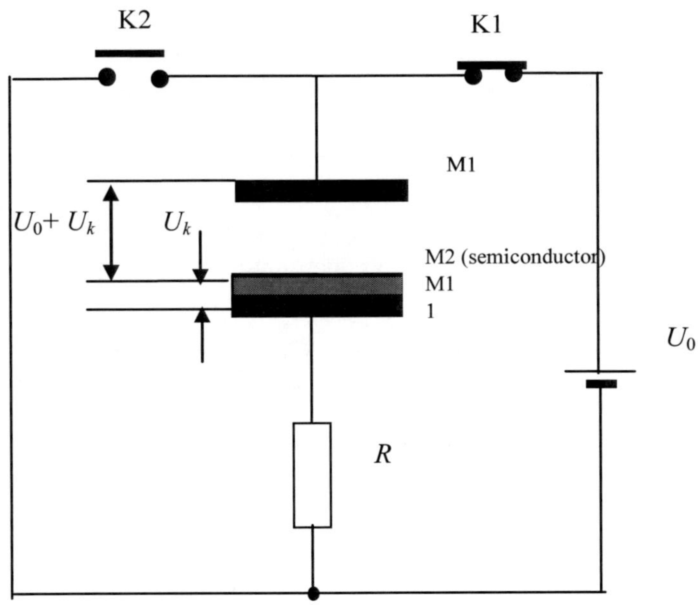

Figure 7. The equivalent circuit of charge-discharge of the thermoelectric capacitor.

The solution of Equation (1) is defined as

$$u(t) = A\exp\left(-\frac{t}{RC}\right) + U_k + U_0 . \tag{2}$$

At the initial moment of time, the capacitor is charged to voltage U_k, therefore $A = -U_0$, and

$$u(t) = U_0\left[1 - \exp\left(-\frac{t}{RC}\right)\right] + U_k . \tag{3}$$

The current passes through the capacitor is determined as $i = (U_0/R)\exp[-t/(RC)]$, and

$$W_C = \int_0^\infty uidt = \int_0^\infty U_0\left[1-\exp\left(-\frac{t}{RC}\right)\right]\frac{U_0}{R}\exp\left(-\frac{t}{RC}\right)dt + \int_0^\infty \frac{U_k U_0}{R}\exp\left(-\frac{t}{RC}\right)dt =$$

$$= \frac{CU_0^2}{2} + CU_k U_0 , \tag{4}$$

where $CU_0^2/2$ is the energy, obtained by the capacitor from voltage source, CU_0U_k is the energy, absorbed from environment, when the direction of the charge current is chosen in such a manner, that via junction "metal 1 – metal 2 (or the semiconductor)" electrons pass from the metal 1 with smaller energy of electrons to the metal 2 (or the semiconductor) with greater energy of electrons. At the current direction, when electrons pass from the metal 1 to the metal 2 (or the semiconductor), they should overcome the potential barrier equal to U_k. It could be only made by most powerful electrons (i.e. "hot" electrons). Therefore, numerous lesser powerful or "cold" electrons remain at the junction [2], and it is cooled, because the electrons come to thermodynamic balance with the crystalline lattice of the metal 1, taking away its energy. This phenomenon is well-known as the Peltier effect and for junction of metal and semiconductor the Peltier factor Π is equal to internal junction potential difference ($\Pi = U_k$), if under the internal junction difference to understand the value of $(E_c - F)/q$ [2], where E_c is the energy at the bottom of the conductivity zone of the semiconductor, F is the Fermi level, q is the charge of electron.

Thus, as it follows from Figure 7, the capacitor obtains energy W_C from the energy sources with voltage U_0 and U_k by means of electrons which transfer this energy to the capacitor. On internal surfaces of plates of the capacitor, the charges are formed. In metals, the penetration depth of these surface charges is lesser than the sizes of atomic lattice. Electrons appear on surface of the negative plate. These electrons transfer all energy, which they obtain from voltage sources, to the electric field of capacitor. Therefore their energy should be the same as the capacitor charge, i.e. their energy is equal to energy of electrons in metallic electrodes of the capacitor (metal 1). If we relate this process to the thermoelectric capacitor (see Figure 7), then energy of electrons of the surface charge should be equal to the energy of electrons in metal 1. In this case, the metal 2 (semiconductor) together with junction to the metal 1 form a source electromotive force with voltage U_k, and the surface charge is defined by the metallic electrode 1 (the black strip over the metal 2 in Figure 7). Thus, at the charging thermoelectric capacitor due to cooling the junction “metal 1 – metal 2”, electrons increase energy, absorbing heat from environment, and then transfer it to electric field of the capacitor, collecting on internal faces of the capacitor. Thus energy of the electrons becomes equal to energy of electrons in the metal 1 because they have the given energy, absorbed in the form of heat from environment, at transition of electric field from the metal 1 to the metal 2. Electrons at transition from the metal 2 to a superficial charge cannot give energy to crystalline lattice (heating) due to it contradicts the law of the energy conservation (the electrons, as it is followed from Equation (4), give all energy, including the energy gotten at transition from the metal 1 to the metal 2 with absorption of heat). As it is appeared from Figure 6, the capacitor is charged to voltage $U = U_0 + U_k$. The fact, that the capacitor is charged to $U = U_0 + U_k$, instead of U_0, means that during process of charging the capacitor obtains an additional energy because of heat absorption at junction. The kinetic energy of thermal movement transforms into potential energy of the charged condenser, as it is fixed by the factor U_0U_kC in the formula (4) [4–7].

The energy arises on the resistor is defined as

$$W_R = \int_0^\infty i^2 R dt = \frac{CU_0^2}{2}. \tag{5}$$

At the capacitor discharging on resistor R (see Figure 7) the key K1 is open, and key K2 is close. The equation, describing this process, has the form:

$$RC\frac{du}{dt}+u=U_k\,. \tag{6}$$

The solution of this equation is determined as $u(t)=A\exp\left(-\frac{t}{RC}\right)+U_k$, and at the initial moment $u(0)=A+U_k=U_k+U_0$, i.e. $A = U_0$ and

$$u(t)=U_0\exp\left(-\frac{t}{RC}\right)+U_k\text{, and } i(t)=\frac{U_0}{R}\exp\left(-\frac{t}{RC}\right). \tag{7}$$

The energy arises on the resistor is equal to

$$W_R=\int_0^\infty i^2Rdt=\frac{CU_0^2}{2}. \tag{8}$$

The energy arises on the metal-semiconductor junction is found as

$$W_k=U_k\int_0^\infty idt=CU_kU_0\,. \tag{9}$$

It is important to note that if originally U_k is negative, then. during charging the thermoelectric capacitor heats, but at discharging one cools. Voltage on the capacitor after charging in this case appears lesser than U_0 by value of U_k, while energy on the resistor at the discharging always the same and equals to $\frac{CU_0^2}{2}$.

The lost energy CU_0U_k arises due to discharging at the junction of the metals 1 and 2 (or semiconductor) because of cooling, i.e. there is a full transformation of thermal energy of environment into electric power. However, it is impossible to consider such way of conversion as effective, because before discharging the capacitor should be charged that will lead to energy expense at the junction, heated on the same value of CU_0U_k. Therefore, during the cycle “charge – discharge”, the increment of energy, caused by expense of the environment thermal energy, will be equal to zero.

3. Description of Cooler Operation

In order to obtain non-zero transformation of thermal energy, the device, shown in Figures 5 and 6, is planned to process in reality. At the capacitor discharging, its heating takes place via junction zone owing to near-surface charge. However, in order to discharge the capacitor, it is necessary a voltage U at which sufficient concentration of electrons arises in near-surface charge, which is required to remain the junction of the electrodes 3. Therefore, we consider the discharging of this capacitor to voltage U through metal electrodes, which are attached to near-surface charge.

The equation, which describes this process, has the form:

$$RC\frac{du}{dt}+u=U_{k1}+U, \tag{10}$$

where u is the voltage on capacitor plates, C is the capacity of the thermoelectric capacitor, U_{k1} is the junction potential difference between metal electrode and near-surface charge, U is the voltage of the power supply.

At the initial moment of time, the capacitor is charged to voltage $U_k + U_0$, therefore the solution of Equation (10) is calculated as

$$u(t)= A\exp\left(-\frac{t}{RC}\right)+U_{k1}+U\ . \tag{11}$$

Then we have

$$A+U_{k1}+U=U_k+U_0\ ;\ A=U_0-U+U_k-U_{k1}=\Delta U+\Delta U_k,$$

where $\Delta U = U_0 - U$, $\Delta U_k = U_k - U_{k1}$.

The current passes through capacitor has the form:

$$i=\frac{\Delta U+\Delta U_k}{R}e^{-\frac{t}{RC}}\ . \tag{12}$$

Further we will consider the capacitor charging from voltage U to voltage U_0 through the ground (lower) electrode.

The equation, describing this process, is written as

$$RC\frac{du}{dt}+u=U_k+U_0\ . \tag{13}$$

The solution of Equation (13) is presented as $u(t)=A\exp\left(-\frac{t}{RC}\right)+U_k+U_0$.

At the initial moment of time, the capacitor is charged to voltage $U_{k1} + U$, therefore, $u(0)=U_{k1}+U$.

Then we have

$$A+U_k+U_0=U_{k1}+U\ ;\ A=-(U_0-U)-(U_k-U_{k1})=-(\Delta U+\Delta U_k),$$

where $\Delta U = U_0 - U$; $\Delta U_k = U_k - U_{k1}$.

The current passes through capacitor has the form:

$$i = \frac{\Delta U + \Delta U_k}{R} \exp\left(-\frac{t}{RC}\right). \tag{14}$$

If now we shall discharge the capacitor to voltage U again via metallic electrodes, which attached to near-surface charge through the resistor, the current, defined by Equation (12) or (14), begins to flow again. If we begin to charge the capacitor again to voltage U_0 through the ground electrode, the current, defined by Equation (12) or (14), will flow through the resistor again.

Thus during the charging-discharging process of the capacitor, the voltage on it varies always from $U_k + U_0$ to $U_{k1} + U$, i.e. the variable amplitude of the voltage is equal to $\Delta U + \Delta U_k$, the charge current is equal to discharge current and defined by Equation (12) or (14).

The energy, which is consumed by the scheme from the power supply (without power on switching) during the charge-discharge cycle, is defined as

$$W_{source} = \int_0^\infty (U_0 - U) i(t) dt = C\Delta U(\Delta U + \Delta U_k).$$

The energy arises on the resistor during the charge-discharge cycle is determined as

$$W_R = 2\int_0^\infty i^2(t) R dt = C(\Delta U + \Delta U_k)^2.$$

Thus, $\Delta W = W_R - W_{source} = C(\Delta U + \Delta U_k)\Delta U_k$, and $\dfrac{\Delta W}{W_{source}} = \dfrac{\Delta U_k}{\Delta U}$.

In the case of $\Delta U = \Delta U_k$, an energy increment on the resistor will be equal to the consumed energy by the scheme from energy sources.

Thus, we have shown, that non-zero conversion of the environment heat into electric energy is possible in MISM structure. Therefore, between the ground electrode 2 and both the drain 6 and the source 7, connected in parallel, there should be potential difference equals to ΔU_k, when the thermoelectric capacitor is charged to voltage U_0 (see Figure 5).

CONCLUSION

In the conclusion, we present the proof of working capacity in the proposed converter (refrigerator). The proof we demonstrate on the base of contradiction. Let us assume that the proposed converter is disabled. It means that either the thermal energy, absorbed at the junction of the ground electrode with the substrate (see Figures 5, 6) doesn't transform into potential energy of the charged capacitor at charging, or this stored energy arises at the junction of metal with near-surface charge during the condenser discharging. However, as it has been shown above, in accordance to the law of the energy conservation, during capacitor charging all energy, stored by electrons, transforms into potential energy of the charged capacitor, including the thermal energy, absorbed at the junction. The near-surface charge energy of electrons is lesser than the corresponding energy in the semiconductor volume. Therefore, the potential barrier, formed at the junction of metal with near-surface charge, becomes lower than the potential barrier between the ground electrode and the substrate (see Figures 5, 6), and the quantity of heat, arisen at the junction "near-surface charge – metal", will be lost and absorbed at the junction "ground electrode – substrate". Hence, the remainder of the absorbed thermal energy does not arise in the form of a heat, and must arise on the resistor. The contradiction to the law of the energy conservation, otherwise, signs that the part of the absorbed thermal energy disappears it is unknown where. This contradiction does not arise, if we assume that the device is efficient. In this case the absorbed thermal energy, which was not arisen at the junction "near-surface charge – metal", will be on the resistor and no contradiction with the law of the energy conservation will arise.

The energy source for the proposed converter is the thermal energy of environment, which absorbs at the junction "ground electrode – substrate" because in the semiconductor and hence in current there is a distribution of electrons on energy. The "hot" electrons are only capable to overcome the potential barrier, but the "cold" electrons remain at the junction and select energy of the crystalline lattice of semiconductor. It leads to the semiconductor cooling together with the junction and heat from environment is directed into substrate. At the junction of substrate with environment the substrate temperature is stabilized, though remains to be lesser than the ambient temperature. If converter has good thermal contact with environment the converter cooling will be insignificant (shares of degree or several degrees) and distribution of electrons on energy remains for all time. It creates conditions for accumulation of the "cold" electrons at the junction, causing an inflow cooling, directed to it due to heat of environment. If heat insulation of the converter will be cooled, the disorder of energy of the electrons, being ever lesser than eventually, will lead to impossibility of accumulation of the "cold" electrons at the junction and, hence, to impossibility to absorb heat from environment. Thus, a moving power in the proposed converter is distribution of electrons on energy, in which the considerable disorder of energy of the electrons will be only at temperature of the semiconductor, changed in certain limits, that is reached by a good thermal contact of the converter with environment.

ACKNOWLEDGEMENTS

The authors are grateful to Prof. Ivan A. Parinov for interest to this work.

REFERENCES

[1] Bell, L. *Accelerating the Commercialization of Promising New Thermoelectric Materials*. Reprinted with permission of ASM International. www.asminternational.org

[2] Stilbans, L. S.; *Physics of Semiconductors*. Nauka, Moscow. 1967, pp 1 - 451.

[3] Agahanyan, T. M.; *Basic of Transistor Electronics*. Energy, Moscow. 1974, pp 1 - 256.

[4] Dneprovski, V. G.; Karapetyan, G. Ya.; Pustovalov, G. A.; Lysenko, M. V. J. *Microsystem Engineering*, 2003, No 9, 38 - 42.

[5] Karapetyan, G. Ya.; Kataev, V. F. *Technology and Designing in Electronic Equipment*, 2006, No 6, 35 - 36.

[6] Karapetyan, G. Ya.; Kataev, V. F.; Dneprovski,V. G. *The Ecological Bulletin of the Black Sea Economic Common Wealth Scientific Center*, 2010, No 4, 44 - 51.

[7] Karapetyan, G. Ya, Kataev, V. F.; Krase, I. V. *Prikaspijsky Journal: Management and High Technologies*, 2009, No 4, 90 – 98.

In: Physics and Mechanics of New Materials …
Editors: Ivan A. Parinov and Shun-Hsyung Chang
ISBN: 978-1-62618-535-7

Chapter 30

NEW DOMAIN REQUIRED FOR THE MODERN ENGINEERING EDUCATION: SCIENCE, TECHNOLOGY AND SOCIETY (STS)

***Wen-Ling Hong*[1,*], *Chitsan Lin*[2] *and Jr-Ping Wang*[3]**

[1]Science Technology and Society (STS) Research Center, National Kaohsiung Marine University, Kaohsiung, Taiwan

[2]College of Ocean Engineering, National Kaohsiung Marine University, Kaohsiung, Taiwan

[3]Department of Naval Architecture and Ocean Engineering, National Kaohsiung Marine University, Kaohsiung, Taiwan

ABSTRACT

In Taiwan, there are more than one quarter undergraduate students studying in engineering departments. In addition, the social resources engineering industries consume may well exceed that ratio. Most of the conflicts arising between engineering industries and society come from using and manipulating materials. There exist great value discrepancies between different societies when it comes to these kinds of conflicts. In the modern democratic society, we expect public participation to carry the responsibility of resolving the issues. Unfortunately, public participants and experts often speak their own languages, one in science and another in social science.

When we look at the current curricula of our engineering higher education, it reflects very little in teaching students with diverse visions. This is however clearly required by the USA Accreditation Board on Engineering and Technology, suggesting the engineering education must cultivate students with global, economic, environmental, and social competence. Present Taiwan engineering higher education does not include these social and humanistic philosophies in our curricula. But engineering teachers who understand the importance and dedicate in multi-discipline teaching often wander in designing syllabus without good guidance and support.

* E-mail: wlhong@webmail.nkmu.edu.tw, wenlinghong@gmail.com.

Since 2007, the Ministry of Education in Taiwan has funded several teaching projects aiming to bring Science, Technology and Society (STS), an inter-disciplinary study, to engineering education. The pioneer STS project at National Kaohsiung Marine University included several courses offered as electives to engineering students. From these experiences, we plan to use STS concepts as a mediator to introduce social and humanistic concepts into engineering education. We hope to see engineering students benefit from discussion of modern engineering issues with STS experts and their term projects. We documented the National Kaohsiung Marine University STS projects carefully and analyze faculties and students' feedback to gain insights of it's effects.

1. INTRODUCTION

The American Engineers' Council for Professional Development defines Engineering as: "The creative application of scientific principles to design or develop structures, machines, apparatus, or manufacturing processes, or works utilizing them singly or in combination; or to construct or operate the same with full cognizance of their design; or to forecast their behavior under specific operating conditions; all as respects an intended function, economics of operation and safety to life and property."[1] Engineering focuses on applying scientific knowledge and findings on design, constructions and implementations. Engineering practices include dealing with objects, and they vary greatly in scales and formats. Traditionally, engineering deals with bigger objects that relate closely to human life, such as combustion engines, bridges and buildings. Until the rapid development of electronic technologies, the rising engineering applications become more delicate and smaller in sizes, taking computers, 3C products and Nano-materials as examples. Disciplines in engineering have very different focuses as mechanical and civil engineering emphasize on mechanics; electrical engineering and computer science is built upon electronics and computer logics. Among all the disciplines, logical analysis, safety issues, optimization, efficiencies are the common values.

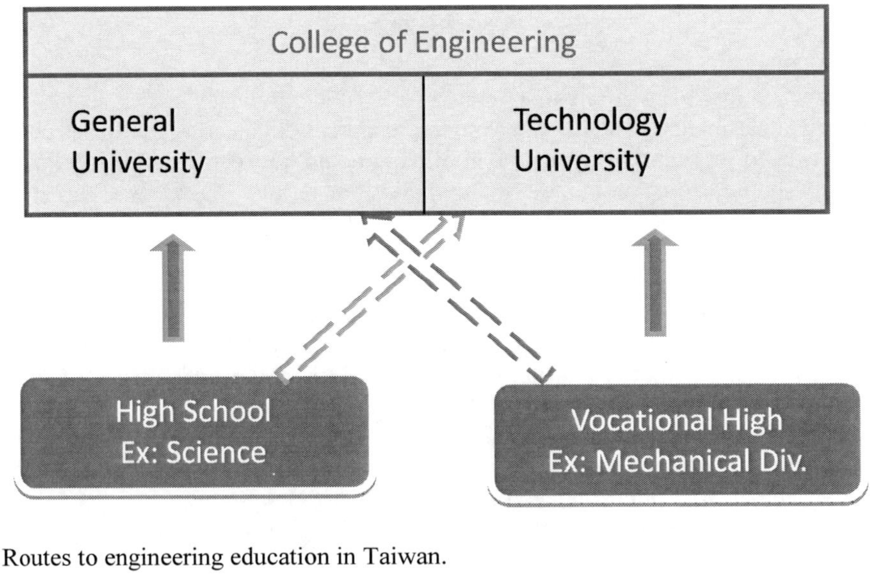

Figure 1. Routes to engineering education in Taiwan.

[1] Cited from http://www.sciencedaily.com/articles/e/engineering.htm.

Engineering is taught mainly at the college level. There are two routes in entering the engineering higher education system. The first one is for science-majoring regular high school students to choose engineering departments while enrolling in colleges. The high school science education focuses on mathematics, physics, chemistry and/or biology. The other is for vocational high school student who majors in technical divisions, then choose engineering departments in technology universities[2]. (There are some crossovers, but small in numbers.) Figure 1 shows the routes to engineering high education in Taiwan. Enlightenment of engineering starts at elementary and secondary school, in science education. The concept of "engineering" is not introduced to students systematically until they enter college.

Engineering sector is one of the largest contributors in Taiwan's economy. National level industrial policies impact social development and citizens greatly. However, Science, Technology and Society (STS) scholars criticize engineering and technology embraces specialty but lacks discussions of complex social issues. Engineering science bases on linearization and simplification of reality, eliminating "higher order" and "unattainable" terms or replacing them with parameters. In facts, the construction of engineering knowledge should be an important research topic, but is indeed overlooked in engineering research. If engineering educators discuss engineering topics with critical awareness, at the same time bringing inter-disciplinary contents to stimulate integrated competency of engineering students, a more flexible and practical engineering professionalism can be expected.

This research is a preliminary study of an experimental project in bringing STS into engineering higher education. The aim is to discourse benefits of such effort and set ground for future promotion of inter-disciplinary content.

2. Engineering Higher Education in Taiwan

Based on national statistics from the Ministry of Education, enrollment in engineering and technology departments still take up the major part of entire student population, as seen in Figure 2. Half of the science-majoring high school students choose to enroll in engineering departments in colleges.

Institute of Engineering Education Taiwan, IEET, established in 2003, offers accreditation to engineering departments. The accreditation has become very popular in Taiwan. Out of 608 engineering departments in Taiwan, as many as 191 departments (about 30%) have passed the accreditation process. The process evaluates the student outcome to ensure education quality of the department. Criteria include: for students to know current social issues; to understand impacts of engineering to environment, society and the globe; to cultivate lifelong learning habit and ability; and to understand professional ethics and social responsibility.

Take a mechanical engineering department as an example, Figure 3 shows a typical curriculum. These courses are very technical oriented. All the above mentioned integrated

[2] In Taiwan, the two kinds of universities, general versus vocational (or technological), cannot be distinguished from their school name.

ability of social responsibility and knowledge of modern issues will have to be provided in general education[3].

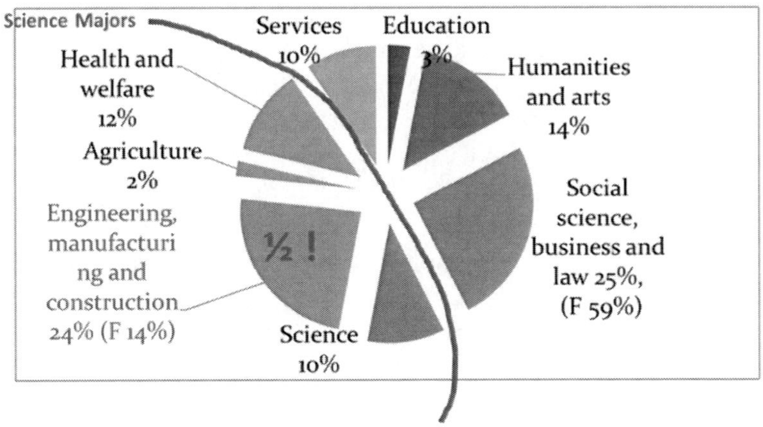

Figure 2. Enrollment in Tertiary by 9 categories defined by the Ministry of Education, Taiwan (by Hong).

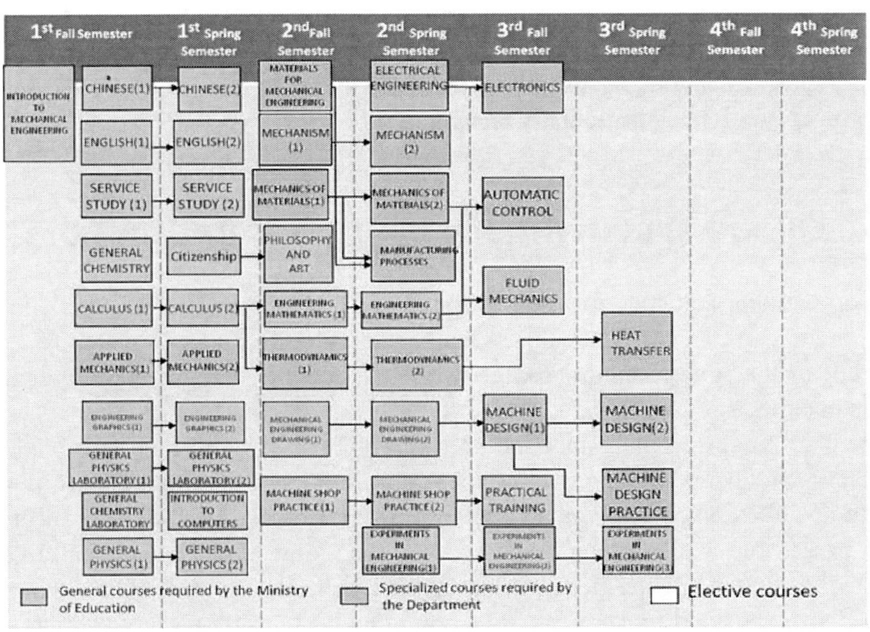

Figure 3 (A). Part of a typical curriculum of a mechanical department.[4]

[3] General education in Taiwan's higher education stands for a series of courses that provide common knowledge in addition to students' major.

[4] From Department of Mechnical Engineering, National Cheng-Kung University, Tainan, Taiwan website, 2011

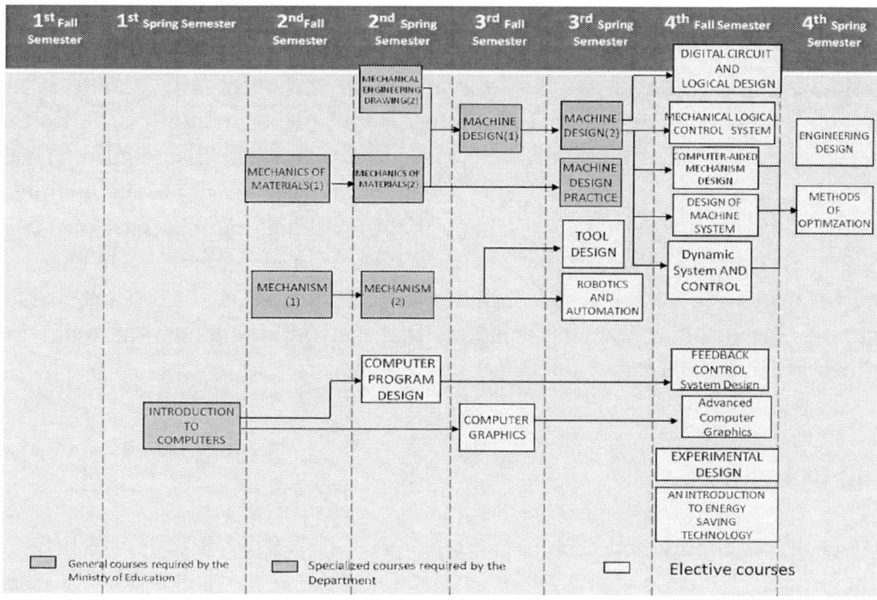

Figure 3 (B). Part of a typical curriculum of a mechanical department (cont).

3. LITERATURE REVIEW OF ENGINEERING EDUCATION

Barrella, Simmons and Buffinton compared first-year professional engineering curriculum in Canada, Australia and Singapore. They discovered most of the courses value spoken and written reporting and team work [1]. Students are encouraged and trained to develop these skills and ability to resolve problems. Moreover, the course training includes required competence of a professional engineer, for example, engineering ethic, social influence, resolving environmental issues and sustainable responsibility. Their research pointed out, Australian universities pay special attention at engineering design and its sustainable development. The courses involve professional engineering practice via many ways, including special talks, case study, guest speaker, or field investigation, etc. Some universities even bring in very creative ways to encourage students to learn about engineering professionalism in practice: Queen's University in Canada offer the "Professional Engineering Skills" course to lead students to understanding of real engineering in the reality of the society; students at the University of Sydney have to compare engineering articles from magazines and news; Bucknell University's students taking "Exploring Engineering"are required to do written reports on engineering books.

Under Taiwan's current education system, professionals in engineering industry receive training from engineering higher education institutes. Chang felt that Taiwan's engineering training lay particular emphasis on theorems, knowledge and skills, engineer's responsibility are overlooked [2]. How to pursue reasonable profit and comply with engineering ethic, environmental ethic and social responsibility at the same time is to be covered in engineering education. Pacheco et al. also raised that, people in general have high expectation on science and technology since technology has strong potential in changing the society [3]. When STS is integrated into engineering education, educators and engineers should use common language to maximize its effect.

Local literatures on engineering education are mostly on observation, analysis and future expectation of engineering general education. Some departments have offered inter-disciplinary courses, encouraging students to take credits in other disciplines. These actions are thought to be the implementation of general education's value [4]. When we think of what can be lacking in engineering education, Chang mentioned that engineering education should strengthen student's ability in carrying out a technical career, alone with basic scientific competence for a general citizen [5]. In additional to Project-Based Learning and Problem-Based-Learning (PBL), other teaching methods that can build students' attitudes in science and technology are to be innovated.

3.1. STS in Education

As Huang, et al. mentioned there are two types of STS in Taiwan, each has its own academic community. Huang named them "Science Education STS" and "Science Study STS" [6]. Lin and Fu are the first to mention such idea of integrated study of technology and social study. Since 1995, many scholars in sociology, history in science, medical sociology, philosophy of science and gender study have participated in STS academic activities, including many workshops, translation camps, etc [7]. Two books which are collections of translated classic papers in STS field were published and remain the most important STS texts in Taiwan, "*Technology Aspires After Society*" and "*Technology Aspires After Gender*" (Wu, et al.) [8]. Generally speaking, "Science Study STS" sees science as a product of human culture, focus on metacognition and inter-disciplinary study. "Science Education STS" is more a revolution of science class or class design concept. "Science Study STS" group of scholars received supports from the Ministry of Education (MOE) in 2004 to build an STS academic community, leading to a big scale STS teaching projects started in 2007 in many universities. Therefore, the present study takes STS as a discipline which promote inter-disciplinary collaboration, uses "Science Education STS" class design concept to enrich the effort in establishing an STS education environment.

Aikenhead mentioned contents of STS courses are independently decided by each state based on common indexes in Canada. They worked for students to learn STS contents, analytical skills, and scientific knowledge and attitudes [9]. "The biggest challenge in Canada continues to be the paradigm shift over who has privilege and the political power to decide what ought to be learned in science classes and what ideologies will prevail -- "science for all" or "science for an elite"?"Aikenhead also discovered "that political power is an issue in both teacher understanding and student learning".

Comparison of STS educations in United State of America (Kumar and Berlin), Canada (Aikenhead) [9, 10] and Taiwan gives interesting finding. The first two countries emphasize on STS science education reform and curriculum implementation, such as different teaching and curriculum lay out improvement. But Taiwan spinned off a "Science Study STS", encouraged educators to leap outside of their original disciplines and embrace a diverse thinking model.

As mentioned before, under the support of the Ministry of Education, a group of Taiwanese scholars started the project to promote the growth of STS academic community

[11, 12]. As one of the phase in this project, the advisory office of the Ministry of Education called for inter-disciplinary teaching proposals in 2007, aiming to bring STS into higher education system. 7 universities received the grant support in 2007; they were National Taiwan University (NTU), National Yang-Ming University (NYMU), National Chiao Tung University (NCTU), National Cheng Kung University (NCKU), Vanung University (VNU) and National Kaohsiung Marine University (NKMU). Other programs were granted during the second year and third year review. Chaoyang University (CYU) of Technology joined in 2008 and Southern Taiwan University (STU) joined in 2009. Each proposal required an inter-disciplinary faculty team to run their own STS program in their university, based on their specific design. Each faculty team consists of both humanity-social science faculty member and other science and technology orientated faculty member, e.g. medicine, engineering, science and agriculture. Each program had to be reviewed twice a year in term of overall performance by the main project office. Out of all the programs granted, the programs from National Chiao Tung University (NCTU), National Cheng Kung University (NCKU), Southern Taiwan University (STU) and National Kaohsiung Marine University (NKMU) were in school of engineering; therefore are the main subject of study for this research. The other programs were in the field of medicine, public health, nursing and others.

The original design of the teaching projects required the STS courses to be offered in the departments, not as the general elective education courses offered by the general education center in the university (names vary in each university). This experimental project ran from 2007-2011. Totally, 180 new inter-disciplinary courses were offered, 8,186 students have enrolled in at least one of the courses, 141 faculty has participated as instructors, speakers and/or workshop participators, according to the final report by the main project office in the advisory office, MOE. All the above statistics are of all the STS programs, not limited to those in engineering.

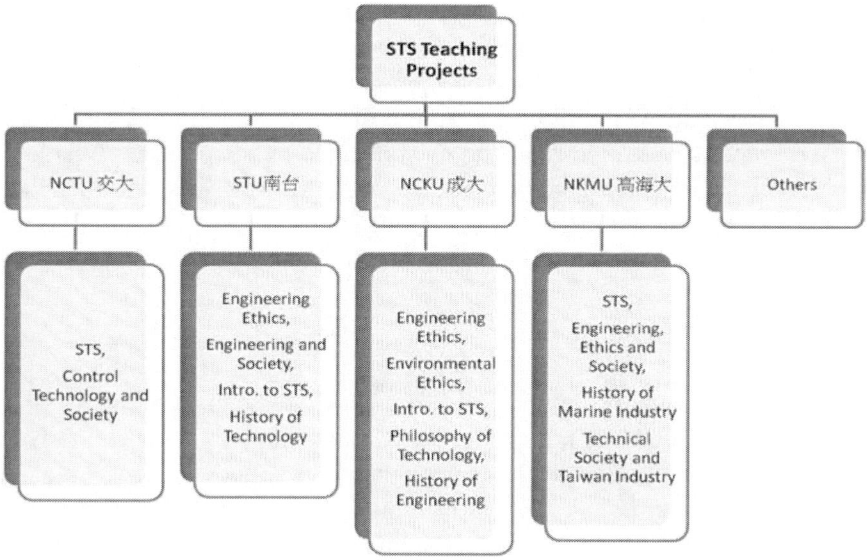

Figure 4. A glimpse of STS courses offered in the MOE supported STS teaching projects in 4 engineering schools.

Overtime, each STS teaching projects developed their own priority and strategies in terms of course placements and funding usage. For example, National Cheng Kung University (NCKU) wanted to ensure high enrollment, "Introduction to STS", "Engineering Ethics", and "History of Engineering" was offered as department electives or general education electives. National Chiao Tung University (NCTU) focused on establishing an STS center and publishing. An inter-disciplinary book "Technology, Society, and Human" was published in 2010 (Yang, et al.) [13]. In National Kaohsiung Marine University (NKMU), STS courses were offered in departments, enrollment was stable. Courses include "STS", "Engineering Ethics and Society", "History of Marine Industry" and "Technical Society and Taiwan Industry". Figure 4 shows a list of STS courses of STS programs in engineering schools in selective universities.

Based on the 4 years efforts in running the STS programs, we want to exam if STS courses nourish the curriculum of engineering higher education with its inter-disciplinary essence? Does this reflect modern society's expectation of engineering professionals successfully?

3.2. Research Questions

The present research focuses on engineering higher education. We want to look at the practical design of STS courses, cases, teaching methods, plus interactive strategies and class placements. The aim is to conclude on effective suggestions in cultivating lifelong care and ability to deal with science related social issues for engineering students.

The curriculum of Taiwan's engineering higher education can be roughly divided into three parts: fundamental engineering science, advance engineering science and design, and other general credits. Among engineering courses, inter-disciplinary is rarely mentioned, such as critical thinking emphasized in social science and caring as in humanity. Instructors themselves may not have such competency, or were reluctant of teaching inter-disciplinary subjects.

The present paper studies the three teaching projects in three engineering colleges. We analyzed the term reports and some class survey conducted by the projects. What did the STS teaching projects achieve? What are the effective strategies and what aren't? What are the changes to students and instructors?

3.3. Objects of Study

For this study, we focus on three STS programs in the school of engineering in three universities in southern Taiwan. *A* university (NCKU) is a big public general university, *B* is a public technical university (NKMU) (vocational track), and *C* is a private technical university (STU) (vocational track). The engineering college in *A* university consists of 34% of students of the entire university, B has 27%, and C has 40%, also the largest college in their university.

We collect detail information of these three STS programs, including midterm and final project reports of each programs, course curriculum and class syllabuses, etc. Some of the courses were followed with before- and after- questionnairess for student's motivations to take the class, expectations, suggestions, etc. We also interviewed scholars participating in the programs, including engineering and STS scholars, for their feedback of the programs, and how the programs affect their professional developments.

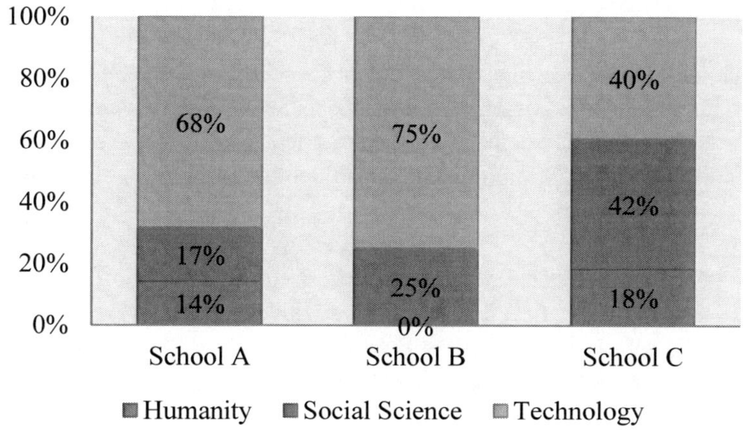

Total student number (undergraduate)
A University □ 11,357
B University □ 4,082
C University □ 11,819

Figure 5. Student's compositions in three universities.

4. Effect Analysis of STS Programs

4.1. Course Placement Matters

Because of its inter-disciplinary nature, when STS program started in *A* university (2007-2008), STS courses were categorized more as humanity and social science courses. These STS courses were offered as dual electives of engineering department/general education; open to all the students in the university. Students can choose to take them as engineering credits or general education credits. "STS" is their core course and is offered every semester by a history background STS scholar. This course aims to encourage student to be familiar with methods and viewpoints from social science, philosophy and history. There are other courses like "Engineering Ethics" and "History of Engineering" given by engineering background faculty members. Students of these courses have mixed backgrounds.

The STS programs in *B* and *C* universities took a different approach. They both offered their STS courses within the engineering colleges. "STS" and "Engineering Ethics and Society" are their core courses. In *C* university, the importance of having engineering ethic courses has already been recognized, partly because of the urge in fulfilling the engineering accreditation requirement of "soft skills". Both STS programs insisted in keeping the STS courses inside the engineering college as electives. These STS courses are given by engineering faculty members. Through a series of workshops, they involved in dialogues and discussion with humanity and social science scholars to enrich their social science knowledge and teaching competence. They felt when the course were taught by engineering instructors to engineering students, the instructors can have better grips on how "inter-disciplinary" the course content get, making it easier for engineering students to follow.

> The teaching of Engineering Ethics under the scope of STS is somewhat more meaningful. This is closer to students' experience. It is easier to guide them to extend their critical thinking about social issues from their professional experience. As an instructor, it enhance my professional research in depth and scope, I am more connected to social issues now. (Ct1)[5]

In order to broaden disciplinary knowledge, *B* and *C* STS programs did not seek collaboration with general education divisions. They developed their own mini series of STS courses and promoted them in the college of engineering.

The course placement really affects student pool and numbers. Having the STS courses as general education electives will almost guarantee good enrollment. Mixed students also enable diversified discussion in STS courses. But with such diversified students, instructors have to adjust contents and keep it from being too "specific". However, when the course was taught too "general", students suggested that the course should be offered in the department to allow deeper discussion to happen. As a student in *A* university stated:

> Maybe because I am not in an engineering related department, I found it very difficult to follow when the instructor talks about engineering this and engineering that. The final project has to be engineering related. But I just don't have that kind of knowledge. It doesn't help even I spend time looking for reference. I wish this course can be more "general". (As9)[6]

4.2. Instructors' Perception

All the three STS programs used instructors from different fields to teach. For *B* university, they do not have humanity and social science colleges in the school. The program was supported by a social science scholar for the first two years and they hired a historian for the latter two years. Guest speakers in the classes were a must to bring in angles from other fields.

Another situation inter-disciplinary courses face is the difficulties in recruiting engineering faculty members. As members in *A* and *C* university program stated:

> Engineering Ethics is probably the only course that can link engineering faculty and STS. College of Engineering has to offer this course to pass the IEET accreditation. Someone has to teach it, but in fact, no one wants to. I actually suggested Engineering Ethics paring with STS, History of Engineering, and Philosophy of Technology, to be

[5] Ct1: C – *C* university, t – teacher

[6] As9: A-*A* university, s-student.

> required electives[7]. If this proposal is passed in December, we can then hold teaching workshop for engineering faculty, penetrate into engineering college, influence engineering faculty. (At2)
>
> It is not enough to offer courses and publish case studies. For STS to generate that kind of effects we hope for in technical universities, there has to be well-thought-out structures (or systems) and feasible operation model in place. (Ct1)

During the first year of the STS programs, the university administrations were very supportive. However, to promote STS course credits to be recognized for graduation faces different challenges in different universities. In general, being dominated by traditional value of professionalism, engineering education is slow to change. Faculty members in research oriented universities devote to research, have little motivation in engaging with STS teaching. Research pointed out, changing engineering viewpoints needs strong support from school administration. Therefore, a reward system has to be in place to encourage and promote faculty's willingness to teach inter-disciplinary courses (Baillie, et al.) [14].

4.3. Changes Within

In the first year of STS teaching, it can be seen from the reports of *B* university students that they still focus on personal errors when think of engineering ethics. Instructors have to spend efforts in breaking the prejudices and re-focus on the blind advocacy of technology.

> A graduate student form the STS class came to talk to me, seriously. He said he can agree with STS's viewpoint from a logical angle, but that will also defeat the entire value about science and technology he grows up upon. He became very concerned about the destruction of that sole standard. I realize, as it is important to promote critical thinking, we still have to leave hope to them. Therefore, we have to emphasize the possibility of democratic participation, of knowledge bringing actions. (At1)
>
> The vocational education in Taiwan emphasizes on repeated and percise execution. Therefore, independent thinking or critical thinking is never part of their education objectives. For those who were "deprived" (quoted from a student) from right to think, it takes much effort to change their habit of little thinking or no-thinking at all. (Bt2)

4.4. Student Feedback

Reasons for students to like STS courses:

A. Non-traditional teaching: instructors used contemporary issues for class discussion, especially local cases. Open discussion set the tone that the class allow and welcome different opinions and ways of thinking. This is very different from traditional lecture of single answer, and students like it very much.

---

[7] Required elective, meaning students have to take the course but it is OK if they did not pass and get the credit.

There are more thinking involved, more interaction. This is not one question one answer anymore. (Bs3)

This is very different from the courses I took before. I like every topics. I now know that engineering is more than engineering, and society is more than society. When people are willing to change, difference can be made, like Ruhr Industrial Base. (Bs8)

B. Conversion of thinking habit: because of credentialism in the past, training to think is not part of education objective. Students may hesitate at first, but find it attractive.

In addition to history, the instructor really wants to make us understand the humanity that was overlooked in traditional engineering value. He tried to explain in many aspects of one single event. The impact to engineering students is very big and positive. (As3)

C. New concept to break the segregation of humanity and engineering: they learn the STS ways to consider an issues jointly with engineering and social science.

I suggest putting STS as department electives. Because this course teaches us to understand the relationship between technology and society, understanding about our mission as engineers. (As6)

STS courses are new to me. I don't know if they offer them in high school or vocational high. If not, I hope they can offer them in high school. This will encourage high school students to think multi-dimensionally and freely, to care more about issues around them. So they will not be so exam-oriented. (Bs13)

4.5. Network Mixed and Widen for Engineering and STS Academic Community

After the execution of STS programs in each university, inter-disciplinary institutes were established in many schools, such as STS Research Center in NKMU and NCTU, and STM (Society, Technology and Medicine) Center in NCKU. They have since played very important and active roles in expending STS academic networks, encouraged social participation in related topics, brought in global issues, and promoted STS knowledge and academic development.

In these engineering STS programs, class instructors often invite social science scholars as guest speakers in the class. Therefore, meeting scholars in other fields became an opportunity to change for these engineering teachers themselves. Figure 6 shows the number of scholars involves in the entire STS teaching projects supported by the Ministry of Education for a four year span. Lin thought that education reform can happen inside the system, it just takes longer time [15]. He took women study in United States of America as an example. When female faculty member were brought in to universities, they started to form social networks, offered courses with gender consciousness, and eventually attracted female students and others' attention. This kind of power seems to be slow and invisible, but is indeed a system reformation turned on by social network operation. With the same concept, engineering scholars involved in STS programs broaden their social networks and extend reading fields. They were able to think holistically and found talks with peers in the same discipline cover deeper and broader aspects, based on their self relections.

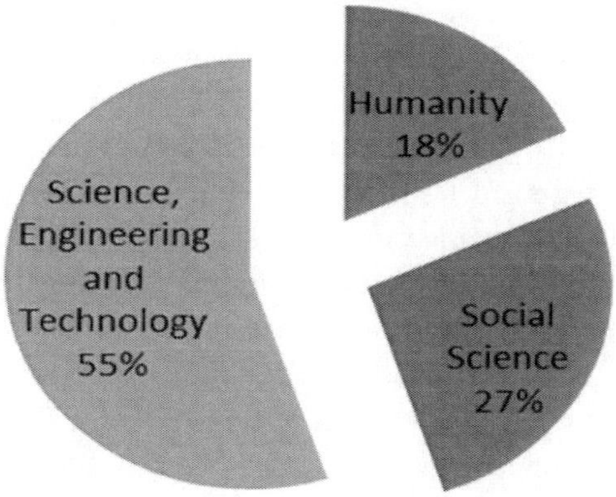

Figure 6. Composition of teachers participated in Ministry of Education's STS teaching projects from 2007 to 2011.

Wang mentioned that practical community is an unofficial organization [16]. Teachers developed common knowledge through interactive mutual learning and shared teaching materials. STS community took exactly this route to enhance professional growth and cultivate personal network. Their methods include:

(1) Develop course pack or case study
(2) Publish STS newsletter
(3) Hold frequent inter-university workshops

In addition to those mentioned above, the first three annual conferences of the newly established Taiwan STS Society (in 2008) were organized by scholars who were deeply involved in these STS programs, and held in three different universities. These cross university academic activities strengthen the lateral connections of STS community, increase channels for STS issues spreading, and set fundamental and positive energy for STS and inter-disciplinary research.

CONCLUSION

We found department's course regulations play important roles in offering STS courses in engineering. Course placement (as in department and general education electives) affects course orientation, potential student pools and enrollment numbers. Class size limits teaching to certain styles. Also, the sense of value of the instructors is the key to successful inter-disciplinary learning. Teaching STS inter-disciplinary courses requires certain level of professional knowledge and ability to conduct constructive discussion with scholars from other fields. They also have to inspire students, who are very exam-oriented, to think outside of their original scope. All of the above demand teaching competency. In the school administration, instructors had to fight for good course placement, recruit inter-disciplinary teaching partners. These are all great challenges for STS and engineering teachers. Passions seems to be the common thread in pulling it together.

When engineering courses integrate STS in case studies of real life issues, students understand the close tie between engineering and social science. Engineering Ethics can be used to explain complex problems, but isn't merely a set of moral codes. For Taiwan's exam-oriented engineering students, STS courses have interesting flavors: 1) teaching style, open-end discussions with no set correct answers; 2) course contecnts, domestic and international case studies; 3) changes from seeing engineering and social science/humanity as dualism to integration. For the scholars of engineering, humanity, social science and history, being involved in STS community nourishes their professional growth and establishes new network.

Overall, bringing in STS is a good way to improve engineering students' professional competence, which includes understanding of social issues and improving liberal arts ability. However, at this moment, finding qualified cross-disciplinary instructors remains the biggest challenge.

REFERENCES

[1] Barrella, Elise; Simmons, John; Keith Buffinton. Professional Engineering Education Best Practice Study for First-year, Multi-disciplinary Courses. *Proceedings of the 2006 American Society for Engineering Education Annual Conference & Exposition* (2006).

[2] 張介耀, 〈工程教育與工程倫理之探討〉 《南華通識教育研究》, 1 (2004) 31 Title translated as Exploring Engineering Education and Engineering Ethics.

[3] Pacheco, Renato Lucas; Bazzo, Walter Antonio; Carlson, Renato; Martins-Pacheco, Lúcia Helena. Science, Technology and Society… of Consumption A Reflection. *Proceedings of the 2004 American Society for Engineering Education Annual Conference & Exposition* (2004).

[4] 陳義揚, 〈工程教育之通識思維〉 《華岡工程學報》, 21 (2007) 1 Title translated as The Liberal Arts in Engineering Education.

[5] 張玉山, 〈科技教育的再思維〉 《生活科技教育月刊》, 44(2) (2011) 1 Title translated as Rethinking the Technology Education.

[6] **黃俊儒、楊文金、靳知勤、陳恒安，〈誰的**STS？「科學教育」與「科學研究」的「同」與「不同」〉 《科學教育學刊》, 6 (16) (2008) 585 Title translated as Who's STS? The Similarities and Differences Between Science Education and Science Study.

[7] 林崇熙、傅大為, 〈**歷史中的台灣科學－關於台灣科學史研究的一些回顧與幾點檢討**〉, **五十年來台灣的歷史研究之回顧研討會，台中** (1995) Title translated as Science in Taiwan's History – Reviews on History of Science in Taiwan.

[8] **吳嘉苓、傅大為、雷祥麟**(編) , 《科技**渴望社會**》**，臺北：群學** (2004) Title translated as Technology Aspires After Society.

[9] Aikenhead, Glen S. STS science in Canada: From Policy to Student Evaluation. *Science, Technology and Society: A Sourcebook on Research and Practice*. Kumar, David D. and Chubin, Daryl E. New York: Kluwer Academic Pub. (1998) 49.

[10] Kumar, David and Berlin, Donna. A Study of STS Themes in State Science Curriculum Frameworks in the United States. *Journal of Science Education and Technology*, 7(2) (1998) 191.

[11] 傅大為, 〈翻譯STS 始末記〉, 《科技**渴望性別**》 **載於吳嘉苓等**(主編), 臺北：群學 (2004) 15。Title translated as Technology Aspires After Gender.

[12] 雷祥麟, 〈代序：相互**渴望的科技與社會〉**, **《科技渴望性別》** **載於吳嘉苓等**(主編), 臺北：群學 (2004) 7 Title translated as Technology Aspires After Gender.

[13] 楊谷洋、陳永平、林文源、方俊育, 《科技社會人：STS跨領域新視界》, 臺北：國立交通大學 (2010) Title translated as Technology, Society, Human: STS the New Interdisciplinary Vision.

[14] Baillie, Caroline; Ko, Edmond; Newstetter, Wendy; and Radcliffe, David F., 2011, Advancing Diverse and Inclusive Engineering Education Practices through Interdisciplinary Research and Scholarship. *Journal of Engineering Education*, 100(1) (2011) 6.

[15] 林南, 〈教育制度與社會資本〉《教育研究集刊》, 50(4) (2006) 1 Title translated as Education System and Social Capital.

[16] Wang, Wei-Kuo. Cultivating Communities of Practice for Curriculum Development and Teacher Professional Development. *Journal of Curriculum Studies*, 2(2) (2007) 41.

INDEX

C

D

E

F

G

H

I

K

L

M

N

T

U

V

W

X

Y

Z